FOREWORD

Two phase flow and heat transfer has continued to receive considerable attention in the past few decades mainly because of its common occurrences in many industrial applications. Some examples of industrial applications of two-phase flow and heat transfer are found in heat exchangers, steam generators, boilers, condensers, evaporators, thermosyphons, nuclear reactors, and others. Efforts to understand fundamental processes and mechanisms in these areas have resulted in achieving great strides and many of the results are fruitfully utilized by industry.

The papers in this volume show a balance of topics and reflect the variety and diversity of two-phase flow and heat transfer problems. The first section begins with M. Ishii's invited paper which considers the concept of interfacial area in two-phase flow modeling and formulation. The second section treats the subject of boiling and convective boiling heat transfer and starts with another invited paper by C. T. Avedisian who gives an authoritative review on nucleation and bubble growth. The third section considers heat transfer in mixtures and refrigerants, a timely topic in light of the environmental concerns about refrigerants. The fourth section presents various topics in condensation. The final section includes seveal interesting special topics of current interest.

These papers were presented at the symposium on Basic Aspects of Two Phase Flow and Heat Transfer held at the 1992 National Heat Transfer Conference. The symposium was sponsored by the Heat Transfer Division's K-13 Committee on Nucleonics Heat Transfer with cosponsorship from the Fluids Engineering Division (Multiphase Flow Committee).

We hope that this volume will add to the growing body of knowledge about two phase flow and heat transfer, stimulate further research in this exciting area, and serve as a timely reference for researchers and practitioners.

We wish to express our special appreciation to the authors for their contributions. Our appreciation is also due to reviewers, session leaders, and other participants of the symposium who collectively contributed to making the symposium greatly successful.

J. H. Kim
R. A. Nelson
A. Hashemi

LIST OF REVIEWERS

C. Allison
J. Ambrose
S. G. Bankoff
A. E. Bergles
B. L. Bhatt
V. P. Carey
F. B. Cheung
S. M. Cho
M. L. Corradini
V. K. Dhir
R. A. Dimenna
S. M. Ghiaasiaan
J. Goodman
R. Gore
S. S. Grossel
T. J. Hanratty
A. Hashemi
J. M. Healzer
H. D. Huang
E. D. Hughes
T. F. Irvine, Jr.
M. K. Jensen
M. Kawaji
M. A. Kedzierski
J. H. Kim
M. Kim
L. D. Koffman
R. T. Lahey, Jr.
R. W. Lyczkowski
E. V. McAssey
G. E. McCreery
A. Majumdar
S. B. Memory
J. D. Menna
B. Mikic
C. D. Morgan
R. A. Nelson
K. Pasamehmetoglu
G. P. Peterson
P. F. Peterson
M. Z. Podowski
J. W. Rose
T. D. Rudy
V. Schrock
F. K. Tsou
K. Tuzla
C. Unal
J. C. Y. Wang
A. S. Wanniarachchi
R. L. Webb
R. Wedel
H. Wong
S. C. Yao

CONTENTS

CONDENSATION

SPECIAL TOPICS

HTD-Vol. 197, Two-Phase Flow and Heat Transfer
ASME 1992

INTERFACIAL AREA IN TWO-PHASE FLOW

M. Ishii
School of Nuclear Engineering
Purdue University
West Lafayette, Indiana

ABSTRACT

In the analysis of two-phase flow, the formulation using a two-fluid model is considered the most accurate model because of its detailed treatment of the phase interaction at the interface. The geometrical effects of the interfacial structures can be modelled through the interfacial area concentration and void fraction. For the interfacial transfer of mass, momentum and energy the interfacial area represents the first order effect of the interfacial structure. A detailed discussion of the interfacial area formulation and recent development in the measurement methods of the local interfacial area concentration is presented.

I. INTRODUCTION

Two-phase flow is characterized by the existence of the interface between phases and discontinuities of properties associated with them. The internal structures of two-phase flow are classified by two-phase flow regimes. Various transfer mechanisms between the mixture and wall as well as between phases strongly depend on these two-phase flow regimes. This leads to the use of flow regime dependent correlations and closure equations together with appropriate flow regime transition criteria. The basic structure of flow can be characterized by two fundamental geometrical parameters. These are the void fraction and interfacial area concentration. The void fraction expresses the phase distribution whereas the interfacial area describes available area for the interfacial transfer of mass, momentum and energy. Therefore, an accurate knowledge of these parameters is necessary for any two-phase flow analysis. This fact can be further substantiated with respect to two-phase flow formulation.

In order to analyze the thermal-hydraulic behavior of two-phase flow, various formulations such as the homogeneous flow model, drift-flux mode (Ishii, 1975; Zuber and Findley, 1965; Wallis, 1969; Ishii, 1977; and Chawla and Ishii, 1980), and two-fluid model (Ishii, 1975; Vernier and Delhaye, 1968; Kocamustafaogullari, 1971; Boure, 1978; Ishii and Mishima, 1980; Chawla and Ishii, 1978; Delhaye and Achard, 1977; Sha et al., 1983), have been proposed. As suggested by Ishii and Kocamustafaogullari (1982), among these models the two-fluid formulation can be considered the most accurate model because of its detailed treatment of the two-phase interactions at the interface.

The two-fluid model is formulated by considering each phase separately in terms of two sets of conservation equations which govern the balance of mass, momentum and energy of each phase. These balance equations represent the macroscopic fields of each phase and are obtained from proper averaging methods. Since the macroscopic fields of each phase are not independent of the other phase, the phase interaction terms which couple the transport of mass, momentum and energy of each phase appear in the field equations. It is expected that the two-fluid model can predict mechanical and thermal nonequilibrium between phases accurately. However, it is noted that the interfacial transfer terms should be modeled accurately for the two-fluid model to be useful. In the present state of the arts, the closure relations for these interfacial terms are the weakest link in the two-fluid model. The difficulties arise due to the complicated transfer mechanisms at the interfaces coupled with the motion and geometry of the interfaces. Furthermore, the closure relations should be modeled by macroscopic variables based on proper averaging. A three-dimensional two-fluid model has been obtained by using temporal or statistical averaging, Ishii (1975). For most practical applications, the model developed by Ishii can be simplified to the following forms:

Continuity Equation

$$\frac{\partial \alpha_k \rho_k}{\partial t} + \nabla \cdot (\alpha_k \rho_k \vec{v}_k) = \Gamma_k \tag{1}$$

Momentum Equation

$$\frac{\partial \alpha_k \rho_k \vec{v}_k}{\partial t} + \nabla \cdot (\alpha_k \rho_k \vec{v}_k \vec{v}_k) = -\alpha_k \nabla p_k + \nabla \cdot \alpha_k (\tau_k + \tau_k^t)$$

$$+ \alpha_k \rho_k \vec{g} + \vec{v}_{ki} \Gamma_k + \vec{M}_{ik} - \nabla \alpha_k \cdot \tau_i \tag{2}$$

Enthalpy Energy Equation

$$\frac{\partial \alpha_k \rho_k H_k}{\partial t} + \nabla \cdot (\alpha_k \rho_k H_k \vec{v}_k) = -\nabla \cdot \alpha_k (q_k + q_k^t) + \alpha_k \frac{D_k}{Dt} p_k$$

$$+ H_{ki} \Gamma_k + \frac{q''_{ki}}{L_s} + \Phi_k \tag{3}$$

Here Γ_k, M_{ik}, τ_i, q''_{ki}, Φ_k are the mass generation rate, generalized interfacial drag, interfacial shear stress, interfacial heat flux, and dissipation, respectively. The subscript k denotes k phase, and i stands for the value at the interface. α_k, ρ_k, v_k, p_k and H_k denote the void fraction, density, velocity, pressure and enthalpy of k phase, whereas τ_k, τ_k^t, q_k, q_k^t and g stand for average viscous stress, turbulent stress, mean conduction heat flux, turbulent heat flux and acceleration due to gravity. H_{ki} is the enthalpy of kth phase at the interface, thus it may be assumed to be the saturation enthalpy for most cases. L_s denotes the length scale at the interface, and $1/L_s$ has the physical meaning of the interfacial area per unit volume a_i. Thus,

$$\frac{1}{L_s} = a_i = \frac{\text{Interfacial Area}}{\text{Mixture Volume}} \tag{4}$$

The above field equations indicate that several interfacial transfer terms appear on the right-hand sides of the equations. Since these interfacial transfer terms also should obey the balance laws at the interface, interfacial transfer conditions could be obtained from an average of the local jump conditions (Ishii, 1975). They are given by

$$\sum_k \Gamma_k = 0 \tag{5}$$

$$\sum_k \vec{M}_{ik} = 0 \tag{6}$$

$$\sum_k (\Gamma_k H_{ki} + q''_{ki}/L_s) = 0 \tag{7}$$

Therefore, closure relations, for M_{ik}, q''_{gi}/L_s, and q''_{fi}/L_s are necessary for the interfacial transfer terms. The enthalpy interfacial transfer condition indicates that specifying the heat flux at the interface for both phases is equivalent to the closure relation for Γ_k if the mechanical-energy transfer terms can be neglected (Ishii, 1975). This aspect greatly simplifies the development of the closure relations for interfacial transfer terms.

By introducing the mean mass transfer per unit area, m_k, defined by

$$\Gamma_k \equiv a_i m_k \tag{8}$$

the interfacial energy transfer term in Eq. (3) can be rewritten as

$$\Gamma_k H_{ki} + \frac{q''_{ki}}{L_s} = a_i(m_k H_{ki} + q''_{ki}) \tag{9}$$

The heat flux at the interface should be modeled using the driving force or the potential for an energy transfer. Thus,

$$q''_{ki} = h_{ki}(T_i - T_k) \tag{10}$$

where T_i and T_k are the interfacial and bulk temperatures based on the mean enthalpy and h_{ki} is the interfacial heat transfer coefficient. A similar treatment of the interfacial momentum transfer term is also possible (Ishii and Mishima, 1980). In view of the above, the importance of the interfacial area concentration, a_i, in developing closure relation for this term is evident. The interfacial transfer terms are now expressed as a product of the interfacial area and the driving force. It is essential to make a conceptual distinction between the effects of these two parameters. The interfacial transfer of mass, momentum and energy increases with an interfacial-area concentration toward the mechanical and thermal equilibrium.

Thus, in general, the interfacial transfer terms are given in terms of the interfacial area concentration a_i and driving force (Ishii, 1975; Ishii and Mishima, 1980) as

$$\text{(Interfacial Transfer Term)} \sim a_i \times \text{(Driving Force)}$$

The area concentration defined as the interfacial area per unit volume of the mixture characterizes the first order geometrical effects; therefore, it must be related to the internal flow-pattern of the two-phase flow field. On the other hand, the driving forces for the interfacial transport characterize the local transport mechanisms such as the turbulence, molecular transport properties and driving potentials. In two-phase flow systems, the void fraction and interfacial area concentration are two of the most important geometrical parameters. The interfacial area concentration should be specified by a closure relation, or by a transport equation. The above formulation indicates that the knowledge of the interfacial area concentration and the interfacial structure classified as the flow regimes are indispensable in the two-fluid model.

II. TWO-PHASE FLOW REGIMES

As explained above, various transfer mechanisms between phases and between a two-phase mixture and structures depends on two-phase flow regimes. The concept of two-phase flow regimes implies the macroscopic characterization of interfacial topology or structure. Because of these, the flow regime transition criteria and regime dependent two-phase flow correlations have been developed in parallel. However, in many practical applications, it is much more difficult to predict the actual two-phase flow regime than to find proper correlations for that regime.

This is due to several reasons as listed below.

1. In the past, two-phase flow regimes have often been identified by subjective experimental methods rather than objective methods.
2. In general, a flow regime transition is a gradual phenomenon and no definite transition point exists.
3. Two-phase flow regimes strongly depend on the geometry of a system. This implies the entrance effects are very important for determining the regime. Furthermore, structural changes such as bends, flow junctions, expansions, contractions and spacers significantly alter or influence the geometrical configurations of the interface.
4. A definition of a flow regime can depend on the scale of the reference volume. A flow regime defined based on a small local volume may be quite different from the one based on a large volume. This occurs often in a system with internal substructures such as rod bundles. For example, a locally annular flow may be a part of a large slug bubble spanning over many subchannels.

For objective determinations of two-phase flow regimes, statistical analyses of local measurements of void fractions, fluid particle sizes and pressure drops become necessary. Gradual transitions of two-phase flow regimes and effects of an entrance geometry may not be easily handled by the concept of flow regimes. In order to describe these effects quantitatively, the introduction of an interfacial area concentration is recommended.

III. INTERFACIAL AREA CONCENTRATION

The geometrical effects of interfacial structures can be modeled through the interfacial area concentration and void fraction (Ishii, 1975; Ishii and Mishima, 1980; Ishii et al., 1982). A number of experimental studies (Sharma and Mashelhan, 1968; Watson et al., 1979; Kasturi and Stepanek, 1974; Akita and Yoshida, 1974; Shilimkan and Stepanek, 1977; Bier et al., 1978; Gregory and Scott, 1968; Shah and Sharma, 1975; Wales, 1966; Proter et al., 1966; Dillon and Harris, 1966; Abdel-Aal et al., 1966; Brugess and Calderbank, 1975; Banerjee and Rhodes, 1970; Kulic and Rhodes, 1974; Robinson and Wilke, 1974; Sridhar and Potter, 1978; Menta and Sharma, 1971; Shende and Sharma, 1974; Sridhar and Sharma, 1976; Sharma and Gupta, 1967) on interfacial area has been published in chemical engineering fields in the past 20 years. Most of these experiments used a chemical absorption technique based on a pseudo-first order chemical reaction. These experiments were performed by using two fluids such as air and water. Then some reacting gas such as CO_2 is added to air and reacting liquid such as NaOH to water. If the reaction is a fast, irreversible, pseudo-first order chemical reaction, the average interfacial area between two sampling points can be measured by applying the surface renewal theory of Danckwerts (1970) in terms of chemical combinations and geometries of systems, see also Veteau (1981), and Veteau and Morel (1981).

Other important techniques for measuring interfacial areas are the light attenuation method (Trice and Rodger, 1956) and photography method Watson et al., 1979) which require a flow channel with transparent walls. The optical method has been reviewed in detail and applied to obtain data for bubbly flow by Veteau (1981) and Veteau and Charlot (1981, 1981a, 1981b). The light attenuation method requires the fluid to be transparent. The attenuation of a light beam crossing two-phase mixture is measured. The particle size should be larger than the incident wavelength and multiple reflections reaching the detector should be negligible. This may limit its application to a low void fraction bubbly flow. The underlining theory and model is described by Calderbank (1958, 1959). and Veteau (1981). This method generally gives a line averaged interfacial area concentration across a flow channel. The advantages and shortcomings of various methods are discussed by Landau et al. (1977).

The chemical method has been the most widely used technique. The value of the interfacial area can be obtained by sampling and chemical analysis of concentrations, however, it can be applied only to a case without phase change and to a steady flow. It is a global measurement over a certain length of a flow channel and therefore it does not give detailed local information on the interfacial area concentration. The experimental set-up is cumbersome and time consuming. Furthermore, the errors associated with this method can be large depending on experimental conditions.

The light attenuation method is simple and cross-sectional area averaged measurement at various axial locations is possible. However, the measurement strongly depends on flow regimes, and it is applicable only when interfaces due to multiple and forward light scattering are negligibly small. Recently the attenuation method of ultrasonic waves has also been used successfully for volume averaged measurement at Grenoble Nuclear Research Center.

The photograph method involves very time-consuming data analysis. One should measure particle sizes in detail from photographs. Furthermore, the method is good only at relatively low concentration of a dispersed phase. The light attenuation and photograph methods are applicable to a flow with phase changes. However, the bounding walls and fluid should be transparent.

Among available experimental data based on the global measurement techniques described above, only one-third of them were for straight tubes or channels, and the ranges covered by these data are far short of being sufficient for general applications. The flows studied fall into the slug, churn and annular flow patterns at moderate liquid fluxes (3 ~ 50 cm/s). The range of the gas flux was 0.4 ~ 30 m/s. It should be noted that no data exist for high liquid fluxes. The diameters of tubes were relatively small (0.6 ~ 2.5 cm). Furthermore, the effect of the density ratio or pressure on the interfacial areas has not been studied experimentally. However, the most important shortcoming of existing data may be the lack of information for developing flows. In view of fundamental difficulties encountered in modeling entrance and rapid transient flow, considerable effort should be made to develop some data base for interfacial areas for such flows.

In spite of its necessity in two and three dimensional analysis, there has been little knowledge on the local interfacial area concentration measurements. The local measurement of the interfacial area concentration has been developed recently by Kataoka, Ishii and Serizawa (1984, 1986). A double-sensored probe based on some statistical assumptions of two-phase flow characteristics has been proposed by them. The method is proven to be practical and quite useful to measure the local interfacial area concentration with considerable accuracy when some statistical characteristics of interfacial motion are known. However, the local probe method based on a double-sensored probe is limited to bubbly two-phase flow in the strict sense. When it is applied to other flow regimes such as the slug, churn-turbulent or annular flow, the data may contain large errors depending on the interfacial geometry. The three double-sensored probe method, which is also proposed by Kataoka et al., (1984, 1986), appears to be the most suitable technique due to the longer interfacial length scale. It can measure the interfacial area concentration, interfacial velocity and wave propagation characteristics.

In view of the above discussion, it can be concluded that, at present, it appears that various measurement techniques such as the chemical, light attenuation, photographic and local probe have a number of shortcomings. Among them, however, the method using one to three double sensors seems to have a real potential in terms of accuracy and wider applicability over two-phase flow regime, fluid, and flow conditions. Since the local probe gives the most detailed information on the interfacial area and geometry, a breakthrough in this area would greatly benefit not only two-phase flow modeling efforts, but also process control and instrumentation in a wide variety of industries such as energy related, chemical process, petroleum, environmental control, etc.

The geometrical relations developed for the interfacial area concentration (Ishii and Mishima, 1980) show the importance of the existence and size of small fluid particles for all flow regimes. This implies that the number density of small bubbles is important in bubbly, slug, and churn flows, whereas in annular-mist, dispersed droplet, or inverted annular film boiling flows, the droplet size distribution (Kataoka et al., 1983; Ishii and DeJarlais (1987), and amount of entrainment (Ishii and Grolmes, 1975; Ishii and Mishima, 1982; Kataoka and Ishii 1982) are important in determining the interfacial area. The entrainment data at higher pressure or for viscous fluids are not sufficient to develop a reliable correlation for these cases. The droplet size distributions and interfacial characteristics in inverted annular film boiling (Ishii and DeJarlais, 1987; Obot and Ishii, 1988) are the key parameters for determining the heat transfer and the relative velocity between phases.

For bubbly, slug and churn flow, the maximum stable bubble size, mechanisms of bubble coalescence, disintegration and nucleation are important. In order to accurately predict the interfacial area concentration for these flow regimes, it may be necessary to introduce a transport equation for the interfacial area (Ishii, 1975). For example, for a boiling flow, a bubble number transport equation can be written in terms of the bulk liquid bubble nucleation rate, the bubble number density sink rate due to coalescence and collapse and generation rate due to bubble disintegration (Kocamustafaogullari and Ishii, 1983). This equation is equivalent to the interfacial area transport equation. The bubble number transport equation and bubble growth model have been applied to predict flashing flow (Jones and Shin, 1984; Riznic et al., 1987).

IV. LOCAL INTERFACIAL AREA MEASUREMENT

A theoretical study carried out at Argonne National Laboratory (Kataoka et al., 1984) gives a method of using electrical resistance probe technique to measure the local interfacial area concentration. Using this theoretically supported method Kataoka et al. (1984, 1986, 1984), Wang and Kocamustafaogullari (1990) and Revankar and Ishii (1992) studied the local interfacial area concentration in air-water bubbly flow using a two-sensor resistivity probe. When using the two-sensor probe to measure the local interfacial area, it is necessary to make certain statistical assumption on bubble parameters. The ANL study (Kataoka et al., 1984) also proposed a four-sensor probe to measure the local interfacial area concentration. This method using a four-sensor probe does not require any statistical assumptions. Multi-sensor probes have been used by Burgess and Calderbank (1975, 1975a, 1975b), and Buchholz et al. (1983) to measure the bubble size and velocity in bubble columns of interest to chemical engineering. These authors used the average void fraction and the mean Sauter diameter measured from these multisensor resistivity probe to find the average interfacial area concentration in bubbly flow. Recently Kataoka and Serizawa (1990) have proposed a correlation method for the four-sensor probe to measure the interfacial area and presented some preliminary measurements on air-water bubbly flow using this method. In this method, the interfacial area is obtained from the measured correlation function of the characteristic functions at two different positions by probe sensors.

The four-sensor probe method for measuring the actual vectorial interfacial velocity to obtain the area has been applied by Ishii and Revankar (1991) to cap bubbly and slug flow regimes. It was possible to measure the local void fraction, interfacial area and interfacial velocity from a single probe. Furthermore, the contributions of large bubbles and smaller follow-up bubbles can be measured separately. Such a detailed measurement method for the local structures of the two-phase flow has never been developed previously. The probe also can give some statistical information such as the Sauter-mean diameter, bubble cord length probability density, number of particles and velocity fluctuation of the interface. The reliability of the data is confirmed by comparing them to direct photographic data as well as to the global data from the pressure transducers through the void fraction.

V. CONCLUSIONS

The importance of the interfacial area concentration in two-phase flow formulation is discussed in detail. The void fraction and the interfacial area concentration characterizes the macroscopic effects of the interfacial geometry. In the two-fluid model formulation, the interfacial area should be specified either by a constitutive relation or a transport equation which describes its evolution. Both approaches are discussed in detail. The interfacial area measurement methods such as the chemical, optical, light attenuation and local probe methods are reviewed. The special highlight is given to the very recent development of the multi-sensor local probe technique and its theoretical background. Both the double sensor and four-sensor methods appear very promising. It is expected that a major breakthrough in this direction is going to be made in the near future.

ACKNOWLEDGMENT

This work was performed under the auspices of the U.S. Department of Energy. The author would like to express his sincere appreciation to Dr. O.P. Manley of DOE/BES for his support of the program.

REFERENCES

1. Ishii, M., *Thermo-Fluid Dynamic Theory of Two-Phase Flow*, Eyrolles, Paris (1975).
2. Zuber, N. and Findley, J. A., "Average Volumetric Concentration in Two-Phase Flow Systems," *J. Heat Trans.*, Vol. 87, pp. 453-468 (1965).
3. Wallis, G. B., *One-Dimensional Two-Phase Flow*, McGraw-Hill Publishing Co., New York, 261-263 (1969).
4. Ishii, M., "One-Dimensional Drift-Flux Model and Constitutive Equations for Relative Motion Between Phases in Various Flow Regimes," Argonne National Laboratory Report, ANL-77-47 (1977).
5. Chawla, T. C. and Ishii, M., "Two-Fluid Model of Two-Phase Flow in a Pin Bundle of a Nuclear Reactor," *Int. J. Heat Mass Transfer*, Vol. 23, p. 991, (1980).

6. Vernier, P. and Delhaye, J. M., "General Two-Phase Flow Equations Applied to the Thermohydrodynamics of Boiling Nuclear Reactor," *Energ. Primaire*, Vol. 4, No. 1, 5 (1968).

7. Kocamustafaogullari, G., "Thermofluid Dynamics of Separated Two-Phase Flow." Ph.D. Thesis, Georgia Institute of Technology (1971).

8. Boure, J. A., "Mathematical Modeling of Two-Phase Flows," *Proc. of CSNI Specialist Meeting*, Banerjee, S. and Weaver K. R., Eds. A.E.C.L., Vol. 1, August 3-4, Toronto, 85 (1978).

9. Ishii, M. and Mishima, K., "Study of Two-Fluid Model and Interfacial Area," Argonne National Laboratory Report, ANL-80-111 (1980).

10. Chawla, T. C. and Ishii, M., "Equation of Motion for Two-phase Flow in a Pin Bundle of Nuclear Reactor," *Int. J. Heat Mass Transfer*, Vol. 21, p. 1057 (1978).

11. Delhaye, J. M. and Achard, J. L., "On the Use of Averaging Operators in Two-Phase Flow Modeling," *Thermal and Hydraulic Aspects of Nuclear Reactor Safety*, Vol. 1, Light Water Reactors, Jones, D. C. and Bankoff, S. G., Eds. ASME, New York, 289 (1977).

12. Sha, W. T., Chao, B. T. and Soo, S. L., "Time Averaging of Local Volume Averaged Conservation Equations of Multiphase Flow," NUREG/CR-3434, ANL-83-49 (1983).

13. Ishii, M. and Kocamustafaogullari, G., "Two-Phase Flow Models and Their Limitations," *NATO Advanced Research Workshop on Advances in Two-Phase Flow and Heat Transfer*, Spitzingsee, BRD, August 31-September 3 (1982).

14. Ishii, M., Mishima, K., Kataoka, I. and Kacamustafaogullari, G., "Two-Fluid Model and Importance of the Interfacial Area in Two-Phase Flow Analysis," *Proc. 9th U.S. National Congress of Applied Mechanics*, pp. 73-80, Ithaca, New York, June 21-25 (1982).

15. Sharma, M. M. and Mashelhan, R. A., "Absorption with Reaction in Bubble Columns," *Inst. Chem. Engrs., Symposium Series*, Vol. 28, Tripartite, Chemical Engineering Conference, Montreal (1968).

16. Watson, A. P., Cormack, D. E. and Charles, M. E., "A Preliminary Study of Interfacial Areas in Vertical Cocurrent Two-Phase Upflow," *Canadian J. of Chem. Engr.*, Vol. 57, 16 (1979).

17. Kasturi, G. and Stepanek, J. B., "Two-Phase Flow-III. Interfacial Area in Cocurrent Gas-Liquid Flow," *Chem. Engr. Sci.*, Vol. 29, pp. 713-719 (1974).

18. Akita, K. and Yoshida, F., "Bubble Size Interfacial Area, and Liquid-Phase Mass Transfer Coefficient In Bubble Columns," *Industrial and Engr. Chemistry*, Process Design and Development, Vol. 13, N° 1, p. 84 (1974).

19. Shilimkan, R. V. anl Stepanek, J. B., "Interfacial Area in Cocurrent Gas-Liquid Upward Flow in Tubes of Various Size," *Chem. Engr. Sci.*, Vol. 32, p. 149 (1977).

20. Bier, K., et al., "Blasenbildung und Phasengrenzflache beim Dispergieren von Gasen in Flussigkeiten an einzelnen Gaszulaufoffnungen, Teil 2.. Einflutz von Systemdruk und Stoffeigenschaften auf die Blasengrobe und die Spezifische Phasengrenzflache," *Warme-und Stoffubertragung*, Vol. 11, p. 217 (1978).

21. Gregory, G. A. and Scott, D. S., "Physical and Chemical Mass Transfer in Horizontal Cocurrent Gas-Liquid Slug Flow," *Proc. Intl. Symposium on Research in Cocurrent Gas-Liquid Flow*, Waterloo (1968).

22. Shah, A. K. and Sharma, M. M., "Mass Transfer in Gas-Liquid (Horizontal) Pipeline Contractors," *Canadian J. of Chem. Engr.*, Vol. 53, p. 572 (1975).

23. Wales, C. E., "Physical and Chemical Absorption in Two-Phase Annular and Dispersed Horizontal Flow," *AIChE J.*, Vol. 12, N° 6, 1166 (1966).

24. Proter, K. E., King, M. B. and Varshney, K. C., "Interfacial Areas and Liquid-Film Mass Transfer Coefficients for a 3-ft. Diameter Bubbble Cap Plate Derived from Absorption Rates of Carbon Dioxide into Water and Caustic Soda Solution," *Trans. Inst. Chem. Engrs.*, Vol. 44, T274 (1966).

25. Dillon, G. B. and Harris, I. J., "The Determination of Mass Transfer Coefficients and Interfacial Areas in Gas-Liquid Contacting Systems," *Canadian J. Chem. Eng.*, Vol. 44, p. 307 (1966).

26. Abdel-Aal, H. K., Stiles, G. B. and Holland, C. D., "Formation of Interfacial Area at High Gas Rates of Gas Flow Through Submerged Orifices," *AIChE J.*, Vol. 12, p. 174 (1966).

27. Brugess, J. M. and Calderbank, P. H., "The Measurement of Bubble Parameters in Two-Phase Dispersions--II, The Structure of Sieve Tray Froths," *Chem. Engr. Sci.*, Vol. 30, 1107 (1975).

28. Banerjee, S., D. S. and Rhodes, E., "Studies on Cocurrent Gas-Liquid Flow in Helically Coiled Tubes, II. Theory and Experiments on Turbulent Mass Transfer with and without Chemical Reaction," *Canadian J. of Chem. Engr.*, Vol. 48, p. 542 (1970).

29. Kulic, E. and Rhodes, E., "Chemical Mass Transfer in Co-Current Gas-Liquid Slug Flow in Helical Coils," *Canadian J. Chem Eng.*, Vol. 52, p. 114 (1974).

30. Robinson, C. W. and Wilke, C. R., "Simultaneous Measurement of Interfacial Area and Mass Transfer Coefficients for a Well-Mixed Gas Dispersion in Aqeous Electrolyte Solutions," *AIChE, J.*, Vol. 20(2), p. 285 (1974).

31. Sridhar, T. and Potter, D. E., "Interfacial Area Measurements in Gas-Liquid Agitated Vessels--Comparison of Techniques," *Chem. Eng. Sci.*, Vol. 33, p. 1347 (1978).

32. Menta, V. D. and Sharma, M. M., "Mass Transfer in Mechanically Agitated Gas-Liquid Contractors," *Chem. Eng. Sci,*. Vol. 26, p. 461 (1971).

33. Shende, B. W. and Sharma M. M., "Mass Transfer in Packed Columns: Cocurrent Operation," *Chem. Eng. Sci.*, Vol. 29, p. 1763 (1974).

34. Sridhar, K. and Sharma, M. M., "New Systems and Method for the Measurement of Effective Interfacial Area and Mass Transfer Coefficients in Gas-Liquid Contactors," *Chem. Eng. Sci.*, Vol. 31, p. 767 (1976).

35. Sharma, M. M. and Gupta, R. K., "Mass Transfer Characteristics of Plate Columns without Downcomer," *Trans. Inst. Chem. Engrs.*, Vol. 45, T169 (1967).

36. Sharma, M. M. and Danckwerts, P. V., "Chemical Methods of Measuring Interfacial Area and Mass Transfer Coefficients in Two-Fluid Systems," *British Chem. Engr.*, Vol. 15, Series 4, p. 522 (1970).

37. Veteau, J. M., "Contribution a l'etude des techniques de mesure de l'aire interfacial dans les ecoulements a bulles," These de docteaur es sciences, University Scientifique et Medicale et Institute National Polytechnique de Grenoble (1981).

38. Veteau, J. M. and Morel, Y., "Techniques de mesure des aires interfacials dans les ecoulements a bulles--II la methode chimique," CEA Report CEA-R-5092, France (1981).

39. Trice, V. G., Jr. and Rodger, W. A., "Light Transmittance as a Measure of Interfacial Area in Liquid-Liquid Dispersions," *AIChE J.*, Vol. 2, N° 2, p. 205 (1956).

40. Veteau, J. M. and Charlot R., "Techniques de sesure des aires interfacials dans les ecoulements a bulles--I Comparison de la methode d-attenuation d'un faisceau lumineux et de la methode photographiques," CEA Report CEA-R-5075, France (1981).

41. Veteau, J. M. and Charlot, R., "Interfacial Area Measurements in Two-Phase Bubbly Flows: Comparison Between the Light Attenuation Technique and a Local Method," European Two-Phase Flow Group Meeting, Eindhoven (The Netherlands), June 2-5 (1981a).

42. Veteau, J. M. and Charlot, R., "Techniques de mesure des aires interfacials dans les ecoulements a bulles---III Comparison de la methode d'attenuation d'un faisceau lumineux et d'une methode locale," CEA Report CEA-R-5122, France (1981b).

43. Calderbank, P. H., "Physical Rate Processes in Industrial Fermentation, Part I and Part II," *Trans. Inst. of Chem. Engrs.*, Vols., 36 and 37, p. 443 and p. 173 (1958, 1959).

44. Landau, J. et al., "Comparison of Methods of Measuring Interfacial Areas in Gas-Liquid Dispersions," *Canadian J. of Chem. Engr.*, Vol. 55, p. 13 (1977).

45. Kataoka, I., Ishii, M. and Serizawa, A., "Local Formulation of Interfacial Area Concentration and its Measurements in Two-Phase Flow," Argonne National Laboratory Report, ANL-84-68, NUREG/CR-4029 (1984).

46. Kataoka, I., Ishii. M. and Serizawa, A., "Local Formulation of Interfacial Area Concentration," *Int. J. Multiphase Flow*, 12, 505-529 (1968).

47. Kataoka, Il, Ishii, M. and Mishima, K., "Generation and Size Distribution of Droplet in Annular Two-Phase Flow," *J. of Fluid Engr.*, Vol. 105, pp. 230-238 (1983).

48. Ishii, M. and DeJarlais, G., "Flow Visualization Study of Inverted Annular Flow of Post Dryout Heat Transfer Region," *Nucl. Eng. and Design*, Vol. 99, pp. 187-199 (1987).

49. Ishii, M. and Grolmes, M. A. "Inception Criteria for Droplet Entrainment in Two-Phase Concurrent Film Flow," *AIChE J.*, 21(2), p. 308 (1975).

50. Ishii, M. and Mishima, Kl, "Liquid Transfer and Entrainment Correlation for Droplet-Annular Flow," *7th Int. Heat Trans. Conf.*, Munich (Sept. 1982).

51. Kataoka, I. and Ishii, M., "Mechanism and Correlation of Droplet Entrainment and Deposition in Annular Two-Phase Flow," NUREG/CR-2885, ANL-82-44 (1982).

52. Obot, N. T. and Ishii, M., "Two-Phase Flow Regime Transition Criteria in Port Dryout Region bared on Flow Visualization Experiments", *Int. J. of Heat Mass Tr.*, Vol. 31, p. 2559-2570 (1988).

53. Kocamustafaogullari, G. and Ishii M., "Interfacial Area and Nucleation Site Density in Boiling Systems," *Int. J. of Heat Mass Trans.*, Vol, 26, pp. 1377-1387 (1983).

54. Jones, O. C., Jr. and Shin, T. S., "Progress in Modelling of Flashing Inception and Critical Discharge of Initially Subcooled Liquids Through Nozzles," *Joint Japan-U.S. Seminar on Two-phase Flow Dynamics*, Lake Placid, NY, July 29-Aug. 3 (1984).

55. Riznic, J., Ishii, M. and Afgan, N., "Mechanistic Model for Void Distribution in Flashing Flow," *Proc. Transient Phenomena in Multiphase Flow*, ICHMT Seminar, May 24-30, Dubrovnik, Yugoslavia (1987).

56. Kataoka, I. and Serizawa, A., "Averaged Bubble Diameter and Interfacial Area in Bubbly Flow," *Proc. 5th Two-Phase Symp. Japan*, Kobe, Japan, Nov. 28029, pp. 77-80 (1984).

57. Wang, Z. and Kocamustafaogullari, G., "Interfacial Characteristic Measurements in a Horizontal Bubbly Two-Phase Flow," *ANS Trans.*, 62, 712-713 (1990).

58. Revankar, S. T., and Ishii, M., "Local Interfacial Area Measurement in Bubbly Flow," *Int. J. Heat Mass Transfer* (To be published) (1992).

59. Burgess, J. M. and Calderbank, P. H., "The Measurement of Bubble Parameters in Two-Phase Dispersions -I. The Development of an Improved Probe Technique," *Chem. Eng. Sci.*, 30, 743-750 (1975).

60. Burgess, J. M. and Calderbank, P. H., "The Measurement of Bubble Parameters in Two-Phase Dispersions -II. The Structure of Sieve Tray Froths," *Chem. Eng. Sci.*, 30, 1107-1121 (1975a).

61. Burgess, J. M. and Calderbank, P. H., "The Measurement of Bubble Parameters in Two-Phase Dispersions -III. Bubble Properties in a Freely Bubbling Fluidized-Bed." *Chem. Eng. Sci.*, 30, 1511-1518 (1975b).

62. Buchholz, R., Tsepetonides, J., Steinemann, J., and U. Onken, "Influence of Gas Distribution on Interfacial Area and Mass Transfer in Bubble Columns," *Ger. Chem. Eng.*, 6, 105-113 (1983).

63. Kataoka, I., and Serizawa, A., "Interfacial Area Concentration in Bubbly Flow," *Nucl. Eng. Design*, 120, 163-180 (1990).

64. Ishii, M. and Revankar, S. T., "Four Sensor Conductivity by Probe Application to Two-Phase Flow," *Proc. Int. Conf. Multiphase Flow*, 91 - Tsukuba, Sep. 24-27 Vol. (1991).

HTD-Vol. 197, Two-Phase Flow and Heat Transfer
ASME 1992

EXPERIMENTAL INVESTIGATION OF AVERAGE PARAMETERS FOR AIR-WATER TWO-PHASE FLOW IN A HORIZONTAL CHANNEL BASED ON *IN-SITU* MEASUREMENTS

Jerry K. Keska, Rohan D. Fernando, and Matthew T. Hamer
University of Nebraska
Lincoln, Nebraska

ABSTRACT

This study focuses on a two-phase air-water flow generated in a small, horizontal, 6.35 mm square channel. The problem of the direct measurement of *in situ* spatial concentration, which has been neglected in the past, is directly addressed. Spatial concentration and film thickness measurements were made with a computer-based capacitive system. Pressure drop over a length of the channel was also measured with electrical pressure transducers. These measurements were made for a variety of flow conditions which encompassed bubble, slug, plug, and annular flow regimes. The results are presented in a form that may be analyzed easily, using forms such as slip ratio, pressure drop and quality. The results are also cast in the familiar Lockhart-Martinelli correlations and compared with the existing literature. A preliminary study is made on the application of statistical methods towards flow pattern prediction. Spatial concentration is shown to be a key parameter in describing the state of the mixture in two phase flow. The differences in the results, when the spatial concentration is not taken into consideration fully or when it is inadequately estimated is displayed, when measured data is compared with experimental data and analytical models existing in the literature.

Key Words: Two-phase, capacitive technique, spatial concentration, statistical analysis, computer-based measurement system, multiphase, histogram, power spectral density, flow pattern, experimental data, air-water mixture flow, square horizontal channel, visual technique, PC-based data acquisition, capacitive sensors, film thickness measurements, time signal analysis.

NOMENCLATURE

Symbol	Description	**UNITS**
A	- Cross-sectional area	m^2
c_v	- *In situ* volumetric spatial concentration, $c_v = \frac{V_W}{V_W + V_a}$	
C	- Capacitance	pF
D	- Pipe diameter	m
h	- Thickness of the liquid film	mm
G	- Mass flux	kg/m^2s
g	- Gravitational constant	m/s^2
L	- Length	m
$\dot{M}$	- Mass flow rate	kg/s
m	- mass	kg
p	- Pressure	kPa
$\frac{\Delta p}{\Delta L}$	- Pressure gradient $(\frac{Dp}{DL})$	kPa/m
$\dot{V}$	- Volume flow rate	m^3/s
V	- Volume, Voltage	m^3,V
v	- Velocity	m/s
S	- Slip Ratio $\frac{v_g}{v_L}$	
t	- Time variable	s
x,y,z	- Axis	
x	- Mass quality	
β	- Relative concentration $= \frac{c_{v_i}}{c_{v_{i\,max}}}$	
α	- Void fraction $= \frac{V_g}{V_g + V_L}$	
γ	- Relative capacitance $= \frac{\Delta C_i}{\Delta C_{max}}$	
ϵ	- Dielectric constant	

Subscripts

a	- Air
c	- Capacitor
g	- Gas
m	- Mixture
L	- Liquid
p	- Plate
rot	- Rotameter
s	- Superficial
std	- Standard
v	- Volumetric
w	- Water
I,II,III	- First, second, third type
1,2,...,i	- 1st, 2nd,...,i-th component or value
< >	- Averaged

INTRODUCTION

The phenomenon of two-phase flow is important in the fields of mechanical, chemical, nuclear, and aerospace engineering. Many studies have been conducted and reported in the literature. Most investigations have been experimental in nature because of the limitations of existing theoretical models. Only a few studies have been concerned with small, horizontal, rectangular channels. Even fewer have reported *in-situ* measurements of spatial concentration and phase velocities. In this situation, any experimental results, with precisely described experimental conditions, are especially valuable. In the description of experimental conditions, the most important parameters are *in-situ* concentration (or its compliment void fraction) and *in-situ* phase velocities. Measurement of either of these parameters allows the determination of the other and results in an unambiguous description of the experimental conditions. This requires the application of a concentration meter and/or a phase velocity meter. Also in two-phase flows, it is necessary to be able to predict *a priori* the flow pattern which will be encountered for given flow conditions. Hence, much experimental and theoretical research has concentrated on this problem. Flow pattern, however, is a subjective parameter and thus various authors have defined flow patterns differently. In addition, the flow structure may be difficult to visualize because of high velocity and translucent liquid films. Flow pattern transitions may be influenced by a number of parameters such as channel geometry, orientation, flow parameters, and physical properties of the fluids. Various analytical, empirical and theoretical methods, which attempt to overcome the problems associated with flow pattern prediction, have been proposed. An attempt is made to correlate recent data obtained from a horizontal gas-liquid flow system with that presented in selected articles.

Flow Pattern Characterization

From a survey of the literature related to flow pattern identification, it is clear that visual flow pattern definitions are ambiguous and vary from paper to paper. Many different definitions have been proposed with varying numbers of flow pattern categories. Obviously, when flow patterns are used, they must be described and defined in detail without ambiguity, with universal definitions, and be related to the experimental conditions in order for different researchers' results to be compared. Empirical, theoretical and statistical methods have been used in attempts to identify flow pattern transitions. Almost all of the literature has studied circular flow channels with most of the diameters being large. Also most of the results in literature are related to superficial parameters such as gas and liquid velocities or quality. Furthermore, these parameters are test-loop or experiment related parameters and they are a function of spatial concentration, flow structure and phase velocities (in-situ). The work by Mandhane et al. [1] identified six major flow patterns. Barnea, Duckler et al. [2] in their work categorize five flow patterns with various subgroups proposing an analytical model for flow pattern prediction in horizontal two phase flows. In another work, Barnea et el. [3] produced flow pattern maps based on the same categories for small diameter pipes. Mukherjee et al. [4] introduced a system of empirical equations for four flow patterns for inclined two phase flow of kerosene and oil. Jepson [5] attempted to model only the transition to slug flow using a physical model. Lin and Hanratty [6] have studied pipe diameter effects on flow pattern in air-water horizontal flow. More recently, Wambsganss et al. [7] examines flow in small horizontal rectangular channels. One important conclusion of their work is that previous flow maps obtained from large pipe data are not directly applicable to small channels. Based on the authors' experience, visual identification of flow pattern is complicated and ambiguous even when techniques such as using high speed CCD cameras with slow motion facilities, "freezing" the motions with a stroboscope or dye injection enhancement, are used in parallel. The transition from one flow pattern to another is influenced by a number of parameters such as channel geometry, channel orientation, rheological properties of the fluids and mixture component composition, content and velocity. In order to eliminate the subjectivity of flow pattern identification, several authors have proposed that flow pattern transitions be defined not by visual observation, but by statistical methods. An early paper produced in this area was by Jones and Zuber [8] which dealt with statistical analysis of void fraction fluctuation in vertical gas-liquid flow. It is significant because it attempts to identify flow patterns not by visual observations, but by frequency distribution of the void fraction signal. Annunziato and Girardi [9] propose a method which uses a number of statistical parameters to recognize (and actually define) flow patterns in horizontal gas-liquid flow. Wambsganss et al. [7] uses the RMS value of a pressure signal to identify the flow pattern.

In this study, six commonly used flow pattern definitions are used. These flow pattern definitions follow those used by Richardson [10]. The characterizations used are as follows:

Stratified - Liquid flows in the bottom portion of the channel with gravity being the governing force. Gas flows along the top portion of the channel. The interface between the gas and liquid phases is smooth. If the liquid layer becomes wavy, and the waves do not touch the top of the channel, the flow is classified as **Stratified Wavy**.

Plug - Elongated bubbles of gas which vary in size, flow in the upper part of the channel. The plugs may be formed by the coalescence of smaller bubbles. Plugs are totally encased by the liquid.

Slug - Slugs of liquid flow in the channel intermittently. Slugs may form from wave flow when the waves reach the upper wall of the channel. Slugs generally flow at high velocity.

Bubble - Bubble flow consists of small, relatively uniform, bubbles flowing in the top of the channel. Bubbles flow with approximately the same velocity as the liquid. Bubbles, which fill the entire channel area, may occur.

Annular - Annular flow consists of a liquid film flowing on the channel walls, and a vapor core. In the core, liquid droplets may be entrained. Film thickness on the channel walls may be relatively uniform, or influenced by gravity, depending upon flow velocity.

These flow patterns are illustrated in fig. 1.

Void Fraction (α)

The void fraction is essential in determining the state of the mixture and its importance is highlighted by the number of correlations that exist in the literature through theoretical modelling, phenomenological curve fitting, and direct and indirect measurements. Mostly analytical and empirical models exist for void fraction prediction and mainly for well defined flow patterns.

Butterworth, in his brief communication [11], compares six models proposed by others and comes to the conclusion that all are similar, but may deviate from the actual situation if flow pattern, etc. are taken into consideration. Chen [12] taking the generalized form of the Butterworth equation gives a simplified form for annular flow. Graphical data for air-water in 50 and 90 mm pipes, measured by conductance probes for liquid slugs, are presented by Andreussi et al. [13]. In this case void fraction has been calculated from a model by assuming the bubble production rate is proportional to the film flow rate. In some reported experimental results, void fraction is calculated using parameters like quality and fluid properties. One equation that is often used is the Butterworth form [11].

$$\alpha = \frac{1}{1+K\left(\frac{1-x}{x}\right)^{p}\left(\frac{\rho_g}{\rho_L}\right)^{q}\left(\frac{\mu_\ell}{\mu_g}\right)^{r}} \tag{1}$$

where K, p, q and r take on values suggested by their respective models, from which the above equation is formulated. It should be noted that this equation does not take the flow regime into consideration.

The Lockhart and Martinelli correlation is obtained by substituting K = 0.28, p = 0.64, q = 0.36 and r = 0.07. Hence,

$$\alpha = \frac{1}{1+0.28\left(\frac{1-x}{x}\right)^{0.64}\left(\frac{\rho_g}{\rho_L}\right)^{0.36}\left(\frac{\mu_L}{\mu_g}\right)^{0.07}} \tag{2}$$

or

$$\alpha = \frac{1}{1+X^{2/3}} \tag{3}$$

expressed in terms of the Lockhart-Martinelli parameter. Chen [12] in his further examination of the void fraction correlations has incorporated an empirical parameter k to account for the deviation of the experimental data from the idealized annular flow model on which equation 3 is based. The modified equation is:

$$\alpha = \frac{k}{k+X^{2/3}} \tag{4}$$

Slip Ratio (v_g/v_L)

A single correlation by itself does not predict the slip ratio satisfactorily. The Butterworth form of the Lockhart-Martinelli correlation [11] is:

$$\frac{V_g}{V_L} = \left(\frac{x}{1-x}\right)\left(\frac{\rho_L}{\rho_g}\right)\left(\frac{1-\alpha}{\alpha}\right) \tag{5}$$

This has shown to hold good for most flow regimes by Spedding et al. [14]. Chen and Spedding [15] have analyzed the general form of the Butterworth equation (empirical) and related it to the various flow patterns by substituting different values for the equation constants. The proposed correlations are presented in graphical form for horizontal air-water data in a 45.5 mm pipe. They also claim that their correlations give good agreement with experimental data of others for pipes ranging from 25 mm to 52.5 mm (5 sizes).

Pressure Drop ($\Delta P/\Delta L$)

The main contributor to the pressure drop in adiabatic horizontal pipe flow is friction. This is the main assumption used in many of the proposed analytical solutions. However, in two phase flow, factors like the phase distribution, stratification and degree of wetting also affect pressure drop.

Lockhart and Martinelli were the first to suggest correlations for pressure drop in their 1949 paper [16], and subsequently many authors have used these same with modifications. The pressure drop in general is a function of a number of parameters including density, spatial

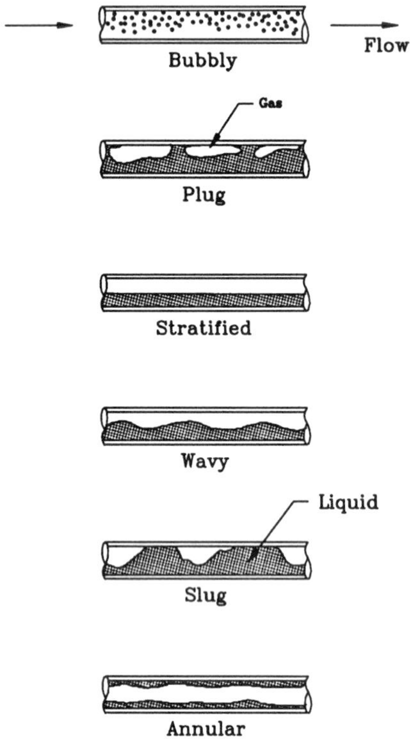

Fig. 1 Flow patterns in gas–liquid mixture flow in a horizontal channel.

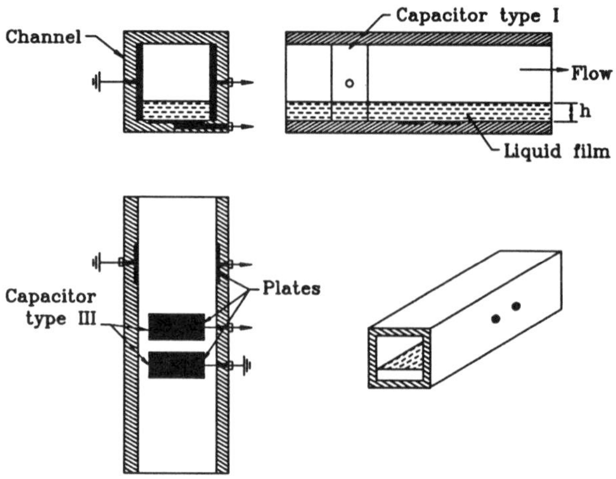

Fig. 3 Capacitive sensor system.

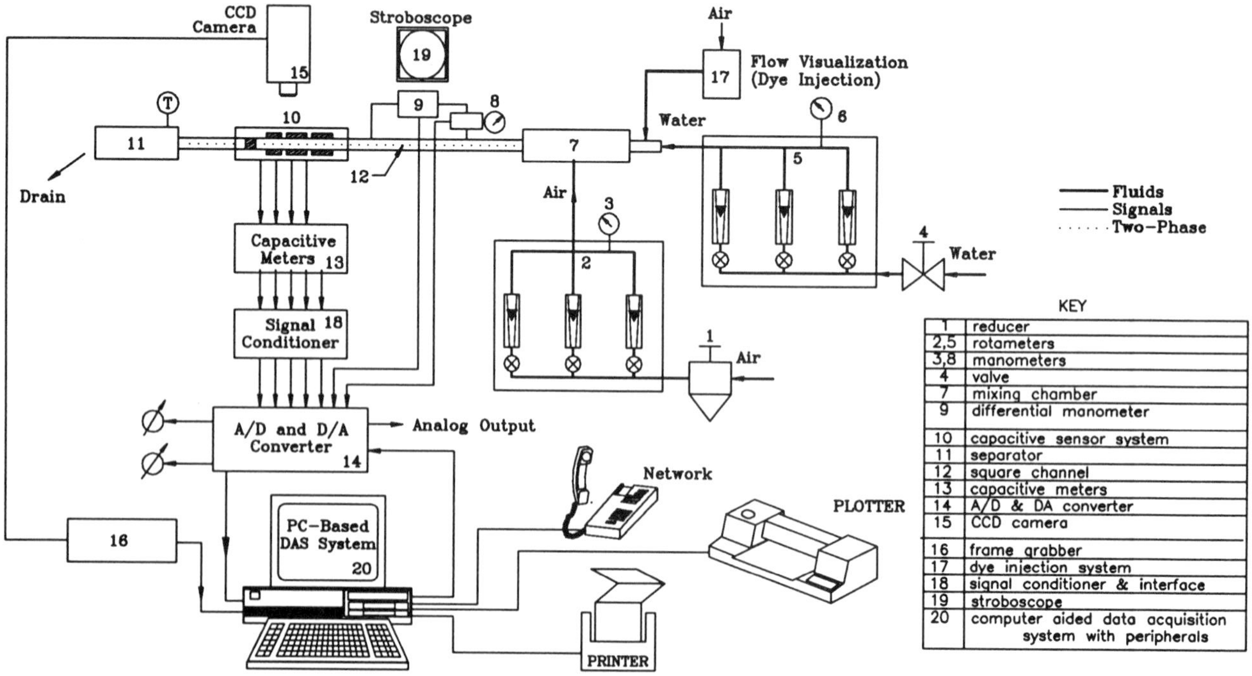

Fig. 2 Experimental apparatus for two–phase flow used in experiments.

concentration, temperature, phase velocities, channel dimensions and flow pattern. The Lockhart and Martinelli correlation introduces the parameters X and ϕ as a means of correlating the pressure drop data. These are related as follows:

$$\phi_L^2 = \frac{(\Delta P/\Delta L)_m}{(\Delta P/\Delta L)_L} \qquad (6)$$

$$\phi_g^2 = \frac{(\Delta P/\Delta L)_m}{(\Delta P/\Delta L)_g} \qquad (7)$$

$$X^2 = \frac{(\Delta P/\Delta L)_L}{(\Delta P/\Delta L)_g} = (\frac{\phi_g}{\phi_L})^2 \qquad (8)$$

Richardson in his thesis [10] expresses the parameter X in terms of the quality and phase constituents thus:

$$X = (\frac{1-x}{x})^{0.875}(\frac{\rho_g}{\rho_L})^{0.5}(\frac{\mu_L}{\mu_g})^{0.125} \qquad (9)$$

Various workers such as Butterworth, Chen & Spedding have adopted the L-M model and tested it against experimental data. While deviations exist, they have found that these are acceptable in relation to that of others that they have compared. The limitations in the L-M correlations are due to the following assumptions:

1. There is no interaction between the phases.
2. The flow pattern does not influence the static pressure drops of the liquid and gas phases which remain equal.
3. There is no acceleration of the phases.

Film Thickness (h)

The film thickness varies along the circumference of the pipe or channel in annular flow. Fukano and Ousaka [17] have examined Laurinat and Lins (1985) models and improved on them to produce a new model which they have supported through experimental data. The experimental data is taken by employing a needle contact method at 7 different circumferential locations in horizontal and near horizontal 26 mm diameter air-water pipe flow. Paras and Karabelas [18] have conducted film height measurements in a 50.8 mm pipe for horizontal air-water flow, using conductance probes, which could be set at 8 circumferential positions. Data was collected in the form of time traces. A summary of results are given for various flow parameters including film thickness. Although the data in these papers were obtained for pipe flow, average values might be comparable to rectangular channel flow.

EXPERIMENTAL APPARATUS

Experimental data is obtained for a two phase flow of air-water mixture generated in a horizontal open loop flow test apparatus as shown on fig. 2. The flow channel is square sectioned. The input parameters are measured at the entrance, the phases mixed in a chamber and flow occurs through a channel, where differential pressure is measured. The flow then passes through a capacitive sensor system and finally exits to waste with phase separation.

The flow channel (12) is constructed from a square sectioned transparent acrylic tube and has internal dimensions of 6.35 mm. Two pressure tappings are set 0.86 mm apart on top of the channel. A mixing chamber (7) is fitted to the entrance end of the channel to intimately mix the air and water. The flow rates are controlled at a metering panel (2,5).

The capacitive sensor system (10) contains the capacitive sensors for the computer-based capacitive concentration meter. The body of the system is machined from a block of acrylic plastic. Two sensors (each consisting of 2 plates) are mounted in this channel flush with the walls, in non-intrusive contact with the mixture flow. In normal orientation these sensors form type I, and type III capacitors as shown in figure 3.

The instrumentation for the capacitive sensor system is based on capacitance meters (13) and consists of two separate systems. Each system consists of a primary electronic part connected to the sensor and secondary electronics (18) where the signal is conditioned and amplified to give an analog signal proportional to the capacitance of a sensor.

The static and differential pressure between the two tappings on the flow channel are measured using electrical pressure transducers (8,9), which provide an analog voltage signal proportional to pressure. All signals in the form of time traces are acquired and stored using computer-based data acquisition systems (DAS) (20). The DAS is composed of a PC compatible computer with an analog/digital converter board (14) and control software.

Analog signals from the system and pressure transducers are digitized using a 12 bit A/D converter installed in the IBM PC compatible system. The capacitance of a type I capacitive sensor bears a linear relationship with the volumetric spatial concentration. Thus calibration may easily be performed *in-situ*. The type III sensor bears a non-linear relationship with the film thickness and this sensor must be calibrated off the flow rig. The flow pattern is observed visually with the help of a stroboscope (19) to 'freeze' the motion and a dye injection system (17) is also used for flow pattern determination. A CCD camera system (15,16) is used to record the flow pattern and video record is analyzed in slow motion.

Results and Discussion

The aim of this study is to report the *in-situ* parameters which describe the flow of air-water in a square horizontal channel. The parameters of concern are, *in-situ* void fraction or spatial concentration which describe the state of the mixture, *in-situ* phase and mixture velocities which are dynamic parameters associated with the flow, and pressure drop which is a measure of the transport energy consumption and which in turn is related to the dynamic and state parameters. Furthermore, the flow pattern, which is an effect resulting from conditions described by these *in-situ* parameters, is reported.

The authors have also presented data in the commonly used parameters in literature such as superficial velocities, quality and Lockhart-Martinelli parameters. This enables comparisons to be easily made to the literature.

A large quantity of data was generated in this experimental study. Various flow conditions defined by the inlet air and water flow rates were set, then spatial concentration, pressure drop, static pressure near the sensor, and film thickness were obtained and stored in the form of time traces by the data acquisition system. Examples of these time traces are shown in fig. 4.

From this large pool of experimental data, 20 data sets, which represent most of the possible flow patterns were chosen for detailed analysis and presentation. Measured and calculated parameters for these data sets are shown in table 2. The results are presented with emphasis on the time averaged values of void fraction, slip ratio, pressure drop, flow pattern, and RMS values.

Flow Pattern

Fig. 5 shows the relative position with respect to flow pattern for experimental data, of the variation of void fraction with the *in-situ* mixture velocity. In this case, the *in-situ* velocity is obtained from the measured data.

Figs. 6, 7 and 8 respectively show the variation of root mean square values (RMS) of the fluctuating components of pressure, spatial concentration and film thickness with quality. Additionally, the mass flux value obtained experimentally is indicated for each point in fig. 6. These 3 graphs show a local maxima of the respective RMS value at a particular quality. Spatial concentration and film thickness both have local maxima at an x value of about 0.002 as shown in fig. 7 and 8 which is significantly different from the local maxima for pressure which occurs around 0.01 for the same experimental conditions as shown in fig. 6. Due to the limited amount of data available in the literature for RMS values of void fraction or spatial concentration and film thickness, the authors were only able to compare their experimental results obtained in this study with that of Wambsganss et al. [19]. Good agreement is seen (fig. 9) for the RMS pressure of the measured data and that of [19] where both the sets of data reach their maximum values at the same quality.

Using in combination, the previously described visual methods of flow pattern recognition and employing the statistical methods of flow pattern identification, fig. 10 shows how the data obtained in this study compares to that in the literature. This figure shows the obtained flow pattern of the measured data superimposed on the flow pattern maps proposed by Wambsganss et al. and Mandhane in the superficial velocity plane. Partial qualitative agreement is seen with results of Wambsganss while quantitative differences are observed with both. The large differences noticed with Mandhane may be caused by the effect of flow channel geometry on the flow pattern.

Void Fraction (α)

Many reported studies use the quality (x) as the only parameter which describe the state of the mixture. However, quality is a composition of the spatial concentration, phase velocities and densities. It is interesting at this point to examine the relationship between the concentration $c_v = (1 - \alpha)$ and gas quality x. From the definition for quality x, by substituting for $\dot{M}_g$, $\dot{M}_L$, $\dot{V}_g$ and $\dot{V}_L$ we get the relation

$$x = \frac{\rho_g V_g (1 - c_V)}{\rho_g V_g (1 - c_V) + \rho_L V_L c_V} \tag{10}$$

Rearranging to express c_v in terms of x we get the following relation

$$c_V = \frac{\rho_g V_g (1 - x)}{\rho_g V_g (1 - x) + \rho_L V_L x} \tag{11}$$

These two equations show the interdependency of c_V and x, when the *in-situ* velocities and phase densities are known. Fig. 11 shows this graphically for the measured data. It can be seen from this graph that the liquid (concentration) has little effect on the quality when the proportion of gas is relatively high indicating that the flow may be treated as a single phase gas flow at high quality which is the case for annular flow. It is worthwhile to examine Eq. 11 for some special cases. If $\rho_g = \rho_L$, (equal phase densities) i.e. a homogeneous mixture, then

$$c_V = \frac{V_g (1 - x)}{V_g (1 - x) + V_L x} \tag{12}$$

which shows spatial concentration to be a function of the quality and *in-situ* velocities only. Furthermore, if $v_g = v_L$ (equal phase velocities) i.e. a single phase flow, or a two phase flow assumed to behave as a single phase flow; then we have the trivial solution

$$c_V = (1 - x) \equiv (1 - \alpha) \tag{13}$$

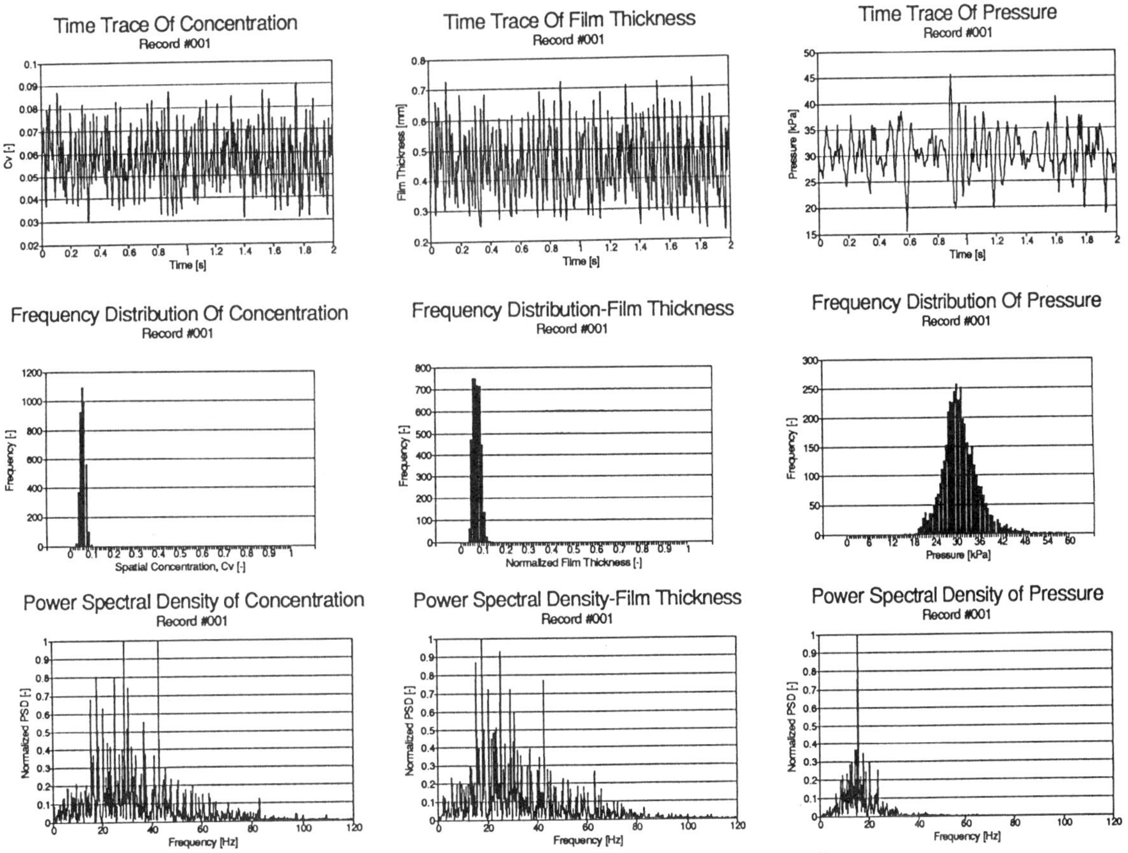

Fig. 4 Annular Flow, $<v_m>$=13.34 m/s; $<c_v>$=0.06, Record #1 (Table 1)

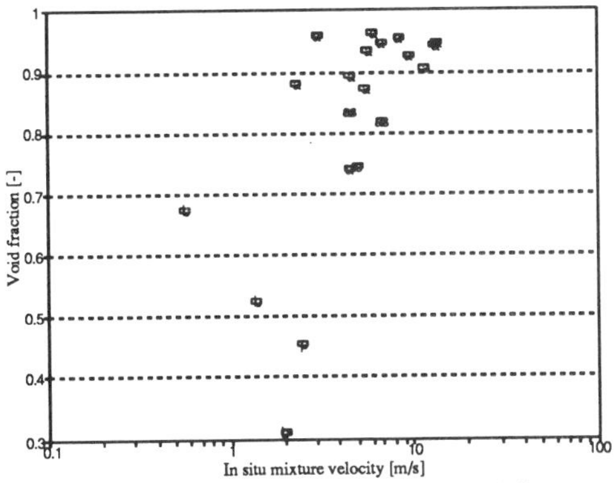

Fig. 5 Void fraction vs in situ mixture velocity as determined experimentally

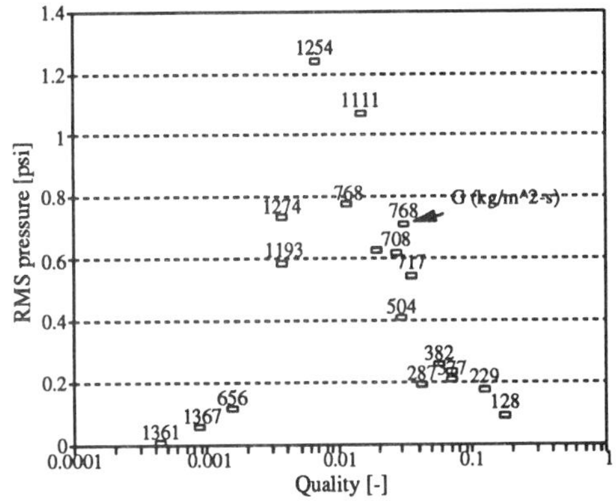

Fig. 6 Root mean square (RMS) static pressure vs quality with given mass flux values for air-water mixture.

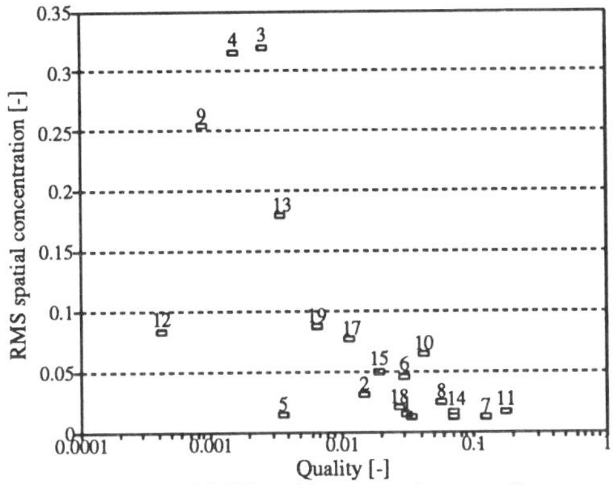

Fig. 7 RMS spatial concentration vs. quality. Number indicates record as in Table 3.

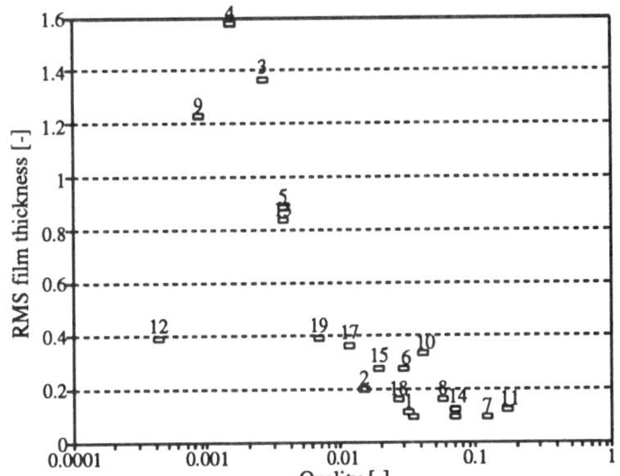

Fig. 8 RMS film thickness. Number indicates record as in Table 3.

where α is the void fraction. The implication of this equation is that determination of the spatial concentration is not required if the two phase flow is assumed to behave like a single phase flow. The same dependency of x on c_V is seen when expressed in the Lockhart-Martinelli parameter (fig. 12) where the change in x is barely noticeable for high gas content mixture (90% of gas).

The void fraction data obtained experimentally is also presented in Fig. 13 along with the idealized annular flow model as described by Chen [12] (represented by equation 3) and data of Richardson [10] and Wambsganss et al. [19]. To be able to represent the measured data in the form of the modified equation 4 by best fit criteria a, value of k = 12 was required; and a value of k = 2.75 was required for data of Wambsganss et al. [19]. Data of Richardson [10] could not be represented by a constant k value however. When this same data is presented in semi-log scale (14), the deviations from the model (Eq. 4) for the bubble/plug regions for the measured data can be clearly seen. These are the regions where the liquid component plays a major part in determining the flow pattern and thus affecting the relationship between void fraction and quality. The data obtained from literature [10] seems to have similar tendency, however the values in between indictate differences that can be generated by different experimental conditions, void fraction determination, flow pattern and other assumptions.

Slip Ratio (v_g/v_L)

The slip ratio is defined as the ratio of the *in-situ* gas velocity to the *in-situ* liquid velocity. The relationship between *in-situ* water and air velocities with indications of the flow pattern for 5 discernable regions are shown graphically in fig. 15. General transitions from bubble to annular flow follow the expected qualitative trends. To be able to compare this data with that from literature, the superficial liquid velocity must be used as an independent parameter and the variation of slip with the superficial liquid velocity is shown in fig. 16, for the measured data and that of Wambsganss et al. [19]. For the case of the measured data, the curve follows the expected trend with the slip tending to one as the liquid component increases. This comparison also indicates that the measured data is in a different range of flow structure to that compared. Further comparison with data in literature can be made in the relationship between slip ratio and void fraction as shown in fig. 17 where for low void fraction the slip tends to 1 indicating that the phases are moving with the same velocity.

Pressure Drop ($\Delta P/\Delta L$)

The obtained pressure drop data shown in table 2 is presented in fig. 18 where the mixture velocity is obtained from eq. 17. Boundary curves for single phase flows i.e. with only water and only air flowing are displayed on the graph to show the relative positions of the two phase data points. Data for these boundary curves were also obtained on the same experimental test rig.

The boundary curve for water is determined by employing the Darcy Weisbach and Blasius relations, and combining them to obtain

$$\left(\frac{\Delta P}{\Delta L}\right) = k v^{1.75} \tag{14}$$

The constant is found to be k = 3 for the measured data with water flowing alone. Hence Eq. 14 can be written as:

$$\left(\frac{\Delta P}{\Delta L}\right)_L = 3 v_L^{1.75} \tag{15}$$

The boundary curve for air is determined by fitting a cubic curve by the method of least squares, to measured data, with air flowing alone. The equation thus obtained is:

$$\left(\frac{\Delta P}{\Delta L}\right)_g = (0.53 v_g^3 - 9.5 v_g^2 + 50 v_g) \cdot 10^{-3} \tag{16}$$

The *in-situ* mixture velocity v_m is calculated by using the conservation of mass principle. i.e.

$$\dot{M}_m = \dot{M}_g + \dot{M}_L$$

Hence:

$$v_m = \frac{\rho_g (1 - c_V)\, v_g + \rho_L c_V v_L}{\rho_m} \tag{17}$$

and this can be seen as the real *in-situ* mixture velocity.

In fig. 19, the same experimental data is displayed, but in this case the *in-situ* mixture velocity is calculated by using the assumption that the total volume flow rate is equal to the sum of the individual volume flow rates ($\dot{V}_m = \dot{V}_g + \dot{V}_L$). This assumption is also deduced from the conservation of mass principle, but by setting the mixture and phase densities equal to each other i.e. a homogeneous single phase-like mixture. In this case the resulting equation is

$$v_m = (1 - c_V)\, v_g + c_V v_L \tag{18}$$

and is seen to be independent of the phase and mixture densities.

The measured data is compared with data from literature [10],[19] in the form of the Lockhart-Martinelli parameters ϕ_L and X. The results compare fairly well as shown in fig. 20. The correlations between the measured and calculated parameter X shows good qualitative agreement as shown in fig. 21. However, the effect of not taking the flow pattern into consideration is shown by the scatter of points at higher values of X where the liquid component has greater influence.

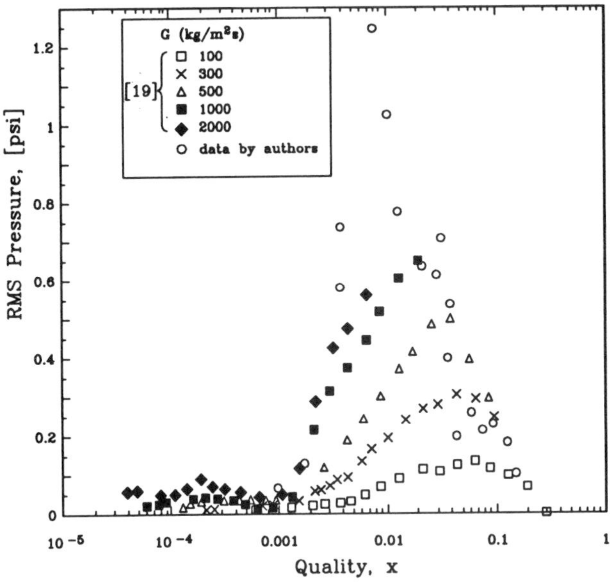

Fig. 9 RMS static pressure as a function of mass quality. Comparison of data obtained in in this experiment with data reported in literature.

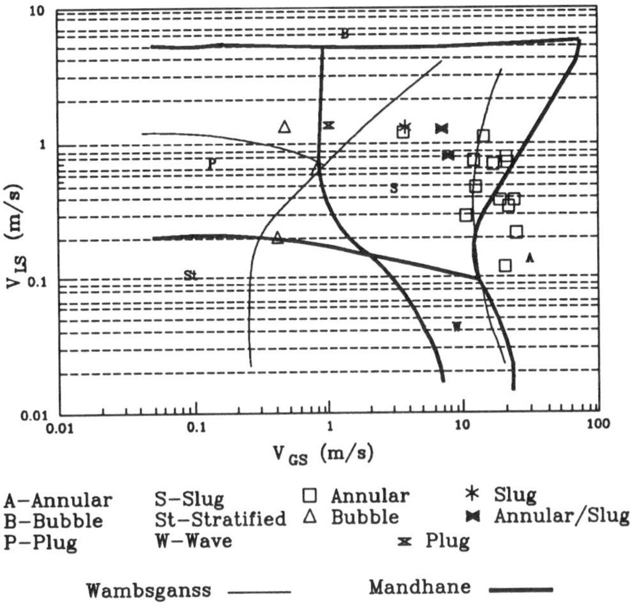

Fig. 10 Comparison of Mandhane (1974) and Wambsganss flow maps with experimental results.

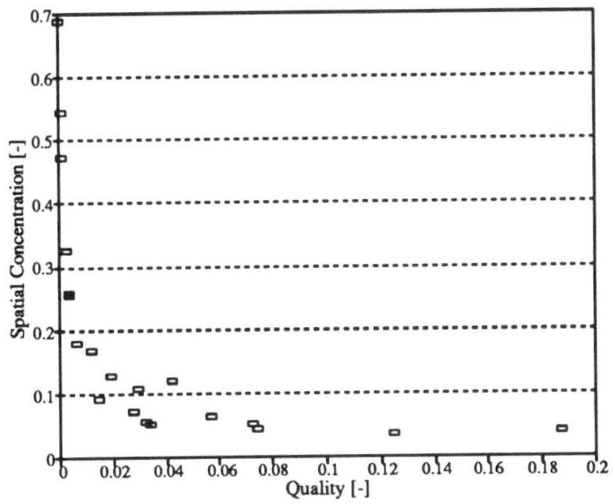

Fig. 11 Spatial concentration of an air-water mixture vs quality as determined experimentally

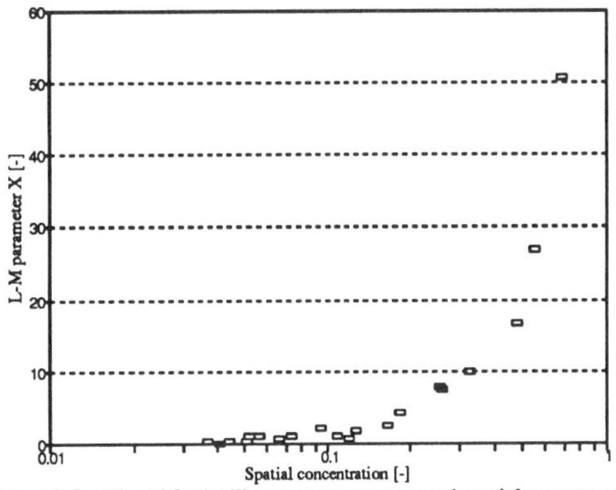

Fig. 12 Lockhart-Martinelli parameter vs measured spatial concentration

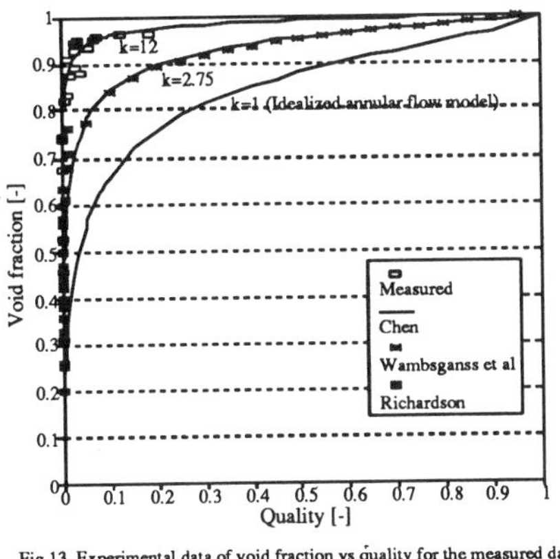

Fig 13 Experimental data of void fraction vs quality for the measured data and that from literature with curves obtained with Chens[10] model

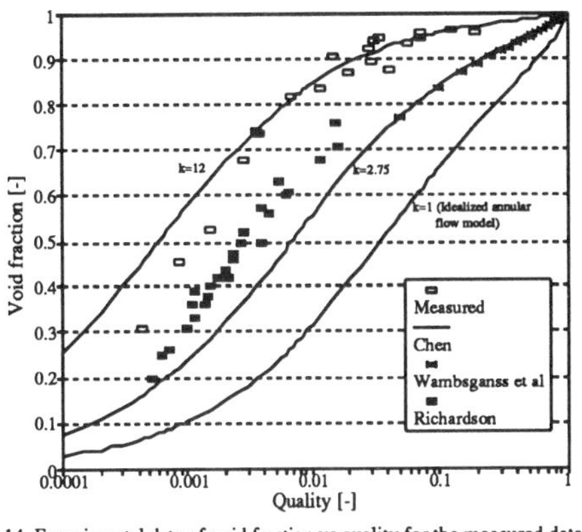

Fig. 14 Experimental data of void fraction vs quality for the measured data and that from literature with curves obtained with Chens[10] model

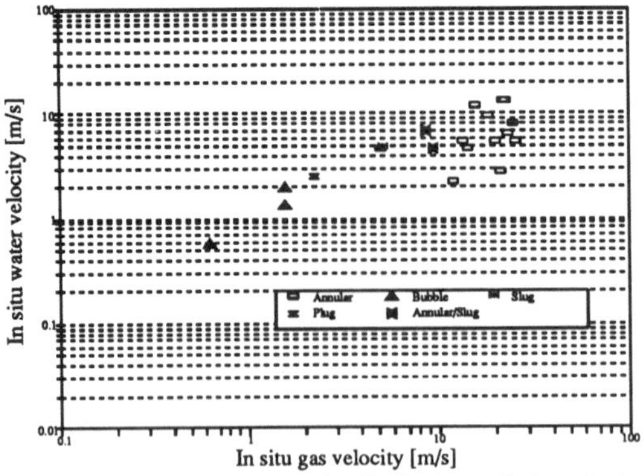

Fig. 15 Experimentally determined in situ water velocity vs in situ gas velocity with flow pattern indicaiton

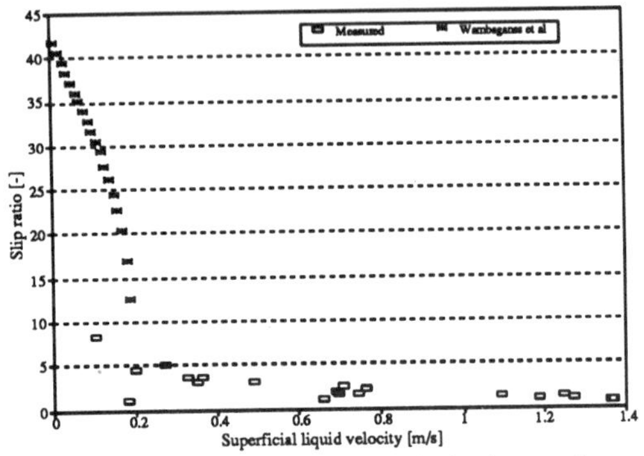

Fig. 16 Slip ratio vs superficial liquid velocity for air-water mixture

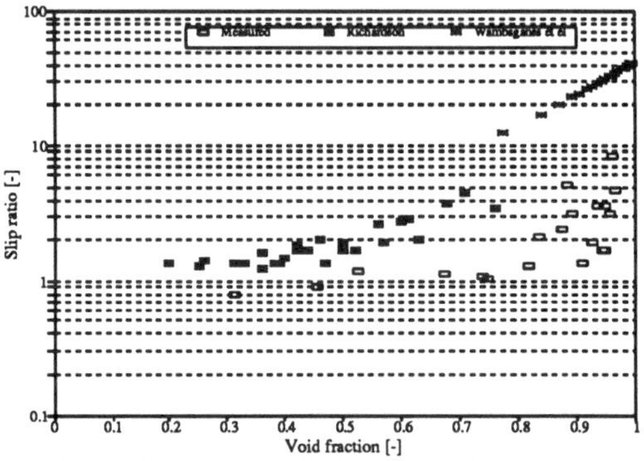

Fig. 17 Slip ratio vs void fraction measured in this experiment and that from literature.

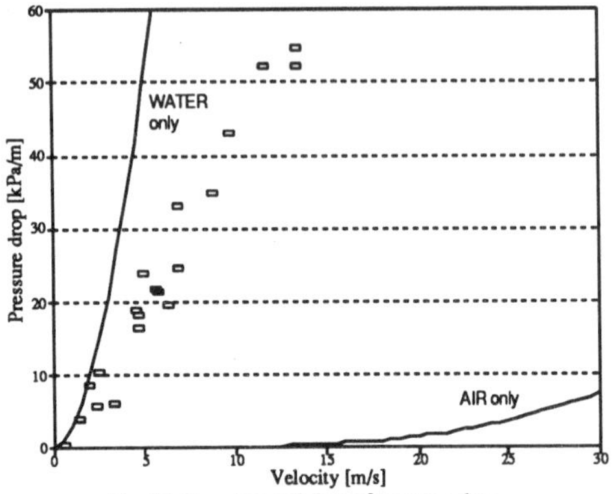

Fig. 18 Experimental data of pressure drop vs in situ velocity for case 1.

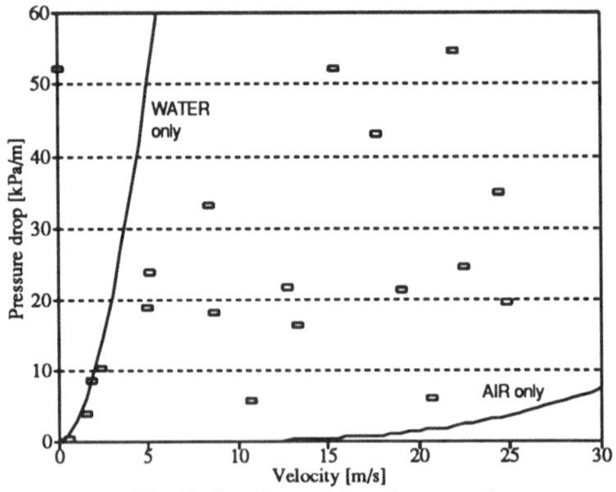

Fig. 19 Experimental data of pressure drop vs in situ velocity for case 2.

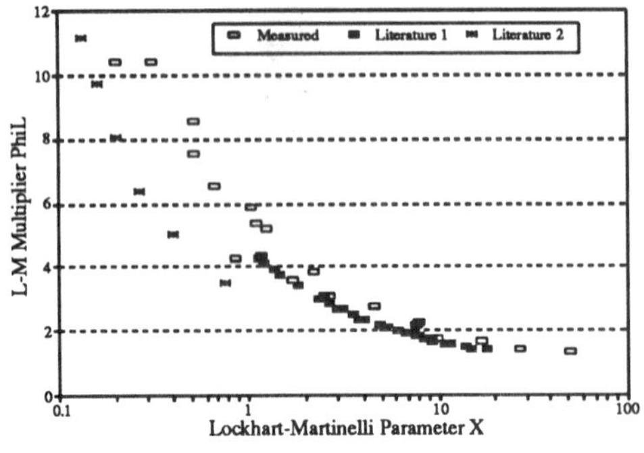

Fig. 20 Pressure drop data expressed in the Lockhart-Martinelli form for the experiment and literature; case 1

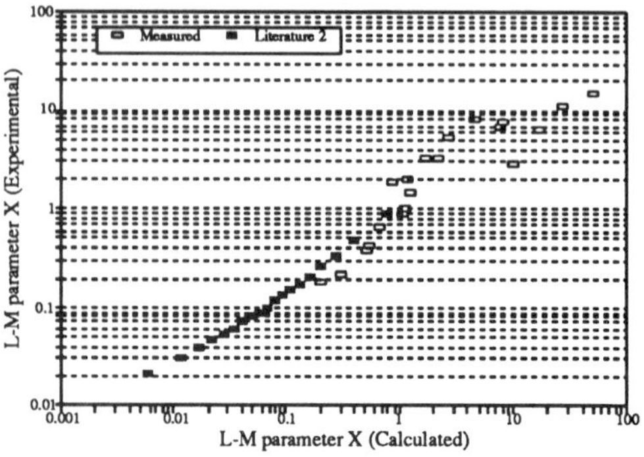

Fig. 21 Comparison of experimental and calculated Lockhart-Martinelli parameters

CONCLUSION

In summary from an investigation of the average *in-situ* components of spatial concentration, film thickness, pressure drop and phase component and mixture velocities, experimental results were obtained using a developed test loop and a computer-based concentration meter for an air-water mixture flow in a square horizontal channel. Analysis was performed on the obtained results and it was found that:

* The RMS values of the pressure, spatial concentration and film thickness are shown to be possible indicators of flow pattern where each of the parameters has a local maxima as a function of quality. The maxima for spatial concentration and film thickness appears at the same value of quality which is significantly different from the maximum of the RMS value of pressure.

* Enhanced techniques used for flow pattern recognition allowed comparison of the detected flow patterns with literature indicating only partial qualitative agreement.
* Quality of the mixture shows a limitation in representing the concentration of that mixture component, especially for a high concentration of gas in the mixture. The *in-situ* measurement of spatial concentration has been shown to be a key factor in describing the state of the mixture in a two phase flow. A general form of equation for void fraction in terms of quality found in literature is shown to fit the measured data in the annular flow regions only as shown in fig. 14.
* Slip ratio values obtained shows differences with that from literature.
* The assumptions made in computing the *in-situ* mixture velocity are shown to have a great effect on subsequent estimation of other parameters associated with two-phase flows.
* Analysis of the pressure drop indicates that the data expressed in the form of Lockhart-Martinelli parameters correlate well with that in literature for conditions close to single phase flow. However, a loss of sensitivity is seen when both phases influence the flow.

REFERENCES

1. Mandhane, J. M., Gregory, G. A. and Aziz, K., "A Flow Pattern Map for Gas-Liquid Flow in Horizontal Pipes", Intl. J. of Multiphase Flow, Vol. 1, pp. 537-553, 1974.

2. Barnea, D., Shoham, O., Taitel, Y. and Dukler, A. E., "Flow Pattern Transition for Gas-Liquid Flow in Horizontal and Inclined Pipes", Intl. J. of Multiphase Flow, Vol. 6, pp. 217-223, 1980.

3. Barnea, D., Luninski, Y. and Taitel, Y., "Flow Pattern in Horizontal and Vertical Two Phase Flow in Small Diameter Pipes", Canadian Journal of Chemical Engineering, Vol. 61, pp. 617-620, 1983.

4. Mukherjee, H. and Brill, J., "Empirical Equations to Predict Flow Patterns in Two-Phase Inclined Flow", Intl. J. of Multiphase Flow, Vol. 11, No. 3, pp. 299-313, 1985.

5. Jepson, W. P., "Modelling the Transition to Slug Flow in Horizontal Conduit", Canadian Journal of Chemical Engineering, Vol. 67, pp. 731-740, 1989.

6. Lin, P. Y. and Hanratty, T. J., "Effect of Pipe Diameter on Flow Patterns for Air-Flow in Horizontal Pipes", Intl. J. of Multiphase Flow, Vol. 13, No. 4, pp. 549-563, 1987.

7. Wambsganss, M. W., Jendrzejczyk, J. A. and France, D. M., "Two-Phase Flow Patterns and Transitions in a Small, Horizontal, Rectangular Channel", Intl. J. of Multiphase Flow, Vol. 17, No. 3, pp. 327-342, 1991.

8. Jones, O. C. and Zuber, N., "The Interrelation Between Void Fraction Fluctuations and Flow Patterns in Two-Phase Flow", Intl. J. of Multiphase Flow, Vol. 2, pp. 273-306, 1975.

9. Annunziato, M. and Girardi, G., "Horizontal Two Phase Flow: A Statistical Method for Flow Pattern Recognition", Proceedings of the 3rd International Conference on Multiphase Flow, The Hague, The Netherlands, Paper F1, 169-185, 1987.

10. Richardson, B. L. (1958), Some Problems in Horizontal Two-Phase Two-Component Flow. Ph.D. Thesis, Purdue University (Argonne National Laboratory Report # ANL-5949).

11. Butterworth, D. (1975). "A Comparison of Some Void-Fraction Relationships For Co-Current Gas-Liquid Flow", Intl. J. Multiphase Flow, Vol. 1, p. 845.

12. Chen J. J. J. (1986), "A Further Examination of Void Fraction In Annular Two-Phase Flow, Int. J. Heat Mass Transfer, Vol. 29 #11, p. 1760.

13. Andreussi, P. Bendiksen, K. (1989), "An Investigation of Void Fraction in Liquid Slugs For Horizontal and Inclined Gas-Liquid Pipe Flow", Intl. J. Multiphase Flow, Vol. 15 #6, pp. 937.

14. Spedding, P. L., O'Hare, K. D., Spence, D. R., Prediction of Holdup In Two-Phase Flow. Paper direct from author, Queen's University of Belfast, Belfast N. Ireland.

15. Chen, J. J. J., Spedding, P. L., Holdup in Horizontal Gas-Liquid Flow. Multi-Phase Flow and Heat Transfer III, Part A - Fundamentals (1984), p. 333, Process Technology Proceedings, 1, Elsevier, Amsterdam, 1984.

16. Lockhart, R. W., Martinelli, R. C. (1949), Proposed Correlation of Data For Isothermal Two-Phase, Two Component Flow In Pipes, Chemical Engineering Progress, Vol. 45, #1.

17. Fukano, T. and Ousaka, A. (1989). "Prediction of the Circumferential Distribution of Film Thickness in Horizontal and Near Horizontal Gas-Liquid Annular Flows", Intl. J. Multiphase Flow, Vol. 15 #3, p. 403.

18. Paras, S. V. and Karabelas, A. J., "Properties of the Liquid Layer in Horizontal Annular Flow", Intl J. of Multiphase Flow, Vol. 17, No. 4, pp. 439-454, 1991.

19. Wambsganss, M. W., *et al.*, "Two-Phase Flow Patterns and Frictional Pressure Gradients in a Small, Horizontal, Rectangular Channel", Report ANL - 90/91, 1990, Argonne National Laboratory, Argonne, IL.

20. Keska, J. K. "Solids Concentration Measurement in Two-Phase Mixture", 3R-International, Vol. 5, 1976, pp. 216-222.

21. Keska, J. K., "Two-Phase Mixture Flow. Theoretical and Experimental Aspects of Measurement, Mechanics and Methodics," Scientific Bulletin of the University of Hannover, Hannover, Vol. 53, 1981, pp. 246-449.

22. Keska, J. K., "Measurements of Two-Phase Flow Patterns and Void Fraction in a Small, Horizontal, Rectangular Channel", pp. 31, 1990, MCT Report, Argonne National Laboratory.

23. Keska, J. K., "Void Fraction Fluctuation as an Indicator of Flow Patterns in Gas-Liquid Flow", ASME-JSME Fluids Engineering Conference Proceedings, pp. 89-95, FED - 110, 1991, Portland, OR.

HTD-Vol. 197, Two-Phase Flow and Heat Transfer
ASME 1992

NON-UNIFORM VOID DISTRIBUTIONS IN AN ANNULAR CHANNEL WITH AIR-WATER DOWNFLOW

J. W. H. Chi
Westinghouse Electric Corporation
Pittsburgh, Pennsylvania

V. Whatley
Westinghouse Savannah River Company
Aiken, South Carolina

The fueled section of a fuel assembly in an SRS production reactor has three main parallel annular coolant channels. Because the channel void and pressure distribution must be characterized to analyze hot channels, void fractions were measured in a single-annulus test rig over a wide range of hydraulic conditions anticipated in postulated LOCAs. Experimental measurements included channel-averaged voids, limited local void fractions, and axial pressure distributions.

Two-phase flow regimes were deduced from the results of the void and pressure measurements and from visual observations in other test rigs. For conditions of critical interest to LOCA analyses (i.e., plenum pressures less than atmospheric and superficial liquid velocities less than 1.8 ft/s), channel flow was found to be predominantly slug and annular (long air bubbles in a continuum of water and air core with a water film against the wall, respectively). An analytical model of the axial void distributions was developed.

The void fractions in slug-annular flow were simulated by a theoretical falling liquid wall film model, and a channel-averaged void fraction equation was derived. The semi-empirical equation yields an excellent correlation of the channel-averaged void fraction data. For hydraulic conditions critical to design-basis LOCA analyses, the model correlates the experimental channel-averaged void fractions to a standard deviation of $\pm$ 8.3%.

NOMENCLATURE

d_e	width of the channel
DP_{ch}	fuel channel pressure drop
DP_{sys}	pressure drop across entire fuel assembly
g	acceleration due to gravity
J_a	air superficial velocity
J_L	liquid superficial velocity
J_t	transition liquid superficial velocity
P_{pl}	plenum pressure
P_{ab}	channel bottom pressure
y	fraction of a channel in slug-annular flow
Z	distance from top of channel
Z_{eff}	effective tank level= Z_{tn} - 23 inches
Z_{tn}	tank level
α_f	void fraction of slug, slug-annular, or annular flow
α_{an}	channel-averaged void fraction
α_H	void fraction for homogeneous bubbly flow
α_S	void fraction for slug flow
δ	time-averaged, local film thickness
ρ_L	liquid density

INTRODUCTION

To support ongoing code-development efforts for postulated loss-of-coolant accidents (LOCA) in SRS production reactors, void fraction was measured in a single annular channel under a wide range of hydraulic conditions. This report describes the results of the void fraction measurements, presents a two-regime model of the non-uniform void distributions, and discusses the excellent correlations with the channel-averaged void fraction data.

EXPERIMENTAL MEASUREMENTS

Two types of void measurements were made. The bulk of the data obtained were channel-averaged values, measured using quick-closing bladder valves. Limited local void fractions were also measured using a gamma-beam densitometer.

Channel-Averaged Void Fraction Measurements

Figure 1 gives a schematic of the test assembly. The test section is a single channel, a reproduction of the outer annulus of a typical fuel assembly. Channel-averaged void fractions were measured by using quick-closing bladder valves located at each end of the channel. In addition to the plenum pressure measurement, pressure taps were placed at five axial locations, as shown, to measure the channel pressure distributions.

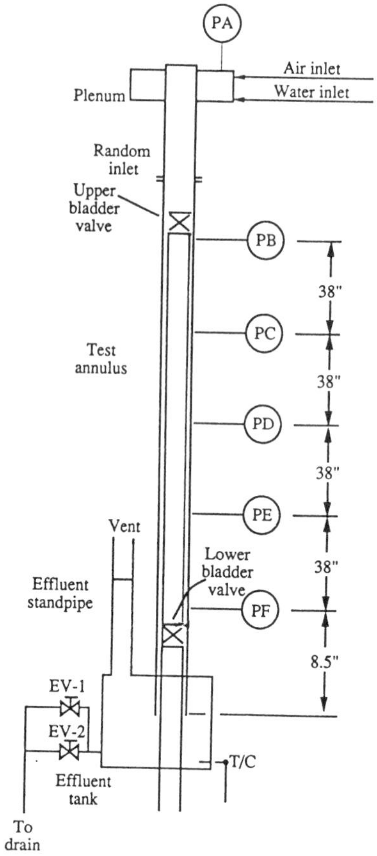

Fig. 1. Schematic of the single annular test channel.

Water, measured by calibrated turbine flowmeters, and air, measured by calibrated mass flowmeters, flowed into a plenum located at the top of the test assembly. The air/water mixture flowed from the plenum downward into a flow distributor and then into the annular channel, simulating the flow path at the top end of a fuel assembly. The test channel is a standard SRS fuel assembly outer annulus. The effluent from the bottom of the annulus discharged into a tank that simulated a low water level in a reactor moderator tank. Each series of tests, at a constant plenum pressure and channel back pressure, was conducted as follows: The channel back pressure was established by setting the "effluent" water level, measured from the bottom of the channel. The desired liquid flowrate was then introduced. The desired plenum pressure level was established by introducing an appropriate air flow.

After steady-state conditions were reached, the two quick-closing valves were actuated to trap the air/water mixture in the annulus. The volume of the water trapped in the annulus was determined by draining and then weighing the water. Pressures were measured by individual pressure gauges connected to the pressure taps shown in Figures 1 and 2. The channel-averaged void fractions so determined are instantaneous values because both valves were actuated simultantaneously. The time-averaged pressures and flowrates were measured every 5 seconds and averaged for 5 minutes. Details on the experimental work can be found in Ref. 1.

Local Void Fraction Measurements

Local void fractions were measured by a gamma densitometer in the test rig at an axial position located 60 inches below the top of the annulus. Azimuthal local void fractions were measured simultaneously with the channel-averaged void fraction determinations by placing the gamma densitometers at different positions. The two gamma densitometer beam positions are illustrated in Fig. 2. Details on these experiments are described in Ref. 2.

Experimental Results

Figure 3 shows typical channel-averaged void fractions plotted against the superficial liquid velocity for a fixed set of plenum pressure and channel back pressure. Figure 4 gives typical axial pressure distributions for a constant plenum pressure and a constant effluent level. Superimposed on Figure 3 are calculated void fractions based on the assumptions of axially uniform homogeneous flow:

$$\alpha_H = \frac{J_a}{J_a + J_L} \qquad (1)$$

and axially uniform slug flow:

$$\alpha_S = \frac{J_a}{1.2\,(J_a + J_L) - 1.5} \qquad (2)$$

The void fractions calculated from homogeneous flow theory and the empirical correlation based on slug flow are in good agreement with the measured channel-averaged void fractions only above a transition liquid superficial velocity, designated J_t.

Below the transition superficial liquid velocity, these models predict void fractions that deviate increasingly from the experimental values with reduced liquid flowrates. Above the transition superficial liquid velocity, there is little difference in the void fractions calculated by the two different models.

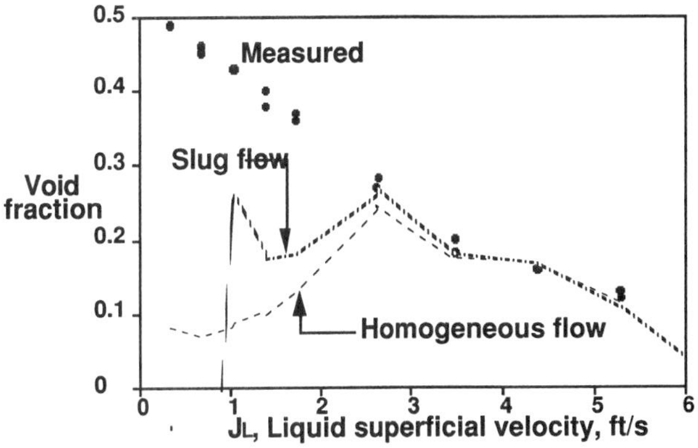

Fig. 3. Comparison of measured channel-averaged void fractions with local void fraction models DPsys=-3.42 psi, Ppl = 0 psig, Ztn = 95 in. (Zeff = 72 in)

From Figure 4, it is noted that, at low liquid superficial velocities, the pressure gradient is non-linear. However, with increased liquid flowrate, a linear axial pressure gradient is approached as the superficial liquid velocity approaches the transition value. Above the transition superficial liquid velocity, the pressure gradients remain uniform for a wide range of liquid flowrates, but the slopes gradually decrease with increased flow. These phenomena suggest that, for liquid superficial velocities less than J_t, the axial void distributions are non-uniform. Thus a generalized void fraction model is needed for assembly flow modelling and analyses.

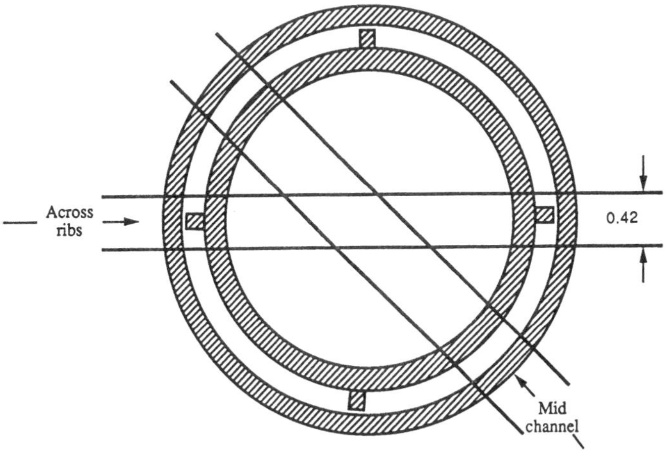

Fig. 2. Local void fraction measurements made with a gamma densitometer.

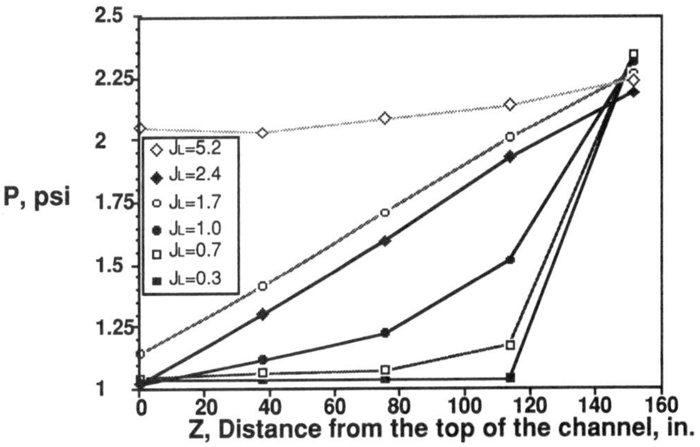

Fig. 4. Typical channel pressure distributions with negative system pressure drops DPsys = -2.42 psi; Ppl = +1 psig; Ztn = 95 in (Zeff =72 in.)

TWO-REGIME MODEL AND DATA CORRELATIONS

To develop a physically significant void fraction model, the expected two-phase flow phenomena were deduced and a physical model of the two-phase flow phenomena was then developed.

Two-Phase Flow Phenomena and Void Distributions

The sequence of two-phase flow patterns through the annuli and fuel assembly was deduced from void and pressure distribution measurements, from visual observations through the transparent sections in the test stand, and from extensive studies on other two-phase flow systems. The idealized flow phenomena are illustrated in Figure 5. Typical assembly flow tests and void measurements were made by setting the moderator tank level. The liquid flowrate was then varied. For a given set of hydraulic conditions, the axial void distribution could be represented by any one of the sub figures shown in Figure 5. Depending upon what parameters were varied, the sequence of two-phase flow patterns and void distributions are expected to follow those indicated by the directions of the arrows shown at bottom of the figure.

Falling Film Model of Slug and Annular Flow

The local void measurements made using a gamma densitometer are reproduced in Figure 6. The local void measurements were compared with the channel-averaged values obtained under the same hydraulic conditions. The comparisons strongly suggested a film flow phenomena, where lower void fractions are measured in the rib region because the gamma beam must traverse through a thicker layer of liquid (parallel to the ribs) and because the thickness of the film in the vicinity of the ribs is expected to be greater than that at the middle of the channel because of the presence of the corners and the effects of surface tension. The fact that these local void measurements bound the channel-averaged values suggest that, under the specific set of hydraulic conditions, the entire 162-in.-long channel was in slug or annular flow, where the former consists of long air bubbles entrained in a continuum of water.

The basis for the deviations between homogeneous flow theory and the measured channel-averaged void fractions at increasingly lower liquid flows appears to be due to the appearance of first slug flow, followed by slug-annular flow, and then annular flow. In both slug and annular flow, the liquid phase is the continuum with liquid continuity provided by the fully wetted wall. Long slugs of air may be regarded as "bridged" annular flow. It is suggested that slug and annular flow may be simulated by a falling liquid wall film model, illustrated in Figure 7.

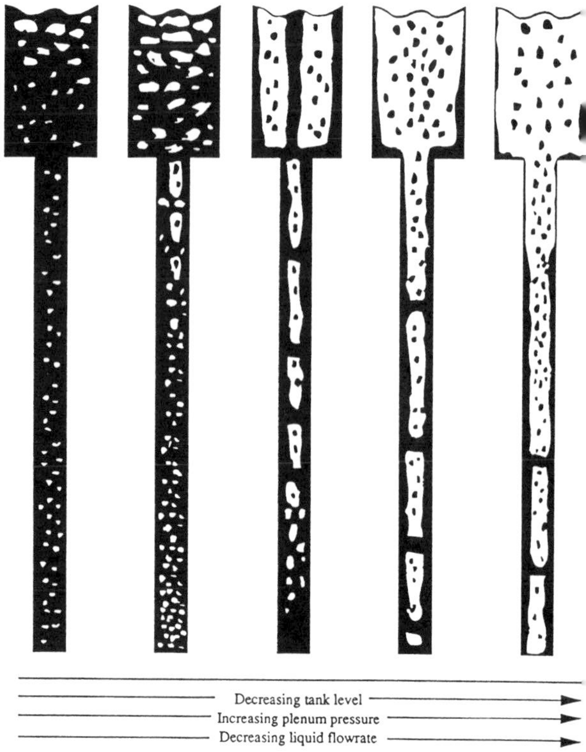

Fig. 5. Idealized sequences of two-phase flow and void distributions in a single channel during a LOCA.

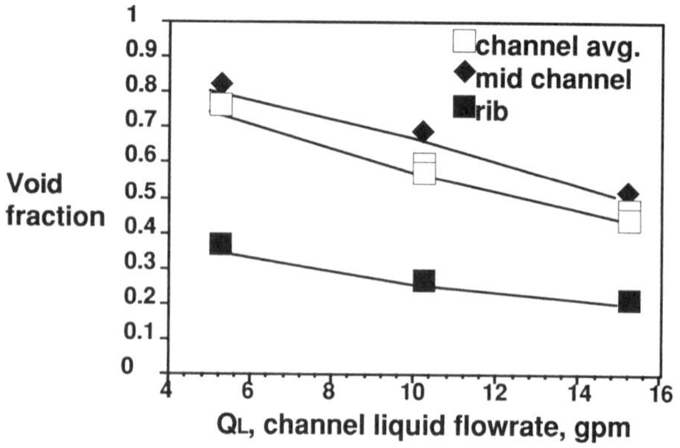

Fig. 6. Effect of liquid flow on local voidfractions.

The void fraction through the middle of the channel is then simply given by the equation

$$\alpha_f = 1 - \frac{2\delta}{d_e} \qquad (3)$$

where δ is the time-averaged, local film thickness and d_e is the width of the channel. The film thickness for a falling film in air/water flow has been studied extensively, and empirical correlations are available in the literature for both laminar flow (3) and turbulent flow (4 and 5). Void fractions calculated using the empirical film thickness equations given in the above references showed that in the laminar flow regime, there is little difference in the calculated local void fractions using either the laminar or the turbulent film flow models. At the lowest channel liquid flowrate studied here (2 gpm, in the laminar film flow regime), the difference in void fractions calculated using laminar and turbulent film flow correlations is 5%. In the turbulent flow regime, there was relatively little difference in the calculated void fractions using the two different correlations.

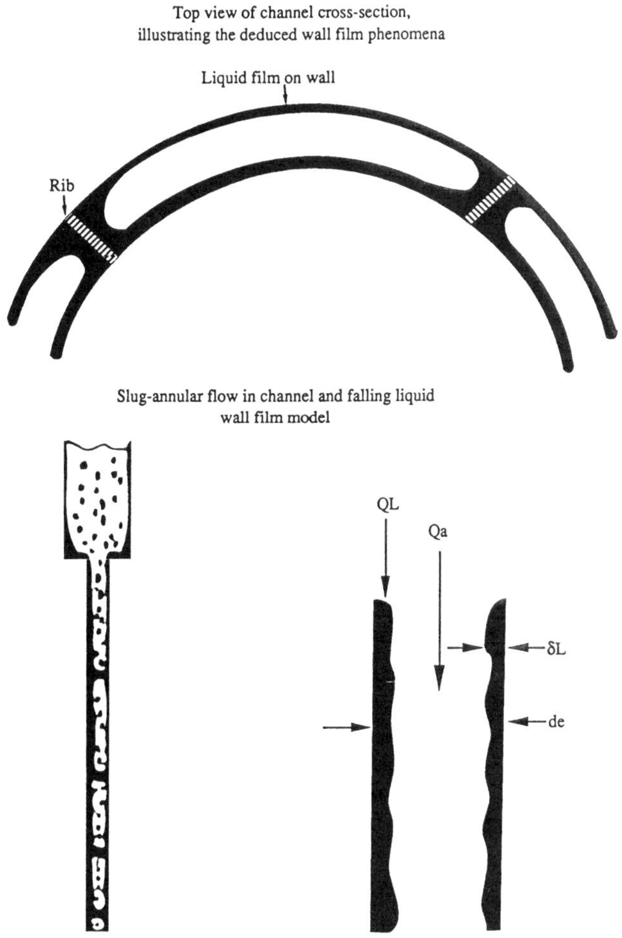

Fig. 7. Falling film model of slug and annular flow.

The semi-empirical correlation given by Belkin (5), based on analogy with turbulent flow between parallel plates (N_{re}>2000) was selected. That correlation yields the following void fraction equation for slug, slug-annular and annular flows for the channel, expressed in terms of the superficial liquid velocity:

$$\alpha_f = 1 - 0.313\, J_L^{0.58} \qquad (4)$$

The local void fractions calculated according to Equation 3 are compared in Figure 8 with the local void fractions measured by the gamma densitometer as well as with the channel-averaged values. From the comparisons, we observe the following:

- The local void fractions (gamma-beam densitometer measurements through the wall) are slightly greater than the channel-averaged values.
- The local void fractions in the vicinity of the ribs are significantly lower than the channel-averaged values.
- The falling wall film model predicts well the effects of the liquid flowrate.
- The falling wall film model yields *local* void fraction values that are in good agreement with the *channel-averaged* values.

These facts suggest that the local, azimuthally averaged void fractions should closely approach the channel-averaged values. It then follows that, under the specific set of hydraulic conditions, the entire channel is in either annular flow or slug flow, where the latter appears as long bubbles of air entrained by a film of liquid on the wall. Moreover, it appears that the falling film model is a good simulation of the two-phase flow phenomena.

Wallis (Ref. 4, p. 307) pointed out that, based on the existing database, transition from annular to slug flow occurs at an approximate void fraction of 0.8, while bubbly flow appears at an approximate void fraction of 0.1. Equation 3 can be used to estimate the transitions from annular to slug flow and from slug to bubbly flow, which occur under the following conditions:

- Transition from annular to slug flow: Q_L = 2.62 gpm or J_L = 0.46 ft/s

- Transition from slug to bubbly flow: Q_L = 35 gpm or J_L = 6.18 ft/s

Our channel-averaged void fraction data were obtained with channel liquid flowrates that range from 2 to 35 gpm. Therefore, based on the above criteria, it appears that slug flow would be the dominant mode of two-phase flow. However, the above limits are estimated on the basis of local void fractions, so that annular, bubbly, or droplet flow may occur simultaneously with slug flow over the 162-in.-long annular channel. This would be consistent with the fact that the slug flow model of the void fraction (Equation 2) correlates well with the channel-averaged void fractions for a wide range of conditions when $J_L > J_t$.

Two Discrete Flow Regime Model of Axial Void Distributions

We measured the channel-averaged void fractions for a range of plenum pressures and tank levels. Comparison of the data with those calculated from the wall film model showed that the model predicts well the monotonically decreasing void fractions with increased liquid flow. The film flow model predicts void fractions that are either higher, the same, or lower than the measured values, depending on the specific set of hydraulic conditions. The predicted values are higher with increased plenum pressures and decreased tank levels. They are lower with increasingly negative plenum pressures (higher vacuums) and higher tank levels. By comparing these trends with the expected sequence of two-phase flow patterns illustrated in Figure 4 and by keeping in mind the fact that the predicted void fractions represent local values, the following general trends may be deduced, based on the assumption that the falling film model correctly simulates slug and annular flow:

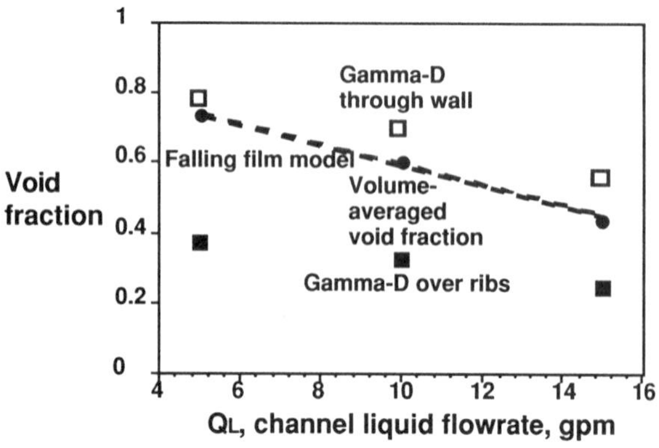

Fig. 8. Comparison of measured void fractions with those predicted by the falling wall film model.

- When the model-calculated void fractions are higher than the measured channel-averaged values, the annulus is in slug and bubbly flow.
- When the model-calculated void fractions are the same as the measured channel-averaged values, the annulus is in either all slug flow, slug-annular flow, or all annular flow.
- When the model-calculated void fractions are lower than the measured channel-averaged void fractions, a portion of the channel is in annular flow, while the upper portion may be in liquid droplet flow, a phenomenon where the walls are either partially or fully dry, with liquid droplets entrained in a continuum of air.

On the basis of these deductions, it is reasonable to postulate that, for any given set of hydraulic conditions, the channel experiences either one, or at most two, principal modes of two-phase flow. Starting from a high liquid flowrate and proceeding to increasingly lower liquid flows (as in a tank draindown process) and starting from the top of the annulus and moving downwards, one, or at most two, discrete flow regimes may be assumed, covering the entire ranges of two-phase flow regimes anticipated.

Accordingly, a simple channel-averaged void fraction model can be postulated to cover all of the two-phase flow regimes. For the general case, the channel average void fraction, α_{an}, is assumed to be made up of two terms: a fraction of the annulus that is in bubbly flow and the remainder in slug-annular flow, such that

$$\alpha_{an} = y\alpha_f + (1 - y)\,\alpha_H \qquad (5)$$

where y is the fraction of the annulus in slug, slug-annular, or annular flow, α_f is the void fraction of slug, slug-annular or annular flow, given by Equation 4, and α_H is the void fraction for bubbly flow, calculated from the homogeneous flow theory (Equation 1). It is reasonable to assume that Equation 1 is applicable to droplet/mist flow as well, where the liquid is entrained in a continuum of air.

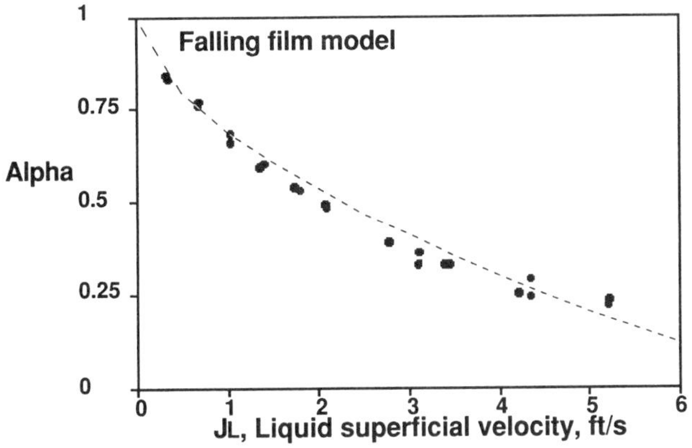

Fig. 9. Comparison of measured void fractions with values calculated from film model.
DPsys = -1.42 psi, Ppl = +2 psig , Ztn = 95 in. (Zeff = 72 in.)

Each set of the channel-averaged void fraction data, where the plenum level and the tank level were held constant, can be correlated by this two discrete flow regimes model, which contains only one adjustable parameter, the empirical constant y. Figure 9 is a representative comparison of the measured channel-averaged void fractions with those calculated from Equations 1, 4, and 5 and best-fit values of y.

General Void Fraction Correlation

The best fit values of y were readily obtained by comparing the model-calculated channel-averaged void fractions with the experimental measurements for each set of data obtained at a constant plenum pressure and tank level. The empirically determined values of y were then plotted against the channel pressure drop defined as the plenum pressure minus the channel bottom pressure ($P_{pl} - P_{ab}$), as shown in Figure 10.

The parameter y was correlated with the system pressure drop simply because plenum pressure and

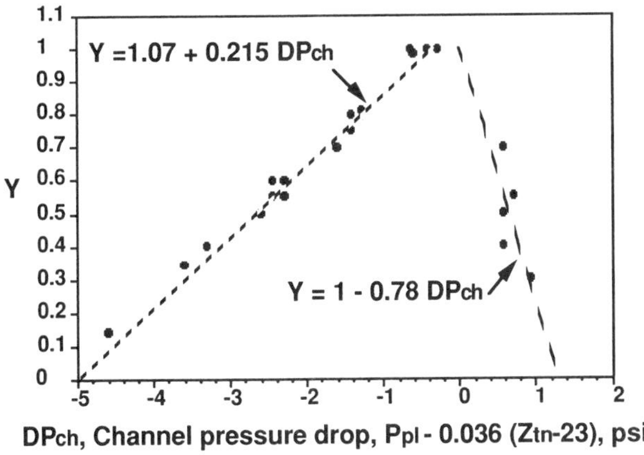

Fig. 10. Effect of channel pressure drop on Y.

tank level are key parameters measured in the hydraulic tests and because this resulted in excellent correlations. Note that the channel bottom pressure is directly related to the tank level: $P_{ab} = \rho_{Lg} (Z_{tn} - 23)$.

From the correlations of y with the system pressure drop and consistent with the expected sequences of two-phase flow patterns, the following are suggested:

- For channel pressure drops less than -0.33 psi, the channel is in bubbly flow with the parameter y given by the following empirical equation:

$$y = 1.07 + 0.215 (P_{pl} - P_{ab}) \quad (6)$$

- At a channel pressure drop of -0.33 psid, the channel is in slug flow only: with y=1.
- For channel pressure drops in the narrow range of -0.33 psi to 0 psi, the channel is in slug-annular flow, again with y = 1.
- For channel pressure drops >0, the channel is in annular-droplet flow. y in this case represents the fraction of the channel in annular flow, with the remaining fraction, (1 - y,) in droplet flow. The parameter y is given by the empirical equation

$$y = 1 - 0.78 [P_{pl} - \rho_{Lg} (Z_{tn} - 23)] \quad (7)$$

Degree of Correlations and Analysis of Errors

The calculated channel-averaged void fractions are compared to the experimental measurements in Fig. 11. Table 1 summarizes the comparison of errors. The experimental errors were estimated from a set of limited data on the reproducibility of the void fraction measurements (P_{pl} = -2 psi, Z_{tn} = 59 in). This is the only set of data that has a sufficiently large number of data points to permit the calculation of reasonable values of the errors in terms of the standard deviations in %. The % error, in contrast to the root-mean-square error, is a more significant measure of the experimental errors, because the void fractions varied over a wide range (from 0.01 to 0.86). The results show that the experimental errors increase sharply with increased flow at high channel liquid flowrates (>25 gpm). This is due to the assembly-limited flow phenomenon where, for the same system pressure drop, assembly air flow decreases with increased liquid flow. With increased liquid flow, the void fraction decreases. With the inherently low void fractions under these conditions, small errors in the

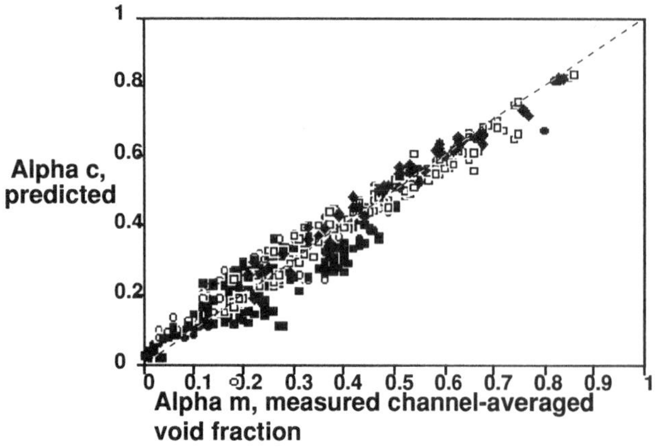

Fig. 11. Comparison of measured and predicted channel-averaged void fractions. All data points, plenum pressures -2 psi to +2 psi .

Table 1. Comparison of experimental errors with errors from the model-calculated void fractions.

DP_{sys}, psi (P_{pl} - P_{tb})	Liquid flowrate, gpm: Single channel	Liquid flowrate, gpm: Assembly	Standard deviation in %: Method A P_{pl} = -2 psi, Z_{tn} = 59 in.	Standard deviation in %: Method B All P_{pl} and Z_{tn} values
> -5.4	<36.8	<104	31.9	37.8
> -5.4	<35	<100	25.5	25.6
> -5.4	<30	< 88	22.7	21.2
> -5.4	<20	< 56	0	19.4
>-4.0	<36.8	<104	0	27.4
>-4.0	<35	<100	0	25.6
>-4.0	<30	< 88	0	17.6
>-4.0	<25	<70	0	15.4
>-4.0	<20	< 56	0	10.4
>-3.0	<36.8	<104	0	22.4
>-3.0	<35	<100	0	22.7
>-3.0	<30	< 88	0	16.5
>-3.0	<25	<70	0	14.8
>-3.0	<20	< 56	0	8.3

Note:
Method A: Based on mean values from reproducibility data.
Method B: Deviations between data and calculated values

void fraction measurements can result in relatively large % errors.

Further examination of the errors in the model-predicted void fractions showed that the deviations between the model-calculated void fractions and the experimental measurements increased with increased tank level and decreased plenum pressure (increased plenum vacuum, increased negative system pressure drop). This is again due to the fact that the void fractions decrease with increased tank level and increased plenum vacuum so that relatively small errors in the void fractions calculated can introduce relatively large % deviations. Therefore, if the two discrete regime model and the associated correlations were used for assembly flow analysis limited to system pressure drops (plenum pressure minus tank bottom pressure) in the range of 0 to -3 psi and assembly liquid flows less than 56 gpm, then the error introduced, within one standard deviation, is less than $\pm$8.3%.

Two-Phase Flow Map

From the correlations of the parameter y with the system pressure drop, a two-phase flow map can be readily constructed. The result is presented in Figure 12. The figure gives the two-phase flow regimes as functions of the plenum pressure and the moderator tank liquid level.

DISCUSSION

The falling liquid film model of slug and annular flow implicitly assumes that the tractive force at the air/water interface is negligible so that air flow has a negligible effect on the liquid film thickness. Zhivaikin (6) studied the effect of air flowrates on the liquid film thickness in co-current air/water downflow. His data showed that air flowrates have a negligible effect on the liquid film thickness for air velocities up to approximately 12 ft/s. Among all our tests, the maximum superficial air velocity was less than 6.7 ft/s. By taking the local void fraction into account, the actual air velocity is estimated to be less than 10 ft/s. Therefore the assumption of negligible surface tractive force is valid.

It should be noted that, the successful development of the void distribution model permits the prediction of the channel-averaged void fractions. This is necessary, but is insufficient for solution of the momentum equation to achieve the basic goal:

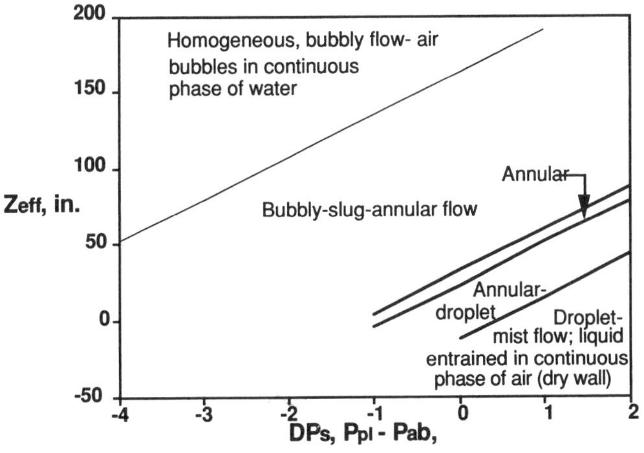

Fig. 12. Flow regime map.

the determination of assembly air flowrates from the system pressure drop. As a consequence, the application of the void fraction model to the prediction of the channel pressure drop and subsequently the system pressure drop needs to be demonstrated.

The two discrete regime model of the non-uniform channel void fractions is consistent with the non-linear channel pressure gradients. The model implies the simulation of the non-linear gradients by two straight lines. A channel frictional loss model consistent with the void fraction model should consist of a fraction of the channel, y, in separated flow, with the remainder in mixed air/water flow.

An examination of Figure 4 and other similar plots showed that the parameter y is a weak function of the assembly liquid flowrate. For an improved correlation, the effect of the liquid flowrate on the parameter y may have to be considered.

CONCLUSIONS

It is concluded that a successful correlation has been developed for the channel average void fractions in the assemblies of SRS reactors under hydraulic conditions critical to double-ended guillotine-break LOCAs. The correlation is within the overall experimental error as determined from reproducibility test data. It is is consistent with the deduced two-phase flow phenomena and the axial pressure distributions observed.

REFERENCES

1. Whatley, V., "Measurements of Void Fractions and Pressure Profiles for Downward Flowing Air/water Mixtures in Single Test Annulus," WSRC-TR-90-328, August 1990.

2. Steimke, J. L., and Fowley, M. D., "Azimuthal Variation of Void Fraction in Ribbed Annulus," WSRC-TR-91-661, February 1992.

3. Cooper, Drew, and McAdams, *Industrial Eng. Chem.*, Vol. 26, 1934, pp. 428–431.

4. Wallis, G. B., *One-Dimensional Two-Phase Flow*, McGraw-Hill, New York, 1969.

5. Belkin, H. H., et al., *J. Am. Inst. Chem. Engrs.*, Vol. 5, 1959, pp. 245–248.

6. Zhivaikin, L. Y., "Liquid Film thickness in Film-Type Units," *Intl. Chem. Eng.*, Vol. 2, No. 3, 1962, pp. 337–341.

TWO-COMPONENT TWO-PHASE FLOW STUDY IN LARGE DIAMETER HORIZONTAL PIPE

D. A. Eghbali
Westinghouse Savannah River Company
Aiken, South Carolina

ABSTRACT

Westinghouse Savannah River Company, Idaho National Engineering Laboratory, and Wyle Laboratory cooperated in a series of single (water) and two–component (nitrogen–water) calibration tests conducted to obtain sufficient information for calibrating ultrasonic flowmeters, to observe flow patterns, and to estimate void fractions. Testing, conducted in large–diameter horizontal pipe (0.38–m I.D.), covered total flows of 0.2 to 1.9 m^3/s and inlet void fractions up to 40%. A flow regime map, constructed using video images of the flow patterns, was compared with maps from the literature, with generally good agreement for interpretation of present flow patterns. In addition, the flow–induced noises on the γ–densitometer signals were analyzed in an attempt to provide another means of defining different flow regimes.

NOMENCLATURE

- C coefficient of discharge for each nozzle
- D pipe inner diameter
- k gas constant
- L_i chordal beam length of ith beam
- $\dot{m}_{N2}$ nitrogen mass flow rate
- P inlet pressure to nozzles
- Q_{N2} nitrogen volumetric flow rate
- Q_W water flow rate
- R pipe inner radius
- R_c ideal gas constant
- S slip ratio
- T inlet temperature to nozzles
- U_{sg} superficial gas velocity
- U_{sl} superficial water velocity
- v_g gas phase velocity
- v_l liquid phase velocity
- Z compressibility factor
- α static void fraction
- β inlet (flowing) void fraction
- ρ_i chordal beam density of ith beam
- ρ_{N2} nitrogen gas density
- ρ_W water density

SUMMARY

In an attempt to simulate and better understand the two–phase flow characteristics of data obtained during in–reactor hydraulic tests conducted at the Savannah River Site (SRS) in 1989, a series of single–component (water) and two–components (nitrogen–water) calibration tests were conducted at Wyle Laboratory in large–diameter (0.38–m I.D.) horizontal pipe. The primary objective was to gather sufficient data for calibrating ultrasonic flowmeters and measuring void fractions. However, because there is little published data on two–phase flow through pipes of these large diameters, the data obtained were used to construct a flow regime map using visual observations of the flow patterns (from video images). The observed flow patterns were defined as bubbly, plug, or slug flow. The regime map generated, was then compared with published flow pattern maps of Simpson et al. (1977) and Weisman et al. (1979).

Simpson's map is based on 127– and 216–mm I.D. horizontal pipes data, while Weisman's map is based on small–diameter horizontal pipe data (12 to 50 mm I.D.) and different liquid properties. Weisman's overall flow pattern map was published for air–water mixture in 25.4–mm I.D. pipe and contains parameters in the coordinates which account for fluid properties and pipe diameter effects. Generally, there was good agreement between the interpretation of flow patterns of our present data and the two reference maps.

The measured void fractions from γ–densitometer reading were plotted against the inlet void fractions to indicate the slip factor present during each flow regime. In addition, the flow–induced noises on the γ–densitometer signals were analyzed in an attempt to provide another means of defining different flow regimes.

INTRODUCTION

The primary objective of the Wyle tests was to obtain sufficient data to calibrate the ultrasonic flow meters used during 1989 in–reactor hydraulic tests. Wyle tests were conducted in a straight, horizontal, 30–m–long section of 0.38–m I.D. pipe. Wyle Laboratories was selected to perform these tests, in part because of their capabilities for large–scale flow testing with two–component flow. Because there is little published data on two–phase flow through relevant pipe diameters, the data obtained were used to construct a flow regime map using visual observations of the flow patterns.

Nitrogen gas and water were metered into a 0.59–m I.D. pipe (mixing region) 1.5 meter before the start of the 0.38–m I.D. test section. The 30–m test section contained a camera viewport to capture the flow patterns and a three–beam γ–densitometer to measure local void fraction. The three–beam γ–densitometer and the video camera were located at L/D of 21 and 71, respectively, from the test section entrance. The γ–densitometer was located upstream of flowmeters to measure the voidage and observe the flow regime before flow entering the flowmeters region. It was assumed that the flow would be fully developed at 21 diameters downstream of the test section. The video camera was mounted at the drag disk location primarily to observe the drag disk behavior throughout the test.

Reference measurements made in the test loop included water and nitrogen flow rates plus applicable pressures and temperatures. Local void fractions were measured using the three–beam γ–densitometer, which operates on the γ–ray attenuation technique. The measured void fractions from γ–densitometer reading were plotted against the inlet void fractions to indicate the slip factor present during each flow regime. In addition, the flow–induced noises on the γ–densitometer signals were analyzed in an attempt to provide another means of defining different flow regimes.

Due to a lack of large pumping capacity, the tests were transient in nature, and quasi–steady state conditions were achieved during a portion of the tests. Sixty tests, with total target volumetric flow rates of 0.2 to 1.9 m^3/s and inlet void fractions of up to 40%, were carried out. A test condition acceptance criteria of ±0.06 m^3/s on flow rates and ±2% on voids was set. The test matrix was designed to simulate the range of flows and void fractions obtained during 1989 in–reactor hydraulic tests and to obtain sufficient data for calibrating ultrasonic flowmeters and measuring void fractions.

Although air–water flow patterns in horizontal pipes have been experimentally studied by a number of investigators, most of these studies were confined to small–diameter pipes. A common technique for map generation of horizontal flows was published by Weisman et al. (1979). Weisman examined several systems with varying fluid properties and pipe diameters (12– to 50–mm I.D.). The fluid properties and pipe diameter effects are inherent in Weisman's overall flow pattern map for 25.4–mm I.D. pipe. Simpson et al. (1977) studied the behavior of the air–water mixture in 127– and 216–mm I.D. pipes.

Smooth stratified flow occurs when the liquid is at the bottom of the pipe and the gas flows along the top with the liquid surface being smooth. Wavy stratified flow is similar to smooth stratified; however, the gas–liquid interface becomes wavy. The intermittent flow is characterized by the liquid bridging the gap between the gas–liquid interface and the top of the pipe. Plug flow is considered to be the limiting case of slug flow where no entrained bubbles exist in the liquid slug. Both slug and plug flows are called intermittent flow regime. In the bubbly flow regime, small gas bubbles are distributed throughout the liquid phase. The transition to bubbly regime is characterized by the gas bubbles losing contact with the top of the tube and becoming uniformly distributed throughout the pipe as liquid flow rate increases.

TEST FACILITY DESCRIPTION

Piping Arrangement

The Wyle facility was designed to accommodate long straight test sections up to 30 meters long. A schematic illustration of the loop configuration is shown in Figure 1. The loop consisted of several

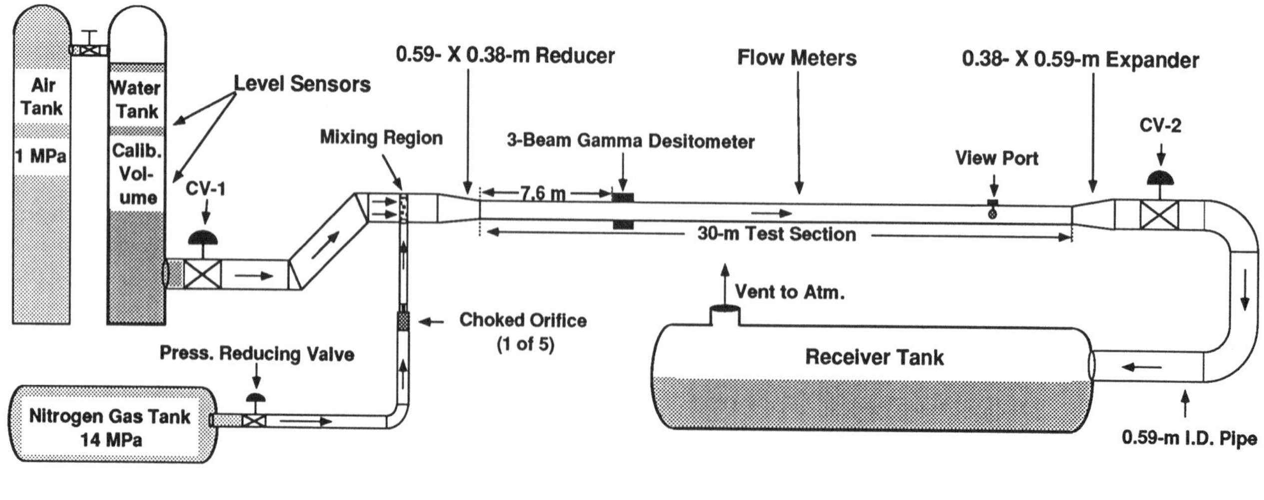

Figure 1. Nitrogen–Water Flow Calibration Loop

tanks to supply water and gas to the test section and to receive the effluent flow. The 95–m^3 water supply tank injected water into the test loop via a 0.59–m I.D. pipe. Water was driven from the water tank under nearly constant head supplied by air through an operator regulating value, from a large pressurized air tank. Two control valves (CV–1 and CV–2), located near the water supply tank exit and receiver tank inlet, respectively, were used for back pressure setting and throttling to obtain the desired flow rates. For the two–component flow tests, nitrogen gas was injected downstream of the CV–1 into the 0.59–m I.D. section (mixing region) through a specially designed cross (see Fig. 2) which consisted of two 76–mm pipes,

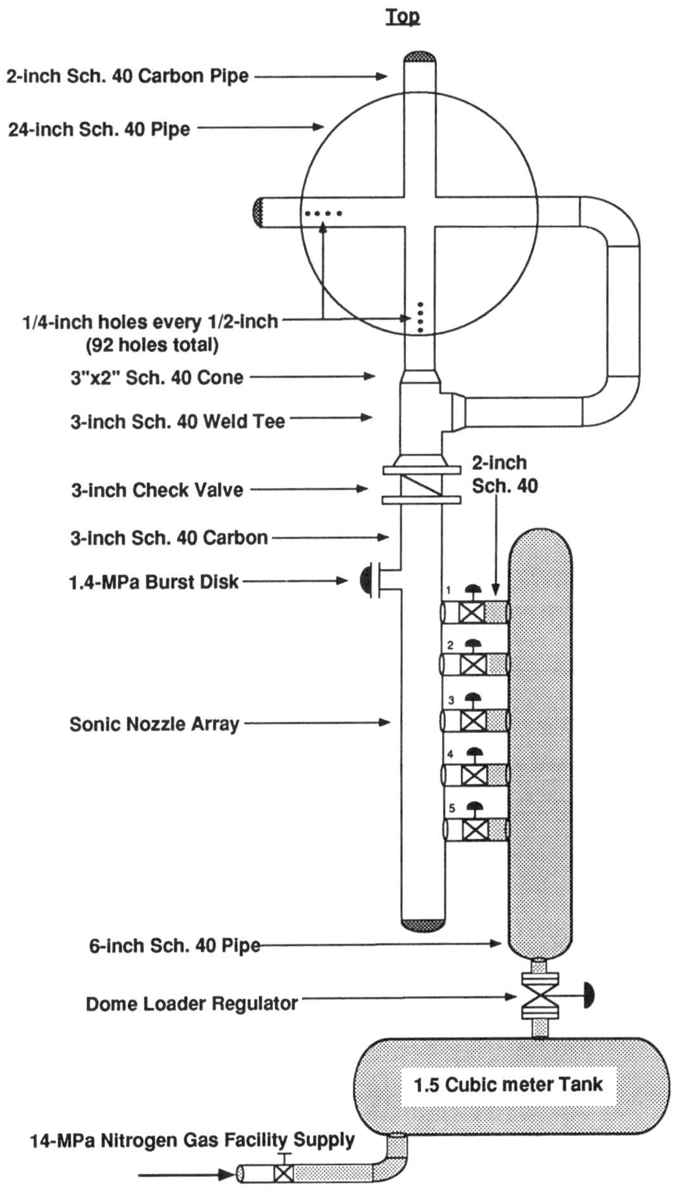

Figure 2. Nitrogen Metering and Ingection Point Cross Section

each with 46 holes of 6.4–mm diameter facing downstream. The nitrogen gas flowrate was determined with up to 5 sonic nozzles using chocked–flow theory. The nitrogen/water mixture flowed through a 0.59– by 0.38–m reducer into a 30–m–long test section where the instruments were mounted. The fluid then passed through a 0.38– by 0.59–m expander and CV–2 before it entered the receiver tank where the air and water separated.

Instrumentation

The 30–m–long non–transparent test section contained a camera viewport to capture the flow patterns. A three–beam γ–densitometer (to measure void fraction) and a video camera (to observe flow patterns) were located at L/D of 21 and 71, respectively, from the test section entrance.

The video camera was mounted at the drag disk upper view port primarily to observe the drag disk behavior throughout the test. The lower port was used for illumination by a high intensity lamp. The three–beam γ–densitometer operated by γ–ray attenuation technique. This instrument used a ^{137}Cs radioactive source to emit three γ–ray beams that passed through the pipe with different angles. The attenuated beams were then measured with three Sodium Iodine (NaI) detectors. Reference measurements made in the test loop included water and nitrogen flow rates and applicable pressures and temperatures. Static pressures were measured at intervals of 0.5 meter along the test section length. Ultrasonic flowmeters were located in the test section to measure average volumetric flow rates.

Data Acquisition System

Data were acquired by an Hewlett Packard–9000 at the rate of 30 samples per second, which establishes an uncertainty in time of ±0.0333 seconds. The basic precision of the data acquisition system was 15 bits plus sign, with a pre– and post–gain range of 1 to 500 for conversion to engineering units.

PROCEDURE

A test run began with the water supply tank filled. The water supply tank was then pressurized to a predetermined pressure (345 to 690 KPa), using air supply tank which was pressurized to 1 MPa. The test loop was filled with water and purged of any remaining air prior to each test. During the course of a test, water supply tank air pressure was manually controlled to maintain constant pressure at the tank outlet. To assure a constant flowrate throughout the test, as the water level in the water supply tank decreased, pressure was added at the top of the tank to maintain constant pressure at the tank outlet.

Target water flow rates were obtained by measuring the time required for the water tank level to drop between two level sensors, using an interval timer. The calibrated volume between the two level sensors is 18.67 m^3. The nitrogen gas flow rates were determined by measuring the pressure and temperature of the compressed gas upstream of the five variously sized sonic nozzles (see Fig. 2). Depending on a desired void fraction, a single nozzle or a combination of up to five sonic nozzles was selected. The sonic nozzles operate in a chocked–flow regime. Following each run the water was pumped through a filter from the receiver tank back to the water supply tank.

RESULTS

In the present tests, the observed flow patterns were defined as bubbly (BUB), plug (PLG), or slug (SLG) flow. Because the test section did not include a transparent pipe segment, video images of the flow patterns, captured through drag disk view port, were the only visual means for observing flow patterns. The interpretations of the observed flow patterns obtained by the video camera were compared with flow pattern maps published by Simpson et. al and Weisman et. al. which are presented in Figures 3 and 4 respectively. These comparisons were limited by the absence of stratified, wavy, and annular flows in the present study. However, a reasonable agreement was indicated between the observed flow patterns and the two reference maps. Simpson's map is based on large diameter horizontal pipe data (127– and 216–mm I.D.) while Weisman's map is based on small diameter horizontal pipe data (12– to 50–mm I.D.) and different liquid properties. Weisman's overall flow pattern map is presented for air–water mixture in a 25.4–mm I.D. pipe, with parameters in the coordinates to account for fluid properties and pipe diameter effects.

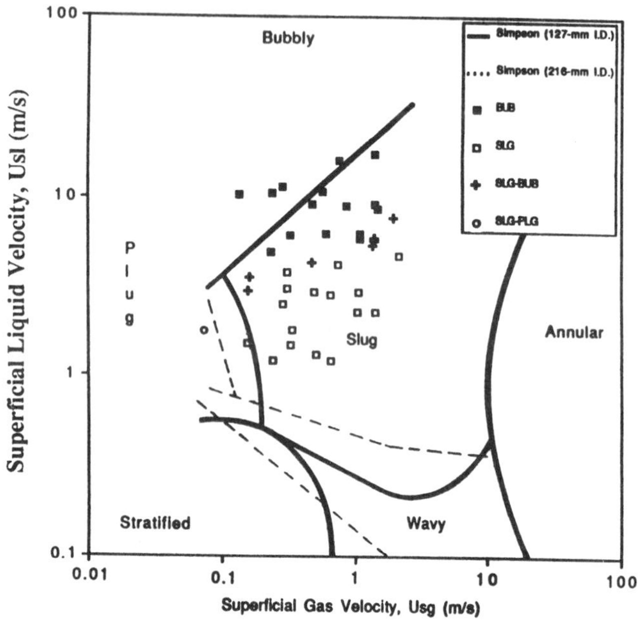

Figure 3. Comparison of Present Data with Simpson's Flow Pattern Map

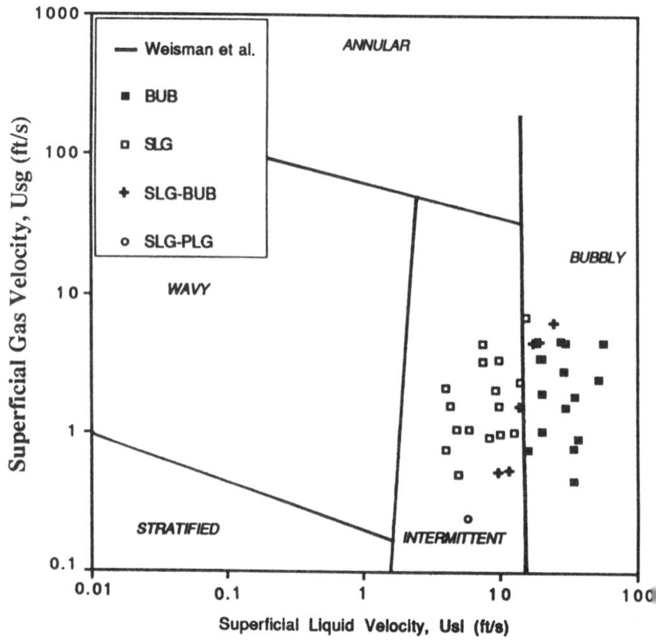

Figure 4. Comparison of Present Data with Weisman's Flow Pattern Map

The pressure measurements in the test section showed significant fluctuations throughout the test and did not supplement the video data for flow pattern identification. The non–uniform addition of pressure (manually controlled valve) at the top of water supply tank could account for most of the pressure fluctuation throughout the loop during the tests. A more uniform pressure drop fluctuations along the test section, however, would be helpful in determination of flow patterns present. Weisman et al. (1979) utilized pressure drop fluctuations technique for flow pattern identification.

The γ–densitometer results were analyzed to estimated local void fractions using three different models: simple stratified model, beam–weighted average model, and Lassahn's model (1983). The detailed derivation of these models are provided in the Appendix.The measured local void fractions using these models were then plotted against the inlet void fractions to indicate the slip factor present during each flow regime. These comparisons are presented in Figures 5, 6, and 7 respectively. These Figures indicate that the slip factor is close to unity for bubbly flow regime and it deviated from unity for other flow regimes. Figure 5 indicates that local void fraction estimated using simple stratified model is not a good representation of true local void fraction due to lack of stratified flow regime in the present data. The beam–weighted average model (Fig. 6) assumes homogeneous flow and therefore is valid only for bubbly flow regime where the two phases move with nearly equal velocities. Figure 7 indicates that local void fraction estimated using Lassahn's model is a good representation of true local void fraction. Whether the γ–densitometer readings are representative of an equilibrium situation at L/D = 21 can be questionable. Some literature documents state that in the absence of bends, an equilibrium situation requires L/D of 60 or greater.

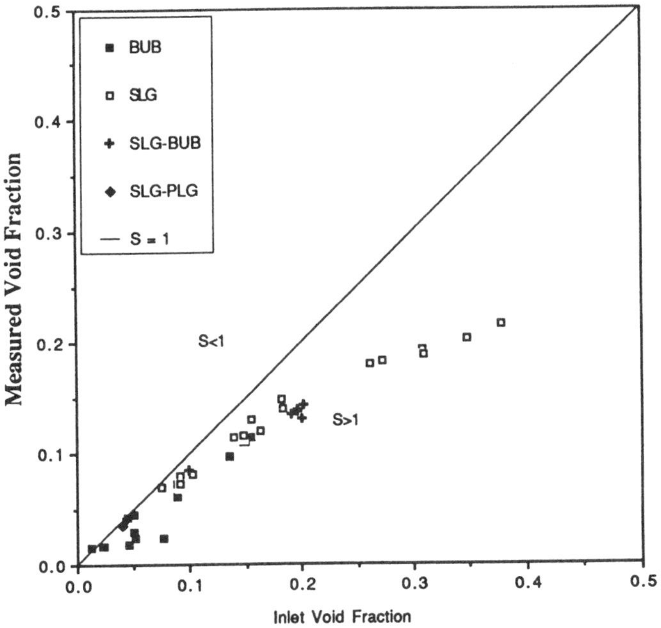

Figure 5. Local Void Fraction Using Smooth Stratified Model vs Inlet Void Fraction

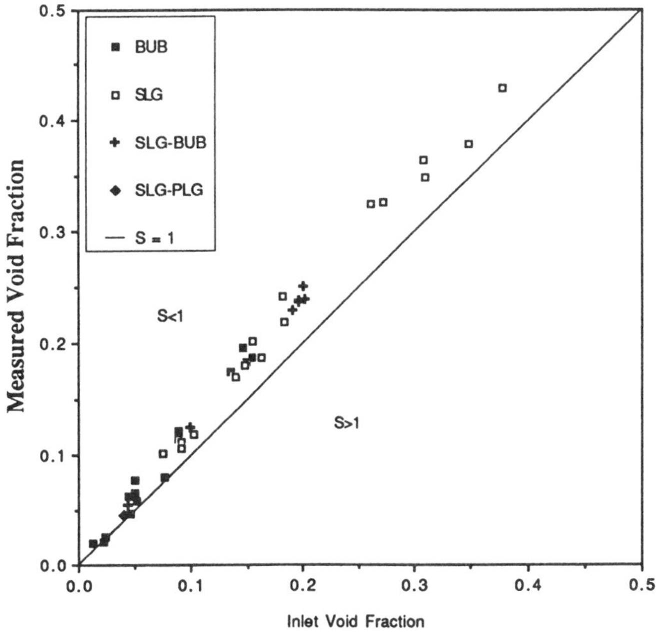

Figure 6. Local Void Fraction Using Beam Weighted Average Model vs Inlet Void Fraction

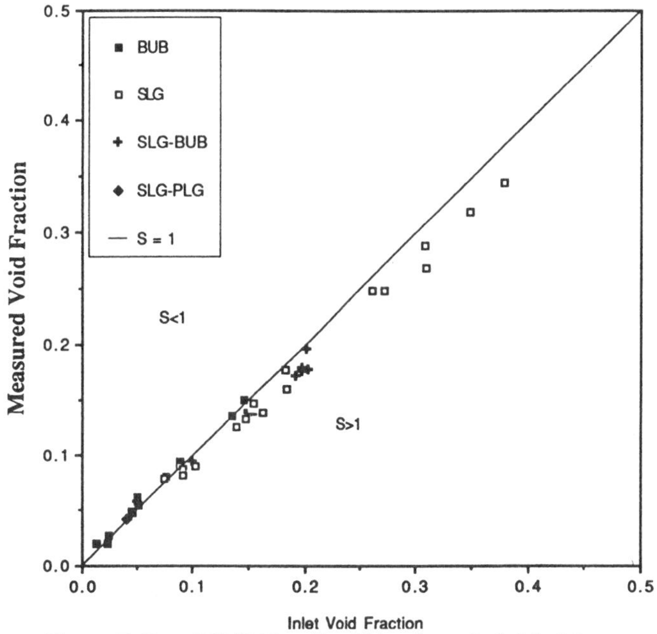

Figure 7. Local Void Fraction Using Lassahn's Model vs Inlet Void Fraction

At sufficiently high Reynolds numbers, the flow will be turbulent, characterized by the random motion of the fluid. Under turbulent two–component flow conditions, the γ–densitometer reading fluctuated about the mean value. The degree of the fluctuation depended upon the inlet water and air flow rates. The instantaneous γ–densitometer reading is an irregularly oscillating function that may be expressed as the combination of the mean γ–densitometer reading and the γ–densitometer reading fluctuation. The coefficients of variation (CV), defined as the standard deviation divided by the mean value of the γ–densitometer readings, were analyzed to provide another means of defining different flow regimes. The plots of CV versus superficial velocity ratio for γ–densitometer upper and middle beam are presented in Figures 8 and 9 respectively. These Figures indicate that the variation from mean increases as superficial gas velocity ratio increases. Analysis of variations in γ–densitometer signals can be helpful in flow regime identification.

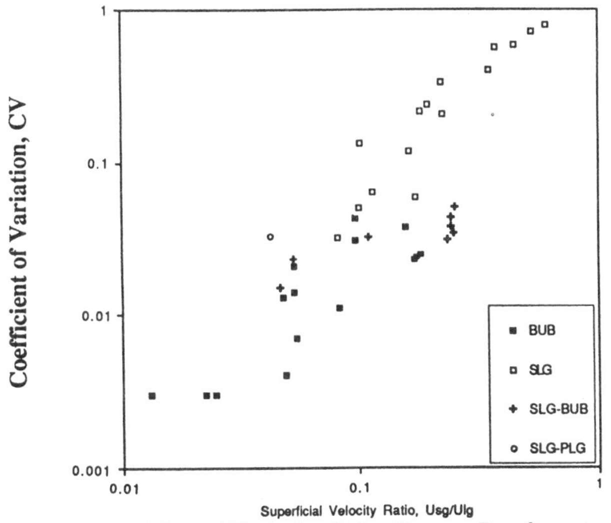

Figure 8. Coefficient of Variation Using Gamma Densitometer Upper Beam vs Superficial Velocity Ratio

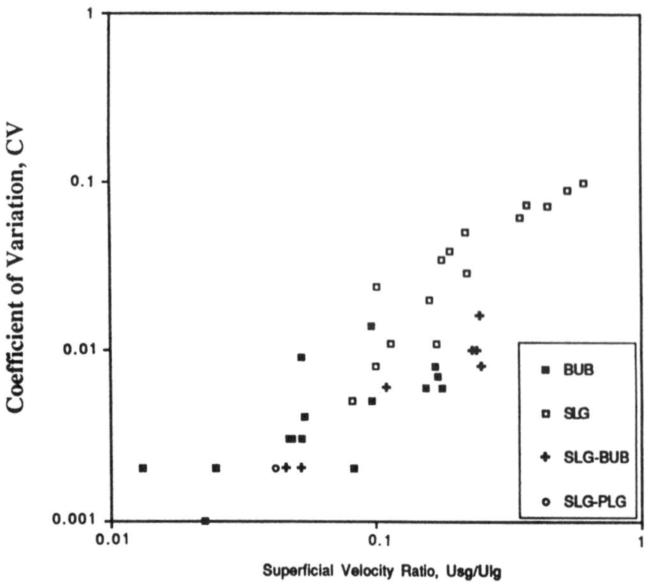

Figure 9. Coefficient of Variation Using Gamma Densitometer Middle Beam vs Superficial Velocity Ratio

UNCERTAINTIES

Evans (1991) reports the uncertainty associated with the γ–densitometer readings as ±0.9, ±1.2, and ±1.0 percent of reading for upper, mid, and lower beams respectively. Uncertainties associated with water and nitrogen volumetric flowrates were ±0.34 and ±0.85 percent of the readings respectively. Uncertainty associated with worst case total volumetric flow rate and inlet void fraction were ±0.33 and ±0.94 percent of reading respectively. The non–uniform addition of pressure at the top of water supply tank caused variation of 1 to 2% in the tank outlet pressure throughout the test. The uncertainty associated with water supply tank thermal expansion was found to be negligible. Use of control valves (CV–1) for adjusting back pressure caused some of the dissolved air to come out of the water/air solution and being detected downstream by γ–densitometer. To avoid this, control valve (CV–1) was kept fully open for the remaining tests.

CONCLUSIONS

With the exception of the bubbly flows, the video images of the flow patterns were difficult to interpret. Because the video data was the only means for observing flow patterns, the flow regime transition boundaries are less obvious. Testing provided a small amount of data for a small portion of the referenced flow regime maps. The transition boundaries on Figures 3 and 4 were defined on the basis of the flow patterns deduced from video camera data, the interpretation of which can be subjective.

The bubbly flow was dominant at low void fraction. However, the high turbulence caused by high liquid flow rates broke up the bubbles and maintained the bubbly flow regime even for higher void fractions.

Figure 7 indicates that the slip factors in bubbly flow regime are nearly equal to 1. In slug flow regime, however, the slip factors are greater than 1 (as expected) and are dependent on air flow rate. Figure 7 also indicates that Lassahn' model for determination of local void fraction using γ–densitometer results best estimates the true local void fraction and should be used for reduction of γ–densitometer results.

Because of limited number of tests, the present data were obtained in the intermittent and bubbly regions of the two referenced flow pattern maps only, limiting the comparison to these regimes. This study indicated that the present data were in good agreement with both referenced flow pattern maps of Simpson et. al and Weisman et. al.

The analysis of variations in the γ–densitometer signals can be helpful in describing the flow patterns present. This technique can be used for flow regime identification where the presence of transparent piping for flow regime observation is not possible.

ACKNOWLEDGMENT

The information contained in this article was developed during the course of work under Contract No. DE–AC09–89SR18035 with the U.S. Department of Energy. This experiment was conducted with cooperation from Westinghouse Savannah River Company, Idaho National Engineering Laboratory, and Wyle Laboratory. The encouragement of D. R. Muhlbaier, and helpful discussions with J. C. Whitehouse are appreciated.

REFERENCES

Collier, J.G., "Convective Boiling and Condensation", 2nd Ed., McGraw–Hill, 1981, pp 5–21.

Eghbali, D.A., "Data Reduction Program Description for 1989 L–Reactor Phase I Test", WSRC–TR–91–456,1991, pp 5–7.

Evans, R.P., "Savannah River Laboratory L–Area Reactor 1989 Phase–I Special Hydraulic Tests Uncertainty Report", EG&G–EE–9496, EG&G Idaho, 1991, p C–5.

Lassahn, G.D., "Mass Flow Rate Estimation Using Subroutines DPR)F4 and EMDOT4", EG&G Idaho, 1983, pp 24–26.

Simpson, H.C., Grattan., D.H.E., and Al–Samarrae, F., "Two Phase Flow in Large Diameter Horizontal Tubes", 1977, European Two Phase Flow Group Meeting, Interim Report, pp 27–31.

Weisman, J., et al., "Effects of Fluid Properties and Pipe Diameter on Two–Phase Flow Patterns in Horizontal Lines", Int. J. Multiphase Flow, 1979, p441.

Whitehouse, J.C., et al., "Measurement of Two–Component Flow Using Ultrasonic Flowmeters", ASME winter meeting,1991, pp 1–2.

APPENDIX

Inlet Void Fraction Measurements

Inlet void fraction was calculated as follows:

$$\beta = \frac{Q_{N2}}{Q_{N2} + Q_w}$$

where

$$Q_{N2} = \frac{\dot{m}_{N2}}{\rho_{N2}} = \frac{f(k,C,P,T)}{f(P,T,R,Z)}$$

and

$$Q_w = \frac{18.67\ m^3}{\Delta T}$$

where ΔT is the time the surface level in the water supply tank takes to pass by two specific level sensors and $18.67 m^3$ is the volume of the calibrated region of the water supply tank.

The static void fraction, α, and the flowing void fraction, β, can be related as follow,

$$\beta = \frac{1}{1 + \frac{V_w}{V_g}\frac{1-\alpha}{\alpha}}$$

and the slip ratio is defined as

$$S = \frac{V_g}{V_l} = \frac{\beta}{(1-\beta)}\frac{(1-\alpha)}{\alpha}$$

Local Void Fraction Measurements using γ–Densitometer

Void fractions at L/D of 21 from the test section entrance were measured by the γ–ray attenuation technique, using a ^{137}Cs radioactive source and three Sodium Iodine (NaI) detectors. Figure 1 presents the geometry and reference angles of the γ–densitometer.

The principle of this technique states that the attenuation of a beam of γ–rays passing through the gas/water mixture, is dependent on the mean density of the tube contents. The mean density and hence the void fraction can be deduced by analyzing the degree of γ–rays beam attenuation after passing through the tube. The intensity of the γ–ray beams after passing through the tube is measured by three NaI detectors. The γ–densitometer apparatus was designed and calibrated by INEL. Theory of γ–ray attenuation gives a straight line relationship between ln (I_x) and void fraction α that is

$$\ln (I_x) = a + b\,\alpha$$

when the tube is full of water, $\alpha = 0$

$$a = \ln (I_{xf})$$

when the tube is full of gas, $\alpha = 1$

$$b = \ln\left(\frac{I_{xg}}{I_{xf}}\right)$$

and hence

$$\alpha = \frac{\ln\left(\frac{I_x}{I_{xf}}\right)}{\ln\left(\frac{I_{xg}}{I_{xf}}\right)}$$

The following models will use γ–densitometer readings to estimate void fraction.

Beam Weighted Average. There are three density values obtained (ρ_1,ρ_2, ρ_3), each measuring along a different chord length in the cross section of the flow. The measured values are bounded by air and liquid densities. The beam–length weighted average of three chordal densities is obtained using the following formula

$$\rho = \frac{L_1\rho_1+L_2\rho_2+L_3\rho_3}{L_1+L_2+L_3}$$

hence,

$$\alpha = \frac{(\rho_w-\rho)}{(\rho_w-\rho_g)}$$

A limitation to this method is due to the fact that it assumes homogeneous liquid–gas mixture model and is valid only for homogeneous bubbly flow.

Smooth Stratified Model. The smooth stratified model utilizes the upper beam density to estimate the void fraction as follows:

$$\alpha = \frac{1}{\pi R^2}\{R^2 \mathrm{Cos}^{-1}(\frac{R-h}{R}) - (R-h)(h(2R-h))^{0.5}\}$$

where

$$h = x + L_{3c}$$

$$x = L_{3c}\frac{(\rho_w-\rho_1)}{(\rho_w-\rho_g)}$$

The γ–densitometer geometry of cord lengths is shown in Figure 10.

Lassahn's Models. Lassahn (1983) states that the readings from the several beams of a typical γ–densitometer cannot uniquely determine the density profile of the fluid in the pipe. This is due to the fact that the γ–densitometer gives no information about the density of the fluid in the space not sampled by the densitometer beams. However, the limited data from a densitometer can be utilized to determine which (if any) of several preselected density profiles actually exists in the pipe. Among all the models for density profile estimation in Lassahn's program, the following two models were applicable to our data.

1. Homogeneous Model: this model checks the three measured chordal average density values. If these values are all equal (or very nearly equal), the other model is not tried, and homogeneous profile is specified as best estimate of the density profile.
2. Stratified Model: the local density profile by this model is given by

$$\rho = \rho_w - \frac{\rho_w-\rho_{N2}}{1+\exp(-4a(z-b))}$$

where a and b are adjustable parameters that specify the width of the liquid to gas transition region and z–coordinate of the middle of the liquid to gas transition region respectively. For more detailed explanation refer to Lassanh (1983).

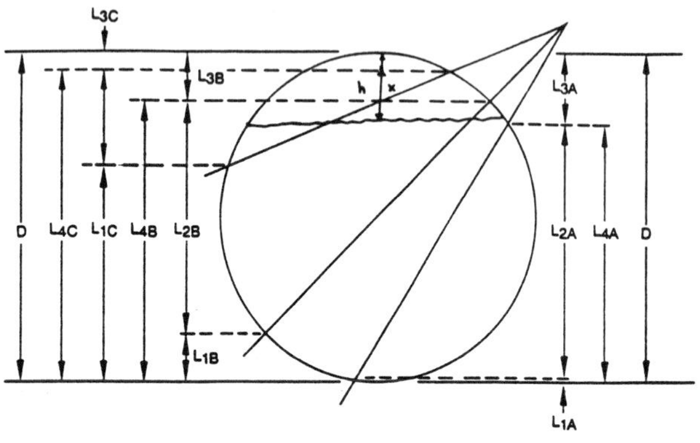

Figure 10. Gamma Densitometer Geometry and Chord Lengths (from Evans)

HTD-Vol. 197, Two-Phase Flow and Heat Transfer
ASME 1992

GAS CORE TURBULENCE AND DROP VELOCITIES IN VERTICAL ANNULAR TWO PHASE FLOW

B. J. Azzopardi
Department of Chemical Engineering
University of Nottingham
Nottingham, United Kingdom

J. C.F. Teixeira
Escola de Engenheria
Universidade do Minho
Braga, Portugal

ABSTRACT

Laser anemometry techniques have been employed to measure gas and drop velocities in annular two-phase flow. Mean velocities and Reynold's stresses in the gas appear to be similar to those in corresponding rough pipes. However, turbulence intensities are higher than the expected values. Drop velocities were 20% below the corresponding local velocities for the gas. Standard deviations of the drop velocities were 10 to 65% higher than those for the gas.

NOMENCLATURE

c_D	Drag coefficient (-)
d	Drop diameter (m)
d_{10}	Linear average drop diameter (m)
g	Gravitational acceleration (m/s^2)
l	Turbulence length scale (m)
r_c	Correlation coefficient (-)
U	Time averaged velocity (m/s)
u'	Axial turbulence intensity (m/s)
v'	Radial turbulence intensity (m/s)
w'	Circumferential turbulence intensity (m/s)
u'v'	Reynolds stress (m^2/s^2)
y	Distance from the wall (m)
z	Axial distance (m)
γ	Density ratio

Subscripts

d	Drop
G	Gas
i	Individual drop
r	Radial
z	Axial

INTRODUCTION

The simultaneous flow of gas and liquid in a duct or pipe can produce a large number of spatial configurations of the phases due to the deformable interface between them. However, for many years it has been realised that for certain combinations of flow rates these configurations can be characterised by a common description. For example, in vertical upflow at high gas rates, part of the liquid flows as a film on the channel walls while the rest is carried as drops by the gas flowing in the centre of the channel. This type of behaviour is usually termed annular or annular-mist flow. This flow pattern is important as it occurs of many pieces of industrial equipment such as boilers and gas/condensate wells. Design of such equipment requires knowledge of first order parameters such as pressure drop. However, there is also a need for more detailed information such as film thickness or the split of liquid between drops and film. These are necessary for models of pressure drop with a sound physical basis as well calculation of heat transfer and processes such as erosion and erosion/corrosion.

Current, physically based, models for annular-mist flow take into account not only the split of liquid between film and drops but also the constant interchange that occurs between them. A recent version of such models, Hewitt and Govan (1990) gives good predictions of experimental data over wide ranges of parameters. Nevertheless, the models are still strongly dependent on empirical correlations for closure. Recently, Abolfadl and Wallis (1985) and Owen and Hewitt (1987) have suggested that knowledge of the turbulence might permit some of the empiricism to be removed. They propose that an effect of the presence of the drops is to suppress the level of turbulence in the gas core which would result in a modified log law for the velocity profile.

An examination of the published literature reveals nothing on turbulence in the gas core of annular flow. A small number of papers were found which report mean velocity profiles, Gill et al (1965), Adorni et al (1961), Subbotin et al (1975) and Kirillov et al (1978). Though there are a number of publication which report drop sizes, hitherto, only the paper by Lopes and Dukler (1987) provides information on drop velocities. This work shows drops to be travelling at velocities significantly below that of the gas. However, the approach used did not consider drops below 100 μm, a size range that could contain a very large fraction of the liquid travelling as entrained drops.

Consideration has also been given to information from other types of two-phase flow. A recent review by Hetsroni (1989) reveals that measurements have been made of the continuous phase turbulence in both gas-solid and liquid-solid flows. Maeda et al (1980), Lee and Durst (1982) and Tsuji et al (1982, 1984) have reported data for gas-solid flow whilst Zisselmar and Molerus (1979) investigated liquid-solid flows. From these it can be seen that turbulence intensities increase or decrease from the single phase values. For smaller particles there is a suppression of turbulence, though the degree of suppression diminishes with particle concentration. For larger particles the turbulence intensity is enhanced the increase being directly dependent on concentration. Gore and Crowe (1989) show that these effects can be quantified by consideration of the ratio between particle size and a characteristic length scale of the turbulence. Ratios below 0.1 result in suppression, higher values correspond to an increase in turbulence. It must be remembered that these two-phase flows differ from annular flow in two fundamental aspects. Firstly, there is no liquid film present. In annular flow this can act as a rough wall. The second difference relates to steady state. Though this is achieved in the fluid-solid flows, in annular flow with its constant creation of drops from the film there can only be a dynamic steady state, i.e., there will always be freshly created drops present.

This paper reports measurements on gas phase turbulence and drop velocities made in vertical annular flow. It brings together and extends material described by Teixeira et al (1989) and Azzopardi et al (1989).

EXPERIMENTAL ARRANGEMENT

Flow rig

Measurements were made in a vertical test section of 0.032 m internal diameter. The rig, shown schematically in figure 1, is supplied from the laboratory compressed air main. This is metered before being mixed with the metered water flow in the test section. Beyond the test section, the two-phase mixture is separated; air is released to the atmosphere whilst the water is returned to the stock tank. The gas is introduced at the bottom of the vertical test section. The liquid is injected through a porous wall section mounted a few diameters from the bottom. Measurements were made at the top of the test section, 120 diameters from the liquid entry point. For the range of flow rates studied, this length was sufficient for the split of liquid between film and drops to be reasonably close to equilibrium values.

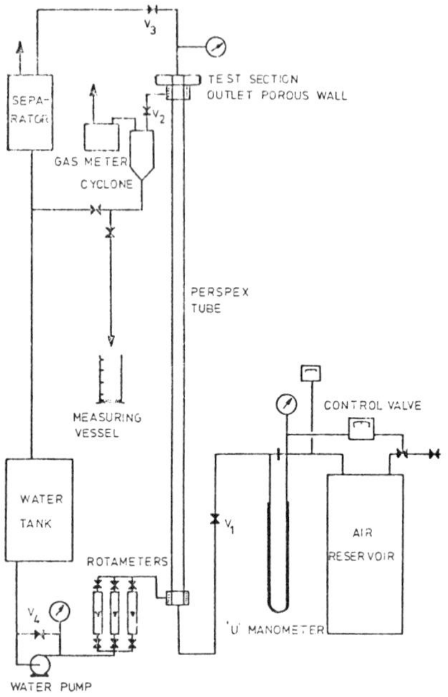

Figure 1 Experimental facility

Special test sections are required to allow for the entry of the light beams and for the exit of the scattered light without their being distorted by the curved tube wall and the highly disturbed film interface. In the case of the phase Doppler, small tubes were inserted through the tube wall and the liquid film as shown diagrammatically in figure 2(a). Once again windows were placed on the outside of this and a purge provided. Extensive tests have shown that these arrangements do not affect the drop flow.

Figure 2(b) indicates the locations within the pipe cross section at which measurements were made.

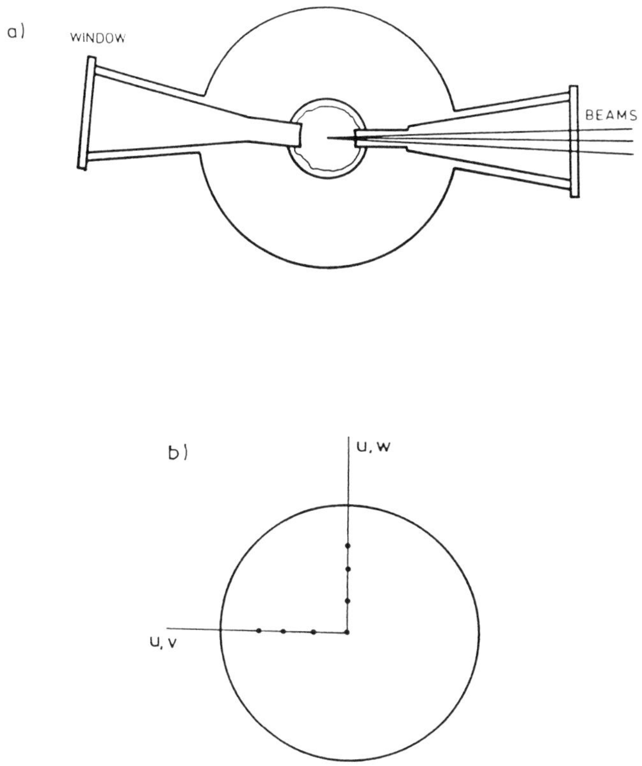

Figure 2 a) Schematic location of positions at which measurement were made with LDA; b) Arrangement for optical access into the gas core

Visibility technique

Gas turbulence data were obtained by inserting 1 μm polystyrene spheres into the gas flow. This size had been chosen as calculations had shown that they would be good flow followers. Signals from these tracer particles were discriminated from those arising from drops by use of the visibility of the doppler bursts (ratio of ac to dc components which is size dependent), see Yeoman et al (1982) for a description of the system and details of the data handling. A suitably configured two-colour system was employed to obtain two components of velocity simultaneously. Axial and radial or axial and circumferential components were studied.

Phase doppler anemometry

Simultaneous measurements of drop size and velocity were made using a Phase Doppler Anemometry technique. The beam of a 15mW HeNe laser was split into two equal intensity beams 15 mm apart. They are focused by a 310 mm focal length lens (crossing angle 2.77°) and the scattered light collected at three positions (at an offset of approximately 30° from the forward direction). This arrangement produces an instrument with a wide dynamic range and from whose signals sphericity can be determined, Livesley (1988). The signals are filtered and processed from which the frequency (and hence velocity) and phase lag (size) can be determined. A microcomputer calculates the velocity and size and sorts and stores the data. For the present optical geometry the size range is 30-650 μm. Because of the gaussian distribution of intensity of the laser beams, the effective probe volume will be larger for bigger drops at a given trigger level. A correction has been made for this as described by Brazier et al (1988) and Teixeira (1988).

RESULTS

For single phase flow, gas velocity and turbulence measurements were made for Reynolds Numbers in the range 57000 to 142100. The data is listed in Table 1. In the case of annular flow, information was acquired for gas mass fluxes in the range 24.5 to 55.6 kg/m²s and liquid mass fluxes between 15.9 and 47.6 kg/m²s. Table 2 contains data on gas velocity and turbulence whilst Table 3 lists drop velocity and size.

Table 1
Gas Velocity and turbulence data - single phase flow

Re_G	U_G	u_*	r/R	u	u'	v'	w'	-u'v'
-	m/s	m/s	-	m/s				m^2/s^2
57000	17.7	0.89	0.00	20.74	0.98	0.85	0.85	0.00
			0.25	20.21	1.07	0.97	0.97	0.13
			0.50	18.80	1.48	0.95	0.86	0.37
			0.69	17.50	1.56	1.07	1.13	0.49
78000	24.3	1.19	0.00	28.28	1.05	0.94	0.94	0.00
			0.25	27.57	1.21	1.05	0.94	0.37
			0.50	25.98	1.60	1.25	1.26	0.52
			0.69	24.03	1.83	1.30	1.36	0.81
99600	30.9	1.48	0.00	36.09	1.40	1.12	1.12	0.00
			0.25	35.21	1.51	1.15	1.29	0.25
			0.50	33.10	1.99	1.45	1.55	0.50
			0.69	31.20	2.39	1.53	1.67	0.69
120900	37.5	1.76	0.00	43.67	1.58	1.29	1.29	0.00
			0.25	42.85	1.91	1.43	1.44	0.65
			0.50	40.50	2.21	1.54	1.71	1.44
			0.69	38.50	2.50	1.70	1.96	1.84
142100	44.1	2.02	0.00	51.53	1.61	1.51	1.51	0.00
			0.25	50.60	2.15	1.46	1.48	0.89
			0.50	48.40	2.64	1.79	1.86	1.63
			0.69	45.80	2.95	1.89	2.13	2.37

Table 2
Gas velocity and turbulence data - two phase flow

G_G	G_L	G_{LE}	r/R	u	u'	v'	w'	-u'v'
kg/m^2s			-	m/s				m^2/s^2
24.5	15.9	0.86	0.00	17.7	1.69	1.30	1.30	0.00
			0.25	17.2	1.96	1.40	1.52	0.39
			0.50	15.3	2.59	1.80	-	0.70
31.8	15.9	0.96	0.00	22.1	1.78	1.49	1.49	0.00
			0.25	21.7	2.09	1.63	1.80	0.53
			0.50	19.5	2.91	1.90	-	1.03
31.8	31.7	2.92	0.00	22.4	2.30	1.87	1.87	0.00
			0.250	21.9	2.59	1.90	2.12	0.68
			.50	19.6	3.18	2.14	-	1.49
31.8	47.6	5.20	0.00	22.7	2.82	2.25	2.25	0.00
			0.25	22.0	3.00	2.15	2.48	0.90
			0.50	19.7	3.40	2.51	-	1.64
43.7	15.9	1.43	0.00	29.8	1.99	1.63	1.63	0.00
			0.25	28.9	2.44	1.74	1.91	1.03
			0.50	26.4	3.30	1.90	-	1.61
55.6	15.9	2.19	0.00	37.4	2.17	1.77	1.77	0.00
			0.25	36.0	2.51	1.85	-	1.21
			0.50	33.1	3.41	2.28	-	2.04

Table 3
Drop velocity and size data

G_G	G_L	r/R	u_d	u_d	-S	K	$-r_c$	d_{32}
(kg/m^2s)		(-)	(m/s)		(-)			μm
24.5	15.9	0.00	13.8	2.1	0.87	4.13	0.55	209
		0.25	12.9	2.2	0.91	3.73	0.52	190
		0.50	7.1	1.4	0.65	3.23	0.38	182
31.8	15.9	0.00	19.6	2.5	1.40	5.81	0.47	144
		0.25	16.1	2.2	0.84	3.86	0.39	140
		0.50	17.2	2.9	0.78	3.52	0.22	131
31.8	31.7	0.00	17.5	2.3	0.63	3.30	0.50	143
		0.25	16.2	2.4	0.60	3.16	0.40	135
		0.50	17.4	3.2	0.66	3.21	0.25	127
31.8	47.6	0.00	17.5	2.6	0.58	3.13	0.49	155
		0.25	16.3	2.5	0.44	3.21	0.43	144
		0.50	17.1	3.3	0.53	3.17	0.23	136
43.7	15.9	0.00	23.1	2.5	1.26	5.26	0.50	127
		0.25	21.1	2.6	1.12	4.46	0.32	117
		0.50	19.5	2.9	0.87	3.93	0.19	115
55.6	15.9	0.00	29.0	2.5	1.49	6.21	0.45	110
		0.25	27.1	3.1	1.28	5.26	0.32	108
		0.50	24.4	3.5	0.78	4.50	0.06	106
55.6	31.7	0.00	30.3	2.7	1.22	5.96	0.49	115
		0.25	28.1	2.9	1.07	4.58	0.34	99
		0.50	24.1	3.6	0.62	3.81	0.06	96
55.6	47.6	0.00	30.7	2.6	1.06	5.29	0.37	114
		0.25	28.5	2.9	0.89	4.47	0.28	99
		0.50	24.3	3.6	0.42	3.30	0.14	89

DISCUSSION

Single phase gas velocity and turbulence

The results for single phase flow were very similar to published data, e.g., Laufer (1954). As in the data of Nikuradse (1932), mean gas velocities were found to fit a power law. Flow rates, obtained from integration of the power law equation, matched those determined from the orifice plate in the inlet line to within +2.8/ -0.8%. Turbulence intensity increased from the centre line towards the wall. The actual values, non-dimensionalised with respect to the friction velocity, were in agreement with the data of Laufer. Wall shear stresses, determined from Reynolds stress measurements, gave interfacial stress values which were in reasonable agreement with those deduced from pressure drop data with the values determined from Reynolds stress being slightly lower than those from the pressure drop or calculated using the standard Blasius equation ($f = 0.079/Re^{1/4}$), figure 3. Figure 4 shows that power law exponents and friction factors from the present data lie on the same line as the data of Nikuradse (1932) and Nunner (1956).

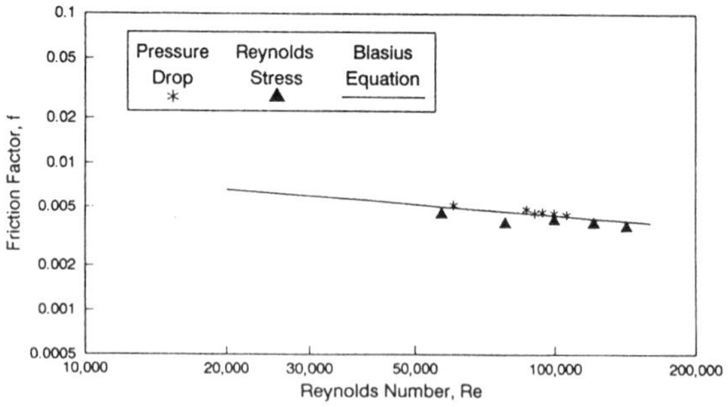

Figure 3 Comparison of friction factors obtained from pressure drop and Reynolds stress

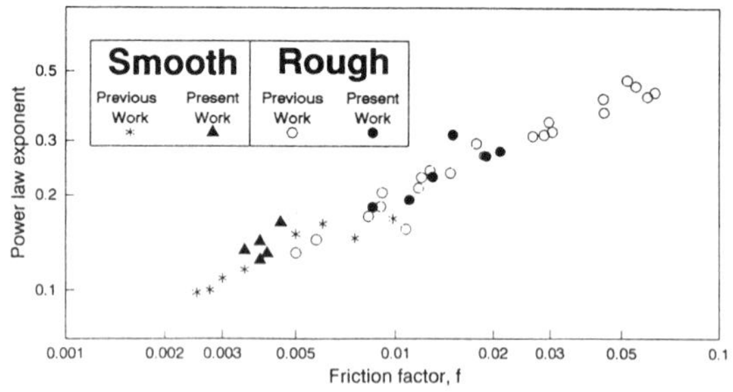

Figure 4 Power law exponent / friction factor relationship for smooth and rough walls

An example of the probability density function of axial velocity is illustrated in figure 5. This shows that the data is well described by a gaussian function. The statistical characteristics, skewness (S) and flatness or kurtosis (K), of the velocity distributions were determined for u', v', w' and u'v'. The calculated values of the fluctuating components of the velocity were found to vary between -0.5 and 0.1 (S) and 2.8 and 3.3 (K). These values, close to those expected for a gaussian distribution (S=0 and K=3), confirm the random nature of the turbulent flow. Similar values are found in Batchelor (1951) and Wood and Antonia (1975).

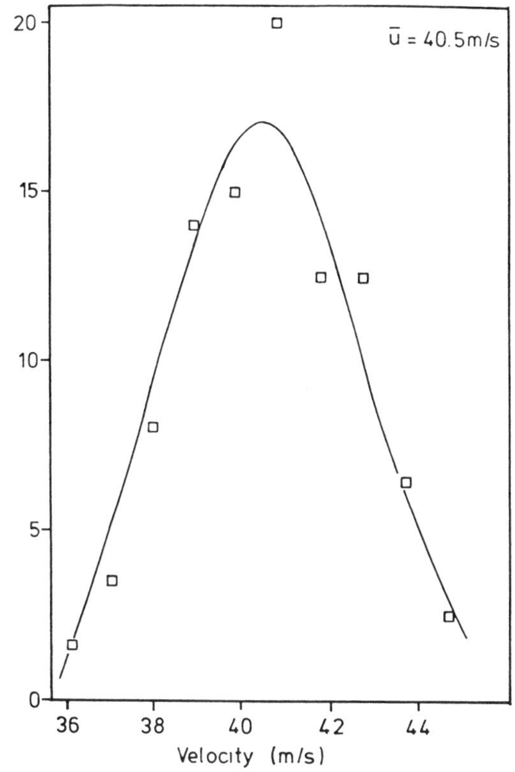

Figure 5 Probability distribution function of the gas velocity: comparison with a Gaussian function

The joint probability distribution function for u'v' was shown by Antonia et al (1973) to be a function of the correlation coefficient u'v'/(uv), when both u' and v' are normally distributed. In such a case, the skewness and kurtosis values would vary between 0 and -2.82 and 9 and 15, respectively. The present values of K range from 9 and 16.3. Deviations from the expected value may be attributed to the actual pdf for u' and v' which are not strictly gaussian.

From the above, it was concluded that the method give accurate measurements for known conditions and that it could be used with confidence in the more novel application of annular two phase flow.

Annular flow gas velocity and turbulence

In the case of annular flow, mean velocity profiles could also be described by a power law. The exponents were larger than the equivalent single phase data values. However, when plotted against friction factor (determined from extrapolation of Reynolds stress data to provide wall shear stresses) the values lie on the same line as data from flow in rough walled pipes, figure 4. In contrast, figure 6 shows that turbulent intensities were higher than the single phase values. Even when non-dimensionalised with respect to the friction velocity, these values were above the levels expected from single phase flow. The higher order moments of the velocity distributions show features similar to those observed in single phase flow: the computed values of the skewness and kurtosis are in the same range as those discussed above single phase flow.

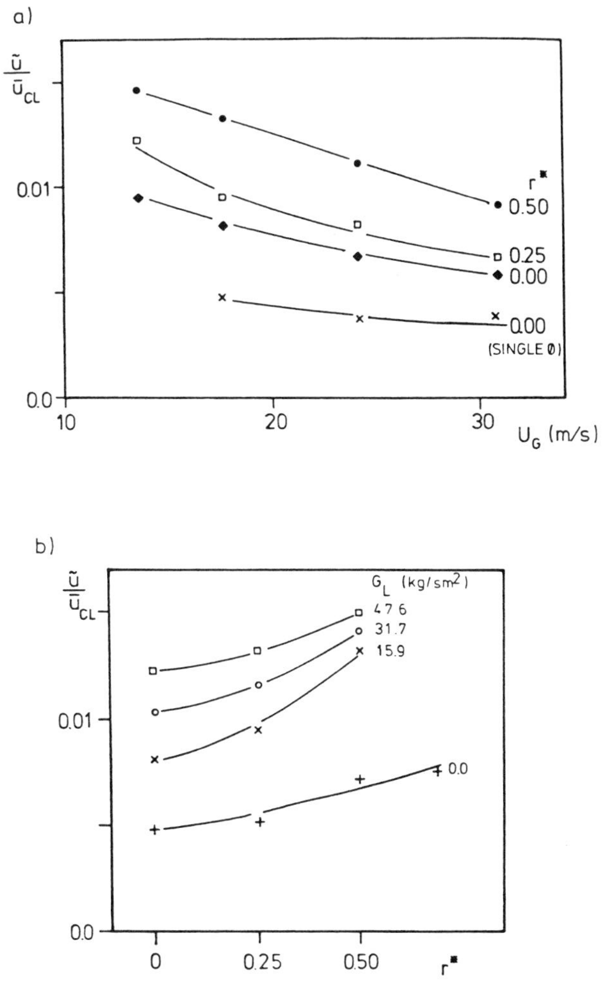

Figure 6 Turbulence intensities in annular flow: a) Effect of the gas velocity (liquid mass flux = 15.9 kg/m²s); b) Effect of the liquid flow rate (gas mass flux = 43.7 kg/m²s)

The change in the turbulence intensity of the gas ([value for two-phase flow minus value for single phase flow]/ value for single phase flow) is plotted against the ratio of drop size to characteristic turbulence length scale in figure 7. The length scale was calculated using the equation of Prandtl.

$$-\overline{u'v'} = l^2 \left(\frac{dU}{dy}\right)^2 \qquad (1)$$

The velocity gradient was determined using the power law profile fitted to the gas velocity data from two-phase flow. Similarly the Reynold's stress was also from the two-phase flow experiments. This approach gave values slightly higher than those from the simple equation suggested by Hutchinson et al (1971).

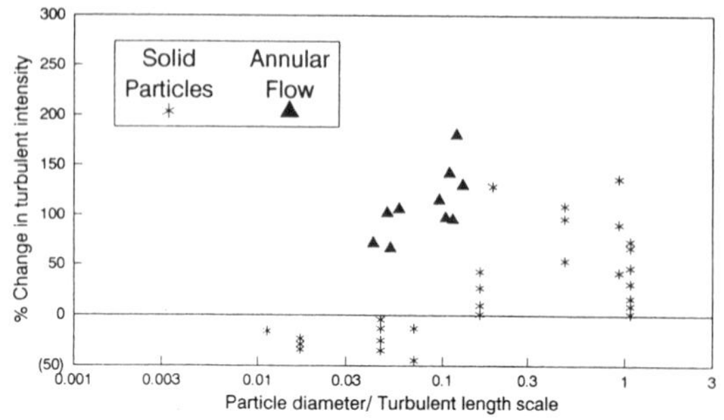

Figure 7 Change in turbulence intensity as a function of the length scale ratio

In contrast to that from gas-solid and liquid-solid flows, the present data shows that there was an increase in turbulence intensity even though the drop size/ length scale ratio was below 0.1. Gore and Crowe (1989) who first proposed this type of plot suggested that for particles, much smaller than the most energetic eddies, will follow the eddy for part of its lifetime and take up energy. thus turbulent energy of the eddy will be transformed to kinetic energy of the particle. The larger particles will tend to create turbulence (in its wake) near the scale of the most energetic eddies. Therefore increasing the turbulence intensity of the gas. In this case energy is transferred from the mean flow, which is moving the particles, to the turbulent kinetic energy. The drops in the present experiment, though they are much smaller than the most energetic eddies, are probably travelling at axial velocities much lower than those of the eddies. Therefore, contact times are small and little energy can be transferred. Also, the large differential velocity will result in drop Reynolds numbers higher than those for gas-solid flows, such Reynolds numbers will result in vortex shedding from the drops, an extra source of turbulence. This idea of turbulence arising from the effect of the particles was first suggested by Kada and Hanratty (1960).

An attempt to quantify the above concepts has been produced by Theofanous and Sullivan (1982). They considered that in a dispersed flow, the shear on the wall and that on the dispersed phase (drops) should both be taken into account. They defined a new friction velocity which depended on the sum of the shear stresses on the wall and on drops. Turbulence intensities were then determined from the assumption that it is equal to the friction velocity. Turbulence intensities calculated in this manner were reasonably close to the measured values. However, the assumptions on which the theory is based require further strengthening.

Further support for the idea that the increase in turbulence intensity is due to newly created drops is provided by the plot of increase in turbulence intensity plotted rate of entrainment, figure 8. Increasing in the amount of new drops increases the augmentation of turbulence.

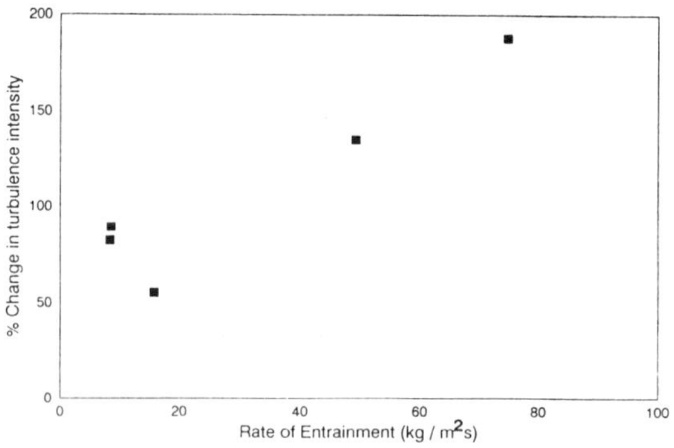

Figure 8 Increase in turbulence intensity as a function of the rate of drop entrainment

Drop size and velocity

Drop size distribution information has been gathered under identical flow conditions using the phase doppler technique and with a laser diffraction technique. Differences are to be expected in the data 'as measured' as the two techniques make different measurements; the diffraction technique measures a time/space average across a pipe diameter whilst the phase doppler measures point values. However, if an allowance is made for spatial variations and for size/velocity relationships, the data can be reduced to the same basis. Teixeira (1988) gives details of the conventions. Such correction results in maximum differences between the Sauter mean diameters from the two methods of 10% with most data showing much smaller differences. Figure 9 shows that there is good correspondence between the data from the two instruments. Similar agreement between the two measurement methods has been observed by Brazier et al (1988), who studied the drops formed by a spray nozzle.

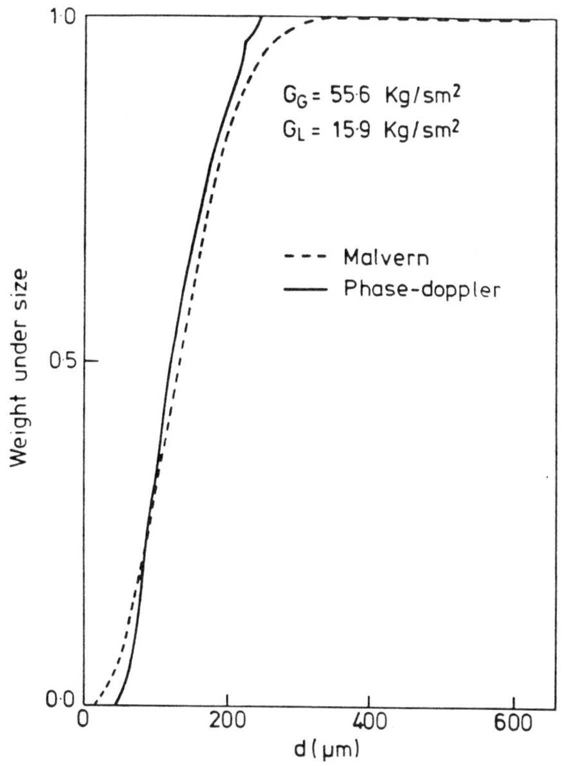

Figure 9 Comparison between the phase Doppler technique and the laser diffraction: cumulative fraction of drops

Mean drop velocities show a similar radial distribution as those for the gas but with values about 20% lower, figure 10. The mean drop velocities are approximately equal to the gas superficial velocity. Similar trends are shown at other flow rates. Figure 11 shows that the standard deviation of the drop velocity (non-dimensionalised by the mean) increases from the centre line towards the interface. Examination of the data indicates that the equivalent values for the gas (turbulence intensity/ mean velocity) are 10 to 65% lower than these values for the drops. Figure 11 also shows a noticeable dependence on gas velocity. The distribution of drop velocities is not differs from a gaussian as the peak is skewed towards the gas velocity. This follows from the fact that drops are created with very low velocities whilst there is an upper limit at the gas velocity. it is interesting to note that the calculations of Schadel et al (1990) predict slip velocities which are about 20% of the gas velocity.

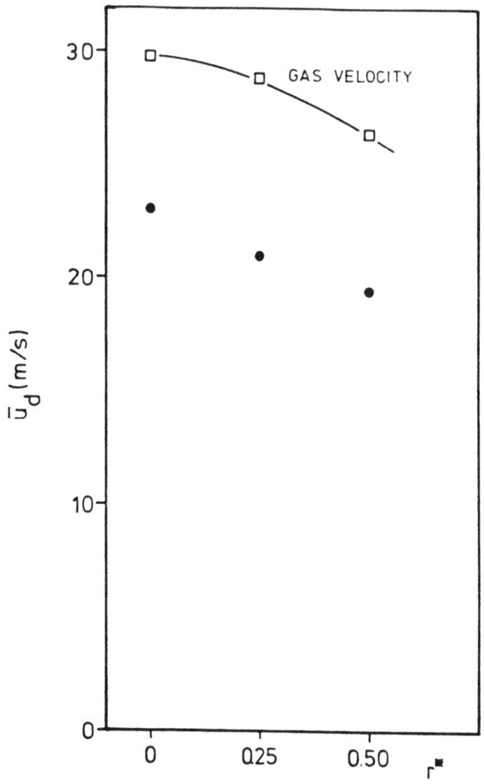

Figure 10 Radial variation of drop velocity: comparison with the local gas velocity (gas mass flux = 43.7 kg/m²s; liquid mass flux = 15.9 kg/m²s).

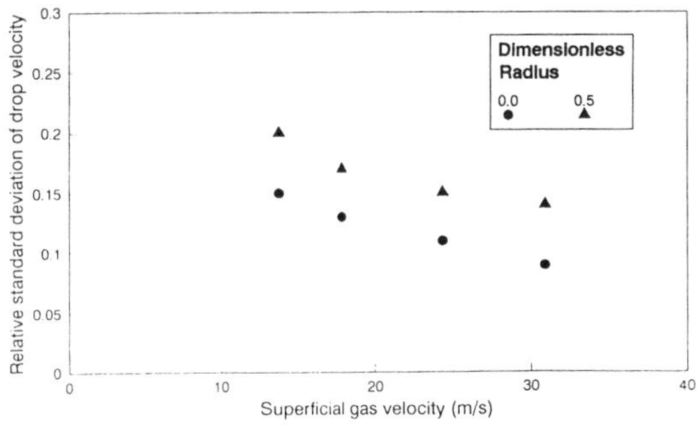

Figure 11 Effect of gas velocity on the relative standard deviation of drop velocity

There is a trend for smaller drops to be travelling at higher velocities, Figure 12. In addition, it can be seen that there is a wider range of velocities at smaller drop sizes. This can be explained by the fact that smaller drops will be most strongly affected by gas turbulence which can cause acceleration and deceleration in the lateral direction and so increase and decrease the time taken for drops to arrive at the probe volume and widen the range of axial velocities that they could achieve. Larger drops, being less susceptible to turbulence will show a narrower range of lifetimes and hence velocities. It is of interest that the only other source of drop velocity data, Lopes and Dukler (1987), shows a completely different trend of drop velocity with drop size; larger velocities were obtained for the bigger drops. An explanation based on the effect of turbulence on the drag coefficient was suggested by those authors.

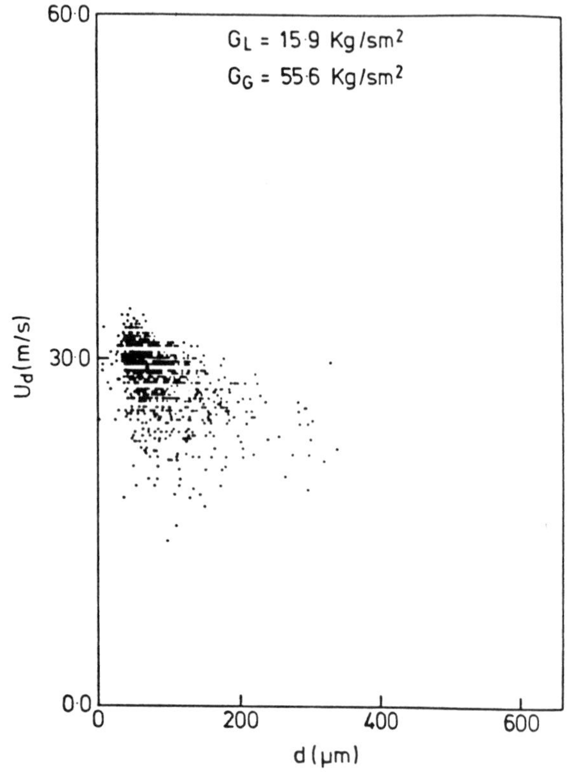

Figure 12 Effect of drop size on drop velocity

Size and velocity are not strongly correlated. The correlation coefficient, defined as

$$r_c = \frac{\Sigma(U_{di} - \overline{U_d})\Sigma(d_i - d_{10})}{(\Sigma(U_{di} - \overline{U_d})^2\Sigma(d_i - d_{10})^2)^{0.5}} \qquad (2)$$

is small and decreases as the interface is approached. It also decreases with gas velocity and, in the centre of the channel, with liquid flow rate.

Calculations have been carried out to establish phenomena which are important in determining the velocities which drops achieve. The motion was considered as two dimensional and the drag equations written as

$$\frac{dU_{dz}}{dt} = \frac{(1-\gamma)}{\gamma}g - \frac{3}{4}C_D\frac{|U_d - U_G|}{\gamma d}(U_{dz} - U_{Gz}) \qquad (3)$$

$$\frac{dU_{dr}}{dt} = -\frac{3}{4}C_D\frac{|U_d - U_G|}{\gamma d}(U_{dr} - U_{Gr}) \qquad (4)$$

where c_d is calculated using the relative velocity $|U_d$-$U_G|$and the equation of Teixeira (1988) which takes account of the effect of gas turbulence on the drag coefficient. γ is the density ratio. The initial value of U_{dr} was calculated from the equation of Andreussi and Azzopardi(1983). The results of Lopes and Dukler (1987) confirm that this equation gives values of the correct magnitude for larger drops. A value of 3 m/s was used for U_{dz} (this was taken from measurements of disturbance wave velocity, Azzopardi(1986), these disturbance waves are the source of drops). The equations were integrated upto the time that the drop reaches the centre line. The predictions give values which pass through the middle of the data except for the smallest drop sizes. The large quantity of small drops with lower velocities is probably caused by their susceptibility to the effect of turbulent eddies which can increase the rate of lateral dispersion of the drops, i.e., either towards or away from the wall, so that some arrive at the centre line earlier than might be expected from newtonian calculations. Consequently they will have lower axial velocities. With larger drops, which are less susceptible to the effect of eddies, this effect is negligible. Predictions obtained assuming that the drops have a constant transverse velocity and ignoring the effect of turbulence on the drag law show significant under prediction at the larger sizes indicating that these effects are important. This, in spite of the experimental finding of Andreussi and Azzopardi (1983) that drops travel with a constant transverse velocity. However, the drops observed in that case were >300 μm and calculations have shown that, though these larger drops do maintain an approximately constant transverse velocity, smaller drops experience a significant transverse retardation, as might be expected.

Estimates indicate that, for the range of flow rates listed in tables 2 and 3, film thicknesses were in the range 0.1 to 0.4 mm.

CONCLUSIONS

From the above it can be concluded that:

(1) Measurements of turbulence in single phase gas flow indicate that the LDA system employed in these tests gave reliable results.

(2) Though mean velocity and Reynolds stress data for annular flow show good correspondence to that for equivalent rough pipes, Turbulence intensities (even when non-dimensionalised by the friction velocity) had values higher than expected. Vortex shedding from slow moving drops was suggested as the source of this extra turbulence.

(3) The phase doppler technique produces drop size distributions similar to those from a laser diffraction instrument. However comparison is only possible if conversions have been made so that both are

measuring under the same conditions of space time averaging.

(4) Mean drop velocities were seen to be about 20% below the corresponding local values for the gas. The averaged drop velocity was approximately the same as the superficial gas velocity. A weak correlation has been found between the velocity and size of drops. Factors which affect this relationship have been identified.

ACKNOWLEDGEMENTS

The experimental work described in this paper was undertaken as part of the Underlying Research Programme of the UKAEA at their Harwell Laboratory.

REFERENCES

Abolfadl, M. and Wallis, G.B. (1985) "A mixing length model for annular two phase flow", Physico-Chemical Hydrodynamics, 6, pp49-68.

Adorni, N., Casagrande, I., Cravarolo, L., Hassid, A. and Silvestri, M. (1961) "Experimental data on two-phase flow: liquid film thickness, phase and velocity distribution, pressure drops in vertical gas-liquid flow", C.I.S.E. report no. R 35.

Andreussi P. and Azzopardi B.J., (1983) "Droplet deposition and interchange in annular two-phase flow", Int. J. Multiphase Flow, 9, pp681-695.

Antonia, R.A., Atkinson, J.D. and Luxton, R.E. (1973) "Comments on statistical characteristics of Reynolds stress in a turbulent boundary layer", Physics of Fluids, 16, p956.

Azzopardi B.J. (1986) "Disturbance wave frequencies, velocities and spacing in vertical annular two-phase flow", Nucl. Eng. and Design, 92, pp121-133.

Azzopardi B.J., Teixeira, J.C.F. and Jepson, D.M. (1989) "Drop sizes and velocities in vertical annual two phase flow", Proc. Int . Conf. on *Mechanics of Two-Phase Flow*, Taipei, Taiwan, 12-15 June, pp261-266.

Batchelor, G.K. (1951) "Pressure fluctuations in isotropic turbulence", Proc. Cam. Phil. Soc., 47, pp359-374.

Brazier K., Gillespie R.F., Dalzell W. and Livesley D.M., (1988) "Bias corrections to size distribution and concentrations in phase-Doppler particle measurement", AERE R13270, UKAEA.

Gill, L.E., Hewitt, G.F. and Lacey, P.M.C. (1964) "Sampling probe studies of the gas core in annular two phase flow II", Chem. Eng. Sci., 19, pp665-682.

Gore, R. and Crowe, C.T. (1989) "Effects of particle size on modulating turbulence intensity", Int. J. Multiphase Flow, 15, pp279-285.

Hetsroni, G. (1989) "Particles-turbulence interaction", Int. J. Multiphase Flow, 15, pp735-746.

Hewitt, G.F. and Govan, A.H. (1990) "Phenomena and predictions in annular two-phase flow", In *Advances in Gas-Liquid Flows*, A.S.M.E. FED vol 99, HTD Vol 155, pp41-55.

Hutchinson, P., Hewitt, G.F. and Dukler, A.E. (1971) "Deposition of liquid or solid dispersion from turbulent gas streams: a stochastic model", Chem. Eng. Sci., 26, pp419-439.

Kada, H. and Hanratty, T.J. (1960) "Effect of solids on turbulence in a fluid", A.I.Ch.E.J., 6, pp624-630.

Kirillov,P.L., Smogalev, I.P., Suvurov, M.Ya., Shumsky, R.V. and Stein, Yu.Yu. (1978) "Investigation of steam-water flow characteristics at high pressures", Proc. 6th Int. Heat Transfer Conf., Toronto, Canada, 7-11 Aug, Vol 1, pp315-320.

Laufer, J. (1954) "The structure of turbulence in fully developed pipe flow", NACA Report 1174.

Lee, S.L. and Durst, F. (1982) " On the motion of particles in turbulent duct flows", Int. J. Multiphase Flow, 8, pp125-146.

Livesley D.M., (1988) "Strengths and limitations of the phase doppler technique for simultaneous measurements of particle velocity and size", AERE R13113, UKAEA.

Lopes J.C.B. and Dukler, A.E. (1987) "Droplet dynamics vertical annular flow", A.I.Ch.E.J., 33, pp1013-1024.

Maeda, M., Hishida, and Furutani, T. (1980) "Optical measurements of local gas and particle velocity in an upward flowing gas-solid suspension", In *Polyphase Flow and Transport Technology*, p 211, Century 2_ETC, San Franciso, Calif.

Nikuradse, J. (1932) "Gesetzmassigkeiten der turbulenten stromung in glatten rohren", VDI-Forschungsheft 356.

Nunner, W. (1956) "Waermeubergang und druckfall in rauhen rohren", VDI-Forschungsheft 455.

Owen, D.G. and Hewitt, G.F. (1987) "An improved annular two-phase flow model", Proc. 3rd Int. Conf. on Multiphase Flow, The Hague, The Netherlands, Paper C1.

Schadel, S.A., Leman, G.W., Binder, J.L. and Hanratty, T.J. (1990) "Rates of atomisation and deposition in vertical annular flow", Int. J. Multiphase Flow, 16, pp363-374.

Subbotin, V.I., Kirillov, P.L., Smogalev, I.P., Suvorov, M. Ya., Stein, Yu.Yu. and Shumsky, R.V. (1975) "Measurement of some characteristics of a steam-water flow in a round tube at pressures of 70 and 100 atm.", A.S.M.E. paper No. 75-WA/HT-21.

Teixeira J.C.F., (1988) "Turbulence in annular two phase flow", PhD Thesis, University of Birmingham, U.K.

Teixeira J.C.F., Azzopardi B.J. and Bott T.R., (1989) "Measurement of turbulence in the gas core of annular two phase flow", Proceedings of the International Symposium on Turbulence Modification of Multiphase Flow, ASME, San Diego.

Theofanous, T.G. and Sullivan, J. (1982) "Turbulence in two-phase dispersed flows", J.Fluid Mech., 116, pp343-362.

Tsuji, Y. and Morikawa, Y. (1982), "LDV measurements of an air-solid two phase flow in a horizontal pipe", J. Fluid Mech., 120, pp385-409.

Tsuji, Y., Morikawa, Y. and Shiomi H. (1984), "LDV measurements of an air-solid two phase flow in a vertical pipe", J. Fluid Mech., 139, pp417-434.

Wood, D.H. and Antonia, R.A. (1975) "Measurements in a turbulent boundary layer over d-type roughness", J. Appl. Mech., 42, pp591-597.

Yeoman, M.L., White, H.J., Azzopardi, B.J., Bates, C.J. and Roberts, P.J. (1982) "Optical development and application of a two-colour LDA system for the simultaneous measurements of particle size and particle velocity", A.S.M.E. Winter Annual Meeting, Phoenix, Ariz., November.

Zisselmar, R. and Molerus, O. (1979) "Investigation of solid-liquid pipe flow with regard to turbulence modification", Chem. Eng. J., 18, p233-239.

HTD-Vol. 197, Two-Phase Flow and Heat Transfer
ASME 1992

BUBBLE-BUBBLE INTERACTON IN MULTI-COMPONENT FLOW

Rizwan-uddin
Department of Nuclear Engineering and Engineering Physics
University of Virginia
Charlottesville, Virginia

ABSTRACT

Bubble-bubble interaction for the simple case of a small spherical bubble in the vicinity of a much larger bubble, both rising in stationary liquid, with no mass transfer between liquid and vapor phases, has been studied. The effect of the motion of the larger bubble on the dynamics of the smaller bubble has been determined both for completely spherical, and for spherical cap shaped large bubbles. The complicated motion of the smaller bubble, as it is entrained in the wake of the larger bubble, is clearly seen. This simple model of bubble-bubble interaction can be used in the individual-particle-tracking method of two-phase flow analysis.

NOMENCLATURE

A	cross sectional area of the bubble perpendicular to flow direction
C	drag coefficient
F_b	buoyancy force
F_d	drag force
R	radius
U	velocity
a	constants
g	gravity
r	distance of the smaller bubble from the larger bubble's center
u	velocity
x	x-location of the smaller bubble in absolute frame
y	y-location of the smaller bubble in absolute frame
α	angle
ρ	density
μ	viscosity

Subscripts

b	bubble
d	drag
l	liquid
v	vapor
1	larger bubble
2	smaller bubble

INTRODUCTION

Stokes (1851) solved the equations of viscous motion with no-slip boundary condition and derived the expression, now known as Stokes law, for the terminal velocity of a *rigid* spherical particle in fluid of infinite extent,

$$u_\infty = \frac{2g(\rho_\ell-\rho_v)\, r_b^2}{9\mu}, \qquad (1)$$

where u_∞ is the steady-state terminal rise velocity, ρ is density, r is radius, μ is viscosity, g is gravity and subscripts ℓ, v and b represent liquid, vapor and bubble, respectively. Application of the boundary layer theory to liquid-vapor interface of a *bubble* led to a slightly different expression for the terminal bubble velocity (Levich, 1949, 1962),

$$u_\infty = \frac{g(\rho_\ell-\rho_v)\, r_b^2}{9\mu} \qquad (2)$$

The difficulties encountered in extending the above results analytically to more general and realistic cases with non-rigid and non-spherical shaped bubbles, led to the introduction and wide spread use of the drag coefficient, defined by the equation

$$F_d \equiv \frac{C_D\, \rho_\ell\, u^2\, A}{2g}$$

where F_d is the drag force on the particle, C_D is the (dimensionless) drag coefficient, A is the cross-sectional area of the bubble perpendicular to the direction of flow and u is the particle velocity relative to the liquid phase. Though there were exceptions (Moore, 1958, 1959, 1963; Hartunian, 1957; Walters and Davidson, 1962) most earlier analyses to determine the dynamics of bubble motion in liquid were carried out experimentally (O'Brien and Gosline, 1935; Pebbles and Garber, 1958; Uno and Kintner, 1956; Harmathy, 1960). Among other analyses these studies determined the range of Reynolds number for the validity of Stokes law applied to gas bubbles in liquids, and experimentally showed that Stokes law, which also can be written as a function of Reynolds number Re, as,

$$C_D = \frac{24}{Re} \quad where \ Re \equiv \frac{2r_b u \rho_l}{\mu},$$

agrees well with data for Re < 2. For other ranges of Reynolds number and other dimensionless parameters, correlations for the drag coefficient also were experimentally obtained, e.g., for 2 ≤ Re < 200 it was found that (Allen, 1900),

$$C_D = \frac{18.5}{Re^{0.6}}.$$

An important factor for the significant functional changes in the drag coefficient's dependence on Reynolds number is that the shape of the rising bubble changes considerably as Reynolds number increases. For air bubbles in water, four general categories of bubble shapes were experimentally determined (Robinson, 1947),

a) Spherical bubbles, Re < 400
b) Oblate spheroids of varying geometric proportion 400 < Re < 1100
c) Oblate spheroids of constant geometric proportions 1100 < Re < 5000
d) Mushroom shaped bubbles with spherical cap 5000 < Re

Spherical cap bubbles are usually described by the radius of curvature and the angle α defined by drawing normals to the cap of the spheres, as shown in Fig. 1. It has been shown, using inviscid flow theory, that α approaches approximately 100° for high Reynolds number (Rippin and Davidson, 1967) and this has been confirmed experimentally for Re $\gtrsim$ 100 (Grace 1970). For intermediate Reynolds number (5 < Re < 100), the relation between α and Re is given by (Davidson et al., 1977),

$$Re = 8\left\{\frac{4}{2-3\cos(\frac{\alpha}{2}) + \cos^3(\alpha/2)}\right\}^{2/3}$$

For mushroom shaped large air bubbles, terminal velocity was experimentally determined and was found to be independent of liquid properties for the two liquids used in experiment, water and nitrobenzene (Davies and Taylor, 1950). Terminal velocity was related to the radius of curvature of the spherical cap r, by

$$U_\infty = 0.667\sqrt{gr}, \tag{3}$$

and radius of curvature r and equivalent spherical radius based on vapor volume were found to be related by (Davies and Taylor, 1950),

$$r = 2.3r_b. \tag{4}$$

Studies on the motion of a stream of air bubbles rising in a liquid column has been carried out experimentally by a few researchers (O'Brien and Gosline, 1935; van Krevelen and Hoftijzer, 1950), only to determine the mean rise velocity as a function of gas flow rate, without measuring the bubble sizes.

The effect on the dynamics of a bubble of the motion of another bubble in the neighborhood has been studied only experimentally and only in the context of coalescence and breakup of bubbles in liquids (Crabtree and Bridgewater, 1971; Otake et al., 1977). It was observed that the leading bubble, due to the wake effect, induced an increment in the rise velocity of the following bubble, hence, making coalescence more probable. Due to the poor understanding of the wake structure behind bubbles, which is necessary for analytical analysis of bubble-bubble interaction, until recently no attempts were made toward the development of models -- to be used in the numerical simulation analysis of two-phase flow by following the dynamics of each individual bubble in a continuum of liquid -- to describe the influence of one or more nearby bubbles on the dynamics of a rising bubble.

With large and fast computers becoming more easily accessible, multi-phase flow analysis via modeling of each individual particle of the discrete phase, its motion, and interaction with other particles and container walls is becoming feasible. In fact, combined Lagrangian-Eularian approaches to model the discrete (bubbles) and continuous (liquid) phases have recently been reported (Stuhmiller et al., 1989; Trap et al., 1991). Modeling the interfacial phenomena accurately is one of the many difficult aspects of such individual-bubble-tracking analyses. This includes the determination of bubble motion in the presence of other bubbles in the surrounding, development of interaction schemes between bubbles and surrounding walls, and development of interaction schemes among individual bubbles as they come close and pass each other or merge into each other to form larger bubbles. Although it is necessary that the bubble dynamics model and interaction schemes are systematically developed using first principles, any interaction scheme based on first principles is likely to be complicated, or at least

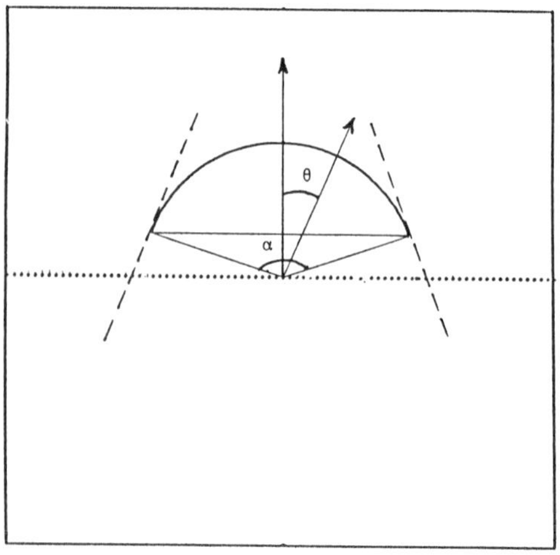

Figure 1. Schematic diagram of a spherical cap bubble and the definition of the angle α defined by drawing normals to the cap of the sphere.

computer intensive in application, for problems of practical use. Hence, simpler interaction schemes based on phenomenological arguments must also be developed to be used in large scale individual-bubble-tracking computer codes. These models must be validated against results from models based on first principles and against experimentally obtained data for single, few, and many bubbles rising and interacting in a liquid column.

Determination of the interaction models among multiple bubbles of various sizes is a very complicated problem, and can only be approached in a step by step fashion. An excellent review of the theoretical results for the motion of solid spherical particles in viscous fluid is given by Feuillebois (1989). Conditions under which these results are applicable to spherical bubbles are also discussed. As a first step toward the development of the interaction schemes, we have carried out analysis of bubble-bubble interaction for the simple case of motion of a small bubble in the vicinity of a much larger vertically rising bubble in stationary liquid with no mass transfer between liquid and vapor phases. The analysis, which has been carried out for different shapes of the large bubble including spherical bubble with no wake, spherical bubbles with wake and spherical cap bubble, does not determine whether coalecence occurs or not.

MODEL

The larger bubble being more buoyant moves faster than the leading smaller bubble and hence, depending upon the values of various parameters and initial conditions, either overtakes the smaller bubble or merges with it. Of course, we assume the smaller bubble is initially close to the vertical line passing through the larger bubble. The motion of the larger trailing bubble, we assume, is not affected as it approaches the much smaller leading bubble and hence the large bubble keeps moving vertically up in a straight line with its terminal velocity, given by Eq. (2) or Eq. (3), depending upon bubble shape. Unless there is contact between bubbles, motion of one bubble affects the dynamics of the surrounding bubbles through the displacement of the liquid that fills the space between them, hence, the smaller bubble, which quickly reaches steady-state and rises with uniform terminal velocity in the absence of the larger bubble, is affected by the relative fluid motion generated by the motion of the larger bubble. The fluid motion increases and exerts a drag force on the smaller bubble as the distance between the larger and smaller bubbles decreases. The dynamics of the smaller bubble, using symmetry around the vertical axis, is determined in a vertical plane that passes through the centers of both bubbles, by solving the bubble dynamics equations in x and y directions,

$$m_b \ddot{x} = -F_d^x \tag{5}$$

$$m_b \ddot{y} = -F_d^y + F_b \tag{6}$$

where m_b is the effective mass of the bubble, F_d is the drag force, F_b is the buoyancy force, and x and y are the coordinates of the smaller bubble in the stationary frame, see Fig. 2. The drag forces in the x and y directions are respectively proportional to the difference in the bubble velocity and velocity of the surrounding fluid -- that is generated due to the motion of larger bubble in the vicinity -- in the x and y directions, and are given by (Lamb, 1945),

$$F_d^x = 12\ \pi\ \mu\ r_2 (u_\ell - \dot{x}) \tag{7}$$

$$F_d^y = 12\ \pi\ \mu\ r_2 (v_\ell - \dot{y}) \tag{8}$$

where r_2 is the radius of the smaller bubble and, $u_\ell(x,y)$ and $v_\ell(x,y)$ are the liquid velocity induced by the larger bubble in the stationary frame, in the x and y directions, respectively. Note that the drag force on bubbles, as shown by Levich (1949, 1962), is twice as much as on solid spherical bodies. Since bubbles of large volume do not retain complete spherical shape, the dynamics of smaller bubbles has been determined both for large complete spherical bubbles and for spherical cap large bubbles.

Models, using various assumptions and approximations, for fluid motion around spheres and around spherical caps have been developed in the past. We use these models, described below, to determine the liquid velocity induced by the motion of the larger bubble at the site of the smaller bubble and solve Eqs. (5) and (6) with appropriate initial conditions to determine the dynamics of the smaller bubble.

<u>Large Spherical Bubble with no Wake</u>

When the fluid motion induced by a moving sphere is determined using non-viscous flow conditions with zero fluid velocity at a distance from the sphere, the resulting flow, symmetric around the axis of bubble motion, is given by the stream function (Batchelor, 1967)

$$\psi = -\frac{1}{2} U \frac{r_1^3}{R} \sin^2\theta \tag{9}$$

or in terms of the velocity potential ϕ, by

$$\phi = -\frac{1}{2} U\, r_1^3\ \frac{\cos\theta}{R^2}$$

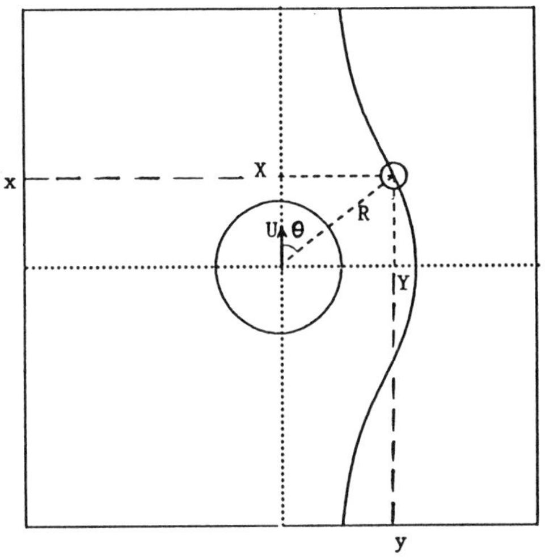

Figure 2. Schematic diagram of a large and a small bubble, and the definition of X, Y, R and θ.

where r_1 is the bubble radius, U is the bubble velocity and R and θ are explained in Fig. 2. Hence, the fluid velocity, as observed by a stationary observer, is given by

$$u_l = -\frac{1}{Y}\frac{\partial\psi}{\partial Y} = \frac{U^1}{2}\left(\frac{r_1}{R}\right)^3 \{1-3\cos^2\theta\} \quad (10)$$

$$v_l = \frac{1}{Y}\frac{\partial\psi}{\partial X} = -\frac{3}{2}U_1 \sin\theta\cos\theta\left(\frac{r_1}{R}\right)^3. \quad (11)$$

However, if the fluid motion induced by the moving sphere is determined using no slip boundary condition at the sphere surface, for small Reynolds number, Re, by ignoring the inertia forces, it is given by the stream function (Lamb, 1945),

$$\psi = \frac{3}{4}Ur_1 R\left(1 - \frac{1}{3}\frac{r_1^2}{R^2}\right)\sin^2\theta. \quad (12)$$

The fluid velocity, as observed by a stationary observer, in this case becomes

$$u_l = U\left(\frac{r_1}{R}\right)\left\{1 + \frac{1}{2}\left[1 - \left(\frac{r_1}{R}\right)^2\right]\left[1 - \frac{3}{2}\sin^2\theta\right]\right\} \quad (13)$$

$$v_l = \frac{3}{4}U r_1 \frac{\sin\theta\cos\theta}{R}\left[1 - \left(\frac{r_1}{R}\right)^2\right] \quad (14)$$

Using the liquid velocities induced by the larger bubble given above, Eqs. 10 and 11 or Eqs. 13 and 14, the

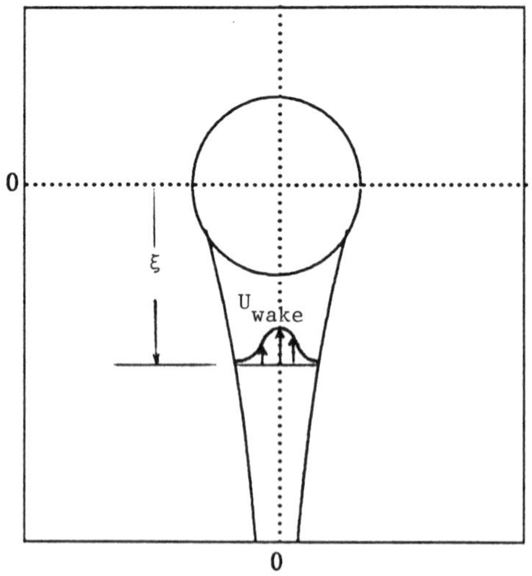

Figure 3. Schematic diagram of the wake behind a bubble. For laminar flow, we assume the width of the wake and the velocity at the center line, decrease exponentially with distance from the bubble's center.

dynamics of the smaller bubble in the vicinity of the larger bubble can be solved using Eqs. 5 and 6. Since the vortex and the wake effect behind the larger bubble is not explicitly included, these models do not include the pulling force exerted by the larger bubble on the smaller bubble in its wake which contributes to coalescence.

Large Spherical Bubble With Wake

As discussed earlier, the wake effect plays an important role in many bubble dynamics and that entrainment in the wake of a leading bubble is considered to be the prime mechanism for bubble coalescence (Crabtree and Bridgewater, 1971; Otake et al. 1977). Though it is known that in laminar flow the width of the wake and liquid velocity decrease with distance, the wake structure behind bubbles is poorly understood as any attempt to measure velocities in the wake also affects the bubble motion. See Fig. 3. We assume that the vertical velocity in the wake can be written as (Stuhmiller et al., 1989)

$$U_{wake}(\xi,\eta) = U_d(\xi)\, e^{-a_1\left(\frac{\eta}{\eta^*(\xi)}\right)^2} \quad (15)$$

where ξ is the distance from the bubble center, η is the perpendicular distance form the line passing through the bubble center, and η^* scales the velocity at different ξ values, and is assumed to be given by

$$\eta^*(\xi) = a_2 e^{-a_3\xi}, \quad (16)$$

which results in exponentially decreasing wake width behind the bubble. The center line velocity in the wake behind the bubble $U_d(\xi)$ is also assumed to decrease exponentially

$$U_d(\xi) = U_o e^{-a_4\left(\frac{\xi}{r_1}-1\right)}, \quad (17)$$

where r_1 is the bubble's radius, U_o is bubble's velocity, and $a_i(i=1,2,3,4)$ are constants. Note that this choice of $U_{wake}(\xi,\eta)$, for a given ξ, results in a Gaussian distribution of velocity in η, where both the peak and spread of the Gaussian decrease exponentially with ξ. Also note that the wake velocity calculated using Eq. 15 is actually super imposed on velocity given by Eqs. (10) and (11) or Eqs. (13) and (14).

Large Spherical Cap Bubble With Wake

It is well known that bubbles effect the dynamics of neighboring bubbles by entraining them in their wake. Length of the wake behind a spherical body has been observed to increase with Reynolds number, but the flow inside the wake is poorly understood. In case of vapor bubbles of spherical cap shape, we assume the bubble is followed by a Hills vortex and hence, the flow inside a closed region right under the spherical cap is given by the stream function (Lamb, 1945)

$$\psi = -\frac{3}{4}UR^2\sin^2\theta\left(1 - \left(\frac{R}{r_1}\right)^2\right) + \frac{1}{2}UR^2\sin^2\theta$$

where the bubble induced velocity outside the vortex is given by Eqs. 10 and 11.

RESULTS

The models described in the section above were used to simulate the fluid motion induced by the rising motion of a bubble. The motion of a small spherical bubble along the path (in the vicinity) of the larger bubble was then determined by solving Eqs. 5 and 6. Results of a typical analysis of the motion of the smaller bubble of 1 mm radius, initially 2 mm from the vertical line passing through the larger bubble, is shown in Fig. 4. Larger bubble's radius is 5 mm. Shown in Fig. 4a is the motion of the smaller bubble in coordinate axes fixed at the center of the (moving) larger bubble. Fig. 4b shows the vertical location of both bubbles as a function of time. The effect of various parameters like bubble radii, liquid viscosity, and various initial conditions were numerically studied for the different models. This is reported first for the case when induced fluid motion is determined using nonviscous flow conditions (Eqs. 10 and 11). Figure 5 shows the motion of the smaller bubble, with respect to the larger bubble, for different initial conditions. As expected, the effect of the larger bubble on the motion of the smaller one significantly decreases if the smaller bubble is initially farther away from the vertical line passing through the larger bubble. The effect of viscosity of the liquid on the bubble-bubble interaction is shown in Fig. 6. An increase in viscosity (by a factor of two) results in an increased drag force and hence, the smaller bubble more closely follows the streamlines generated due to the motion of the larger bubble. The effect of the radius of the larger bubble also was studied. Note that the analysis is valid only for cases where $r_1 >> r_2$, so that the smaller bubble does not influence the motion of the larger bubble. Also, the smaller bubble, for the purpose of drag force calculation, is treated as a point particle. Figure 7 shows bubble motion for two larger bubbles of different radii, 5mm and 7.5 mm. The smaller bubble initially is at distance of 2 mm from the vertical line in both cases.

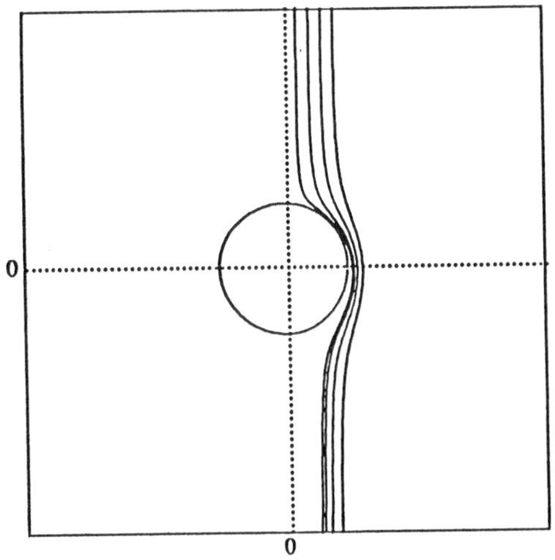

Figure 5. The effect of horizontal distance between the two bubbles on the dynamics of the smaller bubble for induced flow calculated assuming nonviscous flow conditions. Shown are four paths of the smaller bubble starting from initial conditions at 1, 2, 3 and 4 mm from the vertical line passing through the center of the larger bubble.

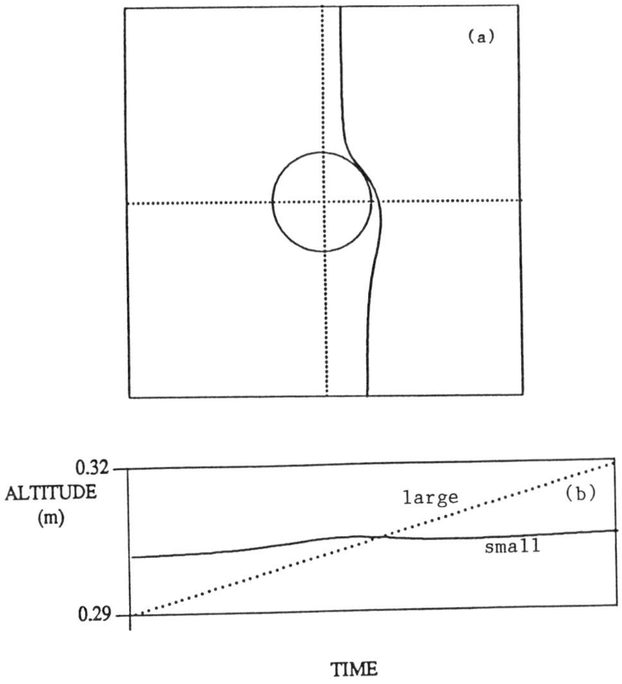

Figure 4. A typical path of the smaller bubble as the larger bubble passes by it. (a) In the frame fixed at the center of the larger bubble; (b) the vertical location of the two bubbles as a function of time.

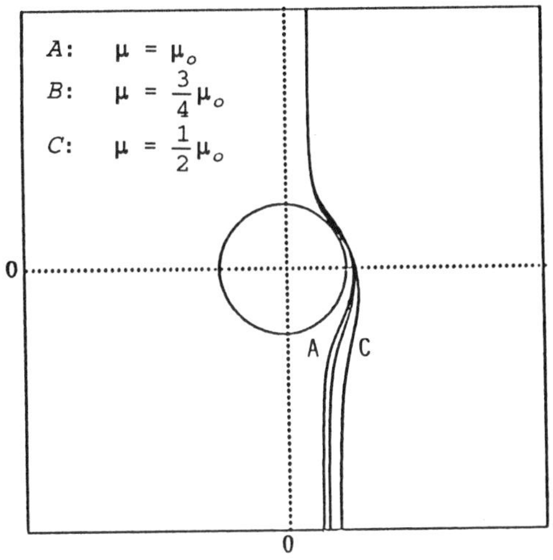

Figure 6. The effect of viscosity on the dynamics of the smaller bubble.

The above analyses were next repeated for case II, in which the larger bubble induced fluid motion is determined by using the no-slip boundary condition, Eqs. 13 and 14. Although more appropriate for solid spheres, it has been suggested that due to contamination in the liquid -- which leads to friction on the bubble surface -- this boundary condition also is relevant for bubble induced flow (Lamb, 1945; Batchelor, 1967). The effect of different initial conditions on bubble dynamics is shown in Fig. 8. A comparison with Fig. 5 shows that in this case bubbles follow the streamlines more closely due to the higher drag force. The drag force, due to high viscosity, necessary because the analysis is valid only for small Reynolds number, is so high that a small change in viscosity does not significantly effect the path of the bubbles. The effect of the size of the larger bubble is shown in Fig. 9. Shown are the paths of the smaller bubble, initially 2 mm away from the vertical line passing through the larger bubble, for larger bubble of 5 mm and 7.5 mm radius.

The dynamics of the smaller bubble was next studied by introducing the wake effect and adding the vertical velocity in the wake as determined by Eq. (15) to the liquid velocity determined by Eqs. (10) and (11). Results for the case of the smaller bubble starting from three different initial conditions (0.5 mm, 1 mm and 2 mm away from the line passing through the center of the larger bubble) are shown in Fig. 10. In the first two cases the smaller bubble gets entrained in the wake and as a result eventually 'hits' the larger bubble which may result in coalescence. In the third case, since the smaller bubble is initially sufficiently removed from the larger bubble, it escapes the wake behind the larger bubble.

For the case of the spherical cap large bubbles, the fluid motion outside the vortex is the same as that given by case I. The motion inside the vortex is given by Hills vortex and for large viscosity fluids, bubbles trapped in the vortex closely follow the circulatory motion in the vortex and, due to the large drag forces, remain trapped. In other cases the smaller bubble leaves the vortex as the larger bubble speeds away, or it may merge with the spherical cap bubble. Since the current model cannot move a small bubble from outside the vortex to inside the vortex, the dynamics induced by the vortex motion is studied by assuming initial conditions for the smaller bubble inside the vortex. Shown in Fig. 11 are two cases of smaller bubble trapped in the vortex (initial condition inside the vortex).

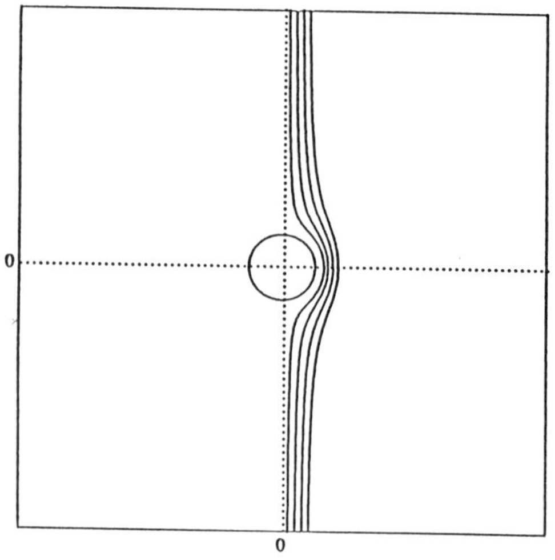

Figure 8. The effect of initial conditions on the dynamics of the smaller bubble for induced flow calculated using no-slip boundary condition (Stokes solution). Shown are four paths of the smaller bubble starting from initial conditions at 1, 2, 3 and 4 mm from the vertical line passing through the center of the larger bubble.

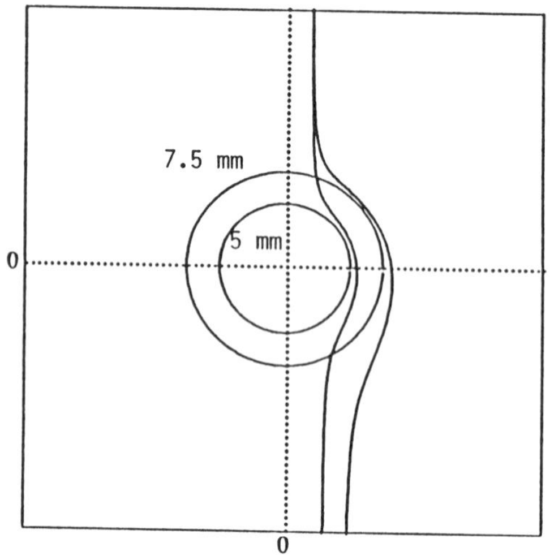

Figure 7. The effect of larger bubble's diameter on the dynamics of the smaller bubble. In both cases, the initial horizontal distance between the two bubbles is 2 mm.

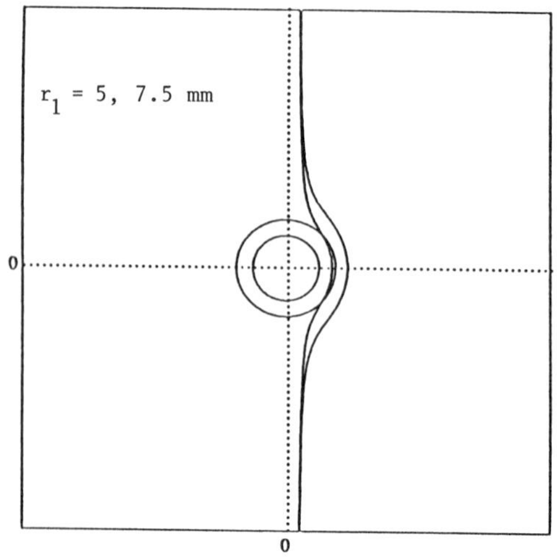

Figure 9. The effect on the dynamics of the smaller bubble of the larger bubbles diameter for Stokes flow around the larger bubble.

For the case of large liquid viscosity, the bubble while slowly drifting out, follows the circulatory motion of Hills vortex, and in the process gets trapped at a location at the bottom of the vortex with zero velocity in the horizontal direction and vertical velocity exactly equal to the larger bubbles' rise velocity. The drag force at this point, due to the relative motion of the smaller bubble and vortex flow, exactly cancels the buoyancy force, and hence, at least theoretically, the smaller bubble keeps following the larger bubble with a constant distance between them, Fig. 11a. This behavior is not observed as the viscosity of the fluid is decreased, and as shown in Fig. 11b, at lower viscosities the smaller bubble is able to 'escape' from the Hills vortex as the larger bubble speeds away.

SUMMARY

Due to increased emphasis on two-phase flow analysis by following the dynamics of individual particles of the discrete phase, we have developed models to study the dynamics of small bubbles in the vicinity of large spherical bubbles with out vortex and in the vicinity of mushroom type spherical cap bubbles. The effects of various physical parameters have been studied and it is found that the numerical analysis in the case of a spherical cap bubble predicts a point behind the large bubble where a small bubble can get trapped and follow the large bubble with zero relative velocity. This behavior was not observed at lower viscosities.

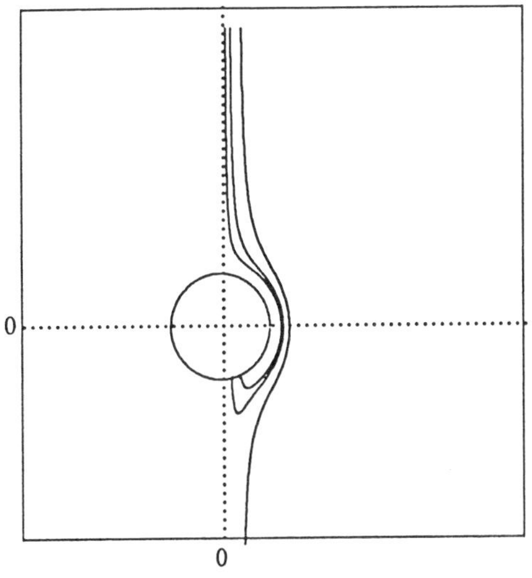

Figure 10. The effect of horizontal distance between the two bubbles when the vertical velocity due to the wake effect behind the larger bubble also is included. Shown are three trajectories of the smaller bubble with initial horizontal distance between the two bubbles equal to 0.5, 1 and 2 mm. The smaller bubble gets trapped in the wake when its initial horizontal distances from the larger bubble is 0.5 mm or 1 mm. The larger bubble's radius is 5 mm.

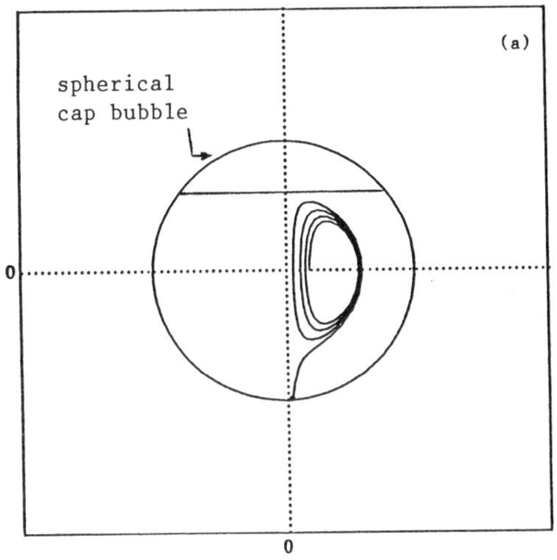

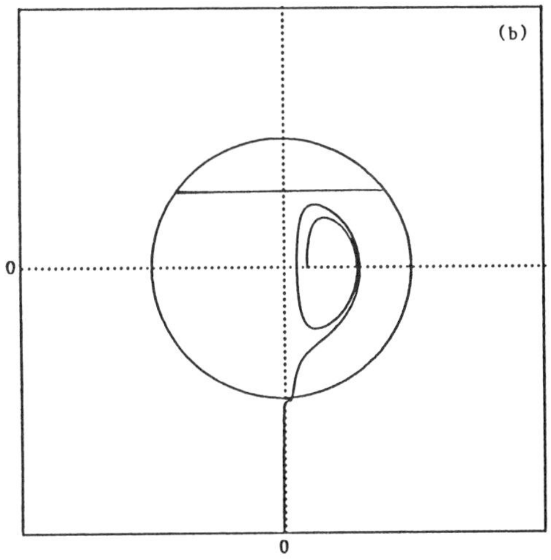

Figure 11. The dynamics of a small bubble inside a Hills vortex behind a spherical cap bubble. Figure 11a shows the case when the smaller bubble gets trapped in the Hills vortex and moves with the larger bubble. As the viscosity is decreased, the smaller bubble, as shown in 11b, is able to escape from the vortex.

REFERENCES

Allen, H.S., 1900, Philosophical Magazine, 50, 323.

Batchelor, G.K., 1967, An Introduction to Fluid Dynamics, Cambridge University Press.

Crabtree, J.R. and Bridgewater, J., 1971, "Bubble Coalescence in Viscous Liquids," Chem. Eng. Scie., 26, 839 (1971).

Davidson, J.F., Harrison, D., Darton, R.C. and LaNauze, R.D., 1977, In Chemical Reactor Theory, A Review, ed. L. Lapidus and N.R. Amundson, Chap. 10, Englewood Cliffs, NJ, Prentice Hall.

Davies, R.M. and Taylor, G., 1950, "The Mechanics of Large Bubbles Rising Through Extended Liquids and Through Liquids in Tubes," Proc. Roy. Soc., Ser. A. 200, 375.

Feuillebois, F., 1989, "Some Theoretical Results for the Motion of Solid Spherical Particles in a Viscous Fluid," in Multiphase Science and Technology, eds. G.F. Hewitt, J.M. Delhaye and N. Zuber, Vol. 4, Hemisphere Publishing Corporation.

Grace, J.R., 1970, Can. J. Chem. Eng., 48, 30.

Harmathy, T.Z., 1960, "Velocity of Large Drops and Bubbles in Media of Infinite and Restricted Extent," AIChE J., 6, 281.

Hartunian, R.A. and Sears, W.R., 1957, "On the Instability of Small Gas Bubbles Moving Uniformly in Various Liquids," J. Fluid Mech., 3, 27.

Lamb, H., 1945, Hydrodynamics, Dover Publications.

Levich, V.G., 1949, Zhur, Eksp. i Teoret. Fiz., 19, 18.

Levich, V.G., 1962, Physiochemical Hydrodynamics, Prentice Hall.

Moore, D.W., 1958, "The Rise of a Gas Bubble in a Viscous Liquid," J. Fluid Mech., 6, 113-130.

Moore, D.W., 1959, "The Rise of a Gas Bubble in Viscous Liquids," J. Fluid Mech., 6, 113.

Moore, D.W., 1963, "The Boundary Layer on a Spherical Gas Bubble," J. Fluid Mech., 16, 161.

O'Brien, M.P. and Gosline, J.E., 1935, "Velocity of Large Bubbles in Vertical Tubes," Ind. and Eng. Chemistry, 27, 1436.

Otake, T., Tone, S., Nakoe, K and Mitsuhashi, Y., 1977, "Coalescence and Breakup of Bubbles in Liquids," Chem. Eng. Sci., 32, 377.

Parlange, J-Y., 1969, "Spherical Cap Bubbles with Laminar Wakes," J. Fluid Mech., 37, 257.

Pebbles, F.N., and Garber, H.J., 1958, "Studies on the Motion of Gas Bubbles in Liquid," Chem. Eng. Prog., 49, 88.

Rippin, D.W.T. and Davidson, J.F., 1967, Chem. Eng. Sci., 22, 217.

Robinson, J.V., 1947, J. Phys. and Coloid. Chem., 51, 431.

Rowe, P.N., 1964, "A Note on the Motion of a Bubble Rising Through a Fluidized Bed," Chemical Eng. Sci., 19, 75.

Stokes, G.G., 1851, "On the Effect of Internal Friction of Fluids on the Motion of Pendulums," Trans. Camb. Phil. Soc., 9, 8.

Stuhmiller, J.H., Ferguson, R.E. and Meiser, C.A., 1989, "Numerical Simulation of Bubbly Flow," Final Report EPRI NP-6557.

Trapp, J.A., Mortensen, G.A., and Ransom, V.H., 1991, "Particle-Fluid Two-Phase Flow Modeling," in Adv. in Math., Comp., and Reactor Phys., Vol. 1, 3.1 3-1 - 3.1 3-12, Am. Nuc. Soc..

Uno, S. and Kintner, R.C., 1956, "Effect of Wall Proximity on the Rate of Rise of Single Air Bubbles in a Quiescent Liquid," AIChE J., 2, 420.

van Krevelen, D.W. and Hoftijzer, P.J., 1950, "Studies of Gas-Bubble Formation--Calculation of Interfacial Area in Bubble Contractors," Chem. Eng. Prog., 46, 29.

Walters, J.K. and Davidson, J.F., 1962, "The Initial Motion of a Bubble Formed in an Inviscid Liquid: Part I. The Two-Dimensional Bubble," J. Fluid Mech., 12, 408.

Wegener P.P. and Parlange, J-Y., 1973, "Spherical-Cap Bubbles," Annual Rev. Fluid Mech., 5, 79.

HTD-Vol. 197, Two-Phase Flow and Heat Transfer
ASME 1992

LOCAL MEASUREMENTS IN THE TWO-PHASE REGION OF SUBCOOLED BOILING FLOW

R. P. Roy, I. A. Choudhury, V. Velidandla, and S. P. Kalra*
Department of Mechanical and Aerospace Engineering
Arizona State University
Tempe, Arizona

ABSTRACT

Accurate measurements of local vapor phase residence time fraction, vapor bubble axial velocity, bubble diameter distribution, and fluid as well as liquid phase temperature in the two-phase region of subcooled boiling flow are reported. The data are appropriate for validation of two-fluid subcooled boiling flow models which are at least two-dimensional.

NOMENCLATURE

D	vapor bubble diameter
$\overline{D}$	average vapor bubble diameter
G	mass velocity of fluid
N_d	total rate of bubble detection
p	pressure
$P_d(D)$	cumulative probability distribution function of detected bubble diameter
p_r	partial pressure of R-113
q_w''	wall heat flux
r	radial coordinate
R^*	nondimensional radial coordinate, $(r-r_i)/(r_o-r_i)$
T	temperature
u_G	vapor bubble axial velocity
x	bubble chord length detected
α_G	vapor residence time fraction
$\rho_c(x)$	probability density function of detected bubbles chord length
$\rho_d(D)$	probability density function of detected bubbles diameter

Subscripts

$bb\ell$	boiling (bubble) layer
i	inner wall of annulus
in	inlet to test section
o	outer wall of annulus
sat	saturation
w	heated wall

† Permanent Address:
Electric Power Research Institute, Palo Alto, CA 94303

INTRODUCTION

Experimental information on the transverse distributions of the residence time fractions of the phases, interfacial area concentration, phase velocities and temperatures, as well as turbulence characteristics is essential for the development of a realistic multidimensional model of turbulent boiling flow. Some of the data may lead to closure relations for the model while others would be appropriate for model validation. However, such information is quite scarce in open literature.

An early experimental study relevant to this information was by Jiji and Clark (1) who carried out transverse temperature distribution measurements in forced convective subcooled boiling of water over a vertical plate in a rectangular channel. Mean and fluctuating temperatures were measured in and outside the boiling layer by means of thermocouples with small time constant. Whether the thermocouples could distinguish between the vapor and liquid temperatures was not discussed although the exposed hot junction diameter (≈ 0.25 mm) appears to have been too large and hence the response too sluggish to allow such a distinction. Walmet and Staub (2) also used thermocouples with very small hot junctions and time constants (≈ 5 ms) to measure the transverse distribution of temperature in subcooled boiling flow of water in a one-side-heated rectangular test section. The local vapor fraction distribution was measured as well by the γ-ray attenuation method.

Shiralkar (3) and Dix (4) reported on the development and application of hot-film anemometer sensors operated in the constant temperature (i.e., overheat) mode for measurement of local vapor fraction in subcooled boiling flow of Refrigerant-114 in an annular test section. Dix also presented qualitative results from direct observation and high-speed film study of subcooled boiling flow. A critical appraisal of the physical mechanisms suggested in models developed by others was also made.

Delhaye et al. (5) made a significant contribution toward local measurements in boiling flow by devising a microthermocouple with a 20 μm diameter hot junction capable of distinguishing between vapor and liquid phase temperatures. The local vapor fraction was obtained from the temperature histograms (probability density functions).

Hasan et al. (6) reported measurements of vapor phase residence time fraction and liquid phase temperature distributions in subcooled boiling flow of Refrigerant-113 (R-113) through a vertical annular channel. The vapor fraction was measured by means of a cylindrical hot-film sensor, 25 μm in diameter and 250 μm long, operated in the constant temperature mode (this being the same technique as Shiralkar (3) and Dix (4)). A chromel-constantan microthermocouple with an exposed disc-shaped hot junction was used in conjunction with a phase-lead compensation circuit to measure the fluid (single- and two-phase) temperature distribution. The measured mean fluid temperature in the two-phase (boiling) region was then equated to the liquid phase temperature on the basis of the argument that the microthermocouple was too slow to respond to the temperature of the flowing small vapor bubbles. However, recent re-examination of the microthermocouple output signals have led to the conclusion that the microthermocouple did respond to the vapor bubble temperature albeit in an attenuated fashion. Depending on the local vapor fraction, the liquid phase temperature was calculated to be up to about 1°C lower (in the high vapor fraction locations) than the mean fluid temperature.

In this paper, we report local measurements of vapor phase residence time fraction, vapor bubble axial velocity, bubble diameter distribution, and fluid as well as liquid phase temperature in subcooled turbulent boiling flow of R-113 through a vertical annular channel whose inner wall is heated and the outer wall insulated. Results are presented for several wall heat fluxes, two mass velocities, and two inlet (to test section) liquid subcoolings. Figure 1 shows, schematically, a typical subcooled flow boiling condition in the annulus exhibiting two distinct regions, one a boiling fluid layer adjacent to the heated wall (region I) and the other an outer all-liquid region (region II).

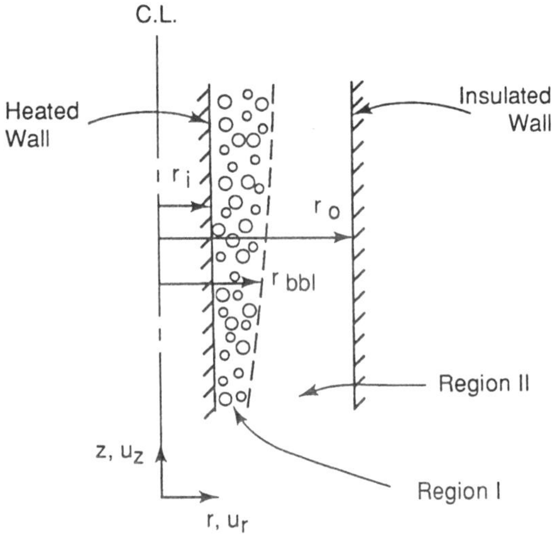

Fig. 1 Boiling layer and adjacent all-liquid region in subcooled boiling flow

THE EXPERIMENTS

The Test Section

Figure 2 shows a part of the vertical annular test section. The inner tube of the test section was of 304 stainless steel (i.d. = 1.46 cm, o.d. = 1.59 cm) which could be resistively heated by direct current. The outer pipe was made of transparent Pyrex glass (i.d. = 3.81 cm, o.d. = 4.70 cm) except for the 0.496 m long *measurement section* which was of quartz† (i.d. = 3.77 cm, o.d. = 4.17 cm). The upper 2.75 m of the total test section length of 3.66 m could be heated to a maximum power of 60 kW. The lower 0.91 m served as the hydrodynamic entrance length.

A dual-sensor fiber-optic probe and a microthermocouple were installed diametrically opposite each other in the measurement section. Four surface thermocouples were installed on the inner wall of the heater tube as well. The heater tube was subsequently filled with aluminum oxide powder insulation. The *measurement plane* shown in Fig. 2 was approximately 1.94 m downstream of the beginning of heated length.

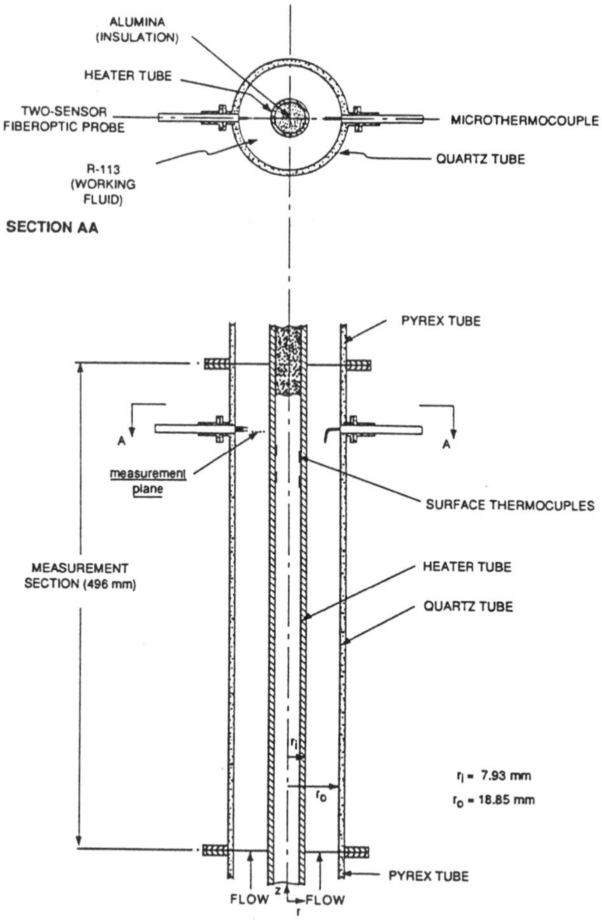

Fig. 2 The measurement section

Measurement Instrumentation

Vapor phase. These measurements were performed by the dual-sensor fiber-optic probe mentioned earlier. This probe (Photonetics) is equipped with two fiber-optic sensors each of which acts as a guide to light emitted by an LED. The sensors are 1.6 mm apart, the measurement diameter of each sensor being about 50 μm. The probe operates in conjunction with an electronic circuit and has a response time of approximately 10 μs.

Figures 3(a) and (b) exhibit typical analog outputs of the two sensors at some specific location in the boiling layer. Each voltage pulse signifies an encounter of a sensor with a vapor bubble. The cross-correlation function between the two signals is shown in Fig. 3(c), with the single dominant peak indicating the average traverse time of the vapor bubbles between the two sensors. The average vapor bubble velocity can now be obtained as the ratio of the distance between the sensors (1.6 mm) and the traverse time.

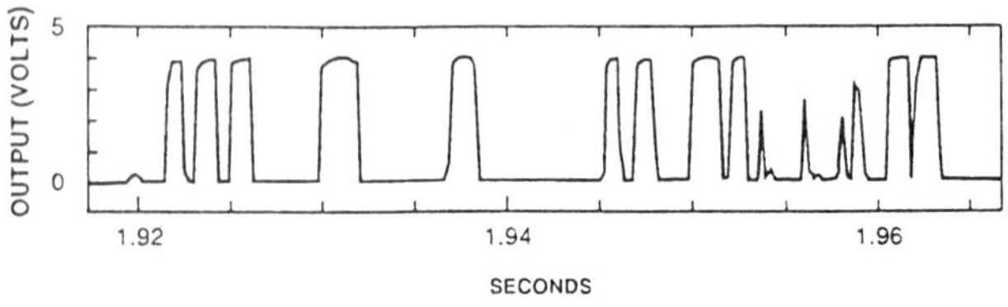

(a) DOWN STREAM SENSOR SIGNAL

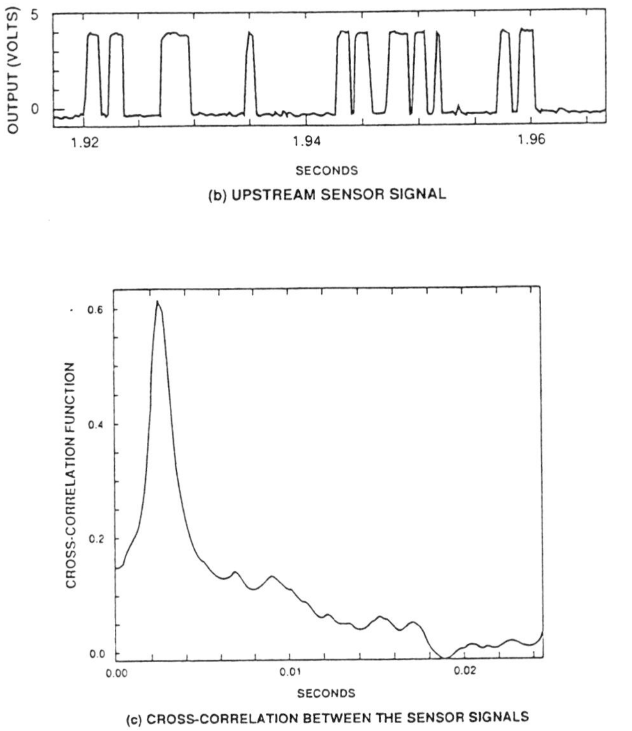

(b) UPSTREAM SENSOR SIGNAL

(c) CROSS-CORRELATION BETWEEN THE SENSOR SIGNALS

Fig. 3 Typical output of fiber-optic sensors and their cross-correlation

Generally, a bubble encountering the sensor tip will not be pierced along a diameter. In other words, a chord length, x, will be sensed as shown schematically in Fig. 4 (Herringe and Davis (7)).

x = Bubble Chord Length

D = Bubble Diameter

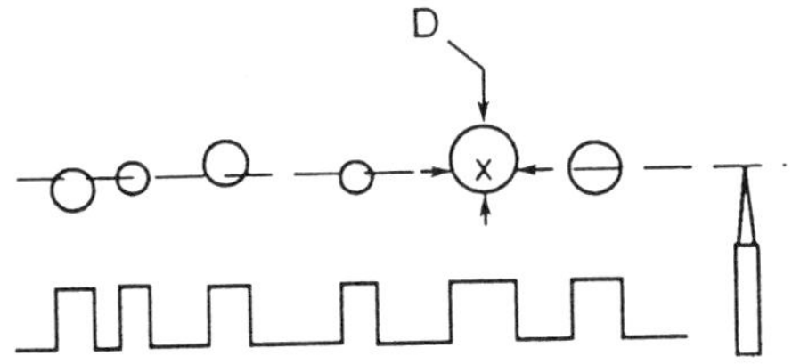

Fig. 4 Bubble chord lengths detected by fiber-optic sensor

One can obtain the probability density function (PDF), $\rho_c(x)$, of the measured chord lengths. It can be shown (Herringe and Davis (7)) that the cumulative probability-distribution function of the diameter of all the *detected* bubbles is

$$P_d(D) = \int_0^D \rho_d(x)\,dx - \frac{1}{2} D\,\rho_d(x = D) \quad (1)$$

The corresponding PDF is

$$\rho_d(D) = \frac{d\,P_d(D)}{dD} \quad (2)$$

Furthermore, the PDF of the diameter of *all* bubbles with centers passing through a unit cross-section area at that location is

$$\rho(D) = \frac{2\,N_d D}{3\,\alpha\,u_G}\,\rho_d(D) \quad (3)$$

where N_d is the total rate of bubble detection and u_G the bubble axial velocity.

It should be noted that the above relations are based on the simplifying assumptions of spherical bubbles and a single bubble velocity irrespective of diameter.

Fluid temperature. A specially constructed chromel-alumel microthermocouple (Beckman et al., (8)) in which the wire elements, 18 µm in diameter, are joined together in a disc-shaped junction of 2.5 µm, thickness and 0.1 mm diameter was used for these measurements. The time constant of the microthermocouple in turbulent single-phase flow of R-113 in the same Reynolds number range as the present experiments was measured to be 4.6 ms. In order to further improve the speed of response, an active phase-lead compensation circuit similar to one suggested by Hishida and Nagano (9) was built. The phase (and gain)-compensated microthermocouple was measured to have a time constant of 3.4 ms. This is an improvement upon the microthermocouple response obtained in our earlier work (Hasan et al., (6)). The present microthermocouple was fast enough to respond reasonable well to the vapor temperature during bubble passages, thus allowing estimation of the local liquid phase mean temperature from the PDF of the fluid temperature signal. We should note, however, that a further reduction in the time constant of the compensated microthermocouple, perhaps to a value ≤ 3 ms, would permit even better distinction between the temperatures of the vapor and liquid phases.

Conduction loss (to the thermocouple wires) correction, usually an important consideration, was considered not to be significant in the present measurements because of the orientation of the thermocouple wires with respect to the axial flow direction (≈ 15 degrees) and the expected isotherms.

Heated wall temperature. The surface thermocouples (Omega, foil thickness = 0.01 mm, junction length = 0.5 mm) were baked on with epoxy onto the inner wall of the heater tube. The tube outer wall (i.e., the heated wall of the annulus) temperature was then calculated analytically from the measured inner wall value by a steady-state heat conduction analysis.

Results

Heat balance calculations based on the imposed wall heat flux, the mean liquid flow rate through the test section, and the rise in its mixed-mean temperature indicated that approximately two percent of the supplied thermal energy was lost to the ambient.

A brief discussion of the importance of careful monitoring of the dissolved air content of the working fluid, R-113, is in order prior to presentation of the results. Air is known to be highly soluble in R-113. Therefore, extensive degassing of the fluid inventory followed by measurement of the residual air content was essential prior to each experiment. Dissolved air measurements were performed by an Aire-Ometer (Seaton-Wilson). On the basis of each measurement and Henry's law, the partial pressure of air in the solution was calculated. This partial pressure was then subtracted from the pressure at the measurement plane to arrive at the local R-113 partial pressure. For the experiments reported here, the air partial pressure was typically 8 kPa out of the measured pressure of 277 kPa.

The effects of dissolved air on the measured values of local vapor fraction and fluid temperature are significant. A high dissolved air content can increase the measured (apparent) vapor fraction by as much as five percent (eg., an apparent vapor fraction of twenty five percent may be measured when the true value is twenty percent). The wall temperature is reduced because of the

† This measurement section material was selected for future laser Doppler velocimetry measurements.

accompanying drop in the R-113 saturation temperature, although the wall temperature reduction was found to be markedly less than the drop in saturation temperature. This is probably due to a decrease in the efficacy of heat transfer at the wall caused by the air coming out of solution. The fluid temperature is correspondingly lower particularly in the vicinity of the heated wall.

Table 1 shows the range of variables over which the experiments were conducted and the associated measurement uncertainties.

Table 1. Range of experiments and measurement uncertainties†

	Range	*Uncertainty*
Wall heat flux	79,400-126,000 W/m^2	± 200 W/m^2
Mass velocity	579, 801 kg/m^{2}s	± 3 kg/m^{2}s
R-113 partial pressure at measurement plane (sat. temp.)	269 kPa (80.1°C)	± 0.7 kPa
Mean liquid temperature at test section inlet	43.0, 50.3° C	± 0.1° C
Wall temperature	95-102° C	± 0.4° C
Vapor local residence time fraction	0-50%	± 1% for 0% < α_G < 10% ± 2% for 10% < α_G < 50%
Sensor radial traverse:		
fiber-optic sensor	0-10 mm	± 30 μm
microthermocouple	0-9 mm	± 40 μm

†The uncertainty estimates are for 95 percent confidence.

Vapor phase residence time fraction--radial distribution. A sampling time interval of 230 μs and a record length of approximately 4.7 seconds were used for each measurement. Each measurement was repeated once and the average of the measured values was taken to be the local value.

Figure 5 shows the radial profile of vapor fraction at the measurement plane at a mass flux of 579 kg/m^{2}s, wall heat flux of 79400 W/m^2, R-113 partial pressure of 269 kPa, and mean liquid temperature at the test section inlet of 50.3° C. The vapor fraction values measured by both the upstream and downstream sensors of the fiber-optic probe are shown. It is apparent from the data that the upstream sensor encounters more vapor bubbles than the downstream sensor. This is reasonable because the downstream sensor is shielded to a certain extent by the sensor upstream. It is then appropriate to adopt the upstream sensor measurement as the more accurate measurement. All vapor fraction data presented subsequently in this paper are based on the upstream sensor measurements.

There are, clearly, two well-defined regions in the flow field (Fig. 5 and the schematic in Fig. 1)--a boiling layer and an all-liquid region. Note also that the closest distance from the heated wall for which data are presented is 0.4 mm.† While it would have been desirable to obtain measurements closer to the wall, this was a risky undertaking since the slightest contact with the wall would have damaged the sensors.

† In some recent experiments, we have ventured closer (0.3 mm) to the wall.

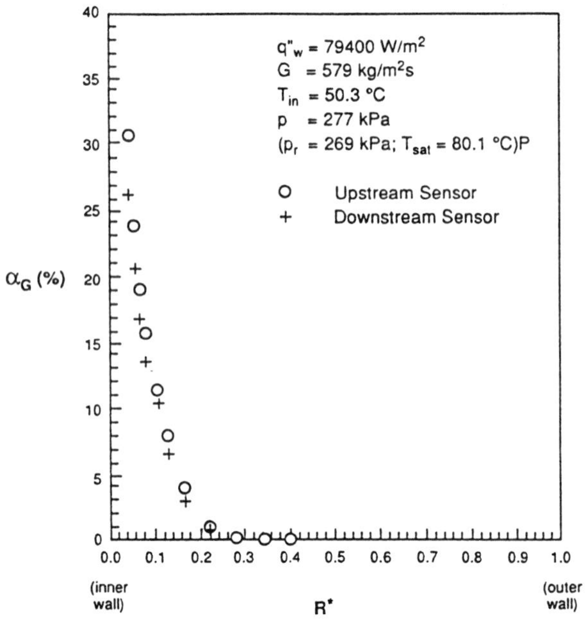

Fig. 5 Typical vapor fraction radial profile measured by the two sensors of the fiber-optic probe

In Fig. 6, we present vapor fraction radial distributions at two mass velocities, 579 and 801 kg/m^{2}s, wall heat flux of 94950 W/m^2, R-113 partial pressure of 269 kPa, and mean liquid temperature at test section inlet of 43° C. The boiling layer is, as expected, thicker at the lower mass velocity. Moreover, the vapor fraction close to the heated wall is significantly higher at the lower mass velocity. Two reasons for this are: the bubbles are larger and the bubbles are more closely packed, the sensor output signals lending credence to both.

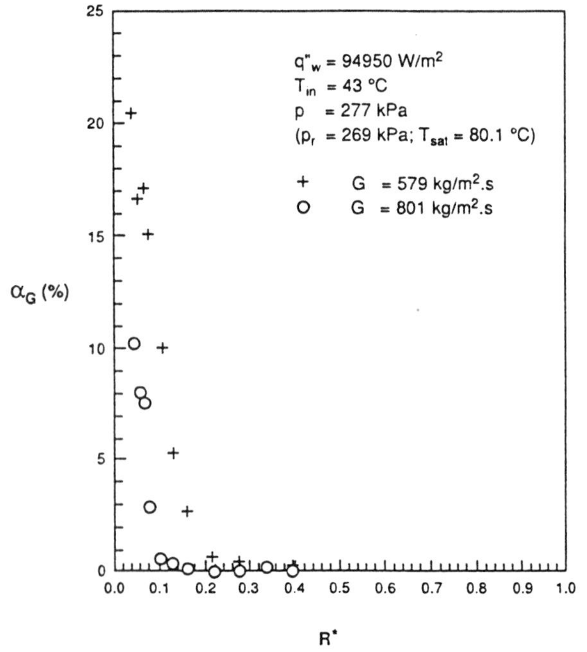

Fig. 6 Vapor fraction radial profiles at two mass velocities

Figure 7 depicts vapor fraction radial profiles at two wall heat fluxes, 79400 and 94950 W/m^2, mass velocity of 579 kg/m^{2}s, R-113 partial pressure of 269 kPa, and mean liquid temperature at test section inlet of 43° C. That the radial extent of the boiling layer is greater at the higher wall heat flux as is the vapor fraction in the wall proximity are also clear from the figure.

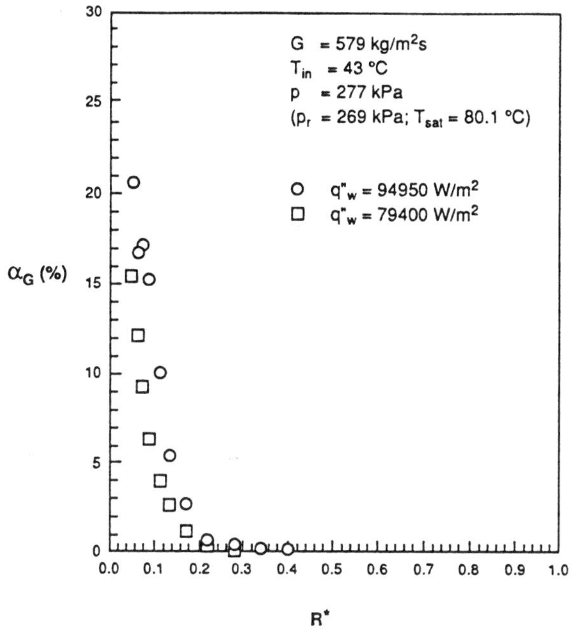

Fig. 7 Vapor fraction radial profiles at two wall heat fluxes

Figure 8 shows the radial profiles of vapor fraction at two inlet liquid temperatures, 43.0 and 50.3° C, mass velocity of 801 kg/m^{2}s, R-113 partial pressure of 269 kPa, and wall heat flux of 115725 W/m^2. At a lower liquid phase subcooling the rate of vapor condensation would be smaller and as such, the transverse extent of the boiling layer will be larger as will be the vapor fraction in the layer. Both features are apparent from the data.

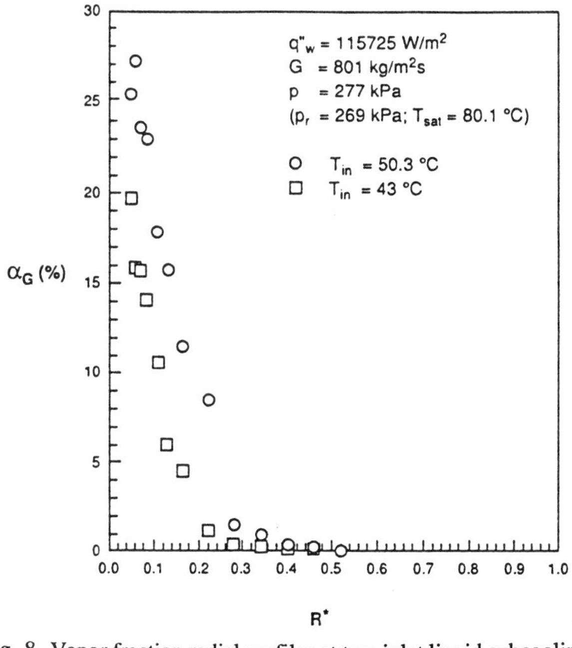

Fig. 8 Vapor fraction radial profiles at two inlet liquid subcoolings

Vapor bubbles--diameter distribution. Figure 9(a) presents the PDF of the detected chord length of vapor bubbles as well as the PDF of the diameter of detected bubbles as calculated from Eq. (1) at an experimental condition for which the vapor fraction distribution was shown in Fig. 8. The vapor fraction at this specific radial location is between 3 and 4 percent. A location such as this is important to investigate because it is near the inter-region (I and II, Fig. 1) boundary. The most probable detected bubble diameter appears to be approximately 1.1 mm.

Figure 9(b) shows the chord length and diameter PDFs of the detected bubbles at the same experimental condition as Fig. 9(a) except that the inlet liquid temperature is now 50.3° C. The corresponding vapor fraction distribution, also shown in Fig. 8, indicates the local (the same radial location as Fig. 9(a)) vapor fraction to be approximately 11 percent. The most probable detected bubble diameter is approximately 1.2 mm. Moreover, nonzero probability persists to larger bubble diameters (> 2.6 mm) in comparison to Fig. 9(a) (≈ 1.9 mm).

That larger bubbles are found when the liquid phase is less subcooled is reasonable. It has already been suggested that the larger bubble size is partly responsible for the higher local vapor fraction, the remaining contribution being from closer packing of the bubbles along the axial direction.

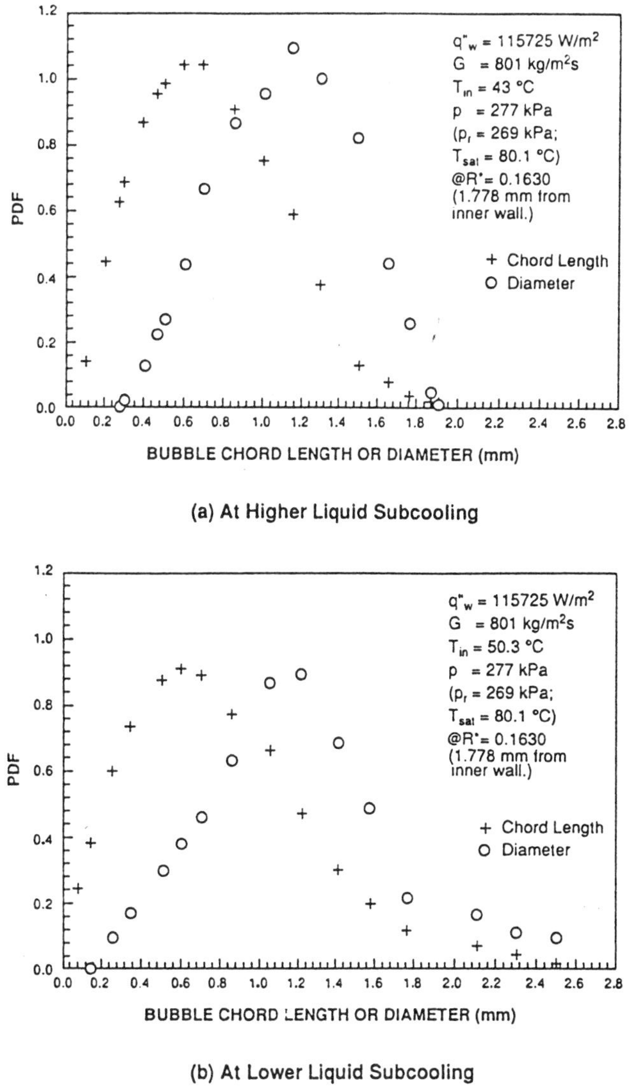

Fig. 9 Vapor bubble diameter distribution at the same radial location for two inlet liquid subcoolings

Vapor bubbles--axial velocity distribution. Vapor bubble axial velocity profiles for the two experimental conditions of Fig. 8 are shown in Fig. 10. The higher axial velocities for the lower subcooling condition can be explained as follows: the local vapor fractions are larger as are the bubble sizes when the liquid is less subcooled, resulting in higher axial velocity for both the vapor and the liquid phase. It should be noted that larger bubbles can be expected to have higher relative velocity with respect to the liquid.

The axial velocity profile in the all-liquid region for the higher liquid subcooling condition was estimated§ from our earlier experiments [Roy et al. (10)] and is shown in Fig. 10. These liquid velocity measurements were performed by means of miniature three-sensor hot-film anemometer sensors.

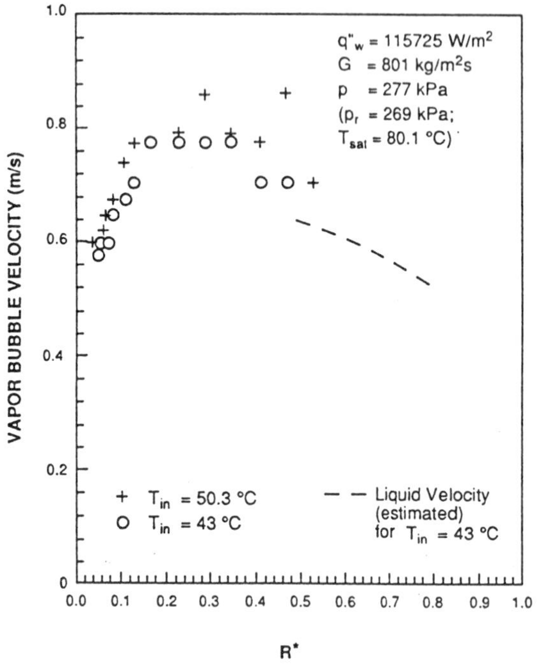

Fig. 10 Radial distribution of vapor bubble axial velocity for two inlet liquid subcoolings

Fluid temperature--radial distribution. By the term "fluid" we mean "liquid" in the all-liquid region and "vapor and liquid" in the boiling layer. It is quite possible for the liquid in the boiling layer to be subcooled and this indeed was the case in our experiments. Figures 11(a) and 12(a) present the mean fluid temperature distributions at the same two experimental conditions as in Fig. 8. The figures also show the mean liquid phase temperature distributions as estimated from the PDF of the microthermocouple output signal. Two such PDFs are shown in Figs. 11(b) and 12(b). For the experimental conditions shown, the mean local liquid temperature was lower than the local fluid temperature by between 0°C (at or very near the boiling layer outer edge) and approximately 2°C (near the heated wall, i.e., the high vapor fraction region). The estimated vapor phase temperature was slightly higher than the local saturation temperature in both cases. Whether this is a physically meaningful attribute or an artifact of the measurement and estimation uncertainties is not clear at this time.

§ The reason an estimate had to be made is that exactly the same experimental condition was not obtained in the earlier work due to the higher dissolved air content of R-113.

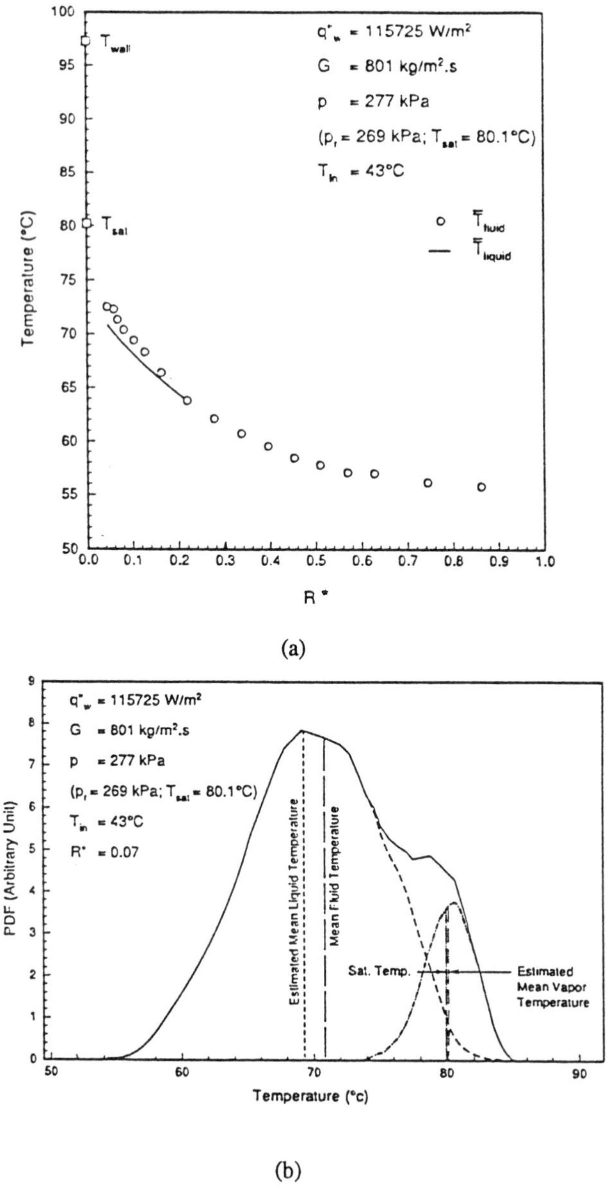

Fig. 11 Radial distribution of fluid temperature and a typical PDF

Discussion

Interfacial area. Considering that the measured local vapor fraction is really the local vapor residence time fraction, it may not be straightforward to relate the concept of *local interfacial area concentration* to the data reported here. One may need to take into account data such as the *bubble encounter (or impaction) rate* and consider the two interface crossings per encounter. It is just before and after these crossings that interfacial phenomena, such as momentum and thermal energy transport, come into play locally.

On the other hand, an appropriate form of spatial averaging could be introduced to estimate the local interfacial area concentration. For example, the expression $6\langle\alpha_G\rangle/D_{sm}$ where $\langle\alpha_G\rangle$ denotes the local volume-average vapor fraction and D_{sm} the Sauter mean diameter has been used for the bubbly flow regime (Delhaye and Bricard, (11)).

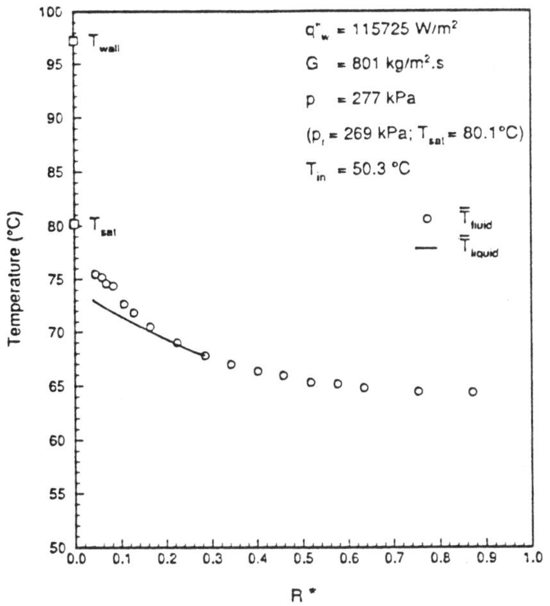

(a)

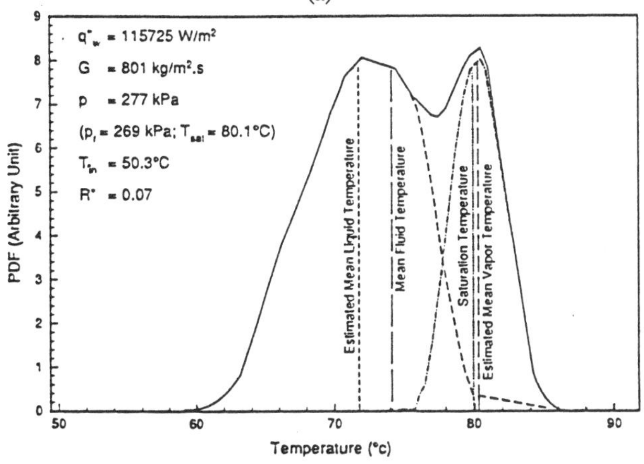

(b)

Fig. 12 Radial distribution of fluid temperature and a typical PDF

Vapor bubbles--shape and velocity. As mentioned earlier, the vapor bubbles were assumed to be spherical in shape. While this is probably a good assumption for small bubbles, the larger bubbles may have more complicated shapes, such as an ellipsoid. However, as stated by Herringe and Davis (7) in the context of air-water flows, the use of nonspherical bubble shapes in the data analysis may not be warranted as yet because detailed information regarding bubble shapes in boiling flows is lacking.

The axial velocity measured by the cross-correlation technique represents an *average* vapor bubble velocity. Clearly, there would be some differences in the velocity of individual bubbles depending on the bubble size.

Vapor and liquid phase temperatures. That the liquid remained significantly subcooled inside much of the boiling layer in subcooled boiling flow was an expected feature, such having been reported by earlier investigators, for example Jiji and Clark (1) and Hasan et al. (6). The reasonably fast response of the microthermocouple did, however, allow us to effectively distinguish between the liquid and vapor temperatures. The small difference between the estimated vapor phase temperature and the local saturation temperature may be physically meaningful. On the other hand, it may be due to measurement uncertainties as well as the somewhat subjective separation of the fluid temperature PDF to the liquid and vapor temperature PDFs.

CONCLUDING REMARKS

The measurements reported here represent a part of our continuing effort to comprehensively study the fundamental character of boiling flow. Only subcooled boiling flow was investigated in this work. Selected data on the distributions of phase residence time fractions, the size and axial velocity of vapor bubbles, and the extent of thermal nonequilibrium between the two phases have been presented. We continue these measurements with the goal of reporting and interpreting data encompassing a wide range of flow conditions.

The dual-sensor fiber-optic probe has, clearly, functioned very well. It would be useful to measure the radial motion (however small) of the vapor bubbles as well by the same technique. Also, a slightly faster microthermocouple would be helpful in distinguishing between the vapor and liquid phase temperatures more accurately especially as we approach saturated boiling condition.

Since the nature of the two regions (boiling and all-liquid) in subcooled boiling flow are quite different, it would not be appropriate to use the reported data for validation of *one-dimensional* models of such flows. A two-dimensional two-fluid model would be more suitable.

Acknowledgement

This research was partly funded by National Science Foundation, Thermal Systems Program, Division of Chemical and Thermal Systems under Grant No. CTS-8918830. Support from Electric Power Research Institute, Nuclear Power Division (RP 2614-60) is gratefully acknowledged. Thanks are also due to Electricité de France for their gift of the fiber-optic instrumentation.

REFERENCES

1 Jiji, L.M. and Clark, J.A., "Bubble Boundary Layer and Temperature Profiles for Forced Convection Boiling in Two-Phase Flow," ASME Journal of Heat Transfer, Vol. 86, 1964, pp. 50-58.

2 Walmet, G.E. and Staub, F.W., 1969, "Pressure, Temperature and Void Fraction Measurement in Nonequilibrium Two-Phase Flow," Two-Phase Flow Instrumentation, ASME, Eleventh National Heat Transfer Conference, Minneapolis, MN, pp. 89-101.

3 Shiralkar,B., "Local Void Fraction Measurements in Freon-114 with a Hot-Wire Anemometer," General Electric Report NEDE-13158, 1970.

4 Dix, G.E., "Vapor Void Fraction for Forced Convection with Subcooled Boiling at Low Flow Rates," General Electric Report NEDO-10491, 1971.

5 Delhaye, J.M., Semeria, R. and Flamand, J.C., "Void Fraction, Vapor and Liquid Temperatures; Local Measurements in Two-Phase Flow Using a Microthermocouple," ASME Journal of Heat Transfer, Vol. 95, 1973, pp.363-370.

6 Hasan, A., Roy, R.P. and Kalra, S.P., "Some Measurements in Subcooled Flow Boiling of Refrigerant-113," ASME Journal of Heat Transfer, Vol. 113, 1991, pp. 216-223.

7 Herringe, R.A. and Davis, M.R., "Structural Development of Gas-Liquid Mixture Flows," Journal of Fluid Mechanics, Vol. 73(1), 1976, pp. 97-123.

8 Beckman, P., Roy, R.P., Whitfield, K. and Hasan, A., "A Fast-Response Microthermocouple," submitted for publication to Review of Scientific Instruments, 1992.

9 Hishida, M. and Nagano, Y., "Simultaneous Measurements of Velocity and Temperature in Nonisothermal Flows," ASME Journal of Heat Transfer, Vol. 100, 1978, pp. 340-345.

10 Roy, R.P., Hasan, A. and Kalra, S.P., "Temperature and Velocity Fields in Turbulent Liquid Flow Adjacent to a Bubbly Boiling Layer," submitted for publication to International Journal of Multiphase Flow, 1991.
11 Delhaye, J.M. and Bricard, P., "Interfacial Area in Bubbly Flow: Experimental Data and Correlations," ANS Proceedings, National Heat Transfer Conference, Minneapolis, MN, 1991, pp. 3-13.

HTD-Vol. 197, Two-Phase Flow and Heat Transfer
ASME 1992

NONLINEAR DYNAMICS OF TWO-PHASE FLOW IN MULTIPLE PARALLEL HEATED CHANNELS

Rizwan-uddin and J. Dorning
Department of Nuclear Engineering and Engineering Physics
University of Virginia
Charlottesville, Virginia

ABSTRACT

Some basic aspects of bifurcation phenomena in two-phase flow and the related nonlinear dynamics of single and multiple parallel, uniformly and nonuniformly heated channels are studied. Specifically, the effects of several two-phase-flow system features on the stability of equilibria are determined. These features include: (1) unheated sections, or risers, at the tops of the heated channels; (2) a return feedback loop that relates the fluid properties at the channel inlets to those at the channel exits at an earlier time; and (3) the flow interaction among multiple parallel channels subjected to fixed total mass flow rate boundary conditions and alternatively to fixed external pressure drop boundary conditions. It is shown that the addition of unheated riser sections at the tops of the heated channels has a destabilizing effect, and that as the lengths of the risers at the tops of otherwise stable heated channels are increased, a supercritical Hopf bifurcation occurs resulting in stable limit cycle density-wave oscillations. It also is demonstrated that the addition of a simple return feedback loop is destabilizing, and that the complex interaction among multiple channels enables one unstable channel to drive the multiple channel system unstable.

INTRODUCTION

Motivated by their many applications in engineering systems such as two-phase heat exchangers and boiling water nuclear reactors, the stability characteristics of two-phase-flow heated channels, has been studied quite extensively in recent years. Most of the earlier analyses were limited to a single channel with a constant pressure drop along its length, no unheated riser section, constant inlet temperature, and spatially uniform and constant (in time) heat flux along the channel length.

Analytical linear stability analyses of single uniformly heated channels have been carried out, and give the marginal stability boundaries, or bifurcation sets, in parameter space (Ishii, M. and Zuber, N., 1970; Achard, J-L, Drew, D.A. and Lahey, R.T., 1981; Rizwan-uddin and Dorning J., 1986a). Subsequent nonlinear analyses (Achard, J-L., Drew, D.A. and Lahey, R.T., 1985; Rizwan-uddin and Dorning, J., 1986a) showed that supercritical Hopf bifurcation (Marsden, J. and McCracken, M., 1976; Guckenheimer, J. and Holmes, P., 1983) occurs, as certain parameters are varied, and thus that stable limit cycles exist in phase space when the bifurcation values of the parameters are exceeded. Similar bifurcation phenomena were observed in nonlinear analyses of single channels that were heated nonuniformly (Rizwan-uddin and Dorning, J., 1986b, 1987). Numerical analyses also have been carried out for single uniformly heated channels to study their dynamic behavior at points in parameter space away from the simple saddle-node and Hopf bifurcation points studied analytically (Achard, J-L, Drew, D.A. and Lahey, R.T., 1985; Rizwan-uddin and Dorning, J., 1986a). In these numerical studies density waves with periodic limit cycle behavior were shown to exist in autonomous heated channels with two-phase flow; and periodic, quasi-periodic and chaotic behavior was shown to exist in nonautonomous heated channels (Rizwan-uddin and Dorning, J., 1988, 1990). Although the results reported so far are important, there is a need to extend the models of two-phase flow heated channels to include unheated riser sections above the heated channels and downcomer feedback recirculation loops that model the external partial return flow region between the riser exit and the channel inlet. Far more importantly, however, since most two-phase flow engineering systems, such as two-phase heat exchangers and boiling water nuclear reactors, have many parallel heated channels, it is essential to analyze single-phase and two-phase flow in multiple parallel channels with different axially varying heat flux distributions in the different channels. Hence, we have extended our earlier numerical model for the analysis of single heated channels with two-phase flow (Rizwan-uddin and Dorning, J., 1988, 1990) to multiple parallel channels with the same or different axially varying heat fluxes along the different channels. We also have included unheated riser sections at the tops of the heated channels and a simple feedback recirculation loop. The model that results makes it possible to study numerically the nonlinear dynamics of multiple parallel heated channels with two-phase flow both as autonomous and nonautonomous (driven) systems. The total static pressure drops, the channel inlet temperatures and the axially varying heat fluxes in different channels can be varied independently and simultaneously as a function of time to simulate different transient conditions. We have numerically solved the equations that result from the model to study the stability of two-phase flow in multiple parallel heated channels with nonuniform heat fluxes, including the effects on bifurcation and stability of unheated riser sections and of downcomer feedback recirculation loops.

MODEL

The system under investigation consists of N_h vertical parallel heated channels of length L connected via a lower header or plenum. There are risers, or unheated sections, at the tops of all the channels, so the fluid leaving the channels passes through the risers all of which are connected via an upper header or plenum. The axial dependence of the heat flux along the length of a channel can be different for each channel, and also can be time-dependent. Subcooled liquid, driven by an externally imposed pressure drop which also can vary in time, enters at the bottom, and leaves -- depending upon the flow rate and amount of heat supplied in the channel -- either as a subcooled liquid or a two-phase liquid-vapor mixture at the top of the channel. The liquid temperatures at the channel inlets also can be time-dependent. Although there are N_h parallel channels, they are connected only via the upper and lower plena; hence each can be modeled separately and coupled to the others using flow or pressure boundary conditions at the inlet and/or exit. Therefore, in this section we shall describe the thermal hydraulics modeling of only a single channel, and we shall discuss the boundary conditions that couple the channels in the next section.

In the general case the fluid exiting a channel is a two-phase mixture, and assuming that mixture can be represented by the homogeneous equilibrium model, the channel at any given time can be divided, into two 'material' regions, a single-phase region and a two-phase region. The single-phase region extends from the channel inlet up to the point where the cross sectionally averaged liquid temperature reaches the boiling temperature for the system pressure. This location, where boiling starts, is called the boiling boundary, and it is denoted by $\lambda(t)$, where $T(z=\lambda(t), t) = T_{sat}$, and in general, it is a function of time (Achard, J-L, Drew, D.A. and Lahey, R.T., 1981, 1985; Rizwan-uddin and Dorning, J., 1986). Of course, in the special cases of various combinations of high flow rate, low heat flux, and high inlet subcooling, the liquid in a channel at a given time may not reach the boiling point before it leaves the heated section. Thus there will be no boiling boundaries in such channels at that time, and there will be only single-phase regions in them. In the more general case there is a boiling boundary in a channel, and the region between it and the channel exit to the upper plenum is called the two-phase region.

A typical channel and the unheated riser above it are divided into K_c and K_r axial (vertical) nodes, respectively. These nodes are fixed-boundary 'geometric' regions. See Fig. 1. Therefore, if at time t the node with the boiling boundary is denoted by index k^λ, then there are k^λ-1 all-single-phase nodes, and K_c-K^λ all-two-phase nodes in the heated channel. Since the riser is not heated, the boiling boundary can not be located in it.

The heated channel is divided into axial computational nodes only to accommodate axially varying heat flux along the channel length, and the heat flux is taken as spatially uniform (but time-dependent) in each of these node. Hence, for the case of axially uniform heat flux along the channel length, a single computational node in the heated channel (K_c = 1) would be sufficient for analysis. Since a riser is not heated, a single node in the entire riser section, K_r = 1, always is sufficient.

In order to represent the heated channel and riser as a nonlinear dynamical system, we start from the set of partial differential equations (PDEs) that describe the flow of a single-phase liquid and a two-phase liquid-vapor mixture, and develop the set of nonlinear functional ordinary differential equations (FDEs) for the all-single-phase heated nodes, the all-two-phase heated nodes, the heated node with the boiling boundary, and the unheated nodes with all-single-phase or all-two-phase flow.

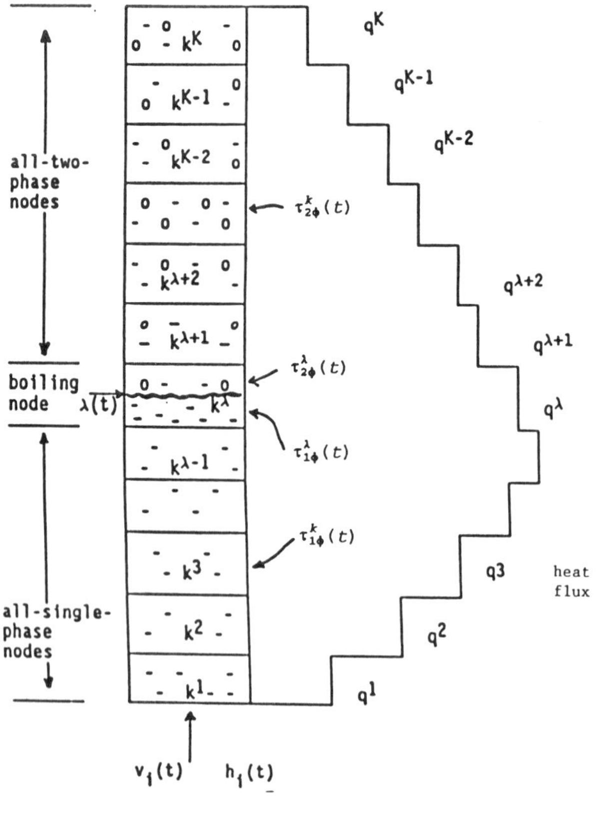

Figure 1. Schematic diagram of a heated channel with piece-wise constant axially varying, time-dependent heat fluxes. Channel is divided into single- and two-phase regions divided by the (time-dependent) boiling boundary. The unheated riser above the heated channel is not shown in the figure.

The exact reduction from the original nonlinear PDEs to the FDEs is achieved by first integrating the mass and energy equations along their characteristics, and then integrating the momentum equation along the node length (after substituting the density and velocity variations along the node length) to determine the nodal pressure drop in terms of node inlet velocity and node residence time -- defined as the time spent within the node by the fluid that is at the node exit at time t. The formulation in terms of the nonlinear FDEs is carried out in such a way that for given channel inlet velocity and node heat fluxes for $t \leq 0$, all the other dependent thermal hydraulic variables can be calculated once the channel inlet velocity and the node residence times are known.

Single-Phase Region ($1 \leq k \leq k^{\lambda}$)

Assuming an incompressible liquid phase, the velocities in the all-single-phase nodes (and in the single-phase region of the k^{λ} node) are equal to the channel inlet velocity,

$$v_k(z,t) = v_i(t), \qquad 1 \leq k \leq k^{\lambda}$$

The residence time for the liquid leaving the k-th node at time t, $\tau_{1\phi}^{k}(t)$, then from its definition is given by

$$\Delta z^{k} \equiv z_{ex}^{k} - z_{in}^{k} = \int_{0}^{\tau_{1\phi}^{k}(t)} v_i(t-\sigma)\,d\sigma \qquad (1)$$

where Δz^k is the height of the k-th node. With the velocity in the node at any given time equal to the node inlet velocity, which in turn is equal to the channel inlet velocity, the energy equation for the all-single-phase nodes is

$$\frac{\partial h^{k}(z,t)}{\partial t} + v_i(t)\frac{\partial h^{k}(z,t)}{\partial z} = \frac{q^{k}(t)}{\rho_{\ell}}, \qquad (2)$$

where $q^{k}(t) \equiv q''^{k}(t)\xi_h/A_f$, and $q''^k(t)$ is the heat flux, ξ_h is the heated perimeter, and A_f is the flow area. Equation (2) can be integrated along its characteristics

$$\frac{dt'}{1} = \frac{dz'}{v_i(t')} = \frac{dh'}{\left(\dfrac{q^{k}(t')}{\rho_{\ell}}\right)}$$

from

$$t' = t-\tau_{1\phi}^{k}(t),\quad z' = z_{in}^{k} \qquad h' = h_{in}^{k}(t-\tau_{1\phi}^{k}(t))$$

to

$$t' = t, \qquad z' = z_{ex}^{k} \qquad h' = h_{ex}^{k}(t)$$

to obtain the following expression for the node exit enthalpy

$$h_{ex}^{k}(t) - h_{in}^{k}(t - \tau_{1\phi}^{k}(t)) = \frac{1}{\rho_{\ell}}\int_{o}^{\tau_{1\phi}^{k}(t)} q^{k}(t-\sigma)\,d\sigma \qquad (3)$$

in terms of the single-phase residence time $\tau_{1\phi}^{k}(t)$. Here $h_{in}^{1}(t)$ is the enthalpy at the channel inlet at time t. The expression for $\tau_{1\phi}^{k}(t)$ given in Eq. (1) above also can be obtained by solving for the characteristics of Eq. (2). Since k^{λ} is not known *apriori*, assuming a channel inlet velocity at time t, and starting from the bottom node and using the time-dependent inlet enthalpy as a boundary condition, the enthalpy increase and exit enthalpies are determined in each of the successive nodes using Eqs. (1) and (3), until the exit enthalpy for a node is found to be greater than h_{ℓ}, the liquid enthalpy at saturation temperature, i.e. $h_{ex}^{k} \geq h_{\ell}$, indicating that the beginning of boiling hence the boiling boundary $\lambda(t)$ is in that node. This is a special node and as noted above will be denoted by index k^{λ}. There are two residence times for this node; the single-phase residence time $\tau_{1\phi}^{\lambda}(t)$ defined as the time spent in this node by the fluid which is at the boiling boundary at time t, and the two-phase residence time $\tau_{2\phi}^{\lambda}(t)$ defined as the time spent in the current two-phase region of this node by the fluid which is at the node exit at time t. The single phase residence time in the node k^{λ}, $\tau_{1\phi}^{\lambda}(t)$, for fluid at the boiling boundary can be obtained from Eq. (3) by realizing that the enthalpy at the boiling boundary is known and is equal to h_{ℓ}, hence $\tau_{1\phi}^{\lambda}(t)$ is given by

$$h_{\ell} - h_{in}^{\lambda}(t - \tau_{1\phi}^{\lambda}(t)) = \frac{1}{\rho_{\ell}}\int_{0}^{\tau_{1\phi}^{\lambda}(t)} q^{\lambda}(t-\sigma)\,d\sigma, \qquad (4)$$

and the location of the boiling boundary, $\lambda(t)$, is given by

$$\lambda(t) - z_{in}^{\lambda} = \int_{0}^{\tau_{1\phi}^{\lambda}(t)} v_i(t-\sigma)\,d\sigma \qquad (5)$$

where z_{in}^{λ} is the inlet axial location of the boiling-boundary node, the node with index $k^{\lambda}(t)$. The pressure drop in the entire single-phase region (from inlet to the boiling boundary) is given by

$$\Delta P_{1\phi}(t) = \rho_{\ell}\int_{0}^{\lambda(t)}\left[\frac{dv_i}{dt} + \frac{f_s}{2D}v_i^2 + g\right]dz + k_i\frac{\rho_{\ell}v_i^2}{2}$$
$$= \rho_{\ell}\left[\frac{dv_i}{dt} + \frac{f_s}{2D}v_i^2 + g\right]\lambda(t) + k_i\frac{\rho_{\ell}v_i^2}{2} \qquad (6)$$

where f_s is the friction number, D is channel hydraulic diameter, g is the gravitational constant, and the last term represents the entrance flow pressure drop at the channel inlet.

Two-Phase Region in the Node k^{λ}

Equation (6) gives an expression for pressure drop in the entire single-phase region of the heated channel which includes all the all-single-phase nodes and the single-phase portion of the node with the boiling boundary, node k^{λ}. Hence, from the node k^{λ} only the pressure drop in the two-phase portion must be added to the total pressure drop equation. Since the boiling boundary is a function of time, the single-phase and two-phase region pressure drop calculation in the node k^λ is a variable domain or moving boundary problem. Hence, the position of the boiling boundary must be determined first, and only then can the two pressure drops be calculated. For a time-dependent heat flux that is uniform along the node length, the homogeneous equilibrium model two-phase flow equations for the volumetric flux j(z,t), the mixture density $\rho_m(z,t)$, and the pressure drop $\frac{\partial P}{\partial z}(z,t)$ are given by

$$\frac{\partial j^k(z,t)}{\partial z} = \left[\frac{\rho_\ell-\rho_v}{\rho_v\rho_\ell h_{fg}}\right]q^k(t) \equiv C_1\ q^k(t) \quad \textbf{(7)}$$

$$\frac{\partial \rho_m^k(z,t)}{\partial t} + j^k(z,t)\frac{\partial \rho_m^k(z,t)}{\partial z} = -\ C_1\ q^k(t)\ \rho_m^k(t) \quad \textbf{(8)}$$

and

$$\frac{\partial P_{2\phi}^k}{\partial z} = \rho_m^k\Big[\frac{\partial j^k(z,t)}{\partial t} + j_k(z,t)\frac{\partial j^k(z,t)}{\partial z} + \frac{f_m}{2D}j^{k^2}(z,t) + g\Big]. \quad (9)$$

The inlet conditions for the volumetric flux and the mixture density in the two-phase region of the node k^λ are

$$j^k(z=\lambda(t),t) = v_i(t)$$
$$\rho_m^\lambda(z=\lambda(t),t) = \rho_\ell$$

Integrating Eq. (7) from the boiling boundary $\lambda(t)$ to z and using the boundary condition, we obtain

$$j^\lambda(z,t) = v_i(t) + C_1[z-\lambda(t)]q^\lambda(t),\quad \lambda(t) \le z \le z_{ex}^\lambda \quad \textbf{(10)}$$

Similarly Eq. (8) can be integrated along its characteristics using the above expression for $j^\lambda(z,t)$ and the inlet condition at the boiling boundary to obtain an expression for mixture density variation in the node k^λ, $\rho_m^\lambda(z,t)$, in terms of fluid residence time in the two-phase region. The two-phase-region pressure drop in the boiling node k^λ, can then be calculated by integrating the two-phase momentum equation, Eq. (9), from $z = \lambda(t)$ to the node exit, z_{ex}^λ. This integration is facilitated by a change of integration variable from z to t_2, the residence time. This development already has been given in the context of uniformly heated channel dynamics, which requires only one node -- the node with the boiling boundary. The results of this analysis which appeared recently (Rizwan-uddin and Dorning, J., 1990) are summarized here. The mixture density in the node k^λ, as a function of $t-t_2$ is given by

$$\rho_m^\lambda(t_2,t) = \exp\left[-C_1^\lambda\int_0^{(t-t_2)} q^\lambda(t-\sigma)\,d\sigma\right] \quad \textbf{(11)}$$

where $t-t_2$ is the time spent in the two-phase region by the fluid element. Two-phase residence time $\tau_{2\phi}^\lambda(t)$ is given by

$$z_{ex}^\lambda = \lambda(t) + \int_0^{\tau_{2\phi}^\lambda(t)}\frac{H(t-\sigma)}{\rho_m(t-\sigma,t)}d\sigma \quad \textbf{(12)}$$

where

$$H(t-\sigma) = \frac{q^\lambda(t-\sigma)\ v_i[t-\sigma-\tau_{1\phi}^\lambda(t-\sigma)]}{\{q^\lambda[t-\sigma-\tau_{1\phi}^\lambda(t-\sigma)] - \rho_\ell\frac{dh_{in}^\lambda}{dt}[t-\sigma-\tau_{1\phi}^\lambda(t-\sigma)]\}}.$$

The pressure drop in the two-phase region of the boiling node is

$$\begin{aligned}\Delta P_{2\phi}^\lambda(t) &= \int_{\lambda(t)}^{z_{ex}^\lambda}\rho_m^\lambda(z,t)\left\{\frac{\partial j^\lambda}{\partial t} + j^\lambda\frac{\partial j^\lambda}{\partial z} + \left(\frac{f_m}{2D}\right)j^{\lambda^2} + g\right\}\\ &= \rho_\ell\{\frac{dv_i}{dt} - C_1\frac{d\lambda}{dt}q^\lambda(t) + C_1q^\lambda(t)\ v_i(t)\\ &\quad + g + \left(\frac{f_m}{2D}\right)v_i^2\}J_1^\lambda(t)\\ &\quad + C_1\rho_\ell\{\frac{dq^\lambda(t)}{dt} + q^\lambda(t)\{C_1q^\lambda(t)\\ &\quad + 2\left(\frac{f_m}{2d}\right)v_i(t)\}\}J_2^\lambda(t)\\ &\quad + \rho_\ell C_1^2\left(\frac{f_m}{2D}\right)q^{\lambda^2}(t)\ J_3^\lambda(t)\end{aligned} \quad \textbf{(13)}$$

where the delay integrals that appear in these equations are

$$J_1^\lambda(t) = \int_0^{\tau_{2\phi}^\lambda(t)} H(t-\sigma)\,d\sigma$$

$$J_2^\lambda(t) = \int_0^{\tau_{2\phi}^\lambda(t)} H(t-\sigma)\int_0^\sigma H(t-\eta)\,d\eta\,d\sigma$$

$$J_3^\lambda(t) = \int_0^{\tau_{2\phi}^\lambda(t)} H(t-\sigma)\left(\int_0^\sigma H(t-\eta)\,d\eta\right)^2 d\sigma$$

and

$$J_4^\lambda(t) = \int_0^{\tau_{2\phi}^\lambda(t)}\frac{H(t-\sigma)}{\rho_m(t-\sigma,t)}d\sigma.$$

All-Two-Phase Nodes ($k^\lambda < k \le K_c$)

The two-phase equations for the volumetric flux $j(z,t)$, the mixture density $\rho_m(z,t)$ and the pressure drop for the all-two-phase nodes are the same as the equations given above for the two-phase region in the node k^λ, except here these equations are to be solved using different inlet conditions. The inlet conditions for the k-th node for $j(z,t)$ and $\rho_m(z,t)$ follow from the continuity of these quantities from the previous node, the (k-1)-th,

$$j_{in}^k(t) \equiv j^k(z = z_{in}^k,\ t) = j_{ex}^{k-1}(t)$$
$$\equiv j^{k-1}(z = z_{ex}^{k-1},\ t)$$
$$\rho_{m_{in}}^k(t) = (\rho_m^k(z = z_{in}^k,\ t) = \rho_{m_{ex}}^{k-1}(t)$$
$$= \rho_m^{k-1}(z = z_{ex}^{k-1},\ t).$$

The solutions for j^k and ρ_m^k, obtained by integrating along the characteristics, are

$$j^k(z,t) = j_{ex}^{k-1}(t) + C_1\, q^k(t)\ (z - z_{in}^k), \qquad (14)$$
$$z_{in}^k < z < z_{ex}^k$$

$$\rho_m^k(z(t_o),t) = \rho_{m_{ex}}^{k-1}(t_o)\ \exp[-C_1\int_o^{t-t_o} q^k(t-\sigma)\,d\sigma], \qquad \textbf{(15)}$$

where z and t_o are related by

$$z - z_{in}^k = \int_o^{t-t_o} \frac{\hat{g}^k(t-\sigma)}{\hat{g}^k(t)}\ j_{ex}^{k-1}(t-\sigma)\ d\sigma,$$
$$z_{in}^k < z < z_{ex}^k,\ 0 < t_o < \tau_{2\phi}^k$$

Hence the k-th node two-phase residence time $\tau_{2\phi}^k(t)$ is given by

$$\Delta z^k = z_{ex}^k - z_{in}^k = \int_o^{\tau_{2\phi}^k(t)} \frac{\hat{g}^k(t-\sigma)}{\hat{g}^k(t)}\ j_{ex}^{k-1}(t-\sigma)\ d\sigma,$$
$$k^\lambda < k \le K \qquad \textbf{(16)}$$

where $\hat{g}^k(t)$ is

$$\hat{g}^k \equiv \exp[-C_1\int q^k(\sigma)\,d\sigma].$$

Integration of the conservation of momentum equation over the k-th all-two-phase node yields

$$-\int_{z_{in}^k}^{z_{ex}^k} \frac{\partial P_{2\phi}}{\partial z}dz = \int_{z_{in}^k}^{z_{ex}^k} \rho_m^k[\frac{\partial j^k}{\partial t} +$$
$$j^k\frac{\partial j^k}{\partial z} + \frac{f_m}{2D}j^{k^2} + g]\,dz,$$

which after a change of integration variable from z to σ, the time spent by a two-phase fluid element in the k-th node, yields the two-phase pressure drop in the k-th node

$$\Delta P_{2\phi}^k(t) = [\frac{\partial j_{ex}^{k-1}(t)}{\partial t} + (\frac{f_m}{2D})\ (j_{ex}^{k-1}(t))^2$$
$$+ g + C_1 q^k(t)\ j_{ex}^{k-1}(t)]\ J_1(t)$$
$$+ [C_1\frac{dq^k}{dt}(t) + 2C_1(\frac{f_m}{2D})\ j_{ex}^{k-1}(t)\ q^k(t)$$
$$+ (C_1 q^k(t))^2]\,J_2(t)$$
$$+ [(\frac{f_m}{2D})\ (C_1 q^k(t))^2]\ J_3(t) \qquad \textbf{(17)}$$

where

$$J_1(t) = \int_o^{\tau_{2\phi}^k} \rho_{m_{ex}}^{k-1}(t-\sigma)\ j_{ex}^{k-1}(t-\sigma)\ d\sigma$$

$$J_2(t) = \int_o^{\tau_{2\phi}^k} \rho_{m_{ex}}^{k-1}(t-\sigma)\ j_{ex}^{k-1}(t-\sigma)\ [\int_o^\sigma \frac{\hat{g}(t-\eta)}{\hat{g}(t)}$$
$$j_{ex}^{k-1}(t-\eta)\,d\eta]\ d\sigma$$

and

$$J_3(t) = \int_o^{\tau_{2\phi}^k} \rho_{m_{ex}}^{k-1}(t-\sigma)\ j_{ex}^{k-1}(t-\sigma)$$
$$[\int_o^\sigma \frac{\hat{g}(t-\eta)}{\hat{g}(t)} j_{ex}^{k-1}(t-\eta)\,d\eta]^2\ d\sigma$$

(For details see Rizwan-uddin and Dorning J., 1990.) Finally, the exit flow pressure drop at the channel outlet is given by

$$\Delta P_{exit}(t) = k_e\ \rho_{m_{ex}}^K(t)\ (j_{ex}^K(t))^2 \qquad \textbf{(18)}$$

Pressure Drop in the Riser

If the fluid does not boil in the heated channel then, of course, as noted previously it remains subcooled throughout the riser, and the fluid velocity in the riser is equal to the channel inlet velocity and the density remains at ρ_ℓ. The enthalpy within the riser can be calculated using the single-phase energy equation with no heat flow (q=0),

$$\frac{\partial h(z,t)}{\partial t} + v_i(t)\frac{\partial h(z,t)}{\partial z} = 0$$

and the pressure drop in the riser is equal to

$$\Delta P^r(t) = \rho_\ell\left[\frac{dv_i}{dt} + \frac{f_s}{2D}v_i^2(t) + g\right]L_r$$

where L_r is the length of the riser. If the fluid boils in the heated channel, then the volumetric flux and mixture density in the riser can be calculated using their respective all-two-phase-node equations with the heat flux equal to zero,

$$\frac{\partial j^r}{\partial z}(z,t) = 0$$

$$\frac{\partial \rho_m^r(z,t)}{\partial t} + j^r(z,t)\frac{\partial \rho_m^r(z,t)}{\partial z} = 0$$

with the continuity of the volumetric flux and mixture density at the riser inlet. The corresponding pressure drop in the riser $\Delta P^r(t)$ can be calculated using Eq. (17) with heat flux q^k set equal to zero. Equating the sum of all the internal pressure drops to the externally imposed pressure drop $\Delta P_{ext}(t)$ yields a nonlinear first order functional ODE for the channel inlet velocity in terms of the node heat fluxes $q^k(t)$,

$$\Delta P_{ext}(t) = \Delta P_{1\phi}(t) + \Delta P_{2\phi}^{\lambda}(t) + \sum_{k=k^{\lambda}+1}^{K_c} \Delta P_{2\phi}^{k}(t) + \Delta P^r(t) + \Delta P_{ex}(t) \tag{19}$$

where $\Delta P_{1\phi}$ is given by Eq. (6). For given heat fluxes $q^k(t)$, external pressure drop $\Delta P_{ext}(t)$ and inlet enthalpy $h_{in}(t)$ the above equation can be rewritten as

$$\frac{dv_i(t)}{dt} = f(q^k(t), \Delta P_{ext}(t), h_{in}(t), v_i(t)) \tag{20}$$

All the other dependent variables, $\tau_{1\phi}(t)$, $\tau_{2\phi}(t)$, $j(z,t)$, $\rho_m(z,t)$, etc., can be evaluated in terms of the inlet velocity. Equation (20), which represents the dynamics of a single heated channel with two-phase flow as an infinite dimensional nonlinear dynamical system (infinite dimensional due to the delay integrals in Eq. (19)), can be solved numerically to determine the steady-state or time-dependent behavior of the channel inlet velocity $v_i(t)$, from which all other thermal hydraulic variables follow.

BOUNDARY CONDITIONS AND SOLUTION PROCEDURE FOR SINGLE AND MULTIPLE PARALLEL CHANNEL MODELS

In order to solve the equations for the multi-node single-channel model described in the preceding section for given initial conditions, boundary conditions related to the original PDEs must be specified, i.e. the external pressure drop to which the channel is subjected and the inlet temperature or enthalpy all of which can be time dependent in general, must be given. The inlet enthalpy, in an extended model in which a return feedback loop is included, is not specified; rather, it is made a function of the enthalpy (or quality) at the channel (or riser) exit at some earlier time, thereby introducing feedback from the channel exit to the channel inlet, such as that which occurs in some two-phase flow systems, e.g. boiling water nuclear reactors. In this case, which models a simple downcomer loop, inlet enthalpy is not a boundary condition but actually becomes a dependent variable. The model of the downcomer loop studied here corresponds to a specified amount of exit vapor being removed (e.g., to drive a turbine or supply process steam) and the rest mixing with the liquid at saturation temperature and returning through the downcomer loop to the channel inlet, mixing on the way with the subcooled liquid returning from the condenser. When a transient occurs which changes the exit quality, the amount of vapor returning through the downcomer changes from its steady-state value, changing the temperature of the fluid entering the channels at a later time.

To transform the single channel model to a multiple parallel channel model, different type pressure or pressure/flow boundary conditions are specified. There are two such types of boundary conditions incorporated as alternatives in the model: (1) a given constant or time-dependent pressure drop is imposed across one of the channels, and the flow in other channels is determined by equating the total (static plus dynamic) pressures at their inlets and exits with the corresponding total pressures in the channel with the specified pressure drop; or (2) a constant total mass flow rate is imposed through all the combined channels, and the total pressures are required at any given time to be the same at all the channel inlets, and to be the same at all the channel exits. (The multiple parallel channel model also includes the possibility of a return feedback loop analogous to the single channel feedback loop, but with the inlet temperatures from the lower header or plenum dependent upon the exit enthalpies (or qualities) of all the channels through their mixing in the upper header or plenum.)

The boundary condition of type (1) requires that

$$\Delta P^1(t) = \Delta P_{ext}(t), \quad and$$

$$\Delta P^i(t) = \Delta P_{ext}(t) + \frac{\rho_\ell}{2}\left[v_{in}^{1^2} - v_{in}^{i^2}\right] - \frac{1}{2}\left[\rho_{ex}^1 v_{ex}^{1^2} - \rho_{ex}^i v_{ex}^{i^2}\right],$$

$$i = 2, 3, \ldots, N_h$$

where ΔP_{ext} is the externally specified pressure drop; the boundary condition of type (2) requires that

$$\sum_{i=1}^{N_h} \dot{m}^i(t) = \dot{m}_{total}(t)$$

and

$$\Delta P^i(t) + \frac{\left[\rho_{in}^1 v_{in}^{1^2} - \rho_{ex}^1 v_{ex}^{1^2}\right]}{2} = \Delta P^{i+1}(t) + \frac{\left[\rho_{in}^i v_{in}^{i^2} - \rho_{ex}^i v_{ex}^{i^2}\right]}{2},$$

$$i = 1, 2, \ldots, N_h - 1$$

where $\dot{m}$ is the mass flow rate.

The steady-state solution, for a given heat flux distribution in all N_h channels, is obtained by specifying a total pressure drop ΔP_o for one of the channels (e.g. channel one) and then iteratively solving the steady-state equations for all the other thermal hydraulic variables in that channel, including inlet velocity v_i^1, the exit volumetric flux j_{ex}^1, and the mixture density $\rho_{m,ex}^1$. The steady-state solutions in all the other channels are then obtained by iteratively solving for inlet velocities that would result in the same total (static plus dynamic) pressures at the inlets and exits as the corresponding total pressures in the first channel. The solution procedure for dynamic problems with pressure boundary conditions of type (1) is the same as for the steady-state except that in time-dependent problems the specified pressure drop in the first channel can be a function of time. In this case, although the dynamics of all the channels are coupled, each channel is solved separately.

For pressure/flow boundary conditions of type (2), the pressure drop is not known in any channel; hence, only N_h-1 equations are obtained by equating the total pressures at all N_h channel inlets and at all N_h channel exits. The final equation is obtained by equating the sum of all mass flow rates into the N_h channels at each time step to the externally imposed total mass flow rate, $\dot{m}_{total}(t)$, which can be a function of time. These N_h equations are solved simultaneously for the inlet velocities to the N_h channels at each time step. The numerical model, thus developed, for the steady-state and dynamic behavior of multiple parallel heated channels with two-phase flow, can be used to determine a steady-state and to determine the stability of this steady-state or fixed-point solution, and also to determine the dynamic behavior of a system as a non-autonomous system, i.e. in response to time-dependent variations in the system parameters. This could include a time-dependent inlet temperature, a time-dependent total pressure drop, or time-dependent heat fluxes in any combination of any nodes in any channels.

RESULTS

The mathematical model with the boundary conditions and the solution procedure described above for the analysis of the dynamics of multiple parallel heated channels with two-phase flow has been solved for various steady-state and dynamic problems. The effects on stability of the length of the risers, the downcomer feedback loop, and the pressure/flow boundary conditions have been studied. The numerical stability analysis of the steady-state solutions or fixed points, was carried out by introducing a short-duration perturbation in the total pressure drop specified, or by introducing a short-duration perturbation in the inlet fluid temperature. In the cases in which the fixed points were unstable these perturbations resulted in system evolving away from the fixed point,in some cases to stable limit cycle oscillations.

Our studies using nonuniform heat fluxes along the channel lengths have shown that, for a fixed inlet mass flow rate, as the length of the riser above a heated channel is increased the steady-state or equilibrium flow that was stable in the original channel becomes unstable through a supercritical Hopf bifurcation when a critical riser length is exceeded. This bifurcation results in stable limit-cycle density-wave oscillations to which the system evolves. Both isolated two-phase heated channels and multiple parallel two-phase heated channels have been analyzed, and the destabilizing effect of the risers was observed in all cases. Figure 2a shows the time evolution of the inlet velocity in a four meter long vertical channel with a half meter riser. The system, following a perturbation in total pressure drop, evolves through damped nonlinear oscillations back to the stable fixed point. Figure 2b shows an analogous time series for the same channel with a 0.75 meter riser added. Again the system returns to the stable fixed point, but more slowly. Figure 2c shows the evolution of the inlet velocity in the same channel with a one meter riser. In this case the Hopf bifurcation as a function of the riser length already has occurred; hence, the fixed point is unstable and the initial condition nearby it evolves away from it to the stable limit cycle. The equilibrium inlet velocity in all three cases is 1.5 m/s.

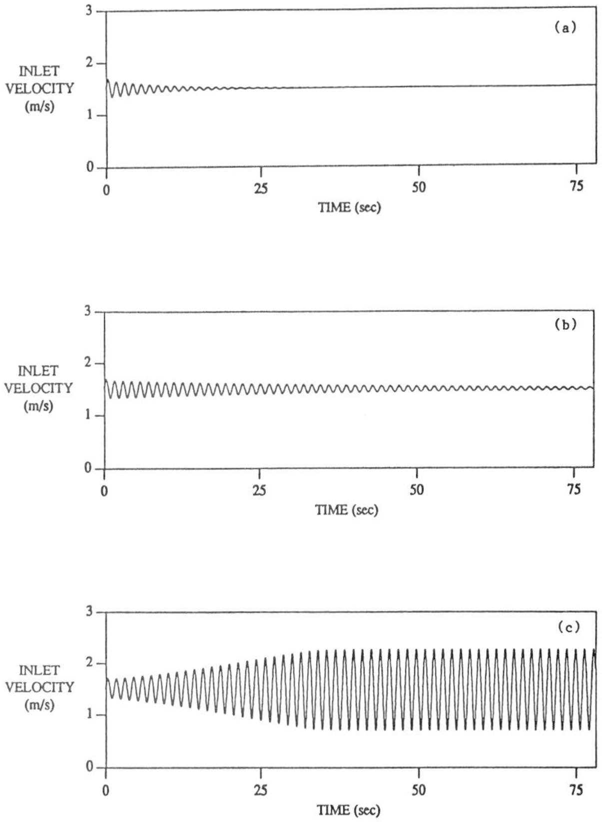

Figure 2. Time evolution of the inlet velocity following a perturbation in total pressure drop in a four meter long vertical heated channel with: (a) a half meter riser; (b) a 0.75 meter riser; and (c) a one meter riser. The system becomes less stable as the length of the riser is increased with the equilibrium losing stability in a supercritical Hopf bifurcation at a riser length between 0.75 meters and one meter, beyond which there exist stable limit cycle density waves.

Analogous behavior was observed in multiple parallel channel systems. Figure 3a shows the time evolution of the inlet velocity in three four-meter long vertical heated channels with half meter risers operating in parallel with the same inlet dynamic pressures and the same exit dynamic pressures. In this system all three channels have bottom-peaked axially varying heat fluxes. However, channel 1 has a higher total heat flux than channel 2 which has a higher total heat flux than channel 3. All three channels, starting from a fixed point, following a perturbation in total pressure drop, evolve back to the fixed point. Figure 3b shows a similar evolution of inlet velocities following a perturbation in the same heated channels with 0.75 meter risers. The equilibrium is again stable. As the length of the risers is increased to 1.0 meter, however the equilibrium becomes unstable, and following a perturbation, as shown in Fig. 3c, the system evolves to a stable limit cycle solution indicating that a supercritical Hopf bifurcation has occurred as the riser lengths were increased from 0.75 meters to 1.0 meter in this three parallel channel system. The equilibrium inlet velocity in channel 1, in all three cases, is 1.5 m/s.

The effect of adding nonlinear feedback to the same three parallel channel system was studied. This was done using three axially varying heat fluxes of the same shape as those used in the previous system, but of lower amplitude. This feedback effect was studied through the addition of a downcomer recirculation loop represented by making the three channel inlet temperatures (same for the three channels) functions of the channel exit qualities (combined in the upper plenum) at an earlier time. Figure 4a shows the evolution of the inlet velocities in the three parallel heated channels with 1.0 meter risers with no recirculation feedback, i.e. constant inlet temperature, following a perturbation in total pressure drop. Clearly the steady-state solution or fixed point, about which the perturbation was introduced, is stable. The recirculation feedback model then was added to the system, and the equations were solved. The resulting evolution of the inlet velocities, following the same perturbation introduced in the model with no feedback, is shown in Figure 4b. It is clear that the same equilibrium for the model with the recirculation feedback loop is less stable, and in this particular case, evolves away from the unstable fixed point toward stable limit cycle oscillations in the three channels. Various multiple parallel channel configurations were studied using the two different kinds of boundary conditions described earlier. In general, for the cases solved, it was observed that the specified total mass flow rate boundary condition resulted in a much more stable system equilibrium than the specified pressure drop boundary condition. This was studied by analyzing the evolution from a steady-state (fixed point) as a result of perturbation in inlet temperature.

The effect of adding risers to the three parallel channel system, and the effect of including a simple downcomer feedback recirculation loop in that same system with lower power, that were described explicitly above, were studied using the specified total pressure drop boundary condition. This boundary condition, which was designated as the type (1) condition in the previous section, is generally related to the analysis of single channels, and is used here and elsewhere in the analysis of flows in multiple parallel channels. However, the type (2) boundary condition, in which the total mass flow rate is specified, is the more relevant one in the analysis of flow in multiple channels. Hence as a final part of this work, the effect of using one or the other of these boundary conditions on the time evolution of the same three parallel channel system with the same three axially varying heat fluxes was studied. This was done even though the two different boundary conditions define two distinct physical systems, and comparisons of their stability properties are meaningful only in a rather general way. Nevertheless, such studies were done as follows. For boundary condition of type (1), the pressure drop specified at the steady-state was kept constant during the transient; and for boundary condition of type (2), the total instantaneous inlet

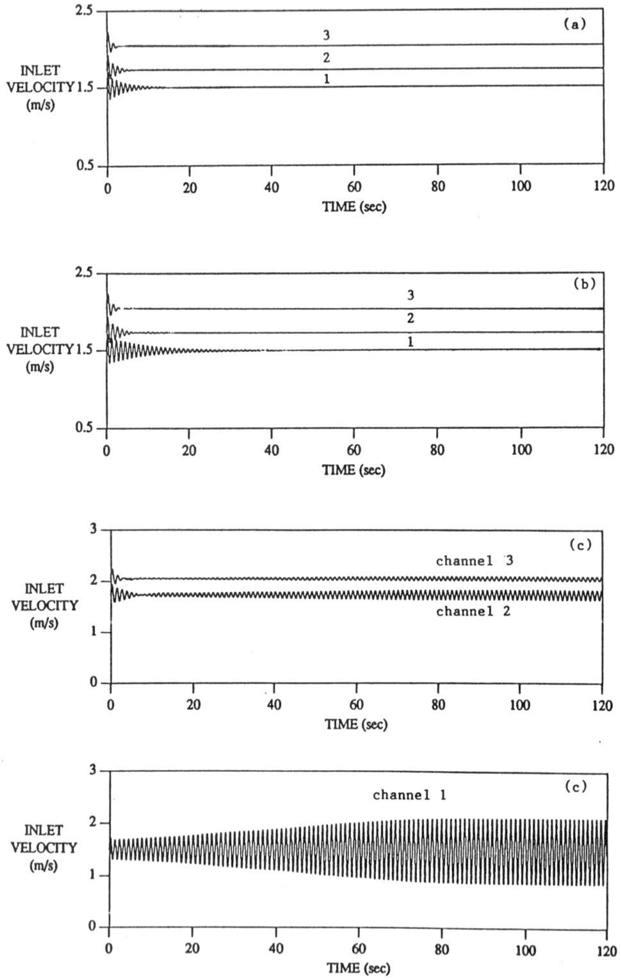

Figure 3. Time evolution of the inlet velocities in three four-meter long parallel heated channels with different heat fluxes, following a perturbation in total pressure drop. The equilibrium (fixed point) for the parallel channels with half meter risers is stable (3a). As the risers length is increased to 0.75 meters the fixed point becomes 'less' stable (3b); the steady-state for heated channels with one meter risers is clearly unstable, and the system evolves to stable limit cycle oscillations (3c).

mass flow rate for all the channels was required during the transient to be equal to the steady-state value. Figure 5 shows, for the constant total mass flow rate boundary condition, the evolution of the inlet velocities resulting from a perturbation in inlet temperature in a three channel system. This system is identical to the first three-channel system with one meter risers described above, but with a smaller inlet pressure drop coefficient which decreases its stability form that indicated by Fig. 3c. The system evolves from nearby the unstable equilibrium to a stable limit cycle. The same fixed point when studied using the constant pressure drop boundary condition in channel 1 was shown to be more unstable in the sense that the amplitudes of the oscillations that resulted form the same perturbation did not saturate at values comparable to the limit cycle amplitudes shown in Fig.5; rather they grew until one of the channel inlet velocities became negative and the channel model ceased to be physically relevant.

SUMMARY

Some basic dynamic bifurcation phenomena and the resulting nonlinear dynamics in single and multiple parallel heated channels have been studied. These included the effects on bifurcation and stability of adding unheated riser sections above the channels, which has been proposed as a way to enhance natural convection in practical two-phase engineering systems, such as so-called inherently safe boiling water nuclear reactors. Although the addition of risers to such systems enhances natural convection, the results presented here show that the presence of such risers will have the adverse effect of decreasing stability. Also studied here was the effect on stability of downcomer feedback recirculation loops which are present in many systems, but sometimes omitted when these systems are modelled. The results show that the equilibria considered became unstable when simple feedback recirculation was added; hence, in systems of this type analyses based on models in which the feedback is omitted would yield nonconservative results with respect to the stability of the equilibrium operating points. Finally, the effect on stability of the boundary conditions used in the calculation of flows in multiple parallel channel systems was studied, and it was shown that the specified total mass flow rate boundary condition, which is more relevant to multiple parallel channel systems, leads to less instability of model equilibria than does the fixed static pressure drop boundary condition.

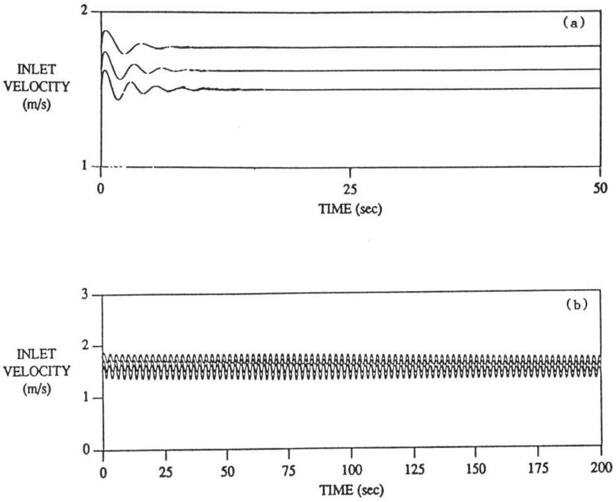

Figure 4. Effect of adding the recirculation feedback to a system with three parallel heated channels with one meter risers. Shown are the time evolutions of the inlet velocities following a perturbation in total pressure drop: (a) no feedback recirculation; (b) feedback recirculation included.

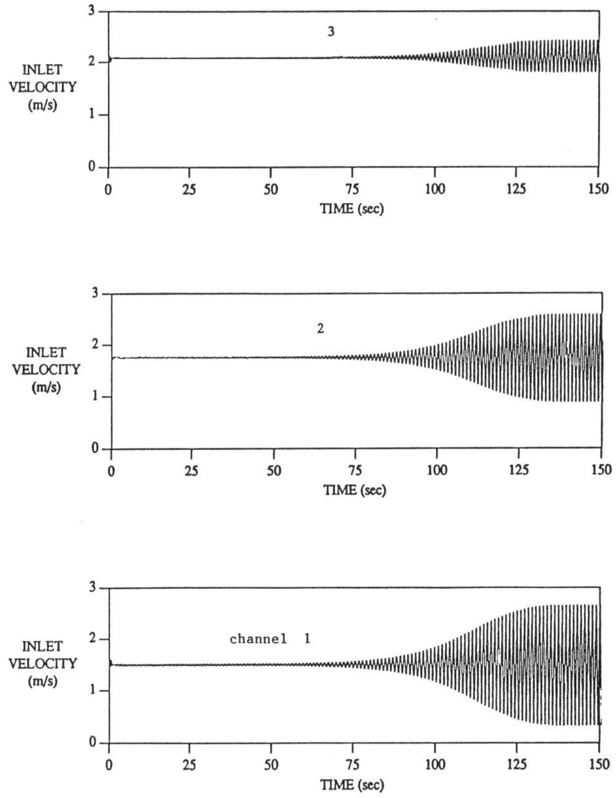

Figure 5. Effect on stability of pressure/flow boundary conditions. Shown are the inlet velocities, which evolve to stable limit cycle oscillations following a perturbation in inlet temperature, in a three parallel heated channel system, with the specified total mass flow rate boundary condition (constant in this case). The amplitude of the oscillations of a system, solved using the constant pressure drop boundary condition, but otherwise identical grows until one of the channel inlet velocities become negative, and the channel model is no longer applicable.

ACKNOWLEDGEMENT

This paper was prepared with the support of the U.S. Nuclear Regulatory Commission (NRC) under Grant No. NRC-04-90-113. The opinions, findings, conclusions and recommendations expressed herein are those of the authors and do not necessarily reflect the views of the NRC.

REFERENCES

Achard, J.L., Drew, D.A. and Lahey, R.T., Jr., 1985, "The Analysis of Nonlinear Density-Wave Oscillations in Boiling Channels," J. Fluid Mech., 155, 213-232.

Achard, J.L., Drew, D.A. and Lahey, R.T., 1981, "The Effect of Gravity and Friction on the Stability of Boiling Flow in a Channel," Chem. Engr. Commun, 11.

Guckenheimer J. and Holmes, P., 1983 Nonlinear Oscillations, Dynamical Systems and Bifurcations of Vector Fields, Springer-Verlag, New York.

Ishii, M. and Zuber, N., 1970, "Thermally Induced Flow Instabilities in Two Phase Mixtures," in Heat transfer 1970, Vol.V, Elsevier Publishing Co., Amsterdam.

Marsden J. and McCracken, M., The Hopf Bifurcation and its Application, Applied Mathematical Sciences, Vol. 18, Springer-Verlag, New York, 1976.

Rizwan-uddin and Dorning, J.J., 1986a, "Some Nonlinear Dynamics of a Heated Channel," Nucl. Eng. Des., 93, 1-14.

Rizwan-uddin and Dorning, J.J., 1986b, "Stability Analyses of Density Wave Oscillations in Channels with Nonuniform Heat Fluxes," Trans. Am. Nucl. Soc., 52, 404.

Rizwan-uddin and Dorning, J.J., 1987, "Nonlinear Stability Analysis of Density-Wave Oscillations in Nonuniformly Heated Channels," Trans. Am. Nucl. Soc., 54, 172.

Rizwan-uddin and Dorning, J.J., 1988, "A Chaotic Attractor in a Periodically Forced Two-Phase-Flow System," Nucl. Sci. Eng., 100, 393-404.

Rizwan-uddin and Dorning, J.J, 1990, "Chaotic Dynamics of a Triply-Forced Two-Phase Flow System," Nucl. Sci. Eng., 105, 123-135.

HTD-Vol. 197, Two-Phase Flow and Heat Transfer
ASME 1992

HOMOGENEOUS BUBBLE NUCLEATION WITHIN LIQUIDS: A REVIEW

C. Thomas Avedisian
Sibley School of Mechanical and Aerospace Engineering
Cornell University
Ithaca, New York

1. Introduction

In this paper the process of bubble nucleation within liquids that is based on overcoming the cohesive forces which hold together the liquid molecular structure is reviewed, and a few applications under which it is likely to be important are discussed. For this case, nucleation is termed "homogeneous" or "spontaneous". Homogeneous nucleation may occur within the bulk of a liquid, at the interface between two liquids in intimate contact, or at a solid surface that is completely wetted by a liquid. The term "heterogeneous" nucleation is commonly used with respect to the latter two situations. In this review, however, the term heterogeneous nucleation will only be used to refer to nucleation by gases trapped in cavities. Nucleation from gases trapped within surface irregularities has been discussed in several excellent reviews (e.g., Cole 1974, van Stralen and Cole 1979, Collier 1981).

Two applications in which homogeneous nucleation may be relevant are the following:

1. two-phase flows during rapid decompression within heated channels or tubes, and

2. vapor explosions.

Bubble formation by homogeneous nucleation during rapid decompression of a liquid (e.g., Hemmingsen 1977, Forest and Ward 1977, Alamgir and Lienhard 1981, Deligiannis and Clever 1990) is used as an example of the importance of the subject for the analyses of two-phase flows. Some applications relevant to vapor explosions include fuel coolant interactions in nuclear reactors (e.g., Witte et al. 1973, Fauske 1974, Bankoff 1978, Nelson and Duda 1982, Corradini 1982, Ciccarelli and Frost 1991), liquid natural gas spills on water (e.g., Reid 1983), the combustion of liquid droplets which contain additives (e.g., Lasheras et al. 1979, Wang and Law 1984, Yang et al. 1990), bubble-jet printing technology (e.g., Allen et al. 1985), and radiation detection by a superheated drop detector (Apfel and Roy 1983). Rapid boiling of liquids related to nuclear reactor accidents and natural gas spills can be potentially hazardous, while it is beneficial in the other applications mentioned.

The essential mechanism of a phase transition that is characterized by homogeneous nucleation, as opposed to "normal" boiling or heterogeneous nucleation, is one of overcoming the force of cohesion between molecules. The energy needed for this comes from the molecular vibrations in the liquid lattice. These vibrations create lower density regions that are termed nuclei, clusters, etc.. They exhibit bubble-like features with a radius and internal vapor pressure (Apfel 1972). For normal boiling processes, such vapor nuclei already exist due to vapor being trapped within surface imperfections when a liquid spreads over a solid (Bankoff 1958, Lorenz et a. 1974). Thus, on the one hand, homogeneous nucleation presumes an absence of such pre-existing gas bubbles within a liquid, whereas on the other, heterogeneous nucleation is based on their presence. The strong cohesion force between liquid molecules generally results in much higher superheats before bubbles form by density fluctuations than when boiling occurs by gas bubbles growing out of the mouth of cavities in which vapor is trapped.

In the following, we review briefly the concept of superheated liquids, discuss the maximum superheat a liquid can sustain from the point of view of homogeneous nucleation, and then illustrate with two examples in which the ideas of homogeneous nucleation can be useful for predicting phase transitions in practical applications. These include two-phase flow during rapid decompression of a liquid and bubble formation within nonburning and burning droplets.

2. Superheated Liquids

All liquid-to-vapor phase transitions occur at temperatures which are bounded on the lower end by the saturation or equilibrium vapor temperature, $T_s(P_o)$ (the assumed functional form imples a single component liquid for illustration) corresponding to the prevailing ambient pressure P_o, and at the upper end by the so-called "spinodal" temperature or "thermodynamic limit of superheat", $T_t(P_o)$ corresponding to the given pressure. $T_t(P_o)$ is determined by the basic extremum principle of thermodynamics which asserts that the entropy of an isolated system is a maximum in a stable equilibrium state with respect to small variations of its natural variables. For a single component liquid, $T_t(P_o)$ is defined by $\partial P/\partial V|_{T_t} = 0$ (Modell and Reid 1983). Within the range T_t-T_s, which can be calculated to be over 100C for many liquids based on a suitable equation of state, the liquid phase can theoretically exist without changing phase.

The term "superheated liquid" refers to a liquid that is at a temperature in the range

$$T_s(P_o) < T(P_o) < T_t(P_o)$$

or alternatively to a liquid which is under a pressure that is in the range

$$P_t(T) < P_o < P_s(T).$$

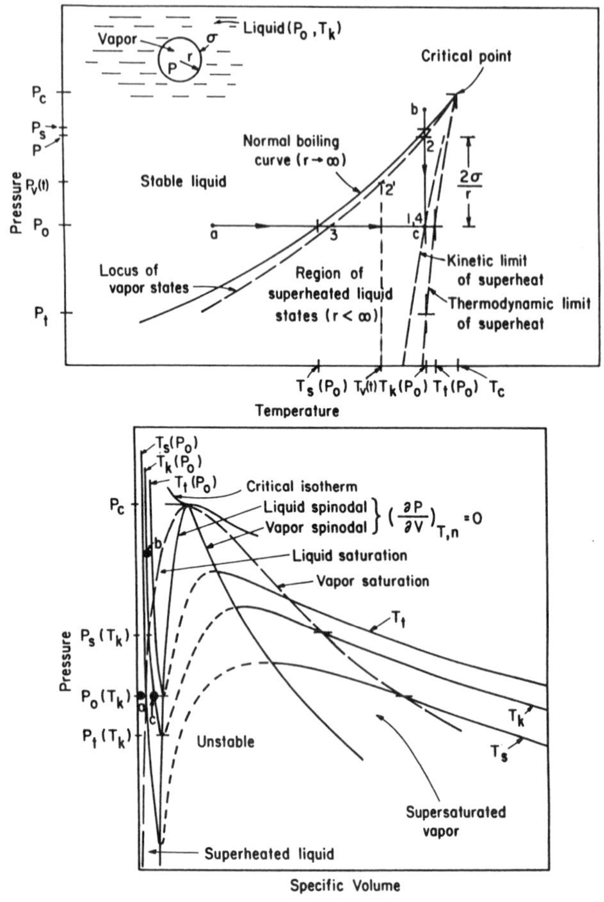

Figure 1 Phase diagram for a single component liquid - pressure/temperature and pressure/volume projections.

Fig. 1 illustrates these temperatures on P-T and P-V projections of a phase diagram for a single component liquid. The "superheat" is defined by the temperature difference T - $T_s(P_0)$ where T is the liquid temperature at pressure P_0.

The thermodynamic state at which a phase transition is initiated between T_s and T_t, and P_t and P_0, can depend on the conditions under which the liquid transgresses its normal (saturation) phase boundary and enters the domain of metastable states. For example, experimental evidence has shown that if a liquid in contact with a solid is rapidly heated (along path a→c in fig. 1), say at the rate of 10^6C/s, temperatures approaching T_t can be sustained (Skripov 1974). On the other hand, if the liquid is heated at a rate which is many orders of magnitude less, then the phase transition typically occurs at temperatures close to T_s.

The precursor to a bulk phase transition will be formation of a bubble that is in thermodynamic equilibrium within a superheated liquid. The static equilibrium state of such a bubble is characterized by a balance of both the forces across the liquid/vapor interface and of the chemical potentials of the two phases. The former requirement yields

$$P - P_0 = 2\sigma/r \qquad 1a.$$

where $2\sigma/r$ can be interpreted in terms of a temperature difference (i.e., "superheat") as shown in fig. 1. This is more clearly seen by writing eq. 1a in terms of temperature with the aid of an integrated form of the Clausius Clapeyron equation:

$$T(P_0) - T_s(P_0) \approx TT_s R \ln[1 + 2\sigma/(P_0 r)]/h_{fg}. \qquad 1b$$

Normal saturation states are conventionally referred to equilibrium across a flat phase boundary, r→∞. In this event, eq. 1 shows that P = P_0 and T=T_s. However, in virtually all practical applications, a liquid is able to sustain some degree of superheat before a phase change occurs, that is $T(P_0)>T_s(P_0)$. From eq. 1b this requires that $r<\infty$ so that the interface between the liquid and vapor must be curved. Hence any vapor present within a superheated liquid must be in the form of bubbles (or slugs, in either event of which the interface curvature is finite) in order to sustain the required pressure difference (cf, eq. 1a) that is commensurate with the superheat. Similar ideas apply to the formation of liquid droplets within vapors that are at a pressure such that $P_0>P_s$, a so-called "supersaturated" vapor (Pound 1972) though that situation is not of direct interest to this review.

The pressure within a bubble that is in equilibrium with a superheated liquid is not precisely the same as the saturation pressure P_s, the difference often being referred to as the "Poynting correction" (Blander and Katz 1975). Equating the chemical potentials between liquid and vapor provides a relation between P and P_s for a single component liquid as

$$P \approx P_s \exp[v_l(P_0-P_s)/RT] \qquad 2.$$

A similar form of eq. 2 applies to multicomponent miscible mixtures (Danilov et al. 1979, Avedisian and Glassman 1981a). The exponential term in eq. 2 is usually very close to unity for liquids at the superheat limit. Fig. 1 compares schematically the vapor states (dotted line) given by eq. 2 to the normal phase boundary for a pure liquid.

While a bubble can exist within a liquid that is superheated, the precise value of the superheat depends on the interface curvature r. This quantity can only be determined from a mechanistic model for bubble formation. In the following, we focus specifically on the case in which bubbles are formed within the bulk of a liquid in which no pre-existing bubbles existed. That is, the bubbles come into existence by overcoming the cohesive forces that bind the molecules together in the liquid, rather than as a result of emergence out of the mouth of surface imperfections in which trapped gases existed. The results will provide a practical lower limit to the bubble curvature, or upper limit to the superheat, that a liquid can sustain below the critical point at a given pressure at which a phase transition must occur.

3. Homogeneous Nucleation

3.1 The Kinetics of Homogeneous Nucleation in the Bulk of a Liquid

A superheated liquid is not quiescent on the microscopic level. Incessant random molecular motion within the loosely structured liquid lattice creates local density fluctuations that momentarily create "holes" or "nuclei" within the liquid. The interior of these nuclei are vapor-like in terms of the density and potential energies of the molecules they contain. The basic problem in nucleation theory is to describe the net rate of forming nuclei - the "nucleation rate" - which are in equilbrium with the surrounding liquid.

The classical view of nuclei formation within the bulk of a liquid is that nuclei grow or decay isothermally in a unimolecular process wherein a molecule "evaporates" or "condenses" on the interface (e.g., Volmer 1939, Frenkel 1946, MacDonald 1963, Skripov 1974, Springer 1978). A molecule entering a nucleus results in incremental growth whereas molecular condensation causes a decrease in the size of the nucleus. This process is visualized by drawing an analogy to a chemical reaction,

$$E_n + E_l = E_{n+1} \qquad 3$$

where n = 1,2,...G and E_G is some arbitrarily large nucleus containing G molecules. One reaction of the type given by eq. 3 applies to every size nucleus containing n molecules. The forward and reverse "rate constants" for the above set of reactions are F_n and R_{n+1}, respectively. The last reaction in the above set is irreversible to allow net conversion of molecules from the liquid to the vapor state: $E_G + E_l \rightarrow E_{G+1}$. R_n is given by the condensation rate of molecules on a surface of area S_n of a bubble containing n molecules and can be expressed in terms of the molecular collision frequency, β, as

$$R_n = S_n\,\beta \quad \text{4a}$$

where for an ideal gas

$$\beta = P/(2\pi mkT)^{1/2} \quad \text{4b}$$

and for a full spherical bubble,

$$S_n = 4\pi r^2 \quad \text{4c.}$$

S_n also depends on P through the ideal gas law (for a liquid droplet S_n is considered to depend only on n). The evaporation rate F_n is not known *a priori* but can be determined as discussed below. A major assumption of nucleation theory is that the F_n are dependent only on the size of the nuclei and not on the vapor density or pressure (Katz and Weidersich 1977).

The rate of forming bubbles - the nucleation rate J - for the above reaction sequence is (Springer 1978)

$$J = F_n f_n - R_{n+1} f_{n+1} \quad \text{5}$$

where f_n is the number density of nuclei that contain n molecules and the rate J is the same for all of the reactions in the steady state approximation.

The specific form of f_n is unknown. The ratio f_n/N_n, where N_n is the number density of nuclei containing n molecules, may be reasoned in the limits $n \rightarrow 1$ ($f_n/N_n=1$) and $n \rightarrow G$ ($f_n/N_n=0$) when J=0. The imposed limit J=0 is a hypothetical constrained equilibrium state (Frenkel 1946, MacDonald 1963). In this limit eq. 5 can be written as

$$R_{n+1} = F_n\, N_n/N_{n+1} \quad \text{6.}$$

Combining eqs. 5 and 6 gives

$$J = F_n\, N_n[f_n/N_n - f_{n+1}/N_{n+1}] \quad \text{7.}$$

The remaining unknown distributions in eq. 7 can be eliminated by summing eq. 7 from n=1 to n=G-1 to yield

$$J = \{\Sigma[1/(F_n N_n)]\}^{-1} \quad \text{8}$$

Finally, by treating n as a continuous rather than a discrete variable the summation in eq. 8 is converted to an integral:

$$J=\{\int[1/(F_n N_n)]dn\}^{-1} \quad \text{9}$$

To evaluate J from eq. 9 both the evaporation rate F_n and equilibrium number density N_n must be known.

The distribution of nuclei in a hypothetical constrained equilibrium state can be determined by assuming that the liquid/nuclei system is an ideal dilute solution of vapor bubbles and single molecules. On minimizing the thermodynamic availability of mixing such a solution, it can be shown that

$$N_n \approx N_0 \exp[-\Phi/kT] \quad \text{10}$$

where Φ is the minimum work required to form a nucleus. To determine this quantity it is assumed that the nucleus is in chemical equilibrium with the surrounding liquid and that the ideal gas law, PV=nkT is valid. Φ can then be expressed as (Avedisian 1986):

$$\Phi = -\,kT(P^*-P_0)/P^*\,n + \sigma[36\pi(kT/P^*)^2]^{1/3}\,n^{2/3} \quad \text{11.}$$

From eq. 11 it is seen that $d\Phi/dn\,|_{n^*} = 0$ at the the state for which a nucleus contains n^* molecules and where eq. 1 is applicable. Also,

$$\Phi_{nn^*} \equiv d^2\Phi/dn^2\,|_{n^*} = -(kT/P^*)^2(P^*-P_0)^4/(32\pi\sigma^3) < 0 \quad \text{12.}$$

The energy of forming a nucleus is then a maximum at the critical size. This maximum energy constitutes an energy barrier to nucleation and hence to phase change that must be overcome before bubble nucleation will occur in the bulk of a liquid. Combining eqs. 11 and 1a with the ideal gas law gives the well-known result

$$\Phi^* = 16\,\pi\,\sigma^3/[3(P^*-P_0)^2] \quad \text{13}$$

It is now desired to determine J from eq. 9. In view of the exponential function that characterizes the constrained distribution in eq. 9, the dominant contribution to the integral will occur in a region around the critical nucleus state, n*. The consequences of this fact are that 1) a series expansion of Φ truncated after the first nonzero term will be sufficiently accurate to represent Φ, and 2) that the limits of integration on n may be extended from $-\infty$ to $+\infty$. Carrying out the integration under these conditions and noting that in the spirit of treating n as a continuous variable, eqs. 6 and 9 can be combined to yield

$$F_n \approx R_n \exp(-\Phi_n/kT) \quad \text{14,}$$

(when $n=n^*$, $\Phi_n \rightarrow \Phi_{n^*}=0$ from eq. 11), the steady state nucleation rate can be written as

$$J = \Gamma\, S_n\, \beta\, N_0\, \exp(-\Phi^*/kT) \quad \text{15a}$$

where

$$\Gamma = [\,|\Phi_{nn^*}|/(2\pi kT)]^{1/2} \quad \text{15b}$$

Γ is a factor that accounts for the possibility that nuclei larger than the critical size may decay. That $\Gamma < 1$ as could be seen by introducing physical properties into eqs. 15 (Avedisian and Andres 1978) implies that not all nuclei that reach the critical size will ultimately continue growing.

Eqs. 15 and 13 show that J varies sharply with T_k because of the strong dependence of vapor pressure and surface tension on temperature. This sharp variation suggests that an effective threshold temperature exists below which J is vanishingly small and above which it is large. The mean temperature in the range of this large change in J is the superheat limit, T_k (i.e., $T \rightarrow T_k$ in eq. 15). This limiting nucleation rate is often defined to correspond to $J \approx 1 cm^3 s^{-1}$. Experimentally measured values of T_k at various nucleation rates have been tabulated for a large number of liquids (Avedisian 1991) and show that $T_k(P_0) - T_s(P_0)$ can be over one hundred degrees for liquids heated at $P_0 \approx 1$ atm. Note that $T_k(P_0)$ must be less than $T_t(P_0)$.

In some applications, it is more convenient to express eq. 15 in terms of temperature if the ambient pressure, P_0, is known or pressure if the temperature is known:

$$T \rightarrow T_k = \Phi^*/k\ [\ln(\Gamma S_{n^*}\beta\, N_0/J)]^{-1} \quad \text{16a}$$

or

$$P_0 \rightarrow P_{0k} = P^* - \{16\,\pi\,\sigma^3/(kT_k)\,[\ln(\Gamma S_{n^*}\beta\, N_0/J)]^{-1}\}^{1/2} \quad \text{16b.}$$

Terms within the logarithm of eq. 16a exert a minor influence on T_k so that eq. 16a could also be written in the form

$$\Phi^*/kT_k = C \quad \text{16c}$$

which is often useful for correlating phase change data that are believed to be governed by homogeneous nucleation. Experimental evidence suggests $10 < C < 75$ for a wide range of applications (Apfel 1971). For example, the rapid isobaric heating of liquids is often characterized by $C \approx 11.5$ (Lienhard and Karimi 1981); for rapid, near isothermal, depressurization $C \approx 28.5$ (Alamgir and Lienhard 1981); and for nucleation at the interface between two immiscible liquids such as might exist within a dispersion of droplets in a field liquid (i.e., an emulsion) $C \approx 66$ (Avedisian and Andres 1978), where in the latter application, the surface tension in eq. 13 is an effective value based on the surface and interfacial tensions between the two liquids.

If the ambient pressure is known, eq. 16a must be solved iteratively because of the strong dependence of temperature on surface tension. If the liquid temperature is known, eq. 16b can essentially be solved explicitly for the nucleation pressure because of the weak contribution of the pressure dependent terms β and S_{n*} which appear in the logarithmic term in eq. 16b.

Depending on the application, eqs 15 or 16 may be more convenient to use. For example, in the rapid depressurization of a liquid in which the temperature and decompression rate are known, the rate of bubble formation appears as a source term in the transport equation for bubble number density in the conservation equations (Deligiannis and Clever 1990); eq. 15 provides the contribution for homogeneous nucleation in such applications (cf, Section 4.1.1). For applications related to vapor explosions, the initial liquid state that triggers vapor explosions (Reid 1983) is more appropriately obtained from eq. 16 in which Γ and J appear in a logarithmic term. Fortunately, T_k or P_{ok} are very weak functions of J, the latter of which can be estimated to within several orders of magnitude accuracy for a range of experimental conditions.

3.2 Modifications to the Classical Theory

Lienhard and Karimi (1981) argued for the substitution of kT in eq. 15a by kT_c on the basis that the ratio Φ^*/kT can be interpreted as the fraction of the molecular energy of a nucleus. The smallest value of this ratio will yield the largest nucleation rate (cf, eq. 15). The resulting thermodynamic state (P_o, T) is then interpreted as coinciding with the spinodal curve. Basing the smallest molecular energy on the potential well value for a Lennard-Jones model of molecular interaction yields $\Phi^*/kT \rightarrow \Phi^*/\varepsilon$ where $\varepsilon \sim kT_c$. This result is phenomenolgical in that it does not appear to emerge naturally from the classic treatment of the kinetics of nucleation, nor on the development which led to the equilibrium distribution of nuclei in eq. 10. While the replacement of T by T_c in eq. 16 may be controversial, the results were successful in correlating measured superheat limit data. Though $T_k \rightarrow T_c$ on the left hand side of eq. 16a, temperature (T_k) still appears prominently through the surface tension and vapor pressure in eq. 13.

Kwak and Panton (1983) based the critical nucleus energy, Φ^*, on the notion that the London disperson force is the only important interaction between molecules in the metastable state. The result was that

$$\Phi^* = 1/2\,(kT)^3/[v_m(P^*-P_o)]^2 \qquad 17$$

where the molecular volume, v_m, is obtained by assuming a FCC packing for the liquid lattice structure. Results using eq. 17 in eq. 16a with $J \approx 10^{30}cm^3\,s^{-1}$ (i.e., at the spinodal limit) yielded close agreement with measured maximum superheat temperatures for liquids that contain dissolved gases.

Katz and Weidersich (1977) discovered that, for the case of formation of liquid droplets by condensation within a supersaturated vapor, the equilibrium state at *saturation* conditions wherein also J=0 can be used to relate F_n to R_n. The saturation state is actually the more appropriate state in which equlibrium should be considered, for from eq. 15a $J \rightarrow 0$ only when $\Phi^* \rightarrow \infty$ which from eqs. 13 and 1a gives $r \rightarrow \infty$ which is the "normal" saturation state of a liquid. For the idea to be useful, it is necessary that the device of summing in eq. 7 lead to a cancellation of all terms except the first and last. Katz and Weidersich (1977) showed how such a cancellation of terms could occur for the case of condensation of liquid droplets. An extension of this idea to the formation of vapor bubbles within a superheated liquid is not straightforward because of the dependence of surface area of a bubble on both the number of molecules it contains and on the internal vapor pressure. However, assuming that $P \approx P_s$ (which is approximately true as shown by eq. 2), then $S_n(P) \approx S_n(P_s) \equiv S_n$. Writing N_{sn} for the distribution of nuclei at saturation ($f_n \rightarrow N_{sn}$) in eq. 5 when J=0 yields

$$F_n = \beta_s\, S_{n+1} N_{sn+1} / N_{sn} \qquad 18$$

where use has been made of eq. 4a and F_n is assumed to be independent of pressure. Substituting eq. 18 into eq. 5, multiplying both sides by γ^n and summing, all terms cancel except the first and last with the result that

$$J = \{\Sigma[\,1/(\beta_s S_n N_n \gamma^{1-n})]\}^{-1} \qquad 19$$

Eq. 18 differs from eq. 8 only in the presence of γ. In view of eq. 2, $\gamma \sim 1$ and the modification to classical nucleation theory based on relating F_n to R_n in a saturated state is small. However, the conceptual interpretation of the process is improved because the notion of a constrained equilibrium distribution of nuclei in a superheated liquid is now unnecessary in the development.

3.3 Homogeneous Nucleation at a Surface

The results of Section 3.1 are readily extended to nucleation at a surface by introducing truncation factors to the volume (ϕ) and surface of the bubble (Δ) due to the presence of the wall:

$$V = 4/3\,\pi\, r^3\, \phi \qquad 20a$$

and

$$S_n = 4\,\pi\, r^2\, \Delta \qquad 20b$$

The specific form of ϕ and Δ depend on the surface structure and many such structures can be envisioned. Taking as an example the case of a conical projection or cavity, geometric arguments can be used to show that the volume and surface truncation factors are

$$\phi(\theta,\xi) = 1/4\{[1-\sin(\theta-\xi)]^2[2+\sin(\theta-\xi)] + \cos^3(\theta-\xi)\cos(\xi)/\sin(\xi)\} \qquad 21a$$

and

$$\Delta(\theta,\xi) = 1/2\,[1-\sin(\theta-\xi)] \qquad 21b.$$

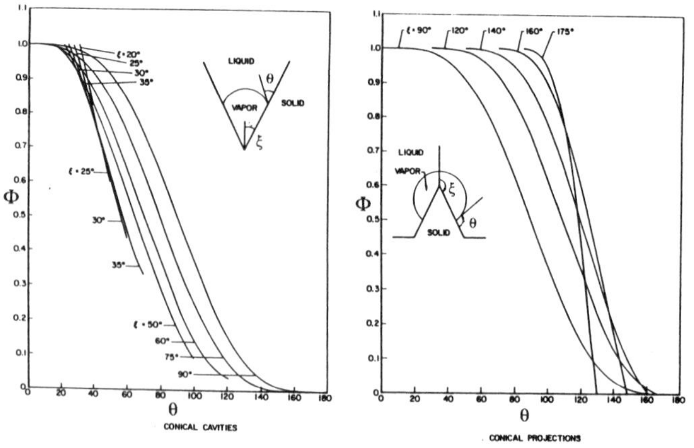

Figure 2 Volume truncation factor (eq. 21a) for a conical cavity and conical projection.

Fig. 2 illustrates the variation of ϕ with contact angle θ. The special case of a flat surface corresponds to $\xi=90°$ as given by many others (e.g., Skripov 1974, Cole 1974, Apfel 1971, Blander and Katz 1975).

Eq. 11 can be extended to determine the minimum work to form a bubble at a surface as

$$\Phi = -kT(P^*-P_o)/P^*\, n + \sigma\, \phi^{1/3}\, [36\pi(kT/P^*)^2]^{1/3}\, n^{2/3} \quad 22$$

from which it follows that

$$\Phi_{nns^*} = \Phi_{nn^*} / \phi$$

where Φ_{nn} is given by eq. 12. The arguments leading up to eq. 13 can be repeated to show that

$$\Phi_s^* = \Phi^*\, \phi \quad 23$$

where Φ^* is given by eq. 13. Finally, noting that the surface density of molecules at a surface (molecules/cm^2) is approximately $N_o^{2/3}$ (Apfel 1970), we can write that

$$J_s = J\, \Delta/\phi^{1/2}\, N_o^{-1/3}\, L \exp[\Phi^*(1-\phi)] \quad 24$$

where J is given by eq. 15a and L is a characteristic length scale (total volume to surface area ratio of the liquid in contact with the solid). An estimate of J_s is required to use eq. 24 and this could be obtained on the basis of the maximum packing density of bubbles at a surface (Alamgir and Lienhard 1981). Eq. 24 could then be inverted for temperature or pressure in a manner similar to that which lead to eq. 16 to provide a means for predicting the superheat limit at a surface. (cf, eq. 16)

It is interesting to compare J_s with J from eq. 15. For example, fig. 3 shows the variation of $\ln[J_s/(JL)]$ with θ for $\xi=90°$ in eq. 21 (i.e., a flat surface) for octane and water at their superheat limits at .101MPa. The interesting fact is revealed that there is a contact angle range below which the probability for homogeneous nucleation is greater in the bulk of a liquid than at a smooth surface ($\ln[J_s/(JL)] < 0$). This is attributed to the competition between a larger number of potential nucleation cites - molecules - existing in the bulk (N_o) compared to the surface ($N_o^{2/3}$) on the one hand with the reduced energy required to form a bubble at the surface because of the contact angle being greater than 0° on the other. Thus, liquids with large contact angles (e.g., water on PTFE (Adamson 1982)) may be more affected by the presence of surfaces than liquids with small contact angles (e.g., many organic liquids on PTFE (Adamson 1982)). Of course, the propensity for vapor trapping increases as the contact angle increases which could promote heterogeneous nucleation.

There are several situations in which homogeneous nucleation can occur in the presence of solid surfaces whether or not they trap gas. These invariably involve a rapid transgression of the normal ($r \rightarrow \infty$) phase boundary, either through rapid decompression (~10^6MPa/s) or rapid heating. In the case of rapid heating, the temperature within the thermal boundary layer increases at a rate that is faster than the time it would take for a bubble to grow and detach at the surface. For many organic liquids and water, this heating rate is on the order of 10^6C/s or larger (Skripov 1974).

4. Two Examples

4.1 Rapid Decompression of an Initially Subcooled Liquid

Rapid depressurization of a subcooled liquid can be a hazard with regard to the so-called "guillotine" break of the recirculation pipe in boiling water and pressurized water reactors (Alamgir and Lienhard 1981). A minimum pressure is reached during the decompression which can be well below P_s (fig. 1) at which nucleation occurs. The resulting large pressure force across the bubbles can lead to an explosive vaporization event. Nucleation induced by a pressure decrease can also occur in a flow system in which the pressure drop is created by the the momentum change of the fluid and wall friction, but the decrease will usually not be great enough to induce homogeneous nucleation for most practical applications. The problem has been analyzed both from the viewpoint of predicting the evolution of pressure and the minimum pressure undershoot.

4.1.1 Two-Phase Flow

An important parameter for predicting the evolution of pressure during rapid blow-down of a liquid filled vessel is the number density of bubbles. This value dictates the interfacial area between the two phases which is an important parameter for predicting the transport properties: pressure, temperature, velocity and void fraction.

For the two-fluid models that are considered to be the most relevant to analyses of phase change precesses during depressurization of fluids, the number density appears in the source terms in the conservation equations. For bubbly flows, the total interfacial heat transfer per unit volume between the liquid and vapor (bubble) phases can be expressed as (Deligiannis and Clever 1990)

$$q_i = N_b^{2/3}\, (6\pi^2)\, \alpha^{1/3}\, Nu\, \kappa\, (T_i - T_j) \quad 25$$

where j = liquid or vapor, and N_b is the bubble number density and T_i is the interface temperature. The bubble radius is related to N_b through the void fraction, $\alpha = N_b\, 4/3\pi\, r^3$.

To provide closure to the conservation equations, a transport equation for the bubble number density was derived by

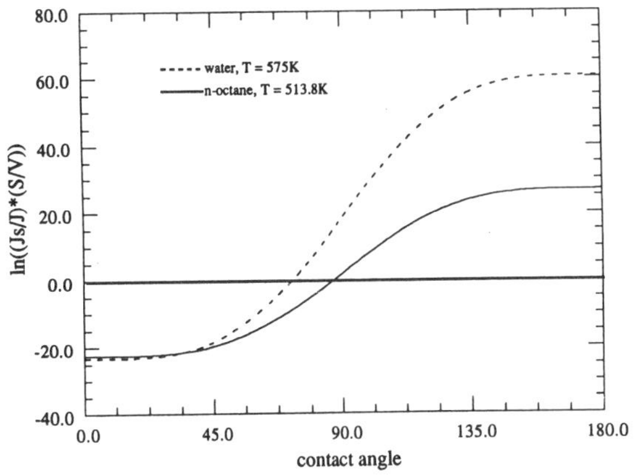

Figure 3 Influence of contact angle, θ, on the relative values of surface (J_s) and bulk (J) nucleation rate for water and octane.

Kocamustafaogullari and Ishii (1983) which illustrates the potential importance of homogeneous nucleation in the problem:

$$\partial N_b/\partial t + \nabla \cdot (\mathbf{v} N_b) \approx \Xi \quad 26a$$

where

$$\Xi = J + J_s + \phi_{wn} - \phi_{cond} \quad 26b$$

and Ξ accounts for the factors by which N_b changes. These include the rate of formation of bubbles per unit volume by gas bubble nucleation from gas-filled cavities, ϕ_{wn}, and a contribution toward condensation, ϕ_{cond}. In the earliest stages of depressurization, ϕ_{cond} can be neglected but it may exert an influence at later times as the nucleated bubbles grow and enter the subcooled bulk liquid. It has often been considered that $\phi_{wn} >> \phi_{cond}$, J, and J_s. The approach, though, appears to require a rather high degree of empiricism because $\phi_{wn} \sim N_{cn}\, f/D$ where D is the channel or tube diameter, f is the bubble generation frequency at the surface, and N_{cn} is the surface nucleation site density, all of which are determined through empirical correlations (Kocamustafaogullari and Ishii 1983).

A recent study (Deligiannis and Cleaver 1990) took advantage of the fact that J_s in eq. 24 could essentially be considered to contain one unknown parameter - the contact angle. By assuming that the dominant contribution to the source terms in eq. 26 comes from J_s, in which case $\Xi \approx J_s$, the problem then contains only one empirically determined parameter.

Fig. 4 compares pressures predicted by a one-dimensional two fluid model (Deligiannis and Clever 1990) with measured values for depressurization of water in a horizontal tube (Edwards and O'Brien 1970). ϕ was selected to achieve agreement with the minimum pressure. Lower values of ϕ than expected were needed to achieve the

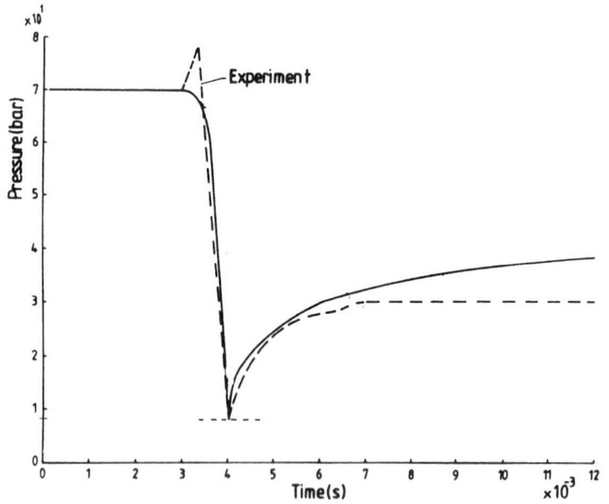

Figure 4 Comparison between measured pressure transients (Edwards and O'Brien 1970) with values predicted by the two-fluid model of Deligiannis and Clever (1990). ϕ in eq. 24 was considered an empirical parameter to fit the maximum pressure undershoot.

agreement shown in fig. 4. This fact could be a consequence of lumping all contributions toward the source term in eq. 26 in ϕ in eq. 24. Agreement is poorer during the recovery stage because nucleation no longer exerts an influence on the change in N_b during that phase of the transient. Furthermore, the influence of surface re-wetting as the nucleated bubbles detach and enter the bulk (Kenning and Thirunaruktarasu 1970) could exert an influence which is also not accounted for in the analyses. Since the change in N_b is considered to be dominated by homogeneous nucleation in the approach of Deligiannis and Clever (1990), agreement between predicted and measured pressure transients during the early stages should be improved as lower minimum pressures are achieved in experiment which was in fact found to be the case.

4.1.2 Maximum Pressure Undershoot

It is often useful to consider ϕ as an adjustable parameter as described above to provide closure to the bubble transport equation. Another application in which ϕ can be used as a parameter is in correlating measured maximum pressure undershoots on the basis of eqs. 16b, 23 and 24 for the purpose of developing a more generalized predictive framwork for the maximum pressure undershoot. Using such data, Alamgir and Lienhard (1981) found that

$$\phi \approx .1058\,(T/T_c)^{28.46}\,[1 + 14\,(dP_0/dt)^{.8}]$$

and $C \approx 28$ in eq. 16c. Combining eqs. 13, 16c and 23 and solving for pressure yields

$$P^* - P_0 \approx .251\,\sigma^{2/3}\,(T/T_c)^{13.73}\,[1 + 14\,(dP_0/dt)^{.8}]^{1/2}/(kT_c)^{1/2} \qquad 27.$$

which was shown to correlate maximum pressure undershoots to within about ±10%.

The influence of initial liquid temperature on the nucleation pressure as predicted from eq. 27 is illustrated in fig. 5 with the depressurization rate as a parameter and the properties of water as an example. The results show that the pressure undershoot increases as

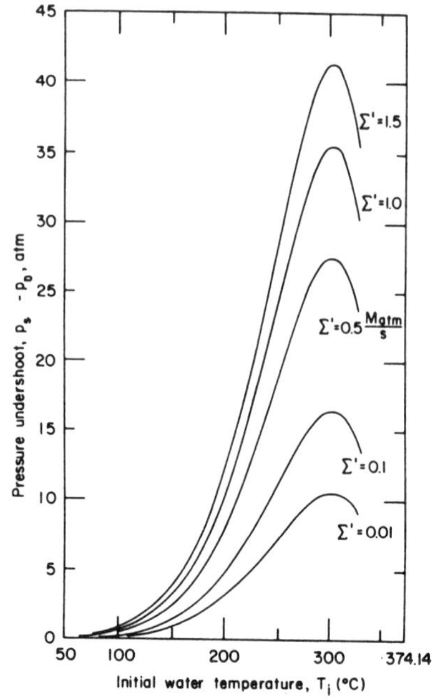

Figure 5 Effect of initial temperature on the minimum pressure undershoot for water at various decompression rates (Alamgir and Lienhard 1981).

the decompression rate increases. This fact may be expected as noted previously in connection with the competition between bringing the liquid into the metastable state at a fast enough rate for homogeneous nucleation to occur before any bubbles which nucleate at surface cavities grow to appreciable size to effect the pressure. That the pressure undershoot should exhibit a maximum and eventually decrease may be expected because as $T \rightarrow T_c$ (374C for water) $r \rightarrow \infty$ and $P \rightarrow P_0$.

4.2 Vapor Explosions of Droplets

4.2.1 Predicting the Superheat Limit

A droplet at the superheat limit can vaporize in an explosive manner in which a bubble grows almost immediately to vaporize the liquid when the superheat limit is reached. In such situations the homogeneous nucleation limit in effect defines the dynamics of the phase change process. For this reason, the homogeneous nucleation limit is often identified with the onset of vapor explosions (Fauske 1974, Bankoff 1978), though the dynamics of the process involve detailed analyses of the equations that govern bubble growth within liquids (e.g., Sideman and Taitel 1964, Bankoff 1966, Vuong and Sadhal 1989).

Predictions of the superheat limit using eq. 16 requires the nucleation rate commensurate with the experimental conditions be known and the ability to predict accurately physical properties. The most important physical properties are the vapor pressure, P_s in eq. 2, and the liquid/vapor surface tension, σ. These properties depend strongly on the temperature, the dependence of which is further enhanced by the way the properties appear in the equations: the surface tension is cubed and the vapor pressure is squared. Furthermore, while the development discussed in Section 1 was restricted to single component liquids, an analogous procedure for miscible liquids can be shown to yield essentially the same form as eq. 16 except for differences in the pre-exponential term in eq. 15a. Thus, composition must be added to the dependence of physical properties. Difficulties with predicting accurately the vapor pressure are more computational than theoretical (Prausnitz 1969). The problem is more severe with surface tension, though methods based on corresponding states (Dickinson 1975) seem to be reasonably accurate for predicting superheat limits.

The final information needed to estimate T_k is the nucleation rate commensurate with the experimental conditions. J in eq. 16a was estimated from its interpretation as a "waiting" time t, which is the mean lifetime of a superheated liquid of volume v: $J = 1/(vt)$. A discussion of more detailed methods for estimating the nucleation rate are provided elsewhere, but generally values in the range of 1 to 10^{22} nuclei/(cm^3-s) characterize a wide range of experimental methods (Avedisian 1985).

One particularly useful method which has been shown to yield reasonably accurate superheat limit data for droplets consists of heating droplets of one liquid in another liquid with which it is immiscible (the field liquid) (e.g., Wakeshima and Takata 1958, Skripov 1974, Holden and Katz 1978, Avedisian and Glassman 1981a). By measuring the temperature and pressure at the location in the column at which the droplet is observed to vaporize, and if thermal equilibrium can be assumed, these conditions also correspond to the droplet. Gas bubble nucleation on cavities is suppressed with this method because the "container" for the liquid is the interface between two liquids which is smooth down to molecular dimensions.

For the method to work, the boiling point of the field liquid must be larger than the superheat limit of the liquid that comprises the test droplet. Additional requirements that the test and field liquids must satisfy are described elsewhere (Avedisian 1986). One system which has been shown to be particularly suitable for this method consists of heptane as the droplet and glycerine as the field liquid. Measured maximum temperatures at which heptane droplets, typically about 1mm diameter, have been observed to vaporize under a range of ambient pressures are displayed in fig. 6a. Data from two sources are

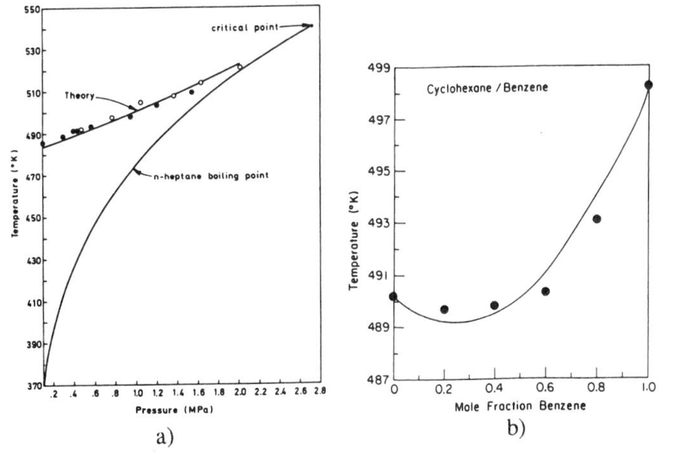

Figure 6 Comparison between measured and predicted superheat limits.

a) Effect of pressure for n-heptane (●-Skripov 1974, o-Avedisian and Glassman 1981a); predictions from eq. 16a.

b) Effect of composition at constant pressure (.101MPa) for cyclohexane/hexane mixtures (o-Holden and Katz 1978); predictions from eq. 29.

shown to illustrate the consistency of the measurements (Avedisian and Glassman 1981a, Skripov 1974). The axes are rotated from fig. 1 to reflect the fact that the pressure was the controlled variable. The trend shown that the superheat limit increases as ambient pressure increases is consistent with all reported limit of superheat data and eq. 16. This trend may be explained by noting, with reference to fig. 1 and eq. 1, that as P_o increases $2\sigma/r$ must decrease because the saturation pressure varies more strongly with temperature than does the nucleation pressure. As the critical point is approached, $P-P_o \rightarrow 0$. The agreement between experiment and predictions ($J \approx 10^6 cm^{-3}s^{-2}$) is evident.

The ability to calculate accurately the superheat limit from eq. 16 is largely dependent on the ability to predict accurately the vapor pressure and surface tension of the liquid. The problem is especially severe for the surface tension because of the way it appears in eq. 16. This has been the motivation for seeking alternative methods for predicting the kinetic limit of superheat. One method is based on eq. 16c. Using the so-called "accentric" factor (Pitzer 1977), defined as

$$\omega = -\log_{10}(P/P_c)\big|_{T/Tc=.7} - 1$$

and assuming that $\sigma \sim (1-T/T_c)^{11/9}$ (Guggenheim 1945), Lienhard and Karimi (1981) developed a correlation for the nucleation pressure on the basis of eq. 16b in the form

$$P_{ro} = P_{rs} - \chi(1-T/T_c)^{1.83}/(\ln(J) - C) \qquad 28a$$

where

$$\chi = 112.82 - 224.24\,\omega \qquad 28b$$

The utility of eq. 28 resides both in relative accuracy and simplicity: no surface tension data are required.

The problem of predicting T_k for mixtures is more complicated because there is no generally acceptable method for predicting the surface tension of a mixture, especially when the temperature is above the critical temperature of one of the mixture components such as will often be the case for miscible mixtures heated to their superheat limits. To overcome this problem, the corresponding states principle was extended to the superheat limit property of a mixture (Avedisian and Sullivan 1984). The result was the following:

$$T_k/T_{cm} = \Sigma\, x_i\, T_{ki} / T_{ci} \qquad 29$$

where the T_{ki} are the superheat limits of the individual components in the mixture evaluated from eq. 16 at the reduced temperature and reduced pressure of the mixture. It is still necessary to be able to predict surface tension to evaluate the T_{ki} in eq. 29, but only single component predictions are necessary. This is a much easier task than to predict the surface tension of a mixture, which does not appear in eq. 29. Fig 6b compares measured maximum temperatures of cyclohexane/benzene droplets heated in glycerine (Holden and Katz 1978) with values predicted from eq. 29. The perfect agreement shown for the mixture components is a consequence of using the measured superheat limits of the mixture components as the reference fluids in eq. 29. The nonlinear variation of temperature superheat with composition illustrated in fig. 6b is a consequence of the particular mixture studied. Linear variations are also predicted and measured for many organic mixtures.

No information on the dynamic aspects of the vaporization process is provided by a superheat limit prediction. Thus, from examining the predictions shown in fig. 6, one may still wonder about the dynamics of the phase change process that triggered the observer or instrumentation to record the data indicated on those figures that were believed to coincide with the homogeneous nucleation limit.

4.2.2 Dynamics of Phase Change at the Homogeneous Nucleation Limit

The question is addressed here using the configuration of droplets for illustration. Larger systems, for example pools of liquids being heated to the superheat limit (e.g., Reid 1983), have also been studied and could have been used to illustrate some of the dramatic aspects of the phase change process at the superheat limit that are revealed in droplet studies.

The behavior of droplets that led to the data illustrated in fig. 6 exhibited a range of dynamic aspects, from a somewhat violent and audible process akin to an explosion at pressures near atmospheric (for example the data in fig. 6a near P_0 = .1MPa to .2 MPa) to a quiescent process in which the droplet merely swelled as the internal

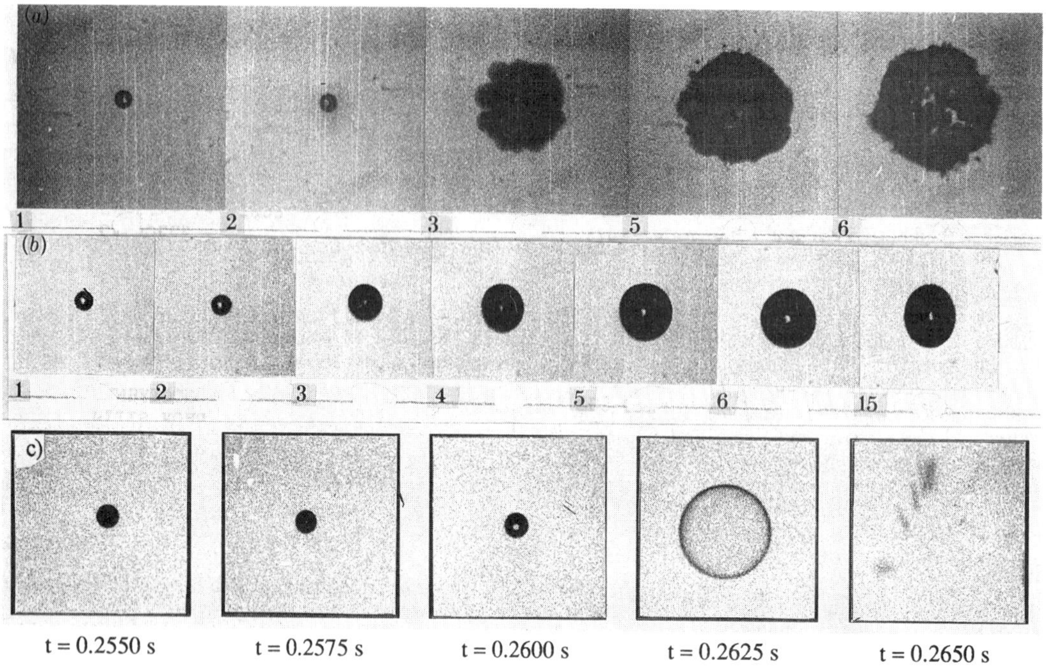

Figure 7 Photographic sequence of bubble nucleation within droplets. Number on picture in (a) and (b) corresponds to the position in the motion picture sequence.

a) n-hexane droplet levitated in a flowing stream of glycerine. Flow is from left to right (Avedisian 1987). Far field pressure is .12MPa. Initial droplet diameter was.72mm. Glycerine (and droplet) temperature is the superheat limit of hexane corresponding to .12MPa (187C). (frames s^{-1}=900).

b) n-hexane droplet levitated in a flowing stream of glycerine. Flow is from left to right (Avedisian 1987). Far field pressure is .44MPa. Glycerine (and droplet) temperature is the superheat limit of hexane corresponding to .44MPa (188.2C). Initial droplet diameter was .76mm. This sequence illustrates the reduction in boiling intensity at elevated pressure. (frames s^{-1}=500).

c) methanol (.75, v/v)/dodecanol(.25, v/v) mixture droplet burning in air (Yang et al. 1990). Number adjacent to each photograph is the time (s) after ignition of the droplet. The vapor explosion is illustrated by the growth of a bubble in the droplet and its subsequent fragmentation. Initial droplet diameter was .58mm.

bubble grew (data in fig. 6a above around P_0 = .6MPa).

The vaporization dynamics of droplets at the superheat limit has been studied at atmospheric (e.g., Shepherd and Strutevant 1982, Frost and Sturtevant 1986, Mori and Komotori 1976) and higher pressures (e.g., Avedisian 1987, Frost and Sturtevant 1986). Liquid composition has also been observed to exert an influence on the intensity of vaporization (Blander et al. 1971, Avedisian 1987). Furthermore, if the interfacial tension between the droplet and surrounding field liquid is comparatively high, a bubble may form at the interface between the droplet and surrounding liquid, and experimental evidence for this type of vaporization has also been reported (e.g., Jarvis et al. 1975, Avedisian and Glassman 1981b).

To illustrate what the vaporization of a droplet at its superheat limit looks like, fig. 7 shows a sequence of photographs of a hexane droplet approximately 1.4mm initial diameter being heated in glycerine. The conditions in the first frame in each sequence corresponded very closely to the predicted superheat limit using eq. 16 (for $J \approx 10^5 cm^{-3} s^{-1}$ where this rate was considered to be appropriate for the conditions of the experiment). For a hexane droplet heated under a pressure close to .1MPa, the entire droplet vaporized in under 3ms. The process was accompanied by an audible "popping" sound. At a pressure of .44MPa, though, the droplet merely swelled as a vapor bubble grew within it. However, the temperature and pressure at which swelling was first detected - the second photograph in fig. 7 - again agreed very closely with predictions from eq. 16. Clearly, the results indicate that vaporization at the superheat limit is not always explosive.

The photographic method used to obtain the pictures shown in fig. 7 was not capable of resolving details of the explosive vaporization process that is shown in the first three pictures in fig. 7a. In a time of under 3ms, the droplet completely vaporized. Additional work by Shepherd and Sturtevant (1982) and Frost and Sturtevant (1986) has revealed new insights on the explosive vaporization of droplets at the superheat limit. They discovered that the explosion process occurs on a *micro*second time scale.

Finally, shown in fig. 7c is a droplet burning in air at .101MPa. The droplet is a mixture of methanol and dodecanol (Yang et al. 1990) and it it is burning in a very low gravity environment (the purpose of which was to create spherically symmetric burning conditions in order to compare results from simple one-dimensional droplet combustion models with the data). The number beneath each frame indicates the time after igniting the droplet that the photograph was taken. The last two frames show the droplet exploding as manifested by growth and bursting of a large bubble inside the droplet. The resulting fragments, of which four are shown in fig. 7c (somewhat blurred as the camera framing rate was about $250s^{-1}$) serve to illustrate why this process is thought to be beneficial in combustion: the larger drop is converted into several smaller ones by the explosion process, and these smaller droplets can burn faster and/or more cleanly due to the increased surface area, collectively, of the smaller droplets.

The initiation of droplet shattering during combustion process seems to be reasonably well predicted based on homogeneous nucleation (Lasheras et al. 1979). The simpliest prediction involves comparing the superheat limit (cf, eq. 16) with the droplet burning temperature that is based on reaching the boiling point of the heaviest component of the mixture. If the droplet temperature is greater than the superheat limit of the mixture, homogeneous nucleation and bubble growth occurs within it and the droplet explodes. The droplet shown in fig. 7c had the requisite composition for it to explode based on this criterion.

The dynamics of droplet explosions clearly are not predictable from knowledge of just the initial conditions needed to form the critical size nucleus within the droplet. These dynamics, for the sequence shown in fig. 7c, involved a bubble growing within the droplet and then bursting, rather like a ballon.

5. Future Directions

The dynamic aspects of the phase change process at the superheat limit appear to have received less attention than methods to predict the superheat limit. Indeed, it may be considered that such methods are reasonably well established. Perhaps what remains concerning the ability to predict the superheat limit of liquids is to expand the data base of physical properties especially in the metastable regime, most notably surface tension, in order to extend the range of fluids for which eq. 16 or one of the approximate methods discussed in Section 4.2.1 can be applied accurately.

Future work may also strive to more clearly understand the dynamic aspects of the phase change process at the superheat limit, whether in two-phase flow applications that occur during rapid decompression of a liquid or in applications involving liquid droplets.

Thanks go to Mr. Gregory Jackson for his assistance in the preparation of some of the figures. The author thanks the National Science Foundation (Division of Chemical and Thermal Engineering), the Department of Energy (Office of Basic Energy Sciences), and the National Institute of Standards and Technology (Office of Standard Reference Data) for support in the past on problems related to superheated liquids.

6. Nomenclature

h_{fg} heat of vaporization
J steady state bulk nucleation rate ($cm^{-3}\ s^{-1}$)
J_s steady state surface nucleation rate ($cm^{-2}\ s^{-1}$)
k Boltzmann constant
m mass of a molecule
n number of molecules in a vapor nucleus
N_o number density of single molecules in the liquid (cm^{-3})
Nu Nusselt number for heat transfer between a spherical bubble and surrounding liquid
P pressure
P_{ro} $\equiv P_o/P_c$
P_{rs} $\equiv P_s P_c$
r radius of a bubble
R gas constant
t time
T temperature
v velocity vector (eq. 26)
v_l molar volume of liquid ($cm^3\ moles^{-1}$)

Greek

β collision frequency
γ $\equiv \beta/\beta_s$
θ contact angle, fig. 2
κ liquid thermal conductivity
ξ cavity cone angle, fig. 2
ϕ_{wn} number density of bubbles created by gas bubble nucleation on surfaces, eq. 26 (cm^{-3})
ϕ_{cond} number density of bubbles condensed, eq. 26 (cm^{-3})
σ surface tension

Subscripts

c critical point
cm critical point of a miscible mixture
i liquid/vapor interface
k kinetic limit of superheat
ki kinetic limit of superheat of component i in a miscible mixture
o liquid in the far field
s saturation ($r \rightarrow \infty$)
t thermodynamic limit of superheat

Superscript

* at the kinetic limit of superheat

7. References

Adamson, A.W. 1982 *Physical Chemistry of Surfaces*, 4th edition, pp. 349-350, New York, John Wiley.

Alamgir, Md., and Lienhard, J.H. 1981 "Correlation of Pressure Undershoot during Hot Water Depressurization," *J. Heat Transf.* **103**, 52.

Allen, R.R., Meyer, J.D. and Knight, W.R. 1985 "Thermodynamics and Hydrodynamics of Thermal Ink Jets," *Hewlett Packard J.* **36**(5), 21.

Apfel, R.W. 1970 "Vapor Cavity Formation in Liquids," Technical Memorandum 62, Harvard University Acoustics Research Laboratory, Cambridge, Mass.

Apfel, R.W. 1971 "Vapor Nucleation at a Liquid-Liquid Interface," *J. Chem. Phys.* **54**, 62.

Apfel, R.W. 1972 "The Tensile Strength of Liquids," *Sci. Amer.* **227**, 58.

Apfel, R.W. and Roy, S.C. 1983 "Instrument to Detect Vapor Nucleation of Superheated Drops," *Rev. Sci. Instru.* **54**, 1397.

Avedisian, C.T. 1986 "Bubble Growth within Superheated Liquid Droplets," *Encyclopedia of Fluid Mechanics*, Chapter 8, Gulf Publishing Co.

Avedisian, C.T. 1987 "High Pressure Bubble Growth within Multicomponent Liquid Droplets Levitated in a Flowing Stream of Another Immiscible Liquid," *Proc. R. Soc. Lond.* **409**, 271.

Avedisian, C.T. 1991 "The Limits of Superheat of Pure Liquids," in *CRC Handbook of Chemistry and Physics*, 72nd edition (D.R. Lide, editor), pp. 6-173 to 6-183, Boston, CRC Press.

Avedisian, C.T. and Andres, R.P. 1978 "Bubble Nucleation within Superheated Liquid-Liquid Emulsions," *J. Coll. Interf. Sci.* **64**, 438.

Avedisian, C.T. and Glassman, I. 1981a "High Pressure Homogeneous Nucleation of Bubbles with Superheated Binary Liquid Mixtures," *J. Heat Transf.* **103**, 272.

Avedisian, C.T. and Glassman, I. 1981b "Superheating and Boiling of Water in Hydrocarbons at High Pressures," *Int. J. Heat Mass Transf.* **24**, 695.

Avedisian, C.T. and Sullivan, J.R. 1984 "A Generalized Corresponding States Method for Predicting the Limits of Superheat of Liquids: Application to the Normal Alcohols," *Chem. Eng. Sci.* **39**, 1033.

Bankoff, S.G. 1958 "Entrapment of a Gas in the Spreading of a Liquid over a Rough Surface," *AIChE J.* **4**, 24.

Bankoff, S.G. 1966 "Diffusion Controlled Bubble Growth," *Adv. Chemical Engineering* **6**, 1.

Bankoff, S.G. 1978 "Vapor Explosions: A Critical Review," *Proc. 6th Int. Heat Transf. Conf.* **6**, 355.

Blander, M. and Katz, J.L. 1975 "Bubble Nucleation in Liquids," *AIChE J.* **21**, 833.

Blander, M., Hengstenberg, D. and Katz, J.L. 1971 "Bubble Nucleation in n-Pentane, n-Hexane, n-Pentane and Hexadecane Mixtures, and Water," *J. Phys. Chem.* **75**, 3613.

Ciccarelli, C and Frost D.L. 1991 "Flash X-Ray Visualization of the Steam Explosion of a Molten Metal Drop," paper presented at the 13th International Colloquium on the Dynamics of Explosions and Reactive Systems, Nagoya, Japan, July 28-August 2.

Cole, R. 1974 "Boiling Nucleation," *Adv. Heat Transf.* **10**, 85.

Collier, J.G. 1981 *Convective Boiling and Condensation*, 2nd edition, New York, McGraw-Hill.

Corradini, M.L. 1982 "Analysis and Modeling of Steam Explosion Experiments," *Nucl. Sci. Eng.* **82**, 304.

Danilov, N.N, Sinitsyn, Ye N. and Skripov, V.P. 1979 "Kinetics of Flashing of Superheated Binary Solutions," *Heat Transf.-Soviet Res.* **11**, 26.

Deligiannis, P. and Clever, J.W. 1990 "The Role of Nucleation in the Initial Phases of a Rapid Depressurization of a Subcooled Liquid," *Int. J. Multiphase Flow*, **16**, 975.

Dickinson, E. 1975 "The Influence of Chain Length on the Surface Tension of Oligomeric Mixtures," *J. Coll. Interf. Sci.* **53**, 467.

Edwards, A.R and O'Brien, T.P. 1970 "Studies of Phenomena Connected with the Depressurization of Water Reactors," *J. Br. Nuc. Energy. Soc.* **9**, 125.

Fauske, H.K. 1974 "The Role of Nucleation in Vapor Explosions," *Trans. Am. Nuc. Soc.* **15**, 813.

Forest, T.W. and Ward, C.A. 1977 "Effect of a Dissolved Gas on the Homogeneous Nucleation Pressure of a Liquid," *J. Chem. Phys.*, **66**, 2322.

Frenkel, J. 1946 *Kinetic Theory of Liquids* Oxford University Press, Oxford.

Frost, D. and Sturtevant, B. "Effects of Ambient Pressure on the Instability of a Liquid Boiling Explosively at the Superheat Limit," 1986 *J. Heat Transf.* **108**, 418.

Guggenheim, E.A. 1945 "The Principle of Corresponding States," *J. Chem.Phys.* **13**, 253.

Hemmingsen, E.A. 1977 "Spontaneous Formation of Bubbles in Gas-Supersaturated Water," *Nature (Lond)* **267**, 141.

Holden, B.C and Katz, J.L. 1978 "The Homogeneous Nucleation of Bubbles in Superheated Binary Liquid Mixtures," *AIChE J.* **24**, 260.

Jarvis, T.J., Donohue, M.D. and Katz, J.L. 1975 "Bubble Nucleation Mechanisms of Liquid Droplets Superheated in other Liquids," *J. Coll. Interf. Sci.* **50**, 359.

Katz, J.L. and Weidersich, H. 1977 "Nucleation Theory without Maxwell Demons," *J. Coll. Interf. Sci.* **61**, 351.

Kenning, D.B.R. and Thirunaruktarasu, S.K. 1970 "Bubble Nucleation Following a Sudden Pressure Reduction in Water," *Proc. 4th Int. Heat Transf. Conf.* **5**, B2.9.

Kocamustafaogullari, G. and Ishii, M. 1983 "Interfacial Area and Nucleation Site Density in Boiling Systems," *Int. J. Heat Mass Transf.***26**, 1377.

Kwak, H. and Panton, R.L. 1983 "Gas Bubble Formation in Nonequilibrium Water-Gas Solutions," *J. Chem. Phys.* **78**, 5795.

Lasheras, J.C., Fernandez-Pello, A.C. and Dryer, F.L. 1979 "Initial Observations on the Free Droplet Combustion Characteristics of Water-in-Fuel Emulsions," *Comb. Sci. Tech.* **21**, 1.

Lienhard, J.H. and Karimi, A. 1981 "Homogeneous Nucleation and the Spinodal Line," *J. Heat Transf.* **103**, 61.

Lorenz, J.J., Mikic, B.B. and Rohsenow, W.M. 1974 "The Effect of Surface Condition on Boiling Characteristics," *Proc. 5th Int. Heat Transf. Conf.* **4**, 35.

MacDonald, J.E. 1963 "Homogeneous Nucleation of Vapor Condensation. II.," *Am. J. Phys.* **31**, 31.

Modell, M. and Reid, R.C. 1983 *Thermodynamics and Its Applications*, Chapter 9, Englewood Cliffs, Prentice-Hall.

Mori, Y.H and Komotori, K. 1976 "Boiling Modes of Volatile Liquid Drops in an Immiscible Liquid Depending on Degree of Superheat," *ASME Paper No. 76-HT-13*.

Nelson, L.S. and Duda, P.M. 1982 "Steam Explosions of Molten Iron Oxide Drops: Easier Initiation at Small Pressurizations," *Nature* **296**, 844.

Pitzer, K.S. 1977 "Origin of the Accentric Factor," *ACS Symp. Ser.* **60**,1.

Pound, G.M. 1972 "Selected Values of Critical Supersaturation for Nucleation of Liquids from the Vapor," *J. Phys. Chem. Ref. Data* . **1**, 119.

Prausnitz, J.M. 1969 *Molecular Thermodynamics of Fluid Phase Equilibria*, 2nd Edition, Englewood Cliffs, Prentice-Hall.

Reid, R.C. 1983 "Rapit Phase Transitions from Liquid to Vapor," *Adv. Chemical Engineering* **12**, 106.

Shepherd, J.E. and Sturtevant, B. 1982 "Rapid Evaporation at the Superheat Limit,' *J. Fluid. Mech.* **121**, 379.

Siedman, S. and Taitel, Y. 1964 "Direct Contact Heat Transfer with Change of Phase: Evaporation of Drops in an Immiscible Liquid Medium," *Int. J. Heat Mass. Transf.* **7**, 1273.

Skripov, V.P. 1974 *Superheated Liquids* New York, John Wiley.

Springer, G.S. 1978 "Homogeneous Nucleation," *Adv. Heat Transf.* **14**, 281.

van Stralen, S.J.D. and Cole, R. 1979 *Boiling Phenomena*, Vols. 1 and 2, New York, Hemisphere.

Volmer, M. 1939 *Kinetics of Phase Formation* ATI No. 81935 (F-TS-7068-RE) from the Clearinghouse for Federal and Technical Information.

Vuong, S.T and Sadhal, S.S. 1989a "Growth and Collapse of a Liquid-Vapour Compound Drop in a Second Liquid. Part 1. Fluid Mechanics," *J. Fluid Mechanics* **209**, 617.

Vuong, S.T and Sadhal, S.S. 1989b "Growth and Collapse of a Liquid-Vapour Compound Drop in a Second Liquid. Part 2. Heat Transfer,"*J. Fluid Mechanics* **209**, 639.

Wakeshima, H. and Takata, K. 1958 "On the Limit of Superheat," *J. Phys. Soc. Japan* **13**, 1398.

Wang, C.H. and Law, C.K. 1984 "Microexplosion of Fuel Droplets under High Pressure," *Comb. Flame.* **59**, 53.

Witte, L.C., Vyas, T.J. and Gelabert, A.A. 1973 "Heat Transfer and Fragmentation During Molten Metal/Water Interactions," *J. Heat Transf.* **95**, 521.

Yang, J.C., Jackson, G.S. and Avedisian, C.T. 1990 "Combustion of Unsupported Methanol/Dodecanol Mixture Droplets at Low Gravity," *23rd Symp. (Int.) Comb.*, 1619.

HTD-Vol. 197, Two-Phase Flow and Heat Transfer
ASME 1992

ONSET OF DRYOUT AND POST-DRYOUT HEAT TRANSFER IN ENHANCED PASSAGE GEOMETRIES FOR COMPACT EVAPORATORS

K. Shollenberger, Van P. Carey, and P. Tervo
Department of Mechanical Engineering
University of California, Berkeley
Berkeley, California

ABSTRACT

Recent studies have demonstrated that enhanced ribbed and finned surfaces used in compact automotive A/C evaporators can exhibit dryout behavior in the latter stages of the vaporization process that is distinctly different from that observed in round tubes. In particular, because of the spatial non-uniformity of the flow field, dryout occurs at first preferentially at a few locations on the passage walls. Dry regions generally get larger with increasing downstream distance until the walls are entirely dry. In this paper, the results of recent experimental studies are summarized which document the dryout characteristics of two idealized enhanced surfaces: a cross-ribbed passage and a large-scale offset fin surface. R-113 was used for the working fluid over the following ranges of flow conditions: $1 \leq p \leq 1.5$ atm, $30 \leq G \leq 115$ kg / m^2•s, $10 \leq q'' \leq 89$ kW / m^2. Data for these two passage types is presented from which the conditions necessary for the onset of dryout are inferred and heat transfer coefficients are correlated for vertical flow boiling.

NOMENCLATURE

A	coefficient in equation
A_o	cross sectional open area of channel
A_p	prime surface area of channel of length L_c
B	cross-stream fin spacing
Bo	Boiling number, $q''/G\,h_{\ell v}$
$c_{p\ell}$	specific heat of liquid at constant pressure
d_h	hydraulic diameter based on wetted perimeter, $4A_o/P_W$
d_{hp}	hydraulic diameter based on heated perimeter, $4A_o/P_H$
F	convective boiling parameter
G	mass flux
h	heat transfer coefficient
$h_{\ell v}$	latent heat of vaporization of coolant
H	dimension of fins or ribs from root to tip
k_ℓ	thermal conductivity of liquid
k_c	thermal conductivity of copper
L	length of fin or spacing between ribs
L_c	length of channel section
m	constant in equation (11)
$\dot{m}$	mass flow rate in passage
n	exponent in equation (7)
Nu	Nusselt number, $h\,d_{hp}/k$
p	pressure
P_H	heated perimeter
P_W	wetted perimeter
Pr	Prandtl number, $\mu\,c_p/k$
q''	surface heat flux
q_c	surface heat transfer rate
q_p	preheater heat transfer rate
r	constant in equation (19)
Re	Reynolds number, $G\,d_h/\mu_{core}$
t	fin or rib thickness
T_i	inlet temperature to test section
T_{pi}	inlet temperature to preheater
T_W	wall temperature of prime surface of channel
T_{sat}	saturation temperature of coolant
∇T	measured local temperature gradient in copper slab
W_c	width of copper slab
We	Weber number defined by equation (12)
x	mass quality
x_i	mass quality at inlet of test section
X_{tt}	Martinelli parameter for turbulent-turbulent flow

Greek symbols

δ	film thickness
η	fin efficiency of fin or rib
μ	absolute viscosity
ρ	density
σ	surface tension of liquid film
τ_i	interfacial shear stress acting on liquid film

Subscripts

core	corresponding to core flow conditions
do	corresponding to dry out conditions
ℓ	liquid properties
v	vapor properties
tp	corresponding to two phase flow conditions

INTRODUCTION

Often in heat exchanger design, enhanced complex channel geometries provide more efficient means for boiling and condensation. Enhanced surfaces, such as offset strip fins and cross-ribbed channels have been used in industrial applications for some time. Chemical, food processing, aerospace, and automotive industries have successfully employed cross-ribbed channels. Offset strip fins have been frequently used in refrigeration, air conditioning, and cryogenic systems. The advantage of these geometries over ordinary round tubes is two-fold. First, since the fins or ribs are intermittent, the thermal boundary layer and/or the liquid film on the surface is broken up and remains thin, thus providing less resistance to heat transfer. Second, the flow patterns caused by the fins and ribs induce turbulence and mixing. The combination of these two effects greatly increases the heat transfer rate which can be ten times or more than that for a conventional shell and tube heat exchanger of equal size. One drawback of the improved heat transfer performance is, of course, that the pressure drop can be significantly higher than in a conventional heat exchanger.

In the design of compact evaporators, precise prediction of the heat transfer rate requires accurate knowledge of the flow regimes present throughout the heat exchanger. Different mechanisms of heat transfer dominate for each flow regime. For low quality two-phase flow in the bubbly or slug flow regimes, both nucleate boiling and forced convection are important. On the other hand, for high quality two-phase flow when the flow is usually annular in configuration (i.e., a thin liquid film flowing on the heat exchanger channel walls with a vapor core), evaporation of the film is the dominate heat transfer mechanism.

In automotive evaporators, the inlet quality is typically 20 - 30% and the inlet void fraction is typically greater than 80%. Consequently, the flow immediately takes on an annular configuration with a liquid film on the wall of the passage, and a core flow that is mostly vapor with some entrained liquid droplets. Vaporization of the coolant occurs primarily as evaporation of the thin liquid film on the passage walls. Nucleate boiling effects, if present, are usually weak in such flows.

For annular flow, as quality increases, the liquid film on the heat transfer surface becomes thinner. As the vaporization process continues, dry spots eventually appear on the surface. The size of the dry areas increases with downstream distance until the liquid film on the channel walls drys out completely. This progressive dryout is very important in the design of evaporators used in automotive and aircraft air conditioning systems because the heat transfer rate drops continuously after the onset of dryout. The ability to predict the onset of dryout conditions and the decrease in heat transfer in the partial dryout regime is required to optimize the design and size of the evaporator.

Dryout for simple round tube passages has been studied by a number of previous researchers (see, for example, Barnard et al, 1974, Shah , 1979, and Sthapak et al, 1976). The dryout characteristics of evaporating round tube flows have been extensively documented in the open literature. For finned or cross-ribbed passage geometries, the mechanisms of dryout are not well understood, and techniques to predict the onset of dryout are not yet available. The main objectives of this investigation were to obtain detailed experimental heat transfer data for two typical enhanced heat transfer surfaces at quality levels spanning the pre-dryout and post-dryout regimes, and to use the data to develop a semi-empirical technique to predict the conditions under which dryout occurs. The present study specifically examined the role of parameters important in dryout, e.g. mass flux, heat flux, and fluid properties, and the development of an analytical model of dryout. The main focus was to investigate the hydrodynamic effects on dryout. Previous researchers such as Bergles (1979), Kitto (1980), and Levitan and Lantsman (1975) have found that at low heat flux levels, heat flux effects are generally negligible compared to hydrodynamic effects. This issue is examined in more detail in a later section.

Our investigation focused on the dryout characteristics of two specific passage geometries: one being an offset fin surface and the other a cross-ribbed passage. In this study we sought to explore the different effects finned and cross-ribbed geometries have on dryout and to visually observe the flow regimes and onset of dryout in both of these geometries. Heat transfer performance in the post-dryout region was also investigated.

EXPERIMENTAL APPARATUS

The experimental portion of this study was carried out by measuring two phase heat transfer performance for the large-scale offset fin and cross-ribbed geometries shown in Figures 1 and 2. The dimensions for these geometries are given in Table 1. The finned geometry is the same as surface #2 used in earlier studies by Mandrusiak and Carey (1989) and the cross-ribbed geometry is the same as geometry #3 used by Cohen and Carey (1989). Pure copper blocks were machined to form the heat transfer surfaces in the test sections shown in Figures 1 and 2. For the finned geometry, the fins were machined parallel to the flow direction. For the cross-ribbed geometry, the ribs were machined at a 60° angle to the flow direction. In addition, for the cross-ribbed geometry, a polycarbonate insert having ribs of the same dimensions and spacing

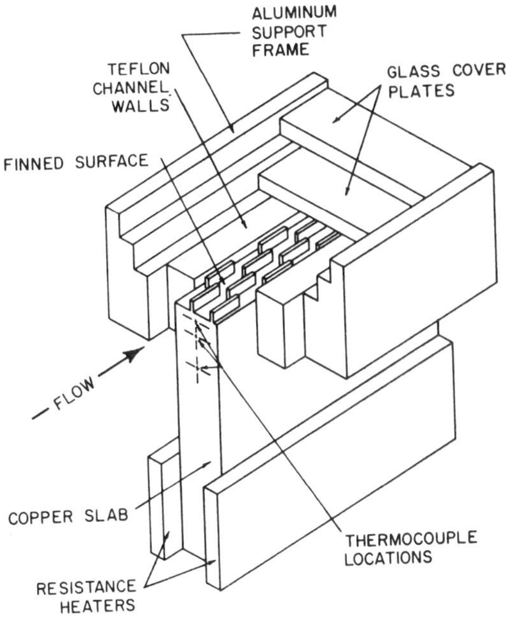

Figure 1. Cutaway view of offset fin test section.

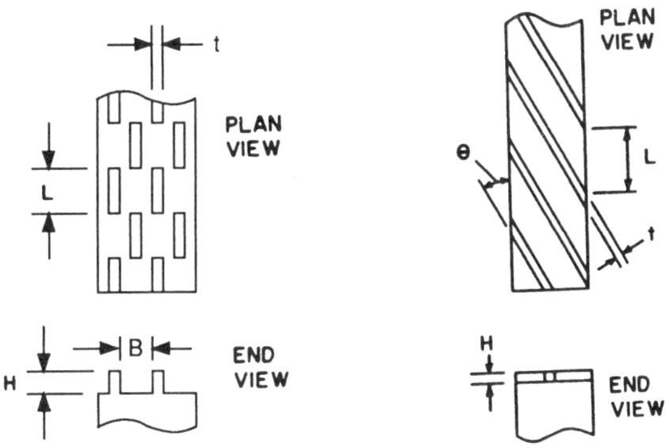

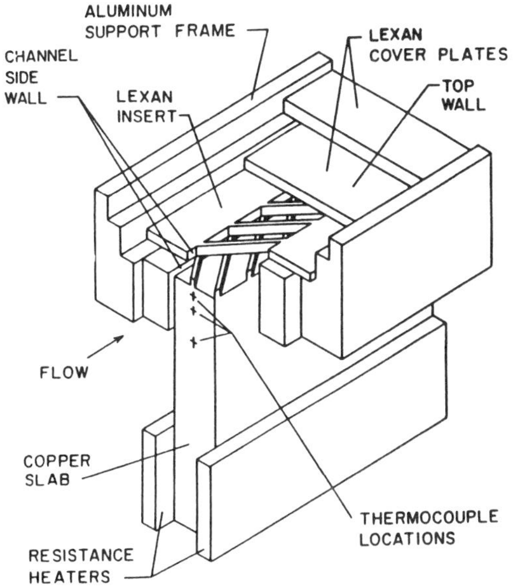

Figure 2. Cutaway view of cross-ribbed test section.

Dimension (mm)	Finned Geometry	Crossribbed Geometry
t	1.91	1.6
H	9.52	3.18
L	12.7	17.1
B	8.26	- - -
q	- - -	60
d_h	8.84	6.83
d_{hp}	11.52	16.7
W_c	25.4	19.05
L_H	762	457
A_o (mm^2)	196.6	108

Table 1. Dimensions of Geometries Tested

was installed in the test section above the copper slab as shown in Figure 2 to form a channel with cross-ribbed walls. The side walls of the flow channel for both geometries were Teflon strips and the top wall was transparent polycarbonate sheet. These materials provided good insulating properties and the clear polycarbonate coverplate allowed direct visual observation of the two phase flow.

Two electrical resistance heaters clamped to the sides of the copper slab provided a virtually uniform heat flux along the length of the test section. Heat was conducted from the heaters through the copper slab to the finned or ribbed surface where it was transferred to the test fluid flowing in the channel. The heat flux was controlled by a rheostat which allowed the voltage input to the heaters to be varied.

The temperature gradient in the copper normal to the axis of the flow passage was measured using three thermocouples imbedded in the copper slab at fixed axial locations along the test section, as illustrated in Figures 1 and 2. From the measured temperature gradient, the surface temperature was determined by linear extrapolation and the heat flux calculated from conduction heat transfer analysis. A fourth thermocouple was installed through the channel wall to measure the local fluid temperature. This local fluid temperature was combined with the local heat flux and surface temperature to compute the local heat transfer coefficient. All temperatures were measured using 30 gauge copper-constantan thermocouples connected to a precision Fluke digital readout through an Omega double-pole selector switch.

The test section was installed in the system illustrated in Figure 3 which provided a steady flow of the test fluid, R-113, through the test section. Liquid flow from the reservoir to the preheater was provided by a centrifugal pump. The flow rate was controlled by the flow control valve and also by varying the voltage supplied to the pump. The flow rate of pure liquid was measured using a Cole-Parmer rotameter which was calibrated for R-113. R-113 was chosen as the test fluid because its fluid properties are similar to R-12 and R-134a, and its normal boiling point of 48 °C made it possible to conduct the experiments at low pressure.

The preheater in the test system vaporized some of the

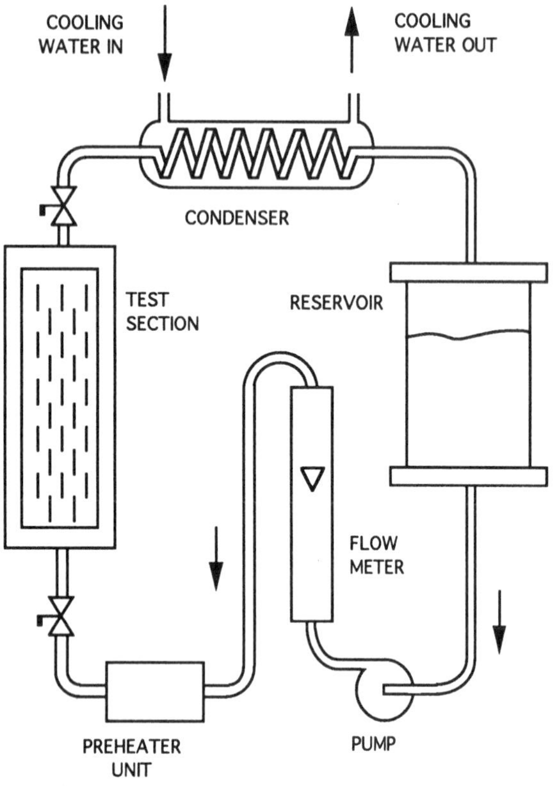

FIGURE 3. Schematic of flow supply loop

refrigerant using an internal electrical resistance heater thereby providing a two phase flow at the inlet of the test section. The preheater was controlled by using a rheostat to vary the voltage supplied to the heater. This allowed direct control of the inlet quality to the test section. The power input to the preheater was calculated from the measured voltage and resistance of the heater. After the refrigerant passed through the test section, it was cooled and condensed back to subcooled liquid in a water cooled condenser and then returned to the reservoir. As Figure 3 illustrates, the test system operates as a continuous closed loop.

EXPERIMENTAL PROCEDURE AND RESULTS

For the experiments done with both finned geometries, the flow passage was in a vertical orientation with upward flow. For all of the experiments, a series of heat transfer measurements were performed for pressures in the range 1 atm $< p <$ 1.5 atm and for mass flux values in the range $30 < G < 115$ kg / m$^2\cdot$ s. For each mass flux setting, the power input to the preheater was adjusted to provide a steady two phase flow to the inlet of the test section. The inlet qualities were typically in the range $0.1 < x < 0.3$. The heat flux to the copper slab was then set so most of the test section was in annular flow with no visible dry spots. The system was then allowed to stabilize for about 10 minutes before measuring the flow rate and temperatures. The system was assumed stable when temperature fluctuations in the copper slab approached zero. This procedure was then repeated keeping the flow rate constant but increasing the heat flux for each run until the temperatures in the copper slab began increasing rapidly. This indicated that the heat transfer coefficient had dropped significantly and critical heat flux condition (CHF) had been reached. The experiments were performed for heat flux values in the range $10 < q'' < 89$ kW / m^2. The range of Boiling numbers (Bo) for this range of mass flux and heat flux values is $.002 < \text{Bo} < 0.008$.

The local flow conditions and heat transfer coefficient were determined at each thermocouple bank location using the methodology employed in the previous investigations of Carey, et al. (1986) and Xu, et al. (1987) who used similar test sections. The inlet quality to the test section was determined from an energy balance on the preheater

$$\dot{m}\, c_{p\ell}\, (T_i - T_{pi}) + x_i\, \dot{m}\, h_{\ell v} = q_p \tag{1}$$

Using the inlet quality x_i determined from the above relation, the quality at each thermocouple bank was determined from an energy balance on the portion of the test section between the inlet and each thermocouple bank

$$(x - x_i)\, \dot{m}\, h_{\ell v} = q_c \tag{2}$$

Following the method used by Carey, et al. (1986) and Xu, et al. (1987) the local two phase heat transfer coefficients were obtained by iteratively solving the energy balance relation

$$W_c\, L_c\, k_c\, \nabla T_c = h_{tp}\, (A_p + \eta\, A_F)\, (T_w - T_{sat}) \tag{3}$$

where, for the finned surface, $\eta = \eta_F$ is the fin efficiency determined by Carey, et al. (1986)

$$\eta_F = \frac{\tanh(MH)}{MH} \qquad M = \sqrt{2h_{tp}(t+L)/k_c tL} \tag{4}$$

For the cross-ribbed surface, the fin efficiency $\eta = \eta_R$ was shown by Xu, et al. (1987) to be given by

$$\eta_R = \frac{1}{M(H+t/2)}\left[\frac{\tanh(MH)+h/k_c M}{1+(h/k_c M)\tanh(MH)}\right] \tag{5a}$$

where
$$M = \sqrt{2h_{tp}/k_c t} \tag{5b}$$

These heat transfer coefficients are the average values over the heated perimeter of the channel and are the values of h_{tp} which correspond to specific values of x and G.

The estimated uncertainty in the experimentally measured values are $\pm 5\%$ for G, $\pm 8\%$ for q'', $\pm 2\%$ for x_i, $\pm 10\%$ for x, and $\pm 13\%$ for h_{tp}. These errors in the measured data would tend to be mostly random; therefore, on average, the errors would be smaller than the values given above.

From the experimental investigation, local heat transfer coefficients were determined for the finned and cross-ribbed

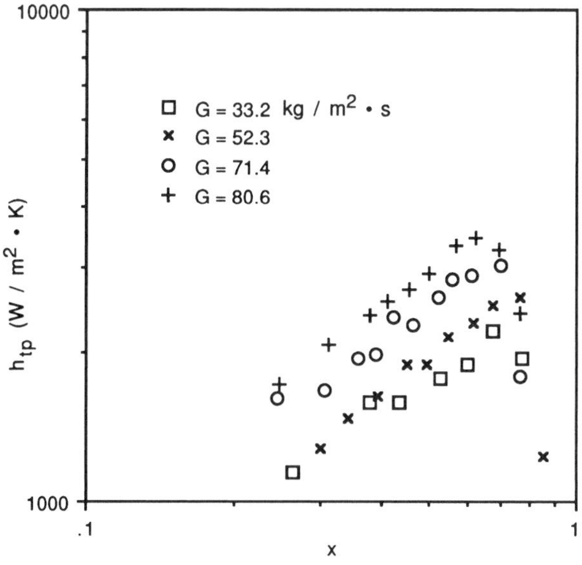

Figure 4. Measured local heat transfer coefficients for vertical flow in the finned geometry

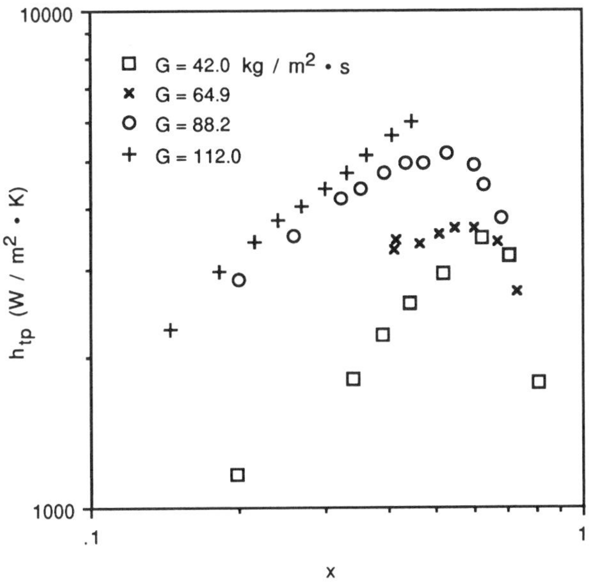

Figure 5. Measured local heat transfer coefficients for vertical flow in the crossribbed geometry.

geometries at various combinations of mass flux and quality. Representative results of this type are shown in Figures 4 and 5. For all but the highest mass flux values, it can be seen that the local heat transfer coefficient h_{tp} increases with quality at low quality, peaking just before the onset of partial dryout. For qualities greater than the onset condition, h_{tp} decreases with increasing quality.

For the data points up to the dryout condition, the flow pattern visually appeared to be annular with a significant amount of liquid entrainment. As the dryout condition was approached, dry spots were observed to appear, and as the quality increased beyond the dryout condition, the surface appeared to be progressively drier. For all of the experiments, bubble nucleation appeared to be completely suppressed. As these figures illustrate, the pre-dryout heat transfer coefficient (h_{tp}) for both surfaces increases with mass flux in a manner similar to that predicted by some correlations for round tubes (Benneet, 1980 and Bjorge, 1982). As these graphs indicate, the peak h_{tp} values were found in the quality range 55% < x < 76% and the exact value of x where the peak occurred varied only slightly with mass flux. In the range 30% < x < 55%, h_{tp} steadily increases with quality even though intermittent small dry spots were frequently observed. These observations indicate that the first appearance of dry spots does not produce an immediate drop in h_{tp}; rather, h_{tp} begins to drop only after dry patches appear on a significant portion of the surface. Qualitatively similar observations were made in the study by Carey and Mandrusiak (1986) for a comparable finned channel using water, methanol, and n-butanol as test fluids. The results in Figures 4 and 5 also indicate that, for a given mass flux, the heat transfer coefficients for the cross-ribbed geometry are 1.5 to 2 times higher than those for the finned geometry. This indicates that the cross-ribbed geometry is more effective from a heat transfer standpoint.

Although observations clearly indicate that the passage walls begin to exhibit intermittent dry patches at qualities lower than that at which h_{tp} is a maximum, we will identify the onset of dryout quality x_{do} with the peak of h_{tp}. For the low boiling number in these experiments 0.002 < Bo < 0.008, the level of heat flux is not expected to play a strong role in determining the dryout condition. Increasing the mass flux is known to increase entrainment in two phase flow in round tubes (see Givan, 1988) for the flow conditions in our experiments. This suggests that varying the mass flow may have a strong effect on the onset of dryout conditions. As indicated in Figures 4 and 5, however, the onset of dryout quality appears to be only weakly affected by varying the mass flux.

At first glance, altering the passage geometry might also be expected to affect the onset of dryout. In complex passages, the complicated flow in the core results in complex entrainment and deposition effects and highly non-uniform cross-sectional variation of the interfacial shear. Altering the geometry could significantly alter one or more of these factors. On the other hand, the data for the two surfaces considered here differs only slightly in the range of dryout qualities observed for comparable mass flux values. A model analysis of the dryout mechanisms is presented in the next section to further explore parametric effects on the onset of dryout condition.

ANALYSIS OF DRYOUT BEHAVIOR

As a first step in exploring the dryout behavior, we consider the pre-dryout heat transfer characteristics of the surfaces. For $x < x_{do}$, we found that the h_{tp} data were well correlated in terms of the dimensionless variables X_{tt} and F defined as

$$X_{tt} = \left(\frac{1-x}{x}\right)^{1-n/2}\left(\frac{\rho_\ell}{\rho_v}\right)^{0.5}\left(\frac{\mu_v}{\mu_\ell}\right)^{n/2} \tag{6a}$$

$$F(X_{tt}) = \left(h_{tp}/h_\ell\right)Pr_\ell^{-0.296} \tag{6b}$$

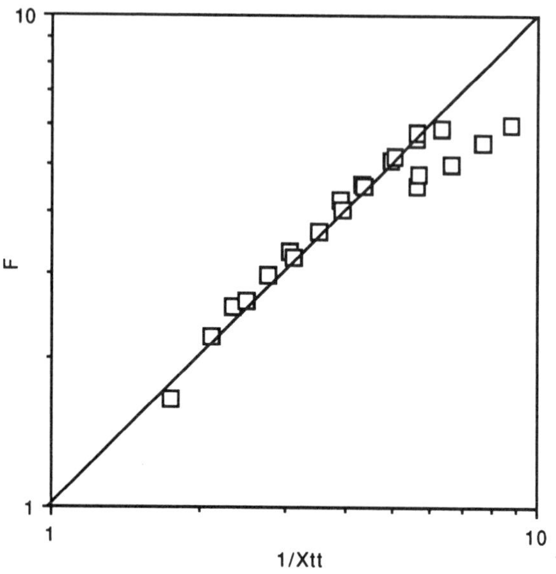

Figure 6. Pre-dryout heat transfer data for the cross-ribbed geometry correlated in terms of F and Xtt

$$\frac{h_\ell}{G(1-x)\,c_{p\ell}} = A\left[\frac{G(1-x)D}{\mu_\ell}\right]^{-n} Pr_\ell^{-2/3} \tag{7}$$

where (A,n) = (0.141, 0.30) for the offset fin geometry and (0.341, 0.30) for the cross-ribbed passage. The data for the cross-ribbed geometry are plotted in terms of F and X_{tt} in Figure 6. Correlations fit to the data for each geometry are

$$F = \left[1 + \frac{1.70}{X_{tt}^2}\right]^{0.450} \quad \text{for the finned surface} \tag{8}$$

$$F = \left[1 + \frac{2.30}{X_{tt}^2}\right]^{0.374} \quad \text{for the cross-ribbed surface} \tag{9}$$

To analyze the conditions that result in the onset of dryout, we will make several simplifying idealizations. The first is that at dryout the film is very thin in comparison to the channel dimensions and that we can approximate the film thickness for annular flow by

$$\delta \cong \frac{k_\ell}{h_{tp}} \tag{10}$$

The resulting values of film thickness calculated at dryout using the above correlations were in the range $10 < \delta < 37$ μm which, if compared to the dimensions given in Table 1, are much less than the channel dimensions.

The heat transfer data in Figures 4 and 5 indicate that in the pre-dryout range $x < x_{do}$, the heat transfer coefficient h_{tp} is a power-law function of quality

$$h_{tp} = \beta\, x^m \tag{11}$$

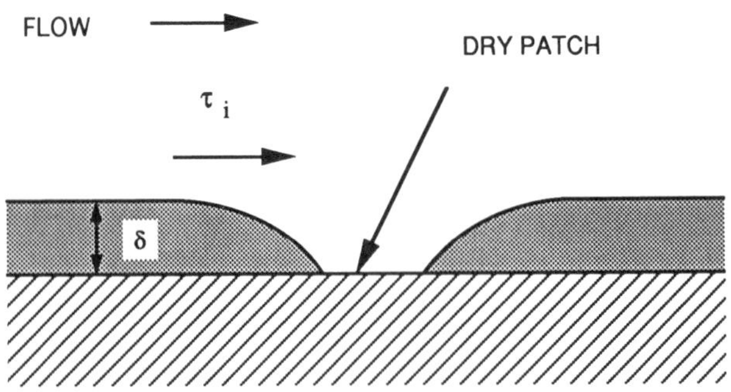

Figure 7. Liquid film at onset of dryout

with the value of the exponent m ranging from about 0.5 to 0.8.

The circumstances at the surface when a dry patch first appears is shown schematically in Figure 7. Note that liquid covers portions of the surface adjacent to the dry patch. Interfacial shear on the film surface due to turbulent Reynolds stresses and/or the downstream motion of the core flow tends to act to rewet the dry spot with liquid. Interfacial tension forces due to the radius of curvature near the contact line of the liquid front tend to resist the re-wetting effect. Postulating that the interfacial shear stresses are proportional to the dynamic pressure of the core flow, the ratio of interfacial to surface tension forces is quantified as a Weber number we defined as

$$We = \frac{G_{core}^2 / \rho_{core}}{\sigma / \delta} \tag{12a}$$

Taking $G_{core} = G$ because very little of the total mass flow is in the film on the walls and approximating

$$\rho_{core} = \rho_\ell(1-x) + \rho_v x \cong \rho_\ell(1-x) \tag{12b}$$

and $$\delta \cong k_\ell / h_{tp} = k_\ell / \left(\beta\, x^m\right) \tag{12c}$$

the above relation can be written as

$$We = \frac{G^2 k_\ell}{\sigma \rho_\ell (1-x) \beta x^m} \tag{13}$$

The variation of We with quality indicated by the above relation exhibits a minimum at a specific x value. Differentiating with respect to x and setting the derivative equal to zero, it can be shown that the minimum value of x is given by

$$x_{min} = \frac{m}{1+m} \tag{14}$$

We expect that the tendency for dry patches to rewet will be the weakest, and hence dryout is likely to become progressively worse, when the above defined Weber number is at its lowest value. We therefore postulate that the onset of dryout will begin at values of quality just beyond the minimum value given by equation (14). We

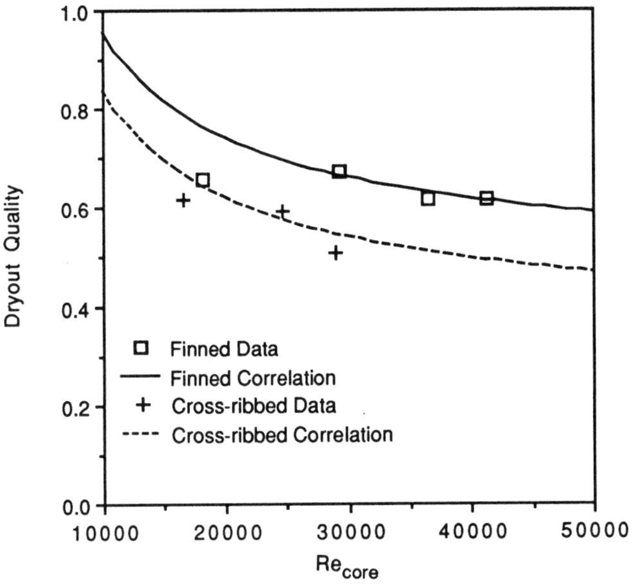

Figure 8. Comparison of onset of dryout data with the prediction of equation (18)

further expect that since entrainment increases with increasing mass flux, that the dryout quality will decrease weakly as the mass flux increases. We therefore postulate the following relation for the dryout quality

$$x_{do} = x_{min} + f(Re_{core}) \tag{15}$$

where

$$Re_{core} = Gd_h / \mu_{core} \tag{16}$$

$$\mu_{core} = \left[\frac{x}{\mu_v} + \frac{1-x}{\mu_\ell}\right]^{-1} \tag{17}$$

and where f (Re_{core}) decreases as Re_{core} increases.

Since the power law exponent m was approximately 0.8 for the finned geometry and 0.5 for the cross-ribbed geometry the corresponding values of x_{min} were 0.45 and 0.33. The function f (Re_{core}) was determined as a best fit to the peak values of h_{tp} in the experimentally determined h_{tp} versus x curves. The resulting relation for x_{do} is

$$x_{do} = x_{min} + 800\, Re_{core}^{-0.80} \tag{18}$$

This relation is compared to the experimentally determined onset of dryout qualities for both surfaces in Figure 8. The experimentally determined onset of dryout data correspond to the peaks in the h_{tp} versus x curves. It can be seen that the agreement with the data for both surfaces is fairly good. The proposed relation (18) provides a means of predicting onset of dryout quality that is reasonably accurate for either of these geometries.

POST-DRYOUT HEAT TRANSFER

When the dryout condition is reached, the heat transfer coefficient begins to drop sharply and approaches the value for a single phase vapor; this value can be less than one tenth the value before dryout has been reached. Since some compact evaporators operate in the post-dryout region, it is important to be able to predict heat transfer coefficients in this region to allow engineers to accurately design compact heat exchangers. Figures 9 and 10 illustrate the behavior of the non-dimensionalized heat transfer coefficient in the post-dryout region versus a normalized quality for the finned and cross-ribbed geometries. As can be seen, there is considerable scatter in the data. However,this is expected since temperatures begin to fluctuate and increase sharply when dryout is reached.

A dispersed turbulent flow of droplets and vapor under saturated conditions exists in the post-dryout region of these experiments . It can be assumed that the complex geometry of the passages that induced this turbulent flow also insures that the heat transfer between the vapor and droplets is sufficiently fast to insure thermodynamic equilibrium. To develop a method which can be used to predict heat transfer performance for this post-dryout region, we begin with a typical single phase relation for turbulent flow in round tubes:

$$Nu_v = A\, Re_v^{1-n}\, Pr_v^{1/3} \tag{19a}$$

where

$$Re_v = \frac{G\, d_h}{\mu_v} \tag{19b}$$

In equation (19), A and n are constants which depend on the geometry of the flow passage and were given for the finned and cross-ribbed geometry in the Analysis of Dryout Behavior section. We now assume that a similar relation can be written for the post-dryout region with the same constants but with fluid properties which are appropriately defined based on an average of vapor and liquid properties:

$$Nu_{tp} = A\, Re_{tp}^{1-n}\, Pr_{tp}^{1/3} \tag{20a}$$

where

$$Re_{tp} = \frac{G\, d_h}{\mu_{tp}} \tag{20b}$$

Dividing equation (20) by equation (19) we obtain:

$$\frac{Nu_{tp}}{Nu_v} = \left(\frac{Re_{tp}}{Re_v}\right)^{1-n}\left(\frac{Pr_{tp}}{Pr_v}\right)^{1/3} = \left(\frac{\mu_v}{\mu_{tp}}\right)^{1-n}\left(\frac{Pr_{tp}}{Pr_v}\right)^{1/3} \tag{21}$$

This can be written in terms of heat transfer coefficients as follows:

$$\frac{h_{tp}}{h_v} = \left(\frac{\mu_v}{\mu_{tp}}\right)^{1-n}\left(\frac{Pr_{tp}}{Pr_v}\right)^{1/3}\left(\frac{k_{tp}}{k_v}\right) \tag{22a}$$

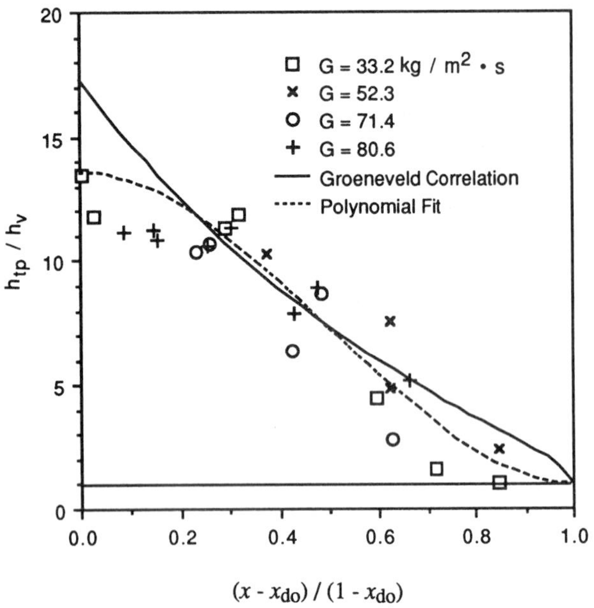

Figure 9. Correlation of post-dryout data for vertical flow in the finned geometry.

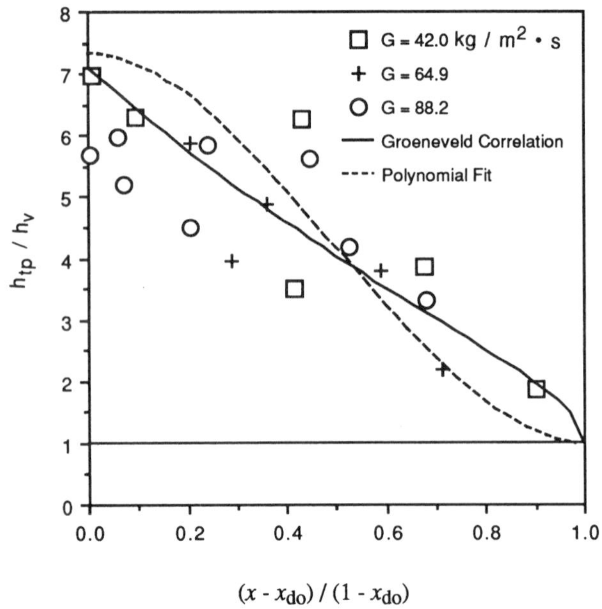

Figure 10. Correlation of post-dryout data for vertical flow in the cross-ribbed geometry.

$$\frac{h_{tp}}{h_v} = f\left(x, \frac{\rho_v}{\rho_{tp}}, \frac{\mu_v}{\mu_{tp}}, \frac{Pr_{tp}}{Pr_v}, \frac{k_{tp}}{k_v}\right) \tag{22b}$$

Thus, the ratio between the two-phase and the pure vapor heat transfer coefficients can hypothetically be correlated in terms of vapor and liquid fluid property ratios and quality. This result appears to be verified by the lack of dependance of h_{tp} / h_v on G in Figures 9 and 10.

One form of f(x) found in the literature that can be extracted from the correlation developed by Groeneveld (1973) for heat transfer in the dispersed flow regime is:

$$f(x) = \left[x + \frac{\rho_v}{\rho_\ell}(1-x)\right]^{1-n} Y^d \tag{23a}$$

$$\text{where} \quad Y = \left[1 - 0.1\left(\frac{\rho_\ell}{\rho_v} - 1\right)^{0.4} (1-x)^{0.4}\right] \tag{23b}$$

In the past, this correlation has been used to fit data for water in vertical round tube and annuli experiments over the following ranges: $0.25 < D < 2.5$ cm, $68 < P < 215$ bar, $700 < G < 5300$ kg•m^{-2}•s^{-1}, $120 < q'' < 2100$ kW•m^{-2} with RMS error of 11.5%. While the dimensions of the passages are comparable, the ranges for P, G and q" are much higher than those for this experiment. If we force the slope of this relation to be zero at x_{do} so that there is a smooth transition between the pre and post-dryout regions, we find that the magnitude of h_{tp}/h_v predicted by f(x) is 5 to 10 times lower than that found experimentally. If we relax the above condition on the slope, we can obtain a correlation with 22.6% and 21.3% RMS error for the finned (d = -3.83) and cross-ribbed (d= -2.78) geometries. These curves are shown in Figures 9 and 10 and indicate that the shape of f(x) predicted by equation (23) is not optimal.

As an alternative to the above correlation, a more generic form of f(x) was developed from the following polynomial:

$$f(x) = C_1\chi^3 + C_2\chi^2 + C_3\chi + C_4 \quad \text{where} \quad \chi = \frac{(x - x_{do})}{(1 - x_{do})} \tag{24}$$

where we impose the following boundary conditions:

$$f(\chi = 1) = 1 \quad \text{and} \quad \left.\frac{df}{d\chi}\right|_{\chi=0} = \left.\frac{df}{d\chi}\right|_{\chi=1} = 0 \tag{25}$$

This results in the following correlation:

$$f(x) = C\left(\chi - \frac{3}{2}\right)\chi^2 + \frac{C}{2} + 1 \tag{27}$$

A best fit to the experimental data resulted in a 23.8% and 26.6% RMS error for the finned (C = 25.15) and cross-ribbed (C = 12.65) geometries. Thus, this correlation shows that although the trend in the data appears to be represented by the above correlation, it is difficult to find a better fit when there is this much scatter in the data. Either equation (23) or (27) can therfore be used to estimate post-dryout heat transfer coefficients for these geometries.

As a final test of all the relations developed, they were combined to predict the heat transfer coefficient over the entire range of quality. The pre-dryout correlations alone were used for $x < x_{do}$. Above x_{do}, the dryout correlation (18) was used to predict x_{do}. Then h_{do} was determined using the pre-dryout correlation evaluated at x_{do} and h_v was determined using the appropriate single

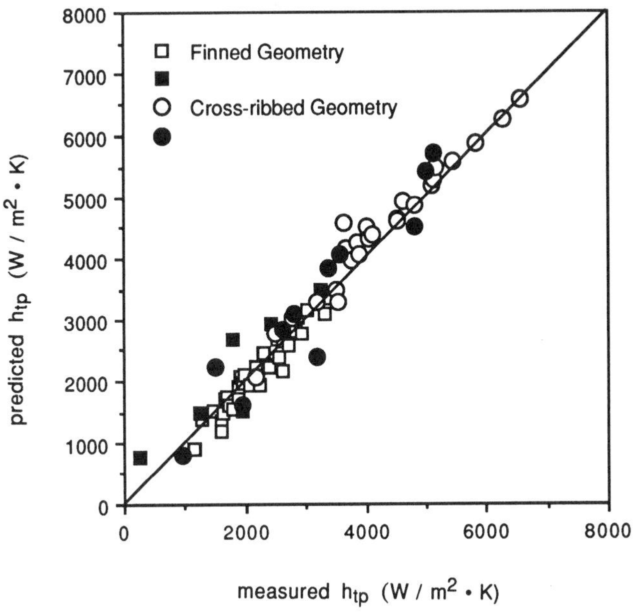

Figure 11. Comparison of heat transfer coefficient data.
(Open symbols denote predryout, closed are for post dryout)

phase correlation for the surface assuming pure saturated vapor flow. Then equation (27) was used to predict h_{tp} for the post-dryout region. The resulting predicted h_{tp} values are compared to the corresponding measured values in Figure 11. It can be seen that the agreement for the region before dyout is good, as expected, and that the amount of scatter for the post-dryout region is reasonable in comparison.

CONCLUDING REMARKS

The good agreement between the measured onset of dryout data for the finned and ribbed surface supports the contention that the postulated balance between surface tension forces and interfacial shear forces play a major role in determining the onset of dryout conditions in complex finned and ribbed evaporator passages. The correlating equation obtained from the analysis developed from this reasoning shows promise as a potential means of predicting the onset of dryout in enhanced evaporator passages of this type.

The scheme for correlating post-dryout heat transfer also shows promise as a means of predicting heat transfer in this regime for enhanced evaporator surfaces. There is, however, a large amount of scatter in the data in this regime, and further evaluation of this correlation method against experimental data is needed before this approach can be used with confidence as a design tool for evaporator development. A further extension of this correlation technique could be to use an existing relation for predicting two-phase properties in equation (22).

ACKNOWLEDGEMENTS

The authors gratefully acknowledge support for this research by the National Science Foundation under grant No. CTS-9024862.

REFERENCES

Barnard, D.A., Dell, F.R., and Stinchcombe, R.A. "R.S. 100: Dryout at Low Mass Velocities for an Upward Boiling Flow of Refrigerant-113 in a Vertical Tube", AERE Report AERE-R 7726, 1974.

Benneet, D.L. and Chen, J.C., "Forced Convective Boiling in Vertical Tubes for satuarted Pure Components and Binary Mixtures," *AIChE Journal*, **26**, pp. 454-461, 1980.

Bergles, A.E. "Burnout in Boiling Heat Transfer. Part III: High-Quality Forced-Convection Systems", *Nuclear Safety*, **20**(6), pp. 671-689, 1979.

Bjorge, R.W., Hall. G.R. and Rohsenow, W.M., "Correlation of Forced Convection Boiling Heat Transfer Data," *Int. J. Heat Mass Transfer*, **25**, pp. 753-757, 1982.

Carey,V.P. and Mandrusiak, G.D. "Annular Film-Flow Boiling of Liquids in a Partially Heated, Vertical Channel with Offset Strip Fins", *Int. J. Heat Mass Transfer*, **29**, pp. 927-939, 1986.

Cohen, M.R. and Carey, V.P., "A Comparison of the Flow Boiling Performance Characteristics of Partially-Heated Cross-Ribbed Channels with Different Rib Geometries", submitted to *Int. J. Heat Mass Transfer*, **32**, pp. 2459-2474,1989.

Givan, A.H., Hewitt, G.F., Owen, D.G. and Bott, T.R., "An Improved CHF Modelling Code," *Proc. Second U.K. Conf. on Heat Transfer*, Inst. of Mechanical ENgineers, London, **1**, pp. 33-48, 1988.

Groeneveld, D. C., "Post-Dryout Heat Transfer at Reactor Operating Conditions," Report AECL-4513, 1973.

Kitto, J.B. "Critical Heat Flux and the Limiting Quality Phenomenon" *AIChE Symposium Series*, **76**(199), pp. 57-78, 1980.

Levitan, L.L. and Lantsman, F.P., "Investigating Burnout with Flow of Steam-Water Mixture in a Round Tube", *Ther. Eng. (USSR) (Engl. Transl.)*, **22**(1), pp. 102-105, 1975.

Mandrusiak, G.D. and Carey, V.P. "Convective Boiling in Vertical Channels with Different Offset Strip Fin Geometries", *ASME Journal of Heat Transfer*, **111**, pp. 156-165, 1989.

Shah, M.M., "A Generalized Method for Predicting CHF in Uniformly Heated Vertical Tubes", *Int. J. Heat Mass Transfer*, **22**, pp. 557-568, 1979.

Sthapak, B.K., Varma, H.K., Gupta, C.P., "Heat Transfer Coefficients in Dry-Out Region of Horizontal Tube Water Heated R-12 Evaporator", *ASHRAE Trans.*, **82**, 47-55, 1976.

Xu, X., and Carey, V.P., "Heat Transfer and Two-Phase Flow During Convective Boiling in a Partially-Heated Cross-Ribbed Channel", *Int. J. Heat Mass Transfer*, **30**, pp. 2385-2397, 1987.

HTD-Vol. 197, Two-Phase Flow and Heat Transfer
ASME 1992

PREDICTION OF THE ONSET OF SIGNIFICANT VOID IN DOWNFLOW SUBCOOLED NUCLEATE BOILING

S. C. Lee
Department of Mechanical Engineering
Yeungnam University
Gyongsan, Korea

S. G. Bankoff
Department of Chemical Engineering
Northwestern University
Evanston, Illinois

ABSTRACT

A model to predict the onset of significant void (OSV) in vertical flow between parallel plates has been developed, which is based upon the influence on vapor bubble departure of the single-phase temperature profile. The model was compared to the experimental data of Whittle and Forgan (1967) and Dougherty et al. (1990), showing excellent agreement. The model was also compared with the Saha-Zuber (1974) correlation, which has been widely used in computer codes for nuclear safety analysis. The present theory is more conservative than this correlation, and further shows that, contrary to this correlation, the Stanton number is not solely related to the Peclet number. This may explain the large error margins required for the Saha-Zuber correlation, and also the scatter beyond the error margins specified by the authors.

KEYWORDS: Onset of Flow Instability, Onset of Significant Void, Forced-Convection Subcooled Boiling, Downflow, Parallel Plates

NOMENCLATURE

A	parameter defined by Equation (15)
A_c	cross-sectional area (m^2)
A_s	surface area (m^2)
A^+	damping constant
C	coefficient
C_p	specific heat (J/kgK)
D	diameter (m)
D_h	hydraulic diameter (m)
F	force (N)
G	mass velocity (kg/m^2s)
g	acceleration due to gravity (m/s^2)
H	distance between the wall and the centerline (m)
K	dimensionless parameter
k	thermal conductivity (W/mK)
L	channel length (m)
Nu	Nusselt number
p	pressure (Pa)
Pe	Peclet number
Pr	Prandtl number
Pr_t	turbulent Prandtl number
q	heat flux (W/m^2)
R	radius (m)
Re	Reynolds number
St	Stanton number
T	temperature (K)
t	time (s)
T_d	bulk temperature at OSV (K)
T_i	inlet temperature (K)
T_o	outlet temperature (K)
u	velocity in x direction (m/s)
u_m	bulk or mean velocity in x direction (m/s)
u^*	friction velocity (m/s)
x	coordinate
y	coordinate

Greek Symbols

α	thermal diffusivity (m^2/s)
ϵ_H	eddy thermal diffusivity (m^2/s)
ϵ_M	eddy viscosity (m^2/s)
θ	dimensionless temperature
ν	kinematic viscosity (m^2/s)
ρ	density (kg/m^3)
σ	surface tension (N/m)

Subscript

B	buoyancy
b	bubble
D	drag
f	liquid
g	gas

S surface tension
sat saturation
w wall

Superscript

* dimensionless
$^+$ dimensionless

1. INTRODUCTION

In a particular reference loss-of-flow accident, the flow through the channels decreases rapidly, owing to the decrease in the pressure difference across the channels resulting from the pipe break. The power also decreases, following the reactor trip. There is thus a race between the decreases in flowrate and power which determines the conditions within the channels, and ultimately the temperature of the fuel elements. Normally, the flow is all-liquid, but as the flow decreases, it is possible to have the onset of nucleate boiling (ONB). This corresponds to the appearance of the first bubbles on the walls. Since the bulk of the liquid at the point of appearance is subcooled, the bubbles usually do not immediately detach from the wall. As the flow decreases further, the point of ONB moves backwards along the wall surface, resulting in a formation of a bubble boundary layer in the vicinity of the wall. The appearance of the bubble boundary layer causes the pressure gradient to increase, owing to the increased apparent roughness of the wall. At some point, owing to the detachment of the individual bubbles or the instability of the bubble boundary layer, there is significant vapor generation and mixing of bubbles with the core liquid. This onset of significant void (OSV) generally signals the onset of flow instability (OFI) in a system with pressure-driven boundary conditions. This is because the sudden decrease in mean flow density results in an increased pressure gradient due, to the combined effects of friction and acceleration. With pressure boundary conditions, rather than specified inlet flowrates, this can lead rapidly to OFI, and consequent dryout and overheating.

Current methods of predicting the onset of significant void, and of the subsequent flow instability, are based on steady-state, small-scale experiments on forced-convection, subcooled nucleate boiling (FCSNB). This is appropriate, since no experimental studies have come to our attention involving transient FCSNB. The current method of choice is to use the empirical correlation of Saha and Zuber (1974), together with a calculation for the velocity and temperature of the bulk liquid as functions of position and time throughout the transient. However, a more mechanistic approach is needed in predicting OFI for the transient case, since the time of the transient is comparable to the residence time of a fluid particle in the channel, and also to the time delays for the development of the thermal and velocity boundaries at the wall.

The objective of the present study is to develop a mechanistic model for OSV, which can be applied to transient, as well as steady-state, flow conditions. The model predicts the OSV heat flux in vertical downwards flow between parallel plates, provided that the operating pressure, the pressure drop and the liquid inlet temperature are prescribed. It is then compared to the available data, and also to the Saha-Zuber correlation. Applicability and limitations of the present theory are discussed. However, only the steady-state case is dealt with in the present study, since no experimental studies on transient OSV or OFI are available at the present time.

2. DEVELOPMENT OF A NEW MECHANISTIC MODEL FOR OSV

2.1 Basic concept

The basic concept of the present model is similar to that of previous models proposed by such investigators as Levy (1967), Staub (1968) and Rogers et al. (1987). However, the present model is based upon a full two-dimensional analysis of fluid flow and heat transfer, whereas the previous models employed a one-dimensional heat transfer equation, although the temperature distribution obtained by Martinelli (1947) was used to evaluate the liquid temperature at the top of the bubble.

Experimental evidence shows that the onset of significant void (OSV) precedes and is very close to OFI (Whittle and Forgan, 1967; Johnston, 1988; Dougherty et al., 1990). It is likely that the OSV phenomenon is connected to the departure of vapor bubbles from the heated surface. Thus, in the present model, attention is focused on the thermal and hydrodynamic conditions under which a bubble starts to leave the surface to predict OSV.

Two requirements are considered in the present model to establish the condition for the departure of an isolated bubble. One is that the resultant of the forces tending to detach the bubble from the surface exceeds the resultant of the forces holding it on the surface at bubble departure. The other is a thermal condition, which states that the temperature at the top of the bubble should be at least equal to the saturation temperature for bubble departure.

Several forces act on the vapor bubble to maintain it in contact with the heated surface in forced-convection boiling flow. In the Levy model (1967) the basic forces are those due to drag, surface tension and buoyancy. The drag and the buoyancy forces act in a direction parallel to the wall in vertical flow, while the surface tension force has components in both tangential and normal directions. Other forces, such as the liquid inertia force, the lift force, the excess pressure force, and the thermocapillary force have been identified (Koumoutsos et al., 1968; Staub, 1968; Winterton, 1984). The liquid inertia force results from the radial growth of the vapor bubble while attached to the wall. It may be very small if the vapor bubble gradually approaches an equilibrium shape, owing to its being surrounded by saturated liquid prior to departure. However, under extreme conditions it can be the principal force causing the bubble to leave the surface, if the bubble grows suddenly into cold liquid in highly-subcooled boiling flow under high heat flux condition. This is a result of the asymmetry of the liquid flow due to the presence of the wall. Likewise, the excess pressure force represents the force due to the difference in pressure between the inside and the outside of the bubble, tending to hold the bubble on to the wall during the period of rapid bubble growth. The thermocapillary force is generated by circulation of the surface of the bubble under a surface tension gradient due to the temperature gradient close to wall. No experimental information is available, but it is expected to be small. The lift force is proportional to the vector product of the vorticity and the relative velocity between the bubble and liquid in an inviscid liquid. For a detached bubble it is a result of circulation around the bubble. In finite-difference form this force has the same form as the drag force. The influence of viscous effects close to the wall is not well understood. All these extra forces act in a direction normal to the wall.

Following Levy (1967) and other investigators (Rogers et al., 1987; Winterton, 1984; Staub, 1968), it is assumed that the forces parallel to the surface are critical in determining the condition of bubble departure. Thus only three major forces will be considered in the model. These are the drag force (which is dimensionally similar to the lift force), the surface tension force (which has components both parallel and normal to the wall), and the buoyancy force (which is only parallel to the wall and is generally very small).

According to Collier (1981), OSV is a transition from partial boiling to fully-developed subcooled boiling, where the heating surface is covered with bubbles. Prior to OSV vapor bubbles cannot survive in the relatively cold core liquid. Prior to, and after, detachment of individual bubbles from the heating surface, a bubble boundary layer builds up at the wall. Hence an alternative description of the present model is that OSV represents the destabilization of this bubble boundary layer, principally by shear forces exerted by the core liquid. This alternative description implies that bubbles may be held in the boundary layer principally by surface tension forces resulting from random contacts of the bubbles with the wall. The lift force exerted on the surface bubbles in the bubble boundary layer by the shear flow is then considered to be the principal destabilizing force. With either interpretation the form of the principal terms in the force balance turns out to be essentially the same.

The temperature profile of the liquid phase is obtained directly from the solutions of the momentum and the energy equation. It is assumed in the present model that the flow is hydrodynamically fully-developed but is thermally developing. This is different from the previous models which employed a temperature profile for fully-developed thermal conditions. The previous models obtained the wall temperature from a one-dimensional heat transfer equation, such as the Dittus-Boelter equation, while the present model gets it directly from the temperature distribution.

2.2 Description of a new model

Figure 1 shows a sketch of the tangential forces acting on the vapor bubble in vertical downwards flow. The bubble is assumed to be a sphere. The force balance around the bubble can be written as follows:

$$F_B + F_S - F_D = 0 \tag{1}$$

where the drag force, F_D, is the component acting to detach the bubble, while the buoyancy force, F_B, and the surface tension force, F_S, are the components tending to hold it on the surface. The buoyancy force is given by:

$$F_B = C_B(\rho_f - \rho_g)g(\frac{4}{3}\pi R_b^3) \tag{2}$$

where C_B is a proportionality constant and may be determined empirically. Similarly, the surface tension force, F_S, can be expressed as follows:

$$F_S = C_S\sigma(2\pi R_b) \tag{3}$$

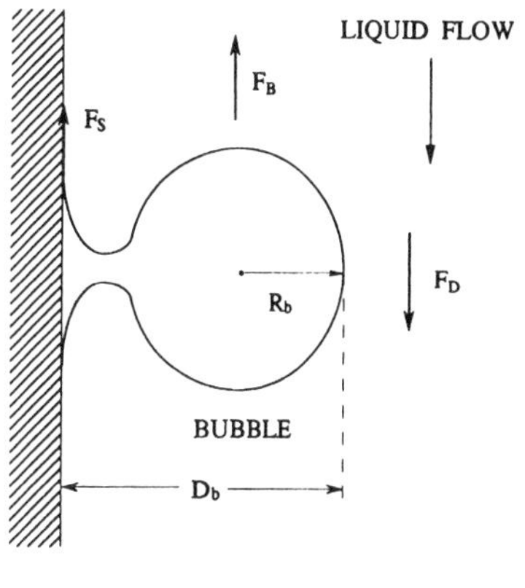

Figure 1. Forces acting on a spherical bubble.

where C_S is a proportionality constant. The drag force is proportional to the projected area of the bubble and the dynamic pressure of the liquid flow, and thus can be written as:

$$F_D = C_D(\frac{1}{2}\rho_f u_m^2)(4\pi R_b^2) \tag{4}$$

Substituting Equations (2), (3), and (4) into Equation (1) and converting it into dimensionless form, one obtains:

$$K_1(R_b^*)^2 - R_b^* + K_2 = 0 \tag{5}$$

where the dimensionless bubble radius, R_b^*, and the dimensionless parameters, K_1 and K_2, are given by:

$$R_b^* = \frac{R_b}{H} \tag{6}$$

$$K_1 = \frac{2}{3}(\frac{C_B}{C_D})[\frac{(\rho_f - \rho_g)gH}{\rho_f u_m^2}] \tag{7}$$

$$K_2 = (\frac{C_S}{C_D})(\frac{\sigma}{\rho_f u_m^2 H}) \tag{8}$$

The solution for R_b^* is:

$$R_b^* = \frac{1}{2K_1}[1 - (1 - 4K_1K_2)^{1/2}] \tag{9}$$

If $4K_1K_2 \ll 1$, the dimensionless bubble radius can be approximated as follows:

$$R_b^* = \frac{R_b}{H} \simeq K_2 \tag{10}$$

Thus, the dimensionless bubble diameter when the bubble starts to depart from the heated wall is inversely proportional to the Weber number:

$$\frac{D_b}{H} = (\frac{C_S}{C_D})(\frac{2\sigma}{\rho_f u_m^2 H}) \tag{11}$$

The temperature profile in the steady-state single-phase turbulent flow can be obtained from the momentum and the energy equation. Under the assumption that the flow is fully-developed, the momentum equation for the flow between parallel plates becomes:

$$-\frac{1}{\rho_f}\frac{dp}{dx}+\frac{d}{dy}[(\nu+\epsilon_M)\frac{du}{dy}]=0 \quad (12)$$

with the boundary conditions:
$u=0$ at $y=0$; $\frac{du}{dy}=0$ at $y=H$
where the pressure term includes the hydrostatic pressure as well as the static pressure. The energy equation may be written as:

$$u\frac{\partial T}{\partial x}=\frac{\partial}{\partial y}[(\alpha+\epsilon_H)\frac{\partial T}{\partial y}] \quad (13)$$

with the boundary conditions:
$-k\frac{\partial T}{\partial y}=q$ at $y=0$; $\frac{\partial T}{\partial y}=0$ at $y=H$; $T=T_i$ at $x=0$

In Equations (12) and (13) ϵ_M and ϵ_H denote the eddy viscosity and the eddy thermal diffusivity, respectively. Several models for the eddy viscosity have been proposed in the past (Launder and Spalding, 1972). A modified Reichardt model suggested by Wilson and Medwell (1971) is used in the present study. The original model for the eddy viscosity proposed by Reichardt (1951) is valid only for regions away from the wall. However, the wall effect was taken into account by multiplying by the van Driest damping factor (1955) in this modified model:

$$\epsilon_M=0.133Hu^*[0.5+(\frac{H-y}{H})^2][1.0-(\frac{H-y}{H})^2][1.0-exp(-\frac{y}{A})] \quad (14)$$

where u^* is the friction velocity and A is defined by:

$$A=A^+(\frac{\nu}{u^*}) \quad (15)$$

In Equation (15), A^+ is the damping constant and is taken to be 40, following Kawamura (1977). For the eddy thermal diffusivity, White (1988) have recommended that the turbulent Prandtl number be taken as a constant.

$$Pr_t=\frac{\epsilon_M}{\epsilon_H}\simeq 0.9 \quad (16)$$

The temperature profile can be obtained numerically from Equations (12) and (13), using Equations (14) and (16). It is noted that the use of a single-phase temperature distribution seems to be reasonable, because it can be assumed that the temperature profile in the liquid core may not be altered by the presence of a very thin vapor bubble layer in the vicinity of the wall.

2.3 Evaluation of the coefficients

Three coefficients emerge in the present model. These should be determined empirically because of the lack of information on physics of the bubble formation and departure.

From Equation (11), the dimensionless bubble diameter can be found without knowing the value of the coefficient, C_B, if $4K_1K_2 \ll 1$. It was found by the present authors that the value of K_1K_2 for the data of Whittle and Forgan (1967) is of order of magnitude between 10^{-3} and 10^{-5}. Physically, this means that the buoyancy force is negligibly small compared to the surface tension and the drag forces. The experimental evidence of Whittle and Forgan (1967), that the flow direction in vertical channel, whether up-flow or down-flow, is not important for the initiation of the onset of flow instability, also confirms the negligibility of the buoyancy force. This result agrees with the argument of Levy (1967) that the buoyancy force exerted on the vapor bubble is negligible in forced-convection boiling flow.

In the previous studies of OSV models, various correlations were employed to calculate the drag coefficient (Levy, 1967; Staub, 1968; Winterton, 1984; Rogers et al., 1987). However, these correlations are either for the friction factor for a rough wall in single-phase turbulent flow or for the drag coefficient for the bubble rising freely under buoyancy. Recently, Avdeev (1986) proposed a relationship for the drag coefficient for the bubbly surface in subcooled boiling flow. Based upon a hydrodynamic analysis for the bubble layer and the Reynolds analogy between the heat and momentum transfer in forced-convection boiling flow, a relationship for the drag coefficient was suggested for the case when the thickness of the bubbly layer is larger than that of the viscous sublayer:

$$C_D=[5.84+16.0log(0.93+0.065K^{0.25})]^{-2} \quad (17)$$

where the dimensionless parameter, K, is defined by:

$$K=\frac{\rho_f u_m C_{pf}(2H)(T_{sat}-T_f)}{k_f(T_w-T_{sat})} \quad (18)$$

Equation (17) is a modified form by the authors, which combines two separate equations suggested by Avdeev (1986).

The coefficient for the surface tension force, C_S, is more complicated to evaluate. The surface tension force in the tangential direction is a result of difference in the contact angle on each side of the bubble. Thus, it would be reasonable to assume that C_S is a function of the contact angle and the liquid velocity. However, neither direct measurement nor theoretical analysis of the contact angle and the surface tension force is possible at the present time. This is the reason why the coefficient, C_S was taken to be an empirical constant in many previous models (Levy, 1967; Staub, 1968; Rogers et al., 1987). However, the data of Whittle and Forgan (1967) and Dougherty et al. (1990) indicate that C_S may be corre-

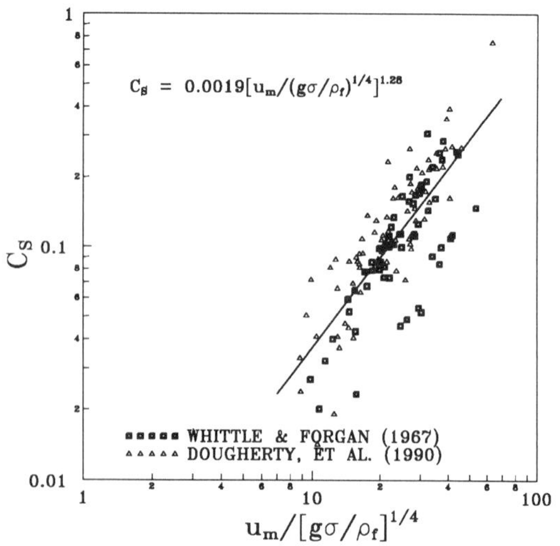

Figure 2. Evaluation of the coefficient for the surface tension.

lated by the following relation:

$$C_S = 0.00193[u_m/(g\sigma/\rho_f)^{1/4}]^{1.28} \qquad (19)$$

where the reference velocity was taken as $(g\sigma/\rho_f)^{1/4}$ which has emerged from the condition that $4K_1K_2 \ll 1$. The result is shown in Figure 2.

2.4 Calculation of the heat flux at OSV

Consider a vertical downwards turbulent flow between heated parallel plates. The channel length is L and the channel depth is $2H$. The calculation process will be described here to find the OSV heat flux when the operating pressure, the mass flux (or equivalently the pressure drop) and the liquid inlet temperature are specified. It may also be noted that OSV always takes place at the bottom of the channel at the OSV heat flux.

Initially the OSV heat flux is assumed. From a simple heat balance the liquid outlet temperature is calculated. This is only to provide the fluid properties at the bulk temperature of the inlet and outlet in the flow and heat transfer analysis. Then, the pressure drop and the velocity profile are obtained from Equation (12), using a numerical integration, to match the given mass flux. If the pressure drop is given instead of the mass flux, the velocity profile can be obtained from Equation (12) and thus the mass flux also can be obtained by integrating the velocity distribution. The energy equation is then solved to find the temperature profile using a finite-difference scheme. The bubble diameter at bubble departure is calculated, using Equation (11) under the given conditions. From the temperature profile obtained from the energy equation, the temperature at the top of the vapor bubble is evaluated. Another heat flux is assumed and this process repeated until the temperature at the top of the bubble reaches the saturation temperature within a desired range of error.

3. COMPARISON

3.1 Description of the available data for comparison

Various experiments have been performed to study OSV or OFI with pressure boundary conditions over the past decades. Several data sets seem to be available for comparison with the present theory. These data sets can be found in Duffey and Hughes (1991). The following should be taken into consideration in selecting the data set for comparison in the present study:

(1) The channel should be heated equally on both plates. Although the present theory has been derived for parallel plates, round tube data may be appropriate if the hydraulic diameter concept is employed. However, annular channel data, or rectangular channel data with unequal heating on both side walls, should be excluded from this comparison.

(2) The flow direction is not an important factor and may be upwards or downwards. This is because of the assumption that the buoyancy force is negligibly small compared to the surface tension and the drag forces.

(3) No restriction may be imposed on the operating pressure. However, low pressure data seems to be preferable because the eddy viscosity model used in the present study emerged from the low-pressure data.

(4) Water is a preferable fluid. The validity of the present model for other fluids seem to be questionable unless the correlations for C_S and C_D are to be tested against those mediums.

The data sets of Whittle and Forgan (1967) and Dougherty et al. (1990) have been chosen for comparison with the present theory in accordance with the remarks mentioned above. These data sets are summarized in Table 1. The

authors	geometry description (m)	pressure (MPa)	heat flux $(\frac{MW}{m^2})$	mass velocity $(\frac{kg}{m^2 s})$	inlet temp. (^oC)
Whittle & Forgan (1967)	rectangular D_h =0.0028 L =0.53 D_h =0.0040 L =0.41 D_h =0.0049 L =0.41 D_h =0.0065 L =0.61	0.12 0.17	0.66 - 3.00	1496 - 9089	35 45 55 60
Dougherty et al.(1990)	circular D =0.0091 L=2.44 D =0.0153 L =2.44 D =0.0155 L =2.44 D =0.0191 L =2.44 D =0.0255 L =2.44 D =0.0284 L =2.44	0.24 0.44	1.06 - 3.17	1296 - 9376	25 50

Table 1. Description of the data set.

data set of Whittle and Forgan (1967) encompasses the OFI data with pressure boundary conditions for vertical upwards and downwards flow in rectangular ducts of four different sizes. The aspect ratio of the channels ranges between 7.9 and 18.2, so that the flow could be considered to be reasonably two-dimensional. The OFI data were determined, under various operating conditions, as a minimum point in the pressure drop- flowrate curve for constant heat flux. Thus, they may be regarded as the OSV point, as explained in Johnston (1988). At a given geometrical condition, the OFI heat flux is presented, using the inlet temperature and the exit pressure as variables.

Dougherty et al.(1990) gives OFI data with pressure boundary conditions for vertical downwards flow in round tubes of six different diameters. The data were determined by a similar method as by Whittle and Forgan (1967).

3.2 Comparison of the present theory with data

The results of comparison of the present theory with the data of Whittle and Forgan (1967) and the data of Dougherty et al.(1990) are shown in Figures 3 and 4, respectively. The calculated OSV heat flux was determined for the corresponding geometrical and operating conditions of the experimental OSV heat flux, using the method outlined in section 2.4. The wall temperature and the liquid bulk temperature in Equation (18) were determined from the two-dimensional temperature profile at the exit. For the data of Dougherty et al.(1990), the length scale, H, was replaced by $D/4$, where D is the tube diameter.

The comparison shows that the present theory is in excellent agreement with both data sets. Most data are within error bounds of $\pm$ 10 %.

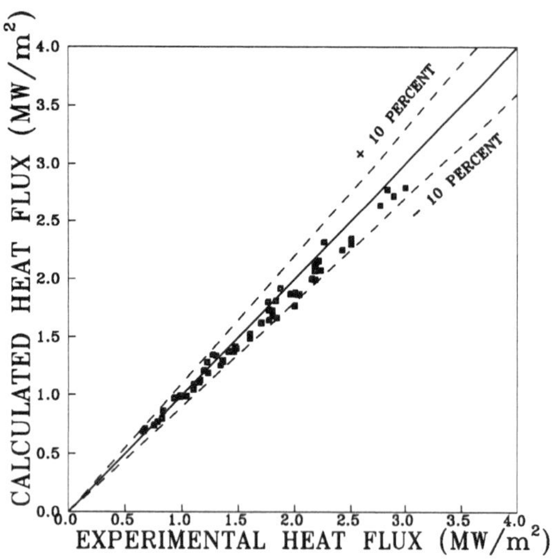

Figure 3. Comparison of the present theory with data of Whittle and Forgan (1967).

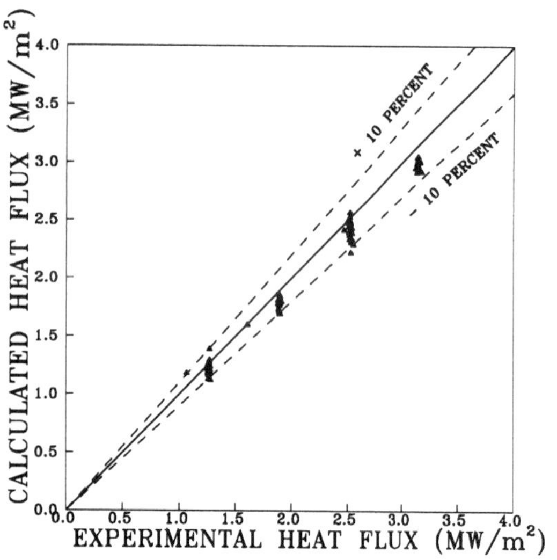

Figure 4. Comparison of the present theory with data of Dougherty et al. (1990).

3.3 Comparison with the Saha-Zuber correlation

Saha and Zuber (1974) proposed a correlation to predict OSV in vertical flow, in which it was argued that the point of net vapor generation is dependent only on local thermal conditions. According to this argument, at low flowrates, corresponding to the thermally controlled region, the local Nusselt number will be the proper similarity parameter, whereas the Stanton number will be appropriate for the high flowrate region, called the hydrodynamically controlled region. Thus, depending upon the Peclet number range, two correlations have been suggested as follows:

$$Nu = \frac{qD_h}{k_f(T_{sat} - T_d)} = 455 \qquad (Pe \leq 70,000) \quad (20)$$

$$St = \frac{q}{GC_{pf}(T_{sat} - T_d)} = 0.0065 \qquad (Pe > 70,000) \quad (21)$$

The Saha-Zuber correlation is very simple and convenient, and has been widely used in computer codes for nuclear safety analysis (Johnston, 1988). However, recent experiments indicate that the discrepancy between the data and the correlation appears to be considerable, with data scatter beyond the error margins described by Saha and Zuber (1974) originally. An example can be found in Figure 5.

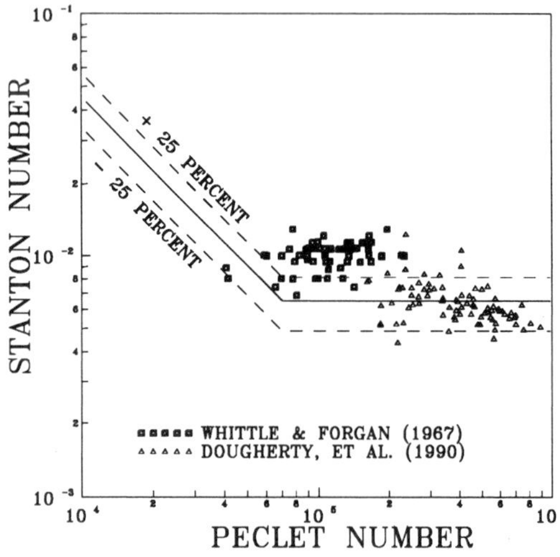

Figure 5. Comparison of the Saha-Zuber correlation with data of Whittle and Forgan (1967) and of Dougherty et al. (1990).

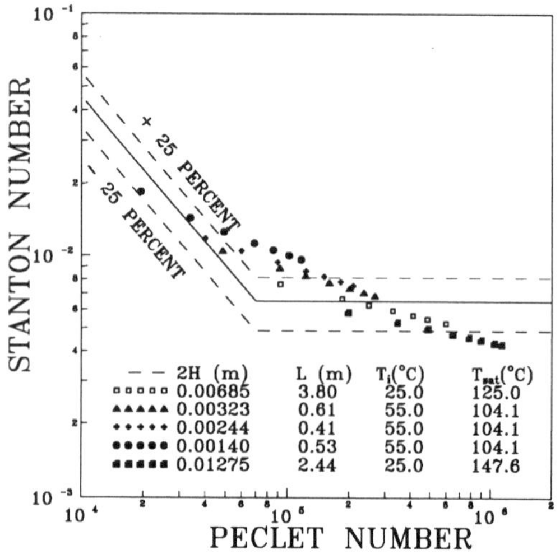

Figure 6. Comparison of the present theory with the Saha-Zuber correlation

The present theory is compared with the Saha-Zuber correlation. This is done by plotting the numerical data calculated from the present theory in the (Pe, St) plane, as shown in Figure 6. Numerical data in this figure have been obtained over a velocity range between approximately 1.0 (m/s) and 7.0 (m/s) for several channel sizes. It is indicated that the present numerical data are fairly well fitted with the correlation. However, the present theory hints that, contrary to the correlation, the Stanton number is not solely related to the Peclet number, showing multiple values of the Stanton number depending upon the geometric dimensions at a single value of the Peclet number. The result may explain the large error margins in the Saha-Zuber correlation, and also the scatter beyond the specified error margins. According to the present theory, the local Stanton number slightly decreases as the Peclet number increases for a specific channel size. This result does not agree with the Saha-Zuber correlation for high Peclet numbers. Considering the physical interpretation of the local Stanton number defined by Saha and Zuber (1974), one obtains the following equation:

$$St = \frac{A_c}{A_s} \frac{T_o - T_i}{T_{sat} - T_d} \tag{22}$$

where the liquid outlet temperature, T_o, is, in general, equal to the local liquid bulk temperature at OSV, T_d. Thus, the Saha-Zuber correlation for high Peclet numbers predicts that the degree of subcooling at the outlet is always the same, regardless of the liquid velocity, under the given geometric and operating conditions. This argument seems to be doubtful, although no physical evidence can be found in the literature yet. This would be another source of data scatter for the correlation. It should be mentioned that the Saha-Zuber correlation is based on Peclet number data up to 400,000. Accordingly, it is questionable that the correlation holds beyond this limit, since the trend of the Stanton number seems to decrease in Figure 5.

4. DISCUSSION

The present theory was derived under the assumption that the buoyancy force is negligibly small. This condition leads to a criterion for a lower limit of the mean velocity for the present theory. This can be calculated from the condition that $4K_1K_2 \ll 1$. Using Equations (7) and (8), one obtains:

$$u_m \gg \left[\frac{8}{3} \frac{C_B}{C_D} \frac{C_S}{C_D} \frac{(\rho_f - \rho_g)}{\rho_f} \frac{g\sigma}{\rho_f}\right]^{1/4} \tag{23}$$

Thus, it will be approximately 0.75 (m/s) at an operating pressure of 0.1 (MPa) assuming that C_B/C_D is $O(10)$ and C_S/C_D is $O(1)$, when $4K_1K_2 \approx 0.05$. For the velocity range smaller than this lower limit, the evaluation of C_B should be made before the present theory is applied. No physical evaluation can be found on the value of C_B in the literature. However, it is expected to be $O(1)$.

It is interesting to note in Figure 6 that the present theory predicts that $St \propto Pe^{-0.20}$. This means that, from the definition of the Stanton number, $Nu \propto Re^{0.80}$, considering the heat transfer from the bubble boundary to the liquid core. This result is exactly identical to the Dittus-Boelter equation for the heat transfer coefficient in single-phase flow.

It should be mentioned that the present theory is still highly empirical, although it contains much more physics than previous models. More experimental information over a wide range, particularly with flow and bubble visualization, is needed to check the details of this model, and to make necessary modifications. Nevertheless, it is encouraging that such a wide range of data can, at the present time, be fitted closely with the choice of a single empirical parameter.

5. SUMMARY AND CONCLUSIONS

A theoretical model to predict OSV in vertical downwards flow between parallel plates has been developed. In the present model, attention was focused on the thermal and hydrodynamic conditions under which a bubble starts to leave the heated surface to predict OSV. A force balance exerted on a vapor bubble was set up to establish a criterion for the onset of bubble departure. From the force balance, a dimensionless distance to the top of the bubble is calculated. The drag coefficient in the force balance was determined from the Avdeev correlation (1986), and the proportionality constant for the surface tension force was evaluated empirically from the available data. It turns out that the buoyancy force is negligibly small compared to the other forces in forced convective boiling flow. This indicates that the force balance is identical regardless of the flow direction, and thus the present model can be applied to vertical upflow without any modification. It is assumed in the model that the liquid temperature at the top of the bubble should be at least equal to the saturation temperature when the bubble leaves. Thus, the momentum and energy equations are set up to solve the temperature distribution in fully-developed, single-phase, turbulent flow between parallel plates. The eddy viscosity model used in the present study is the Reichardt model (1951), multiplied by the van Driest damping factor (1955). The equations are solved numerically to obtain the temperature profile under the given operating conditions. The heat flux at OSV was then determined by checking whether the liquid temperature at the top of the bubble is equal to the saturation temperature or not.

The model was tested against the OFI data of Whittle and Forgan (1967) and Dougherty et al. (1990). It is based upon the fact that the onset of significant void (OSV) precedes and is very close to OFI with a pressure-driven system. For both data sets nearly all data are within ± 10 % agreement with the theory.

The present theory was also compared with the Saha-Zuber correlation. The result show that the theory is in fairly good agreement with the correlation. However, the present theory indicates that, contrary to the correlation, the Stanton number is not solely related to the Peclet number. This may explain the large error margins in the Saha-Zuber correlation and also the scatter beyond the error margins specified by the authors.

ACKNOWLEDGMENT

This work was supported by Westinghouse Savannah River Company. We thank the project monitor, B. S. Johnston for helpful comments. S. C. Lee received partial support from the Ministry of Education, Republic of Korea.

REFERENCES

Avdeev, A. A., 1986, "Application of the Reynolds Analogy to the Investigation of Surface Boiling in Conditions of Forced Motion," *High Temperature*, Vol. 24, pp.100-108.

Collier, J. G., 1981, "Convective Boiling and Condensation," McGraw-Hill, New York, Chapter 6, pp.178-205.

Dougherty, T., Fighetti, C., Reddt, G., Yang, B., Jafri, T., McAssey, E. and Qureshi, Z., 1990, "Flow Boiling in Vertical Down-flow," *Proceedings of 10th Int. Heat Transfer Conf.*, Vol. 6, pp.9-14.

van Driest, E. R., 1955, "On Turbulent Flow near a Wall," *Heat Transfer and Fluid Mechanics Institute Symposium*, University of California, Los Angeles, Paper No. 12.

Duffey, R. B. and Hughes, E. D., 1991, "Static Flow Instability Onset in Tubes, Channels, Annuli, and Rod Bundles," *Int. J. Heat Mass Transfer*, Vol. 34, pp.2483-2496.

Johnston, B. S., 1988, "Subcooled Boiling of Downward Flow in a Vertical Annulus," Report No. DPST-88-891, Savannah River Laboratory, Aiken, SC.

Kawamura, H., 1977, "Experimental and Analytical Study of Transient Heat Transfer for Turbulent Flow in a Circular Tube," *Int. J. Heat Mass Transfer*, Vol. 20, pp.443-450.

Koumoutsos, N., Moizsis R. and Spryidonos, A., 1968, "A Study of Bubble Departure in Forced Convection Boiling," *J. Heat Transfer*, Vol. 90, pp.223-230.

Launder, B. E. and Spalding, D. B., 1979, "Mathematical Models of Turbulence," Academic Press, London Lecture 2, pp.23-45.

Levy, S., 1967, "Forced Convection Subcooled Boiling: Prediction of Vapor Volumetric Fraction," *Int. J. Heat Mass Transfer*, Vol. 10, pp.951-965.

Martinelli, R. C., 1947, "Heat Transfer to Molten Metals," *Trans. ASME*, Vol. 69, pp.941-947.

Reichardt, H., 1951, "Vollstandige Darstellung der Turbulenten Geschwindigkeitsverteilung in Glatten Leitungen," *Z. Angew. Math. Mech.*, Vol. 31, p.208.

Rogers, J. T., Salcudean, M., Abdullah, Z., Mcleod, D. and Poirier, D., 1987, "The Onset of Significant Void in Up-flow Boiling of Water at Low Pressure and Velocities," *Int. J. Heat Mass Transfer*, Vol. 30, pp.2247-2260.

Saha, P. and Zuber, N., 1974, " Point of Net Vapor Generation and Vapor Void Fraction in Subcooled Boiling," *Proceedings of 5th Int. Heat Transfer Conf.*, Vol. 4, pp.175-179.

Staub, F. W., 1968, "The Void Fraction in Subcooled Boiling: Prediction of the Initial Point of Net Vapor Generation," *J. Heat Transfer*, Vol. 90, pp.151-157.

White, F. M., 1988, "Heat and Mass Transfer," Addison-Wesley Publishing Co., Reading, Massachusetts, Chapter 6, pp.315-332.

Whittle, R. H. and Forgan, R., 1967, "Correlation for the Minima in the Pressure Drop versus Flow-rate Curves for Subcooled Water Flowing in Narrow Heated Channels," *Nuclear Engineering and Design*, Vol. 6, pp.89-99.

Wilson, N. W. and Medwell, J. O., 1971, "An Analysis of the Developing Turbulent Hydrodynamic and Thermal Boundary Layers in an Internally Heated Annulus," *J. Heat Transfer*, Vol. 93, pp.25-32.

Winterton, R. H. S., 1984, "Flow Boiling: Prediction of Bubble Departure," *Int. J. Heat Mass Transfer*, Vol. 27, pp.1422-1424.

HTD-Vol. 197, Two-Phase Flow and Heat Transfer
ASME 1992

A CORRELATION FOR FORCED CONVECTION FILM BOILING HEAT TRANSFER FROM A HORIZONTAL CYLINDER

Q. S. Liu, M. Shiotsu, and A. Sakurai
Institute of Atomic Energy
Kyoto University
Kyoto, Japan

ABSTRACT

Forced convection film boiling heat transfer from a horizontal cylinder in water or Freon-113 flowing upward perpendicular to the cylinder under saturated condition was measured for the flow rates ranging from 0 to 1 m/s at the system pressures ranging from 100 to 500 kPa. The cylinders made of platinum with the diameters ranging from 0.7 to 5 mm were used as the test heaters. Bromley's correlation for forced convection film boiling heat transfer could not well describe the experimental data obtained by the authors, especially under pressurized conditions. By modifying an approximate analytical solution for a laminar flow film boiling model derived by the authors to agree better with the experimental data, a new forced convection film boiling heat transfer correlation was given. It was confirmed that the correlation represented not only the experimental data at various pressures obtained by the authors within ±15% for flow rates below 0.7 m/s, and within −30% to +15% for higher flow rates, but also the data by Bromley et al. for diameters ranging from 9.83 to 16.2 mm and for flow rates up to 4.4 m/s in various liquids at atmospheric pressure within ±20% .

NOMENCLATURE

A = non-dimensional quantity
B = non-dimensional quantity
B_N = proportional constant
C = non-dimensional quantity
c_p = specific heat capacity, $J/(kgK)$
D = diameter of the cylinder heater, m
D' = $D[g(\rho_l - \rho_v)/\sigma]^{1/2}$, non-dimensional diameter of the cylinder heater
d = the departing bubble diameter, m
E = non-dimensional quantity
Fr = $U^2/(gD)$, Froude number
Gr = $g(\rho_l - \rho_v)D^3/(\rho_v \nu_v^2)$, Grashof number
g = acceleration due to gravity, m/s^2
h = heat transfer coefficient, $W/(m^2K)$
h_{co} = heat transfer coefficient if there were no radiation, $W/(m^2K)$
h_r = radiation heat transfer coefficient for parallel plates, $W/(m^2K)$
$In(Fr)$ = an integration function of Froude number
$In_1(Fr)$ = an integration function of Froude number
$In_2(Fr)$ = an integration function of Froude number
J = radiation parameter
K = ρ_l/ρ_v, the ratio of the density of liquid to the density of vapor
$K(D')$ = a function of D', non-dimensional quantity
K_d = a factor
k = thermal conductivity, $W/(mK)$
L = latent heat of vaporization, J/kg
L' = $L + 0.5c_{pv}\Delta T_{sat}$, latent heat plus sensible heat content of vapor, J/kg
M = non-dimensional quantity
N = non-dimensional quantity
Nu_v = average Nusselt number
P_l = the pressure in liquid along the cylinder surface, Pa
Pr = Prandtl number
P_v = the pressure in vapor boundary layer along the cylinder surface, Pa
q = heat flux, W/m^2
R = $[\rho_v \mu_v/(\rho_l \mu_l)]^{1/2}$, non-dimensional
r = radius of the cylinder heater, m
Sp = $c_{pv}\Delta T_{sat}/(L' Pr_v)$, non-dimensional superheat
Sp_r = $c_{pv}\Delta T_{sat}/(L Pr_v)$, non-dimensional superheat
T = temperature, K
U = liquid velocity, m/s
u = $u_1/(gr)^{1/2}$, x component velocity of vapor, non-dimensional
u_1 = x component velocity of vapor, m/s
v = $v_1[r/(g\nu_v^2)]^{1/4}$, y component velocity of vapor, non-dimensional
v_1 = y component velocity of vapor film, m/s
x = coordinate along the heater surface, m
y = $y_1[g/(\nu_v^2 r)]^{1/4}$, coordinate normal to the heater surface, non-dimensional
y_1 = coordinate normal to the heater surface, m
α = absorptivity of liquid (taken to be unity in this paper)
ΔT_{sat} = $T_w - T_{sat}$, heater surface superheat, K
δ = $\delta_1[g/(\nu_v^2 r)]^{1/4}$, thickness of laminar vapor layer, non-dimensional
δ_1 = thickness of laminar vapor layer, m

δ_D = average vapor film thickness over the cylinder surface, non-dimensional
δ_F = vapor film thickness upstream the separation point, non-dimensional
δ_{FS} = vapor film thickness at the separation point, non-dimensional
δ_N = vapor film thickness downstream the separation point, non-dimensional
λ' = $L(1 + 0.4c_{pv}\Delta T_{sat}/L)^2$, latent heat plus sensible heat content of vapor in Eqs.(2) and (4) presented by Bromley
λ_m = the most dangerous wavelength of the vapor-liquid interface
μ = viscosity, $kg/(m\ s)$
ν = kinematic viscosity, m^2/s
ϕ = x/r, the angle measured from the front stagnation point of the cylinder heater
ϕ_s = $\arccos[-1/(8Fr)]$, the angle of the separation point
ρ = density, kg/m^3
σ = surface tension, N/m
σ_s = Stefan-Bolzman constant
ε_w = emissivity

Subscripts

l = liquid
r = radiation
sat = saturation
v = vapor
w = wall

1 INTRODUCTION

The knowledge of forced convection film boiling heat transfer for relatively low flow rates under atmospheric and pressurized conditions is important as the basis of understanding the reflooding phenomenon during the emergency cooling in a water-cooled nuclear reactor under a loss of coolant accident. Bromley et al. (1953) first carried out a systematic experiment for the forced convection film boiling at atmospheric pressure: they presented a semi-empirical correlation based on their experimental data. However, because there are few experimental data under pressurized conditions up to the present, there exists uncertainty as to whether or not the correlation is applicable to those for the pressures higher than atmospheric one.

The purpose of this study is, first, to obtain the experimental data of film boiling heat transfer from a horizontal cylinder in water or Freon-113 for the wide ranges of flow rate, surface superheat, system pressure and cylinder diameter, second, to make clear the limit of the applicability of the Bromley's correlation to the data, and third, to derive a new correlation applicable to various kinds of liquids at various pressures by modifying an approximate analytical solution of a simple laminar flow film boiling model obtained to agree better with the experimental data.

2 EXPERIMENTAL APPARATUS AND METHOD

2.1 Forced Convection Film Boiling Heat Transfer Apparatus

Figure.1 shows a schematic diagram of the forced convection film boiling heat transfer experimental apparatus. Distilled and deionized water or Freon-113 of more than 99.5 % purity is circulated by a pump at a flow rate up to 300 l/min and is heated to the desired temperature level by a preheater. The flow rate is controlled by a flow-regulating valve and measured by a turbine flow meter. The liquid flows upward through the test section. This test section made of stainless steel with a rectangular cross section of 120 mm × 40 mm and a height of 1,312 mm was used. The maximum mean velocity at this test section is 1 m/s. Test heater is installed horizontally in this test section. A stainless steel screen(with a mesh of 2 mm) is installed in the upstream of the test heater to straighten the velocity distribution across the test heater. Liquid and steam are separated in a separator expansion tank. The steam will then be collected and liquefied by a condenser. A cooler is used to cool down circulating liquid to a desired temperature if necessary.

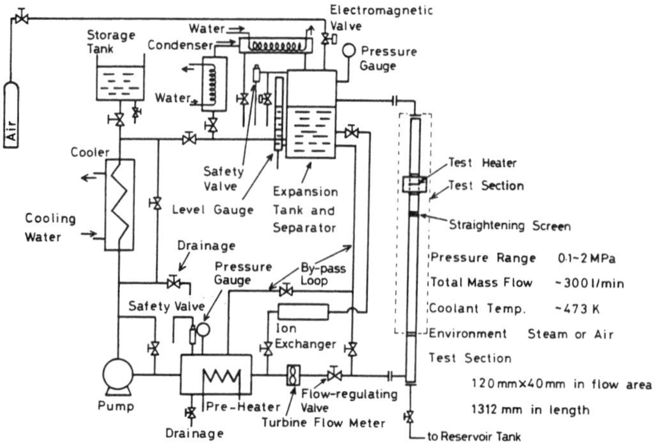

Fig. 1 Schematic diagram of the experimental apparatus

2.2 Heated Section

Figure 2(a) shows a cross-sectional view of the heated section with a quartz glass window to allow visual observation, and Fig.2(b) shows a sketch of the heater configuration. As shown in Fig.2(b), The test heater is bent to a U-shape consisting of horizontal and vertical heater parts. The lower horizontal part

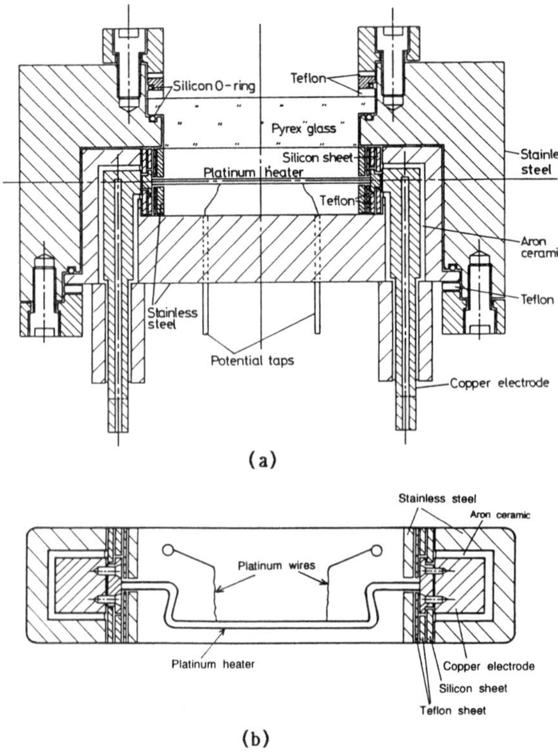

Fig. 2 Cross-sectional view of the heated section and a sketch of the heater configuration

was about 80 mm and the upper horizontal part was about 40 mm in length and the two vertical parts were about 20 mm in height, respectively. U-shaped cylinder heater was used to keep its homogeneity of the axial temperature distribution and also to prevent it from deformation due to thermal expansion at film boiling with a high superheat. Two fine 0.030 mm-dia. platinum wires were spot welded at the position 15 mm from each end of the lower horizontal part of the test heater. The effective lengths of the heaters between the potential taps on which film boiling heat transfer was measured were around 50 mm. Each heater was annealed and its electrical resistance versus temperature relation was calibrated in water and glycerine baths before using it in the experiments. As shown in Fig.2(a), the test heater was installed horizontally in the heated section, both ends of which were connected to the copper electrodes. The margin between copper electrode and stainless steel body was filled with ceramics, and silicon sheets and teflon sheets were used to provide a high pressure seal. Each electrode was heated by two microheaters to keep its temperature at about 200 to 250 °C to realize a relatively flat temperature distribution along the cylinder heater axis.

2.3 Measurement Apparatus

The test heater is heated by direct current from a power amplifier. The test heater temperature is measured by resistance thermometry by using a double bridge circuit including the test heater as a branch. The heater surface temperature is calculated from the measured average temperature and heat generation rate by solving conduction equation in the heater. Experimental error is estimated to be ± 1 K in the heater surface temperature and ±2% in the heat flux.

2.4 Experimental Method and Procedure

Film boiling state was realized with the procedures as follows: first, the liquid temperature in the experimental apparatus was raised to a saturation temperature and was kept constant for about an hour for degassing. Then, by closing the flow-regulating valve upstream of the test section, the liquid was forced to circulate through a by-pass loop parallel to the test section and the liquid in test heater section was drawn to a reservoir tank outside the test loop. The cylinder test heater was heated by the direct current up to well above the current corresponding to the minimum film boiling temperature in the vapor phase over the liquid surface. By slightly opening the flow-regulating valve upstream of the test section, the liquid flew upward slowly; the test heater was immersed. At first, stable film boiling was realized at very low flow rate and then the forced convection film boiling to be measured was realized by adjusting the heating current and the flow rate step by step up to the desired values. A high speed video camera was used to observe the dynamic behavior of film boiling.

2.5 Experimental Conditions

The experimental conditions for water and Freon-113 experiments are shown in Table 1.

Table 1

Water		
	cylinder diameter	0.7, 1.2, 3.0, 5.0 mm
	system pressure	101 ~ 490 kPa
	maximum superheat	~ 800 K
	flow rate	0 ~ 1 m/s
Freon-113		
	cylinder diameter	1.2, 3.0, 5.0 mm
	system pressure	106 ~ 490 kPa
	maximum superheat	~ 400 K
	flow rate	0 ~ 1 m/s

3 EXPERIMENTAL RESULTS

3.1 Film Boiling Heat Transfer Coefficients

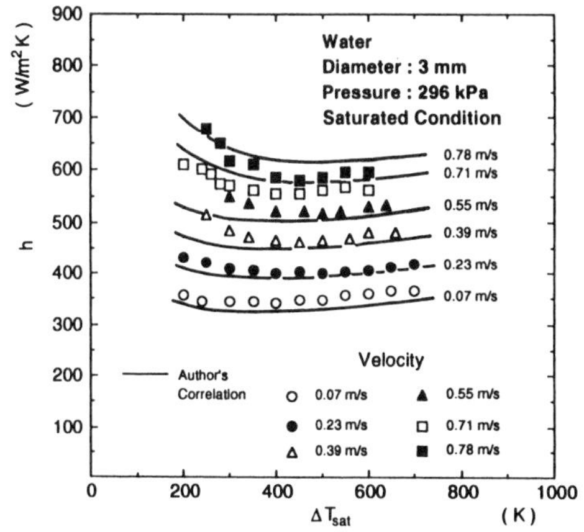

Fig. 3 Film boiling heat transfer coefficients for various flowing velocities on 3 mm diameter horizontal platinum cylinder in saturated water at the pressure of 296 kPa

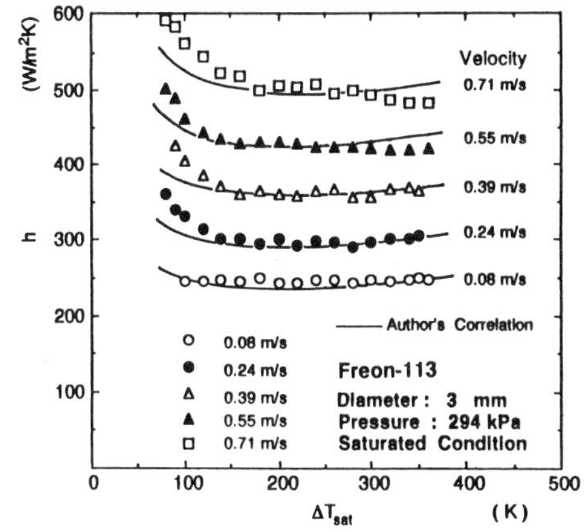

Fig. 4 Film boiling heat transfer coefficients for various flowing velocities on 3 mm diameter horizontal platinum cylinder in saturated Freon-113 at the pressure of 294 kPa

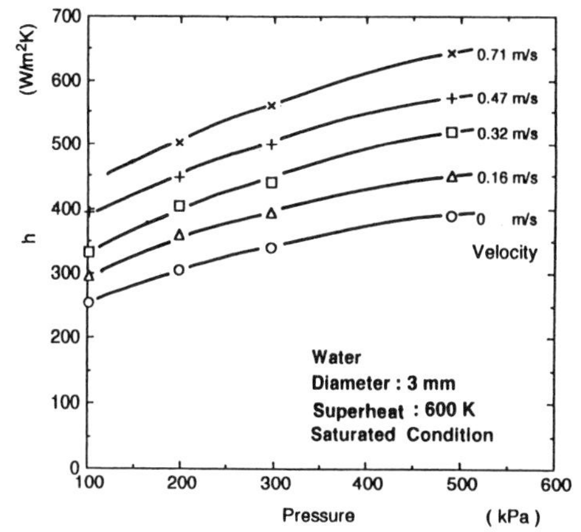

Fig. 5 Effect of the system pressure on saturated film boiling heat transfer coefficients

Film boiling heat transfer coefficients from horizontal cylinders in water and in Freon-113 flowing upward perpendicular to the cylinders were measured for the experimental conditions shown in Table 1. Figures 3 and 4 show the typical data of film boiling heat transfer coefficients, h, versus cylinder surface superheats, ΔT_{sat} , for 3 mm diameter cylinder in water and in Freon-113, respectively, with flow rate as a parameter under system pressures of 296 kPa for water and 294 kPa for Freon-113. As shown in these figures, heat transfer coefficients are weakly dependent on cylinder surface superheats but significantly dependent on flow rates; they are higher for higher flow rates.

Figure 5 shows the heat transfer coefficients at a surface superheat of 600 K in water versus system pressures with the flow rate as a parameter. It can be seen that the heat transfer coefficients at a constant flow rate are higher for higher system pressures. Figure 6 shows the heat transfer coefficients versus flow rates with the cylinder diameter as a parameter. The heat transfer coefficients at each flow rate are higher for smaller cylinder diameters.

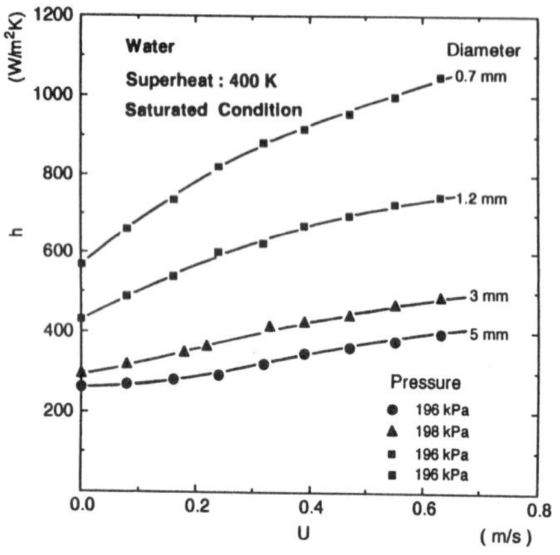

Fig. 6 Effect of the cylinder diameter on saturated film boiling heat transfer coefficients

3.2 Comparison with Bromley's Correlation

Using horizontal graphite cylinders of the diameters 9.83, 12.6, and 16.2 mm, Bromley et al. (1953) carried out the forced convection film boiling heat transfer experiments on the cylinders in four kinds of liquids such as ethyl alcohol, benzene, n-hexane, and carbon tetrachloride at atmospheric pressure. They presented the following correlations based on their data.

$$h = h_{co} + \frac{3}{4}h_r \quad (1)$$

$$h_{co}[\frac{\mu_v D\Delta T_{sat}}{k_v^3 g\rho_v(\rho_l - \rho_v)\lambda'}]^{\frac{1}{4}} = 0.62 \qquad for\ \ U' < 1 \quad (2)$$

$$h = h_{co} + \frac{7}{8}h_r \quad (3)$$

$$h_{co}[\frac{D\Delta T_{sat}}{\rho_v U k_v \lambda'}]^{\frac{1}{2}} = 2.7 \qquad for\ \ U' > 2 \quad (4)$$

here U' is the non-dimensional velocity given by $U' = U/(gD)^{\frac{1}{2}} = Fr^{\frac{1}{2}}$, Fr is Froude number. Equations (1) and (2) are the expression for the non-dimensional velocity U' smaller than unity, and Eqs.(3) and (4) are the expression for U' larger than two. There is no expression for the U' between one and two. Figures 7(a) and 7(b) show the comparison of author's experimental data with the solid lines given by Bromley's correlations. We can see from Fig.7(a), that the experimental data for $U' < 1$ are not constant but increase gradually with U'. For $U' > 2$, we can see from Fig.7(b), that the values of Eq.(4) obtained from the experimental data are a little dependent on U' and the values at near the atmospheric pressure are almost in agreement with the predicted values by the Bromley's correlation. However, the experimental values become lower than the predicted values with the increase in the pressure and they are about 50 % lower than the predicted values at 490 kPa. It is concluded by this comparison that Bromley's correlation can't express our experimental data at higher pressures.

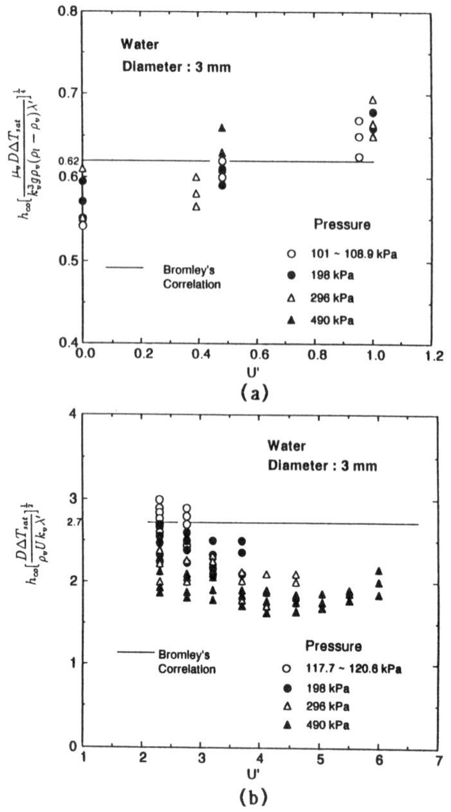

Fig. 7 Comparison of the author's typical experimental data with Bromley's correlations given by Eqs.(1) to (4)

4 ANALYTICAL SOLUTION FOR FORCED CONVECTION FILM BOILING HEAT TRANSFER IN SATURATED LIQUIDS

4.1 Physical Model

Film boiling heat transfer from a horizontal cylinder to saturated fluid flowing upward perpendicular to the cylinder is considered based on the physical model and coordinates shown in Fig.8. The following assumptions are made to obtain the analytical solution.

1) Laminar vapor layer exists around the cylinder with a uniform temperature T_w, and the liquid flowing outside the vapor layer with saturation temperature T_{sat} is assumed to be potential flow.

2) The inertia terms in vapor film momentum equation and the convection terms in the vapor film energy equation can be neglected.

3) The vapor-liquid interface is smooth, and pressure difference due to vapor- liquid interface tension is negligible.

4) All thermal physical properties of the liquid and its vapor are assumed to be constant and may be evaluated at the arithmetic average temperature ($T_w + T_{sat}$)/2 for the vapor. The density of vapor ρ_v is negligible compared with the density of liquid ρ_l .

5) For the liquid flow rate at which the separation of the vapor film from the cylinder surface occurs, the heat transfer at the downstream region of the separation point is evaluated by the solution of the natural convection heat transfer with a condition of a continued vapor film thickness δ_1 at the separation point.

6) Radiation heat transfer is neglected.

4.2 Basic Conservation Equations

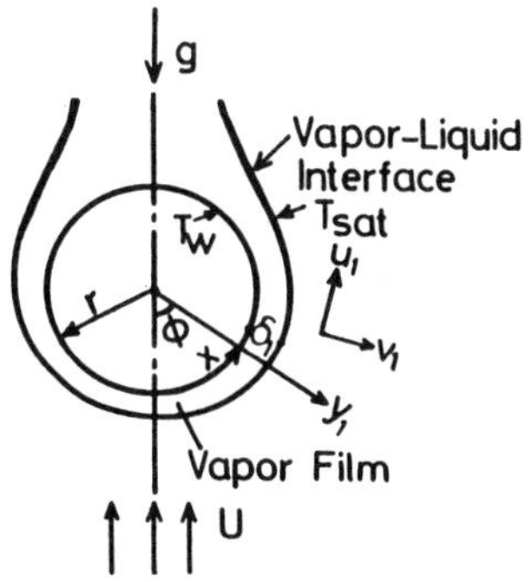

Fig. 8 Physical model and coordinates

For the laminar vapor boundary layer, mass, momentum and energy conservation equations take the forms, respectively.

$$\frac{1}{r}\frac{\partial u_1}{\partial \phi} + \frac{\partial v_1}{\partial y_1} = 0 \tag{5}$$

$$-g\rho_v \sin\phi - \frac{1}{r}\frac{dP_v}{d\phi} + \mu_v \frac{\partial^2 u_1}{\partial y_1^2} = 0 \tag{6}$$

$$k_v \frac{\partial^2 T_v}{\partial y_1^2} = 0 \tag{7}$$

here, ρ_v is the density of vapor, μ_v is the viscosity of vapor, k_v is the thermal conductivity of vapor film and T_v is the temperature inside vapor film. As the external liquid flow is assumed to be potential flow, the pressure gradient near the cylinder surface can be written as

$$-\frac{1}{r}\frac{dP_l}{d\phi} = \rho_l (g \sin\phi + \frac{2U^2}{r}\sin 2\phi) \tag{8}$$

According to the assumption 3), we have $P_v = P_l$, and combining Eqs.(6) and (8), we can obtain the following equation.

$$\nu_v \frac{\partial^2 u_1}{\partial y_1^2} + \frac{g(\rho_l - \rho_v)}{\rho_v}\sin\phi + \frac{2\rho_l}{\rho_v}\frac{U^2}{r}\sin 2\phi = 0 \tag{9}$$

here, ν_v is the kinematic viscosity of vapor film.

The boundary conditions are: at $y_1 = 0$,

$$u_1 = v_1 = 0, \qquad T_v = T_w \tag{10}$$

at $y_1 = \delta_1$, velocity boundary condition will be considered as follows. As this model is not considering the liquid boundary layer with some velocity distribution, vapor velocity at the vapor-liquid interface can't be determined theoretically. Then, the following two extreme cases are considered as the boundary conditions for vapor velocity.

i) In the case of no shear stress at the boundary

$$\mu_v \frac{\partial u_1}{\partial y_1} = 0 \tag{11}$$

ii) In the case of zero vapor velocity at the boundary

$$u_1 = 0 \tag{12}$$

At $y = \delta_1$, other boundary conditions are:

$$T_v = T_{sat} \tag{13}$$

$$-k_v \frac{\partial T_v}{\partial y_1} = k_v \frac{T_w - T_{sat}}{\delta_1} = \frac{L'\rho_v}{r}\frac{d}{d\phi}(\int_0^{\delta_1} u_1 dy_1) \tag{14}$$

4.3 Method of Solution

Now, using the dimensionless variables and parameters defined by Eq.(15) through Eq.(21), the fundamental equations and boundary conditions are expressed as the following dimensionless forms such as Eq.(22) to Eq.(27).

$$y = y_1 (\frac{g}{\nu_v^2 r})^{\frac{1}{4}} \tag{15}$$

$$\delta = \delta_1 (\frac{g}{\nu_v^2 r})^{\frac{1}{4}} \tag{16}$$

$$u = u_1 (\frac{1}{gr})^{\frac{1}{2}} \tag{17}$$

$$v = v_1 (\frac{r}{g\nu_v^2})^{\frac{1}{4}} \tag{18}$$

$$Sp = \frac{c_{pv}(T_w - T_{sat})}{L' Pr_v} \tag{19}$$

$$K = \frac{\rho_l}{\rho_v} \tag{20}$$

$$Fr = \frac{U^2}{gD} \tag{21}$$

here, r is the radius of the cylinder, Sp is non-dimensional superheat, Fr is Froude number, K is the ratio of the density of liquid to the density of vapor. Eq.(5) is transformed to

$$\frac{\partial u}{\partial \phi} + \frac{\partial v}{\partial y} = 0 \tag{22}$$

Eq.(9) changes to ($\rho_l \gg \rho_v$ is considered according to the assumption 4))

$$\frac{\partial^2 u}{\partial y^2} + K(1 + 8Frcos\phi)sin\phi = 0 \tag{23}$$

Eq.(14) changes to

$$\delta \frac{d}{d\phi}(\int_0^{\delta} u dy) = Sp \tag{24}$$

Eq.(10) changes to

$$u = v = 0, \qquad T_v = T_w \tag{25}$$

Eq.(11) changes to

$$\frac{\partial u}{\partial y} = 0 \tag{26}$$

Eq.(12) changes to

$$u = 0 \tag{27}$$

4.3.1 The Solution for Zero Shear Stress at y = δ.

The solution for Eq.(23) to satisfy the boundary conditions given by Eq.(25) and Eq.(26) is,

$$u = \delta[K(1+8Fr\cos\phi)\sin\phi]y - \frac{1}{2}[K(1+8Fr\ cos\phi)\sin\phi]y^2 \tag{28}$$

by substituting Eq.(28) into Eq.(24), we have

$$\frac{d\delta^4}{d\phi} + \frac{4}{3}\frac{\cos\phi + 8Fr\cos 2\phi}{(1 + 8Fr\cos\phi)\sin\phi}\delta^4 = \frac{4Sp}{K(1 + 8Fr\cos\phi)\sin\phi} \tag{29}$$

This is a one-order differential equation with respect to δ^4. Let the solution of the above differential equation be δ_F, we can obtain

$$\delta_F = \left(\frac{4Sp}{K}\right)^{\frac{1}{4}}\left[\frac{\int_0^{\phi}\{(1+8Fr\cos\phi)\sin\phi\}^{\frac{1}{3}}d\phi}{\{(1+8Fr\cos\phi)\sin\phi\}^{\frac{4}{3}}}\right]^{\frac{1}{4}} \tag{30}$$

The condition for the occurrence of the vapor film transforming into a thick vapor wake (separation point) is given by

$$\frac{\partial u}{\partial y}\Big|_{y=0} = \delta K(1+8Fr\cos\phi_s)\sin\phi_s = 0 \tag{31}$$

Where ϕ_s is the angle of the separation point measured from the front stagnation point of the cylinder heater.

$$\phi_s = \arccos\left(-\frac{1}{8Fr}\right) \tag{32}$$

According to Eq.(32), the separation does not occur for $Fr < \frac{1}{8}$. The vapor film thickness for $\phi > \phi_s$ is assumed to be given by the following equation which is similar to that for pool film boiling (Nishikawa et al., 1972).

$$\delta_N = B_N\left[\frac{\int_0^{\phi}\sin^{\frac{1}{3}}\phi\, d\phi}{\sin^{\frac{4}{3}}\phi}\right]^{\frac{1}{4}} \tag{33}$$

here, B_N is a proportional constant. The vapor film thickness δ_{FS} at the separation point ϕ_s is obtained from Eq.(30) as,

$$\delta_{FS} = \left(\frac{4Sp}{K}\right)^{\frac{1}{4}}\left[\frac{\int_0^{\phi_s}\{(1+8Fr\cos\phi)\sin\phi\}^{\frac{1}{3}}d\phi}{\{(1+8Fr\cos\phi_s)\sin\phi_s\}^{\frac{4}{3}}}\right]^{\frac{1}{4}} \tag{34}$$

According to the assumption 5), δ_F and δ_N should be equal at $\phi = \phi_s$, so that,

$$\begin{aligned}\delta_N &= \left(\frac{4Sp}{K}\right)^{\frac{1}{4}}\left[\frac{\int_0^{\phi_s}\{(1+8Fr\cos\phi)\sin\phi\}^{\frac{1}{3}}d\phi}{\{(1+8Fr\cos\phi_s)\sin\phi_s\}^{\frac{4}{3}}}\right]^{\frac{1}{4}} \\ &\quad\times\left[\frac{\sin^{\frac{4}{3}}\phi_s}{\int_0^{\phi_s}\sin^{\frac{1}{3}}\phi\, d\phi}\right]^{\frac{1}{4}}\left[\frac{\int_0^{\phi}\sin^{\frac{1}{3}}\phi\, d\phi}{\sin^{\frac{4}{3}}\phi}\right]^{\frac{1}{4}}\end{aligned} \tag{35}$$

The vapor film thickness averaged over the cylinder surface δ_D is,

$$\begin{aligned}\frac{1}{\delta_D} &= \frac{1}{\pi}\left[\int_0^{\phi_s}\frac{1}{\delta_F}d\phi + \int_{\phi_s}^{\pi}\frac{1}{\delta_N}d\phi\right] \\ &= \left(\frac{K}{4Sp}\right)^{\frac{1}{4}}[In_1(Fr) + In_2(Fr)] \\ &= \left(\frac{K}{4Sp}\right)^{\frac{1}{4}}In(Fr)\end{aligned} \tag{36}$$

where

$$In(Fr) = In_1(Fr) + In_2(Fr) \tag{37}$$

$$In_1(Fr) = \frac{1}{\pi}\int_0^{\phi_s}\left[\frac{\{(1+8Fr\cos\phi)\sin\phi\}^{\frac{4}{3}}}{\int_0^{\phi}\{(1+8Fr\cos\phi)\sin\phi\}^{\frac{1}{3}}d\phi}\right]^{\frac{1}{4}}d\phi \tag{38}$$

$$\begin{aligned}In_2(Fr) &= \frac{1}{\pi}\left[\frac{\{(1+8Fr\cos\phi_s)\sin\phi_s\}^{\frac{4}{3}}}{\int_0^{\phi_s}\{(1+8Fr\cos\phi)\sin\phi\}^{\frac{1}{3}}d\phi}\frac{\int_0^{\phi_s}\sin^{\frac{1}{3}}\phi\, d\phi}{\sin^{\frac{4}{3}}\phi_s}\right]^{\frac{1}{4}} \\ &\quad\times\int_{\phi_s}^{\pi}\left(\frac{\sin^{\frac{4}{3}}\phi}{\int_0^{\phi}\sin^{\frac{1}{3}}\phi\, d\phi}\right)^{\frac{1}{4}}d\phi\end{aligned} \tag{39}$$

Average heat transfer coefficient h_{co} is given by

$$h_{co} = \frac{q}{T_w - T_{sat}} = k_v\frac{1}{\delta_D}\left(\frac{g}{\nu_v^2 r}\right)^{\frac{1}{4}} = k_v\left(\frac{K}{4Sp}\right)^{\frac{1}{4}}In(Fr)\left(\frac{g}{\nu_v^2 r}\right)^{\frac{1}{4}} \tag{40}$$

Nusselt number is given by

$$Nu_v = \frac{h_{co}D}{k_v} = D\left(\frac{K}{4Sp}\right)^{\frac{1}{4}}\left(\frac{g}{\nu_v^2 r}\right)^{\frac{1}{4}}In(Fr) = 2^{-\frac{1}{4}}In(Fr)(Gr/Sp)^{\frac{1}{4}} \tag{41}$$

where Gr is Grashof number, $Gr = D^3 g(\rho_l - \rho_v)/(\rho_v \nu_v^2)$. This is the solution of film boiling heat transfer for zero shear stress at the vapor-liquid interface.

4.3.2 The Solution for Zero Vapor Velocity at y= δ.

The solution for Eq.(23) to satisfy the boundary condition given by Eq.(25) and Eq.(27) is,

$$u = \frac{\delta}{2}[K(1+8Fr\cos\phi)\sin\phi]y - \frac{1}{2}[K(1+8Fr\cos\phi)\sin\phi]y^2 \tag{42}$$

By substituting Eq.(42) to Eq.(24), we have

$$\frac{d\delta^4}{d\phi} + \frac{4}{3}\frac{\cos\phi + 8Fr\cos 2\phi}{(1+8Fr\cos\phi)\sin\phi}\delta^4 = \frac{16Sp}{K(1+8Fr\cos\phi)\sin\phi} \tag{43}$$

Eq.(43) is different from Eq.(29) only by the coefficient at the right hand term. Then, the average Nusselt number is given by

$$Nu_v = 8^{-\frac{1}{4}}In(Fr)(Gr/Sp)^{\frac{1}{4}} \tag{44}$$

This is the solution for zero vapor velocity at the vapor liquid interface.

4.4 Approximate Analytical Solution

Our experimental data for forced convection film boiling heat transfer under wide ranges of conditions in water and Freon-113 are compared with the analytical solutions for two extreme cases of boundary conditions: they are compared with Eq.(41) and Eq.(44) on the $Nu_v/(Gr/Sp)^{\frac{1}{4}}$ vs. $Fr^{\frac{1}{2}}$ plane in Fig.9. As the radiation emissivity of the platinum cylinder used in the experiment is around 0.12, the radiation contribution in the data is negligible. As shown in the figure, almost all of the data are within these two limits presented by the upper and lower dashed curves. It is considered from this fact that the principal mechanisms of the forced convection film boiling are described by this simple model. However, the experimental data are somewhat different in the trend of dependence on the Fr number from the analytical solutions: the experimental data monotonously increase with the increase in the Fr number, but the analytical results are almost constant for $0 < Fr^{\frac{1}{2}} < 0.3$, then begin to decrease and have

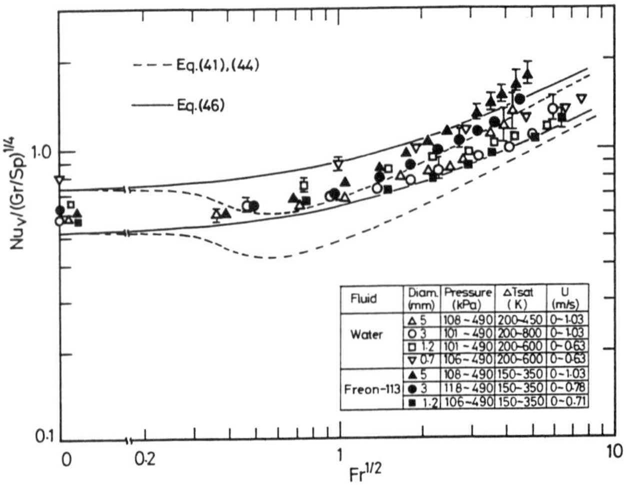

Fig. 9 Comparison of the author's experimental data with the solutions for a laminar flow film boiling model given by Eqs.(41), (44), and Eq.(46)

a minimum at $Fr^{\frac{1}{2}} \approx 0.5$, and then increase with the increase in the Fr number. Several causes can be supposed for the difference: the vapor film separation angle ϕ_s may be affected by the oscillation of the vapor liquid interface or the departure of vapor bubble or the heat transfer at the region downstream of the separation point may be different from that for the pool film boiling. This problem needs further study.

The term $In(Fr)$ in Eq.(41) and Eq.(44) is a very complicated function of Froude number($Fr = U^2/(gD)$) which can be solved only by numerical integration. As Fr number approaches 0, $In(Fr)$ approaches a constant of 0.8658, and it approaches the curve proportional to $Fr^{\frac{1}{4}}$ with Fr number grows very large. By considering those two asymptotes, the approximate equation for $In(Fr)$ is fitted to be as follows based on the experimental data.

$$In(Fr) = 0.8658(1 + 0.68Fr^{\frac{2.5}{4}})^{\frac{1}{2.5}} \tag{45}$$

Substituting this equation to Eqs.(41) and (44) yields the following equation.

$$Nu_v = K_d(1 + 0.68Fr^{\frac{2.5}{4}})^{\frac{1}{2.5}}(Gr/Sp)^{\frac{1}{4}} \tag{46}$$

Where, $K_d = 0.728$ for zero shear force and $K_d = 0.515$ for zero vapor velocity at the vapor liquid interface. The curves given by Eq.(46) are also shown in Fig.9 in comparison with the experimental data. Experimental data are almost within these two curves and the trend of dependence on Fr number is similar to these curves.

To derive approximately an analytical solution for forced convection film boiling heat transfer, the analytical solutions for two extreme cases obtained are compared with the analytical solution for pool film boiling heat transfer for the boundary conditions of equal vapor and liquid velocities and shear stresses at the vapor-liquid interface already derived by the authors (Sakurai et al. 1990a). The latter solution of the heat transfer coefficient with no radiation, h_{co}, for two-phase laminar boundary layer pool film boiling model is expressed as

$$Nu_v = 0.612M^{\frac{1}{4}} \tag{47}$$

M is a function as follows.

$$M = [Gr/Sp][E^3/\{1 + E/(SpPr_l)\}]/(RPr_lSp)^2$$

where,

$E = (A + C\sqrt{B})^{1/3} + (A - C\sqrt{B})^{1/3} + (1/3)Sc^*$

$A = (1/27)Sc^{*3} + (1/3)R^2SpPr_lSc^* + (1/4)R^2Sp^2Pr_l^2$

$B = (-4/27)Sc^{*2} + (2/3)SpPr_lSc^* - (32/27)SpPr_lR^2 + (1/4)Sp^2Pr_l^2 + (2/27)Sc^{*3}/R^2$

$C = (1/2)R^2SpPr_l$, $L' = L + 0.5c_{pv}\Delta T_{sat}$, $R = [\rho_v\mu_v/(\rho_l\mu_l)]^{1/2}$

$Sp = c_{pv}\Delta T_{sat}/(L'Pr_v)$, $Sc^* = 0.93Pr_l^{0.22}Sc$, $Sc = c_{pl}\Delta T_{sub}/L'$

Sp is the non-dimensional superheat, Sc is the non-dimensional liquid subcooling, and Sc^* is the modified non-dimensional subcooling.

For saturated boiling, $Sc = 0$, and Eq.(47) reduces to,

$$Nu_v = 0.612[\{0.5 + 2E/(SpPr_l)\}/\{1 + E/(SpPr_l)\}]^{\frac{1}{4}}[Gr/Sp]^{\frac{1}{4}} \tag{48}$$

If liquid viscosity approaches infinity which corresponds to zero velocity at the vapor-liquid interface, the value of $E/(SpPr_l)$ in the second term of Eq.(48) approaches zero and the value of the combined first and second terms approaches 0.515. On the other hand, if the liquid viscosity approaches zero, which corresponds to the zero shear stress at the vapor- liquid interface, $E/(SpPr_l)$ approaches infinity, and the combined terms approaches 0.728. It is easily understood that the solutions of Eq.(46) are consisting of those for pool film boiling heat transfer for the two extreme cases and the term $(1 + 0.68Fr^{\frac{2.5}{4}})^{\frac{1}{2.5}}$ expressing the effect of flow rate. This means that the approximate analytical solution for forced convection film boiling heat transfer under saturated condition can be obtained by substituting the first and third term of Eq.(46) for the right hand terms of Eq.(47). The approximate analytical solution of h_{co} thus obtained is as follows.

$$Nu_v = 0.612(1 + 0.68Fr^{\frac{2.5}{4}})^{\frac{1}{2.5}}M^{\frac{1}{4}} \tag{49}$$

In film boiling, radiation contribution cannot be neglected generally as the surface temperature can become very high. Sakurai et al.(1990a) has given the following analytical solution for pool film boiling heat transfer coefficients h including radiation contribution from a horizontal cylinder.

$$h = h_{co} + Jh_r \tag{50}$$

$$h_r = \frac{\sigma_s}{1/\varepsilon_w + 1/\alpha - 1}\frac{T_w^4 - T_{sat}^4}{\Delta T_{sat}} \tag{51}$$

$$J = F + (1 - F)/(1 + 1.4h_{co}/h_r) \tag{52}$$

$F = 1 - 0.25\exp(-0.13Sp_r)$ $\quad$ $for \quad F \geq 0.19$

$F = 0.19$ $\quad$ $for \quad F < 0.19$

where, h_{co} is the heat transfer coefficient if there were no radiation, J is a radiation parameter, h_r is a radiation heat transfer coefficient for parallel plates, σ_s is the Stefan-Bolzman constant,ε_w is the emissivity of cylinder heater, α is the absorptivity of liquid(taken to be unity in this paper), and $Sp_r = c_{pv}\Delta T_{sat}/(LPr_v)$.

It is supposed in this paper that Eqs.(50) to (52) are also applicable to forced convection film boiling under saturated condition, though it needs the verification by experimental data. Then, the analytical solution for forced convection film boiling heat transfer coefficient h is given by Eqs.(50), (51), (52) and (49).

5 A CORRELATION FOR FORCED CONVECTION FILM BOILING HEAT TRANSFER

Sakurai et al.(1990a, 1990b) have already made clear that the experimental data of pool film boiling heat transfer coefficients for the non-dimensional cylinder diameter $D' = 1.3$ are in good agreement with the analytical solution for the two-phase laminar boundary layer film boiling model and that the data become higher than the theoretical values as D' increases or decreases from the value. They presumed that the disturbance of the vapor liquid interface induced by growth and release of the vapor bubbles on the top of the cylinder might affect the film boiling heat transfer and make it dependent on D'. They qualitatively explained the diameter dependence by supposing that the disturbance might be proportional to d/λ_m, which is minimal at $D' = 1.3$ by their observation using a high speed video camera, where d is the departing bubble diameter and λ_m is the most dangerous wavelength of the vapor-liquid interface. They had presented the general correlation of pool film boiling heat transfer including the effects of cylinder diameter, by slightly modifying their analytical solution given by Eq.(47) based on the experimental data for various liquids under wide ranges of experimental conditions. According to their correlation, saturated pool film boiling heat transfer coefficients without radiation contribution, h_{co}, is given by the Nusselt number ($Nu_v = h_{co}D/k_v$) in the following equation.

$$Nu_v/(1 + 2/Nu_v) = K(D')M^{\frac{1}{4}} \tag{53}$$

where

$K(D') = 0.44D'^{-\frac{1}{4}}$, $\quad$ $for \quad D' < 0.14$

$K(D') = 0.75/(1+0.28D')$, $\quad$ $for \quad 0.14 \leq D' < 1.25$

$K(D') = 2.1D'/(1+3.0D')$, $\quad$ $for \quad 1.25 \leq D' \leq 6.6$

$K(D') = 0.415D'^{\frac{1}{4}}$, $\quad$ $for \quad D' > 6.6$

The denominator in the left hand side of Eq.(53) is explained by them to be the compensation factor for the similarity transformation by which the analytical solution was derived becoming inappropriate with the decrease in cylinder diameter.

Substituting the terms of analytical solution for pool film boiling in Eq.(49) for Eq.(53) yields.

$$Nu_v/(1+2/Nu_v) = K(D')(1+0.68Fr^{\frac{2.5}{4}})^{\frac{1}{2.5}}M^{\frac{1}{4}} \tag{54}$$

It is expected that the correlation given by Eq.(54) combined with Eqs.(50), (51), (52) can describe well the forced convection film boiling heat transfer at least in the low flow range where the boiling phenomena are similar to those in pool film boiling. Fig.10 shows our experimental data of forced convection film boiling heat transfer coefficients on various diameter cylinders in water and in Freon-113 plotted on $H = Nu_v/(1+2/Nu_v)/K(D')/M^{\frac{1}{4}}$ vs. $Fr^{\frac{1}{2}}$ graph in comparison with the dashed curve given by Eq.(54). The values of h_{co} in Nu_v are obtained by substracting the values of Jh_r from the experimental data of h, referring to Eq.(50), though the radiation contribution in heat transfer coefficients here obtained by using a platinum cylinder heater are almost negligible. As shown in the figure, the data for the non-dimensional diameter D' of about 1.3 (D = 3 mm in water, and D = 1.2 mm for Freon-113) are almost in agreement with the curve of Eq.(54) for the range of Fr number here tested. For the D' smaller or larger than the value, the data become smaller or higher than the dashed curve, respectively, with the increase in Fr number, though all the data almost agree with the curve for $Fr < 1$.

Equation (54) was modified as follows to agree better with our experimental data.

$$Nu_v/(1+2/Nu_v) = H(Fr,D')K(D')M^{\frac{1}{4}} \tag{55}$$

where

$$H(Fr,D') = (1+0.68Fr^{\frac{2.5}{4}})^{\frac{1}{2.5}} + 0.45\tanh\{0.04(D'-1.3)Fr\} \tag{56}$$

The curves on Fig.10 are the values calculated from Eq.(56). They are almost in agreement with the values obtained from the experimental data. The second term in Eq.(56) is considered to be the modification term representing the effect of flow rate on the thermal disturbance; this effect seems to be negligible at the minimum disturbance point of D' = 1.3. The correlation for forced convection film boiling heat transfer from a horizontal cylinder is thus derived as a combination of Eqs.(50),(51),(52),(55), and (56).

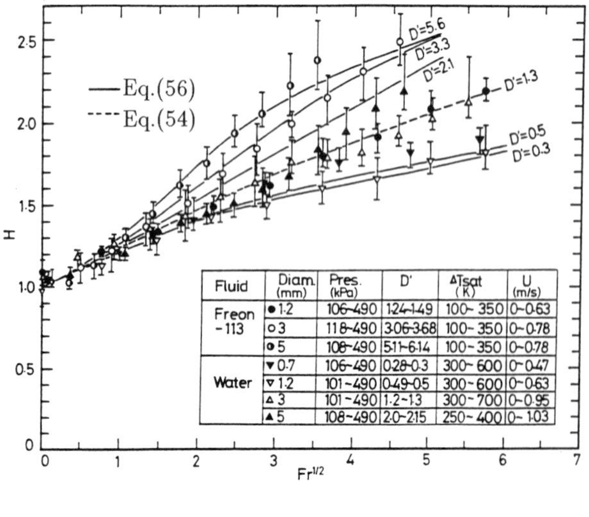

Fluid	Diam. (mm)	Pres. (kPa)	D'	ΔTsat (K)	U (m/s)
Freon -113	● 1.2	106~490	1.24~1.49	100~350	0~0.63
	○ 3	118~490	3.06~3.68	100~350	0~0.78
	◐ 5	108~490	5.11~6.14	100~350	0~0.78
Water	▼ 0.7	106~490	0.28~0.3	300~600	0~0.47
	▽ 1.2	101~490	0.49~0.5	300~600	0~0.63
	△ 3	101~490	1.2~1.3	300~700	0~0.95
	▲ 5	108~490	2.0~2.15	250~400	0~1.03

Fig. 10 Effect of flowing velocities on saturated film boiling heat transfer for various of non-dimensional cylinder diameters

6 COMPARISON OF THE CORRELATION WITH THE EXPERIMENTAL DATA

Rearranging Eq.(55), yields,

$$N = Nu_v/(1+2/Nu_v)/K(D')/H(Fr,D') = M^{\frac{1}{4}} \tag{57}$$

Typical experimental data for the film boiling heat transfer from horizontal cylinders in water and in Freon-113 are shown on N versus M graph in Figs.11(a), (b), (c) to 13(a), (b), (c) in compar-

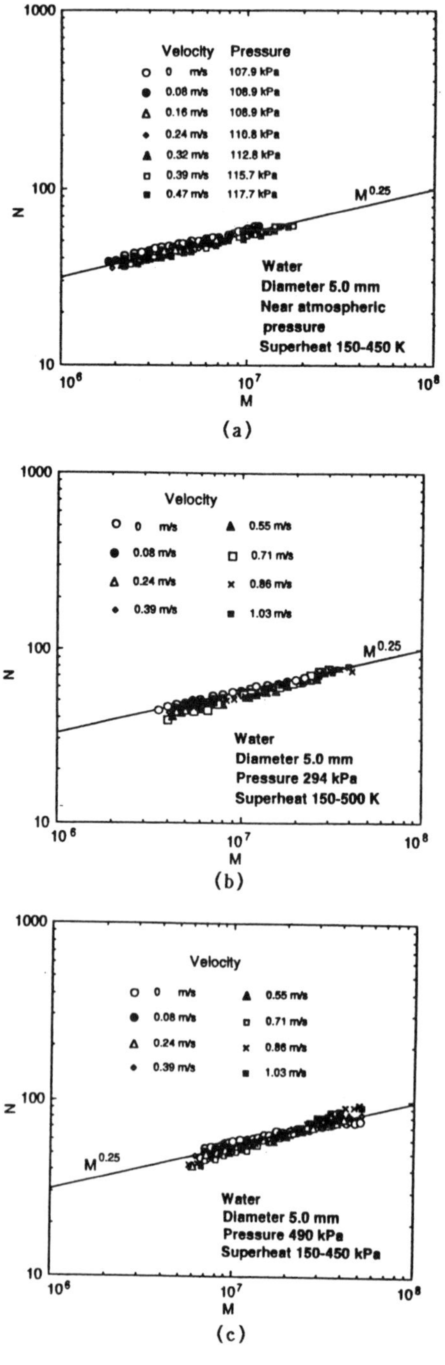

Fig. 11 Comparison of the forced convection film boiling heat transfer correlation with the experimental data on 5 mm diameter cylinder in saturated water at pressures: (a) near atmospheric pressure, (b) 294 kPa, (c) 490 kPa

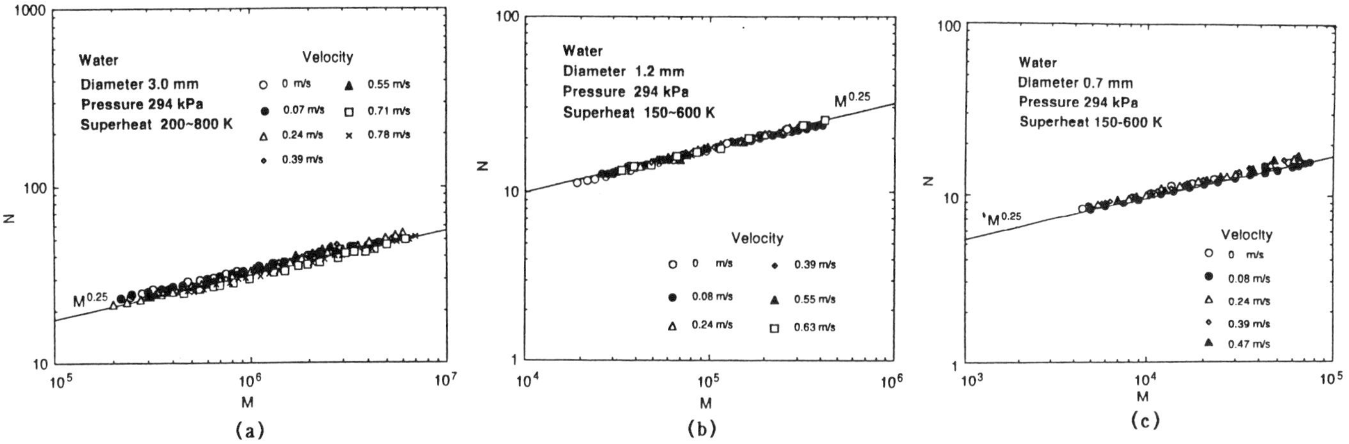

Fig. 12 Comparison of the forced convection film boiling heat transfer correlation with the experimental data on 3, 1.2, and 0.7 mm diameter cylinders in saturated water at the pressure of 294 kPa

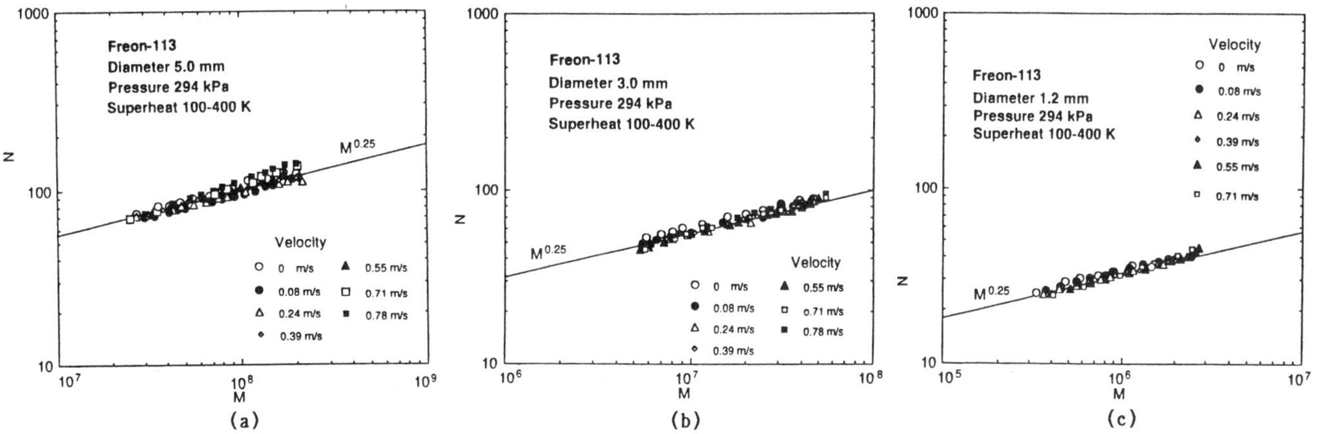

Fig. 13 Comparison of the forced convection film boiling heat transfer correlation with the experimental data on 5, 3, and 1.2 mm diameter cylinders in saturated Freon-113 at the pressure of 294 kPa

ison with the correlation. The values of h_{co} in Nu_v are obtained by substracting the values of Jh_r from the experimental data of h, referring to Eq.(50). It is seen from these figures that the correlation agrees with the experimental data for the cylinder diameters ranging from 0.7 to 5 mm for the liquid velocities ranging from 0 to 0.7 m/s at atmospheric to 500 kPa within $\pm15\%$, and within -30% to $+15\%$ for the higher velocities. The solid curves shown in Figs.3 and 4 are the film boiling heat transfer coefficients at various velocities calculated from the new correlation of forced convection film boiling heat transfer given by Eqs.(50), (51), (52), (55), and (56). They agree well with the experimental data as shown in these figures.

As mentioned in section 3.2, Bromley et al.(1953) carried out the forced convection film boiling heat transfer experiments under saturated conditions with horizontal graphite cylinders of diameters 9.83, 12.6, and 16.2 mm in four kinds of liquids such as ethyl alcohol, benzene, n-hexane, and carbon tetrachloride at atmospheric pressure. The flow rates ranged from 0 to 4.4 m/s, and the surface superheats ranged from 200 to 650 K. The radiation emissivity of the graphite cylinder heater is 0.80 which is higher than that of the platinum heater used in our experiments. The ratios of radiative effects to total heat transfer coefficients, Jh_r/h, are estimated from Eq. (50) to be between about 4% to 30%: the ratios are higher for higher surface superheat and lower flow rate. On the other hand, the ratios of radiative effects to total heat transfer coefficients are about 5% at surface superheat of 800 K in the case of our experiments. Bromley et al.'s data (1953) with the radiative effect substracted are compared with the correlation in Figs.14(a),(b),(c),(d), respectively. As seen from these figures, their data are within $\pm20\%$ of the author's correlation. It seems that the correlation can generally express the forced convection film boiling heat transfer coefficients from horizontal cylinders in various kinds of saturated liquids including the radiation from the cylinder for wide ranges of surface superheats, system pressures and cylinder diameters.

7 CONCLUSIONS

1) Film boiling heat transfer coefficients are weakly dependent on the cylinder surface superheat but are dependent on flow rate, system pressure and cylinder diameter; the heat transfer coefficients are higher for higher flow rate, higher system pressure and smaller cylinder diameter.

2) Bromley's correlation of forced convection film boiling heat transfer can not express our experimental data at higher pressures.

3) A correlation of forced convection film boiling heat transfer from horizontal cylinder was presented. This correlation can express not only the author's experimental data for wide ranges of cylinder diameters and system pressures in water and Freon-113 within $\pm15\%$ for flow rates ranging from 0 m/s(pool boiling) to 0.7 m/s and - 30% to + 15% for the higher flow rates up

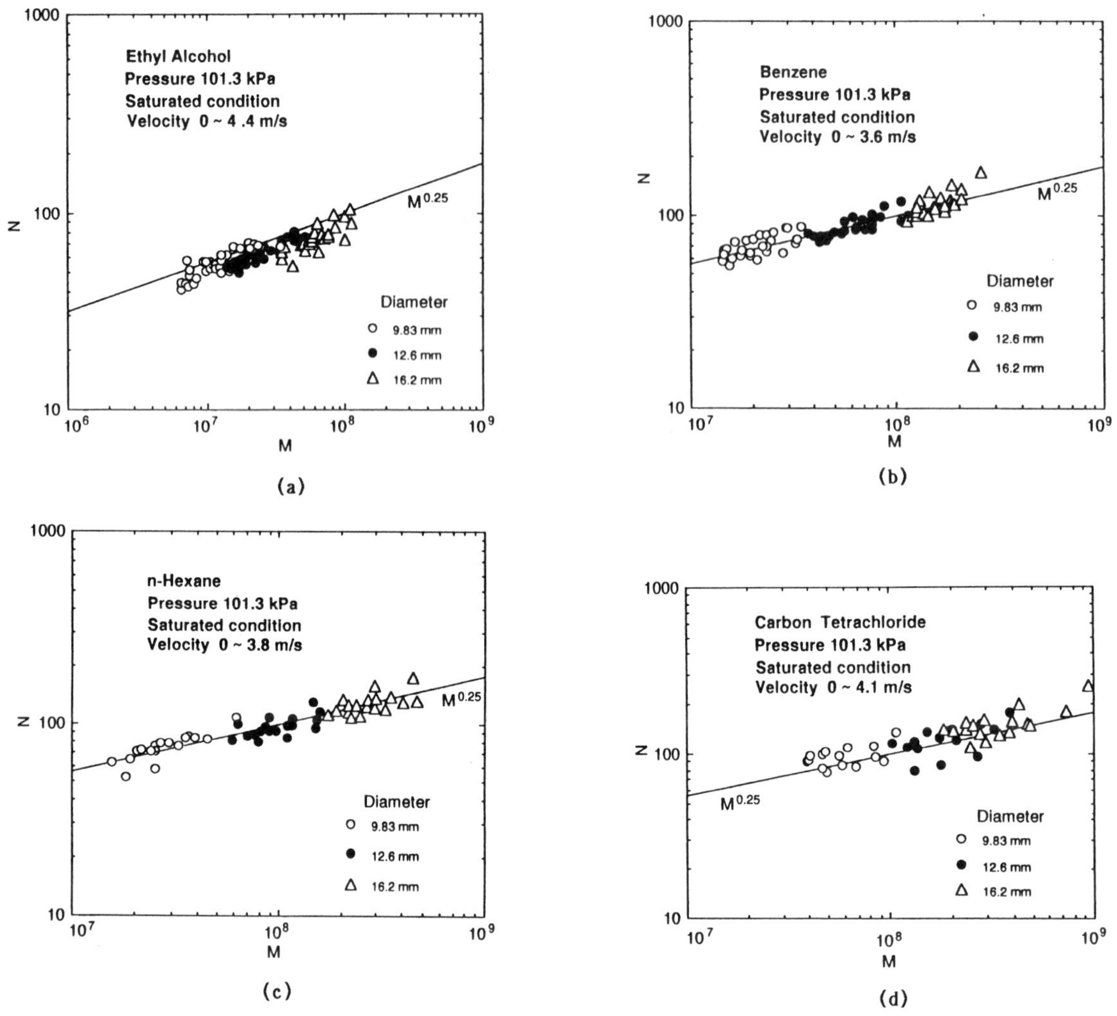

Fig. 14 Comparison of the forced convection film boiling heat transfer correlation with Bromley's experimental data on 9.83, 12.6, and 16.2 mm diameter cylinders in ethyl alcohol, benzene, n-hexane and carbon tetrachloride near atmospheric pressure

to 1 m/s, but also Bromley's data for larger diameter cylinders with higher radiation emissivities at the flow rates up to 4.4 m/s within ±20% difference.

4)It was confirmed that this correlation can generally express the forced convection film boiling heat transfer coefficients including radiation effects from horizontal cylinders in various kinds of saturated liquids under wide range of system pressure, surface superheat and cylinder diameter.

REFERENCES

Bromley, L. A., Leroy, N. R., and Robbers, J. A., 1953, "Heat Transfer in Forced Convection Film Boiling," *Ind.Eng.Chem.*, 45, 2639-2646.

Nishikawa, K., Ito, T., Kuroki, T., and Matsumoto, K., 1972, "Pool Film Boiling Heat Transfer from a Horizontal Cylinder to Saturated Liquids," *Int. J. Heat Mass Transfer,* Vol 15, pp.853-862.

Sakurai, A., Shiotsu, M., and Hata, K., 1990a, "A General Correlation for Pool Film Boiling Heat Transfer from a Horizontal Cylinder to Subcooled Liquid: Part 1 - A Theoretical Pool Film Boiling Heat Transfer Model Including Radiation Contributions and its Analytical Solution," *J. Heat Transfer,*Trans. ASME, Series C, 112, 430-440.

Sakurai, A., Shiotsu, M., and Hata, K., 1990b, "A General Correlation for Pool Film Boiling Heat Transfer from a Horizontal Cylinder to Subcooled Liquid: Part 2 - Experimental Data for Various Liquids and its Correlation," *J. Heat Transfer,*Trans. ASME, Series C, 112, 441-450.

HTD-Vol. 197, Two-Phase Flow and Heat Transfer
ASME 1992

MECHANISM OF BOILING HEAT TRANSFER IN NARROW CHANNELS

Chunlin Xia
Department of Power Engineering
Nanjing Aeronautical Institute
Nanjing, People's Republic of China

Zengyuan Guo and Weilin Hu
Department of Engineering Mechanics
Tsinghua University
Beijing, People's Republic of China

Abstract

While the literature contains many references to pool boiling heat transfer, there are as yet little information and no sucessful explanation on heat transfer enhancement offered by boiling in a confined narrow space. Experiments have been performed for saturated R−113 at heat flux very low to critical in narrow channels. The channels consist of a pair of flat, either transparent or porcelain, and isoflux plates. On the basis of our experimental results of boiling heat transfer in narrow channels with the gap sizes of 0.8, 1.5, 3.0, 5.0, and ∞ mm, an analysis of boiling heat transfer mechanism in narrow space is developed as a modification and extension of previous analyses and a model of boiling heat transfer is provided as following: For the channels with larger gap sizes, two−phase heat transfer is assumed to be a superposition of pool boiling and liquid convective component; For narrower channels, since each elementary portion of the wall surface is periodically washed either by a thin liquid film or a liquid plug, the heated surface is first cooled by evaporation for the time duration during which it is covered by the coalesced bubble; liquid flows into the space immediately after the bubble passes. Flow along the heated surface is taken into account. The results predicted from the model are consistent with the experimental heat transfer data. The influence of gap sizes, plate heights, and heat fluxes on heat transfer is given and discussed. There exists an optimum channel size of heat transfer enhancement.

Nomenclature

Symbol Quantity

- D gap size, mm
- f_w viscous drag, N / m^3
- g gravitational acceleration, m / s^2
- G mass flux, $Kg / m^2 \cdot s$
- h heat transfer coefficient,
- H height, mm
- H_{fg} latent heat of vaporization, J / kg
- j mixture mean velocity, m / s
- k thermal conductivity, W / m · ℃
- P′ effective pressure, $P-P_\infty$, N / m^2
- P_c critical pressure, N / m^2
- P_r Prandtl number
- Q heat flux, W / m^2
- Re Reynolds number
- T temperature, ℃
- U average velocity, m / s
- u_{vj} drift velocity, m / s
- x quality, dimensionless
- X_{tt} Martinelli parameter, dimensionless
- y coordinate in vertical direction, m
- ρ density, kg / m^3
- $\triangle T$ superheat, ℃
- σ surface stress, N / m
- μ dynamic viscosity, Kg / m · s
- v kinetic viscosith, m^2 / s
- δ_o initial thickness of liquid film, m
- Γ vaporation rate per unit volumn, $Kg / m^3 \cdot s$
- τ time, s
- φ weighting factor in equation (1)
- γ weighting factor in equation (1)

Subscripts

- c convection
- f liquid phase
- g vapor
- w wall
- v vaporation
- e exit

Introduction

The trend of increasing number of microelectronic devices at the chip level is continued to accelerate. Now, for the ultra large scale integration (ULSI), the number of devices per chip is kept increasing rapidly to 5×10^5 and more. In many cases, circuit power has also been increased to increase circuit speed. The overall result has been a trend towards ever higher heat fluxes at the module level. And also, the space available for cooling device in such module is getting smaller and smaller. The continued growth in functional density of microelectronic devices is today constrained by available thermal design and heat removal techniques.

In order to achieve even higher cooling rates for the electronic components, the use of liquid phase–change in the form of narrow space liquid film boiling and evaporation has been considered.

In recent years, the reduced size and the enhanced performance offered by natural convective boiling in a confined narrow space have made it increasingly attractive for wide application, although the critical heat flux (CHF) is reduced with decreasing gap sizes of the narrow channels. Microelectronic device cooling and plate–type nuclear fuel element cooling are just a few examples of the applications motivating research in this area,where miniaturization and minimization of packaging volume and weight are of paramount important. In other cases, it is very useful for resisting erosion and controlling heat transfer property in various heat transfer equipment.

According to the mechanism of the heat transfer enhancement offered by boiling on a horizontal tube bundle in shell–and–tube type evaporators, some studies [1, 2] were made employing a part of a horizontal bundle to make it clear, but were unable to elucidate the mechanism. The mechanism of the enhancement due to bubbles passing through a narrow channel with asymmetric heating can be explained from two different points of view: First, the transient thermal conduction model by Ishibashi and Nishikawa [3] in which 70% of the total heat transfer is transported by convection and the remainder by latent heat. It is assumed that the surface temperature remains constant in spite of the temperature fluctuations caused by passing bubbles. In addition, the analysis differs from reality because evaporation of the surface liquid film depends on the heat flux and bubble frequency. These effects should cause differences between theoretical and experimental results. And second, the evaporation model by Nakashima [4] that states that the thermal transports mainly by evaporation of the liquid film which appears when a bubble passes by the surface. This model derives a correlation predicting the heat transfer coefficient, $\bar{h}$, for a mean duration time of $t_b = 0.0205$s, during which the passing bubbles cover a heat tube, based on the assumption that the liquid film is between $28 \sim 47\ \mu$m thick.

Kusuda et al [5] studied the mechanism of enhanced heat transfer due to passing bubbles, using a model introduced by measuring the relationship between a change in the heated surface temperature and period at which the rising bubbles pass through. The heat transfer coefficients theoretically obtained are smaller than those measured in the experiment because the evaporation on the interface of the remaining liquid film is not taken into account.

Monde [6], Monde and Mitsutake [7], Monde and Kusuda [8], and Monde et al [9] experimentally and theoretically studied heat transfer enhancement due to bubbles passing through a vertical narrow channel with asymmetric heating. The study accounted for both the effects of evaporation and conduction. The mechanism of enhancement is: heat is first transported by a latent heat of evaporation and is followed by sensible heating of the succeeding liquid. The latent heat transport is calculated using an integral method and the sensible heating is determined exactly on the model of heat conduction. For low heat fluxes, the calculated heat transfer coefficients are in fairly good agreement with the experiments; at high heat fluxes, however, the calculated values are approximately half the experimental values, since flow along the heated surface is not taken into account.

Yao and Chang [10] studied pool boiling in a vertical annuli with a closed bottom for various gap sizes, fluids, and heat fluxes. Fujita et al. [11] investigated pool boiling heat transfer for saturated water in a confined narrow space bounded by a heated and an opposed unheated parallel rectangular plates. Within a narrow parametrical rang for boiling from vertical arrays of parallel, densely packaged Printed Circuit Boards, Bar–Cohen and Schweitzer [12] reported boiling heat transfer data obtained experimentally from a pair of flat, isoflux plates immersed in saturated water. A theoretical model of slip flow for liquid flow rate throgh the channel is successfully developed and used as a basis for correlating the rate of heat transfer from the channel walls. The proposed correlation agrees well with their experimental data. But the correlation is not widely valid.

While the literature contains many references to boiling heat transfer in a confined narrow space, especially in an annular narrow space (or / and) under asymmetric heating, little work has been conducted in the narrow channels between vertical parallel plates with symmetric heating (for boiling heat transfer from vertical arrays of parallel, densely packaged Printed Circuit Boards (PCB) with narrow spacing). Therefore, the further investigations are required to explain the mechanism of boiling heat transfer in these channels.

In our study, systematical experiments have been performed for saturated R–113 in narrow channels between a pair of flat, smooth, either transparent or porcelain, and isoflux plates. The boiling characteristics in narrow channels, including the influence of gap sizes on the heat transfer, was in details concerned. We were particularly interested in mechanism of heat transfer in narrow channels. The present study will account for both the effects of evaporation and convection (flow) on the basis of the experimental results. The theoretical results will be compared with the experimental data.

Experimental Apparatus and Procedures

Fig.1 depicts a schematic of the experimental apparatus. It consists of main power, the laser dropler anemometer (LDA), the recording system of boiling flow pattern, the signal processing system, a metal tank, and test section.

The tank includes four glass windows on the side walls to facilitate visual observation of the boiling phenomena, as can be seen in Fig.2, and encloses four auxiliary heaters used to heat liquid from room temperature to a saturated state and compensate heat losses to the surroundings in a saturated state. On the top of the tank is a water cooler which condenses freon vapor generating from the boiling, in order to maintain the saturated state in the tank under the constant pressure. The tank is large (500mm × 300mm × 900mm) enough to be expected that the walls and the auxiliary heaters would have no effect on the boiling flow in the channel. It was made of tin plated to protect from the erosion of freon. To minimize heat losses to the surroundings, outer walls of the tank are thermally insulated. The tank is connected to atmosphere through a thin tube in which a small cooler is used to condense freon vapor and send it back to the tank.

The test section is installed in the middle of the boiling tank by a

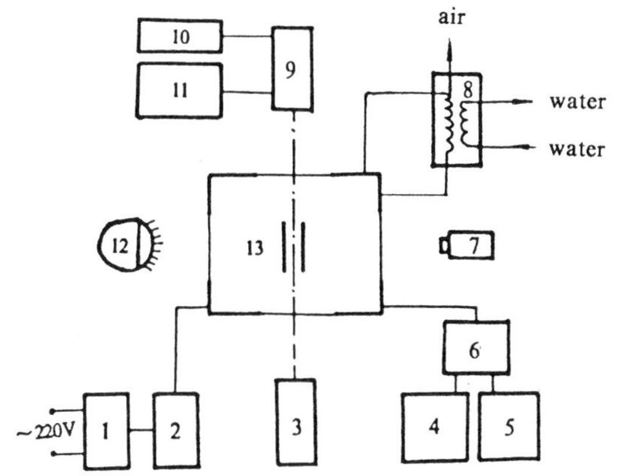

Fig.1 Schematic representation of the experimental system
1) constant–voltage supply 2) transformer 3) laser transmitter 4) multimeter 5) UCAM 6) thermocouples selector 7) camera 8) condenser 9) receiver 10) oscilloscope 11) computer 12) light source 13) test vessel

support frame which is easily adjusted and firmed. The location of the channel inlet was at least one and a half times farther away from the bottom of the tank than the channel heights, lest the tank bottom should have any effect on the flow in the inlets of the channels. Since the mass flux resulting from natural convection depends on the liquid level in the tank, this level always stays constant and the distance between the level and the inlet of the channel is kept to be 500mm in all the courses of present experiments. Gap sizes from 0.8mm to ∞ (pool boiling on a single vertical plate) of the channels are given accurately with the aid of glass or teflon spacers.

The heating plates shown in Fig.3 were made by means of plating electric heating film by a new technique. The material of the film belongs to semiconductive oxide and does not react on R–113 each other. R–113 liquid has almost the same contact angle on the film as on

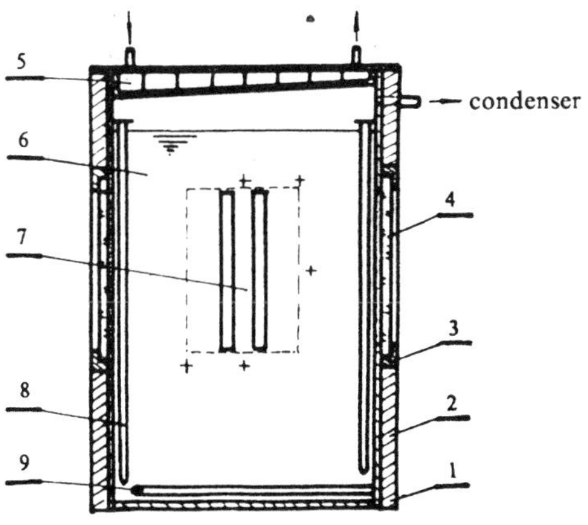

Fig.2 Geometry of boiling vessel
1) thermally insulated material 2) steel vessel 3) window cover 4) window glass 5) cooler 6) saturated water 7) test channel 8,9) auxiliary heater

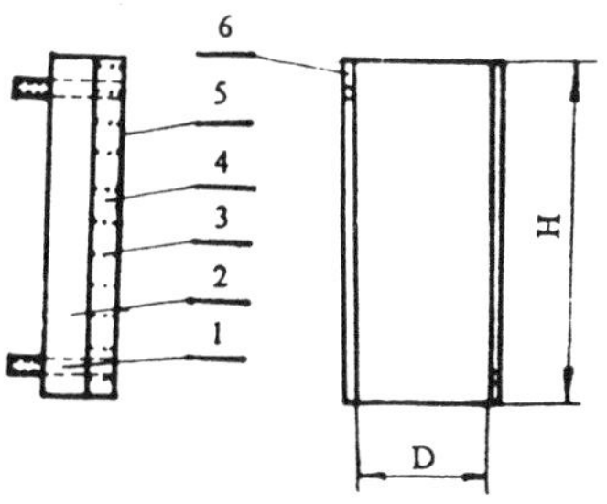

Fig.3 Geometry of the heating plate
1) wire pole 2) thermally insulated material 3) thermocouples 4) porcelain base 5) electric heating film 6) electric pole

the surface of normal carbon steel. The film has a width D of 45mm, lengths H of 56, 88, 128, and 197mm. It is transparent, very thin (less than 1 μm), uniform, and smooth. Its actual thickness, depending on the total electrical resistance of the film that is required for present experiments, does not vary across the heater surface . The required electrical resistance is adjustable in plating process. The electric poles arranged along the direction of the plate height are led to the back of the plate, having no effect on the flow field in the channels. The bases are of two types, quartz and porcelain (95% Al_2O_3). The transparent quartz plates were used for the observation and photography of the flow pattern, and the porcelain plates for the measurement of temperature distribution on the boiling surface. Before the bases were kilned, the holes had been drilled on the porcelain base for positioning thermocouples. In order to attenuate heat losses from the base to the ambient, thermally insulated material plates are attached to the back of the plates.

As a result of this design and careful assembly of heating plates, it was found feasible to obtain clear photographs of the boiling flow pattern with the help of the transparency of the quartz heating plates. In addition the uniform heat flux is as high as 100W / cm^2 across each heating surface so that flow states are easily varied from single–phase natural convection to boiling crisis.

In the course of this experiment, the heat flux across the boiling surface is determined by assuming it to be equal to the electric power input minus heat loss from the back of the heating plates. The heat loss is evaluated by the temperature gradient on the base thickness. In fact, heat transfer coefficient on the boiling surface in this experiment is very high and thermal insulation on the back sides of the heating plates is so efficient that the heat loss is less than 2% of the electric power input. Taking into consideration the uncertainties of the current and volt meters, the error thus determined the heat flux is less than 1%.

The temperature distribution on the boiling surface is measured by Copper–Constantan thermocouples of 0.1mm in diameter, inserted into the heating plate at eleven distinct elevations. At each elevation, a single thermocouple was placed at the center of the plate into a depth of about 1mm from the boiling surface and two thermocouples at different depths on two neighbour lateral sides to extrapolate the boiling surface temperature. It should be noted that by virtue of effective insulation measures on the back side of the base, the total temperature difference is always less than 2 ℃ in the direction of the base thickness in

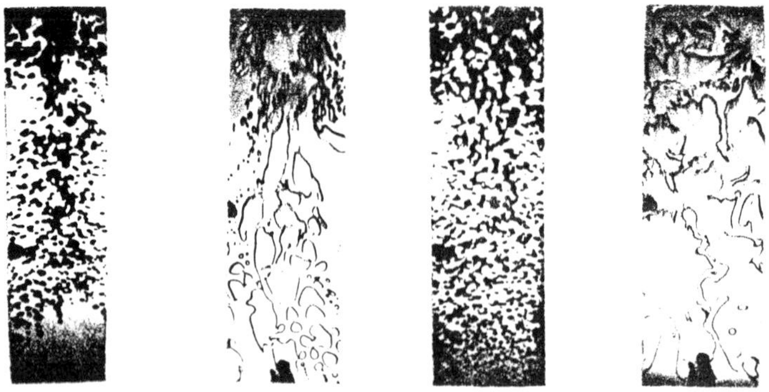

(a) D = 5.0mm
Q = 15.5Kw / m^2
$\triangle$T = 13.5 ℃

(b) D = 0.8mm
Q = 15.3Kw / m^2
$\triangle$T = 11.3 ℃

(c) D = 5.0mm
Q = 15.3Kw / m^2
$\triangle$T = 11.2 ℃

(d) D = 0.8mm
Q = 15.3Kw / m^2
$\triangle$T = 9.5 ℃

Fig. 4 The boiling flow pattern with the variety of gap size and heating process. (a, b) heating with increasing power (c, d) heating with decreasing power

all range of the experiments. The differences between temperatures on the boiling surface and at the centric measured point vary in the range from 0.1 to 0.2℃. Therefore, extrapolated temperature values have an accuracy within 0.1℃, with the uncertainties of thermocouples taken into account. The temperature distribution of the liquid in the tank, especially in the inlet of the channel, was monitored and measured by a group of thermocouples (an accuracy within 0.1℃). The voltage, current, and thermo–emf from the thermocouples were monitored, measured, and dealt with UCAM–8A / 8B (Universal Digital Measuring System, grade 0.1).

Since many interesting phenomena are related to the mass flux in the channels, the one–dimensional LDA was used to measure the velocity distribution of two–phase flow. The mass flux in the channel is evaluated by the numerical integration of the inlet velocity over the cross section of the channel inlet. The results will be discussed in [13].

All of the experimental data reported in this paper were obtained using saturated R–113 as the test fluid at atmospheric pressure. In our experiments, first is to set up saturated state. During this start–up stage, four auxiliary heaters were set at a moderate power level. The temperature distribution in the tank was monitored. As the temperature increased to near–saturated state, the power input to the heaters was slowly reduced until the saturated temperature was reached. In the meantime, the cooler went slowly into run. Finally, the saturated state and constant pressure were maintained under very low power input to the auxiliary heaters. This start–up procedure usually took about an hour to complete. After the saturated state had been maintained for ten minutes, the dissolved gas was completely degassed in the tank and the power input was conducted to the boiling surface and adjusted carefully.

Experimental Results and Discussions

Heat transfer characteristics are closely related to boiling behavior. Before the measurement of heat transfer data, boiling visualization photographs were taken using a RX–10 camera fitted with a 20～200 mm lens and a normal flash unit under the same boiling conditions.

Effects of gap sizes on boiling flow pattern. Fig.4 illustrates the boiling flow patterns in the channels with the variety of gap sizes. Photographs of the two boiling flow patterns are presented in this study.

It is found that there exists a critical gap size 3mm.

When gap size is 3mm or larger, the boiling flow pattern is very similar to that of upward forced convective boiling in vertical tubes with the larger diameters. With the increase in heat fluxes, many small sphere bubbles generated from the upper region of the boiling surface are observable first, then the region spreads down to cover all the surface. Fig.4a shows typical bahaviour of bubble generation in the nucleate boiling (the isolated bubble region).

When gap size is smaller than 3 mm, few isolated bubble generation can be observed, but large squeezed coalesced bubbles are generated within the channels, for example, D = 0.8 mm, as shown in Fig.4b, and the behavior becomes quite different from that of the isolated bubble region described above. An ebullition cycle includes nucleation, bubble growth, departure, and rise. When the boiling space is small relative to bubble sizes, bubble growth and departure will be affected by the walls. For examples, bubbles will be squeezed by the walls and flatten out. Bubble diamerters become so large that adjacent bubbles coalesce.

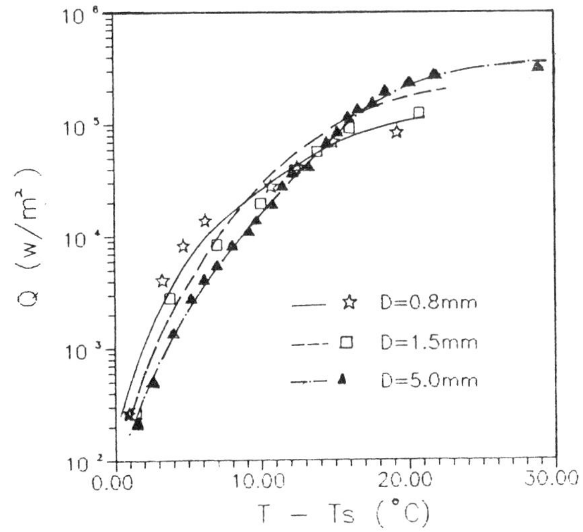

Fig. 5a Influence of gap size and heat flux on wall superheat, and comparison between experimental results and calculated results from the relation(1), point: experimental value, line: calculated result

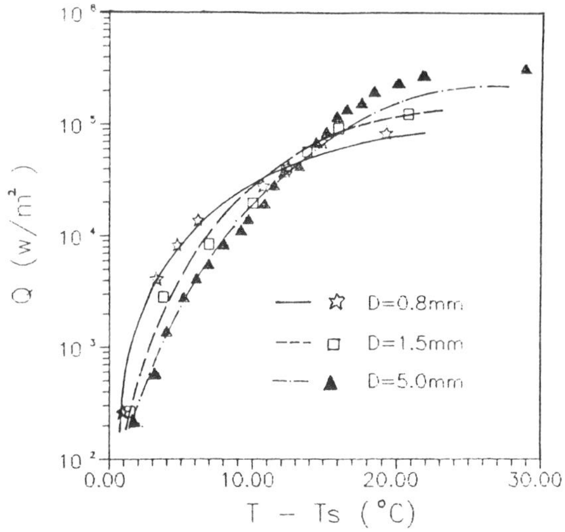

Fig. 5b Influence of gap size and heat flux on wall superheat, and comparison between experimental results and calculated results from the relation(13), point: experimental value, line: calculated result

This critical gap size is different from those reported by Yao and Chang [10] and Fujiti et al.[11] who indicated the critical gap size as 5mm. The difference is mainly caused from the different heating conditions, i.e., symmetric and asymmetric heating.

When coalesced bubbles are generated, the existence of a thin liquid film on the heating surface occupied by the bubble, which can be concluded from the heat transfer enhancement, is visible at the moment the liquid film is torn to pieces and forms the dry patches.

The boiling flow pattern depends on not only gap sizes but also the heating processes. With different heating processes (increasing power or decreasing power), the boiling flow patterns obtained is very different each other in Fig.4. Under the approximately same heat flux, wall superheat is decreased and more nucleation sites are more evenly distributed on the boiling surface in the heating process with decreasing power than increasing power.

Effects of gap sizes on heat transfer. The boiling hysteresis exists in the experiments. Heat transfer data were taken with heat flux going down step by step to avoid the boiling hysteresis effects.

With a given channel, superheat on the boiling surface increases with wall heat flux. It may be arqued that for channel spacings greater than 3 mm, the heat transfer data reflect nearly the behavior of the usual natural convective boiling in a vertical channel. When the heat flux is less than an assumed heat flux value where two boiling curves meet, the wall superheat at constant imposed heat flux decreases as the channel narrows, which can be seen for example in Fig.5. Narrow channel may enhance heat tranfer by a factor of five. When the heat flux is larger than this assumed value, in contrast for narrow channels whose spacings are less than 3mm, wall superheat increases. The curves tend to horizontal when the heat flux is enough high. This trend is connected with the phenomena that the mass velocity in the channel falls off sharply due to heavy heating and the dry patches form.

Theoretical Analysis

On the basis of our experimental results of natural convective boiling heat transfer in narrow channels between two parallel plate with symmetric heating, a model of boiling heat transfer is provided as following:

I **Channel with larger gap sizes, D$\geqslant$3mm.** When gap size is 3mm or larger, the bubbles generated in the channel will not be deformed or flatten little. The boiling flow pattern is very similar to that of upward forced convective boiling in vertical tubes with larger diameters. Nucleate boiling is predominant in this channel.

Various correlations have been proposed for forced convective boiling heat transfer in vertical tubes, but the correlations available in the literature are not good to predict present experimental data. Recently, Kilmenko [14] and Gungor Winterton [15] obtained their generalized correlations for two−phase forced flow heat transfer from a large quantity of experimental data, some data for small−diameter tubes. The predicted results from their correlations are less than present experimental data. This may be induced from the fact that the mechanism of flow and heat transfer in natural circulation is different from that in forced circulation. The reason for this difference is uncertain at present. Thus, the existing models of forced convective boiling is not directly applicable as stated above.

From the measurement of the velocity distribution, the fact that the velocity is nearly uniform in the core and sharply changing near the wall suggests that a conventional forced convection correlation may be applicabe by modification. The successful and the most widely used correlation of the heat transfer rate in flow boiling by Chen [16] and Bjorge, Hall, and Rohsenow [17], appears to validate reliance on a superposition of convective and boiling heat transfer mechanisms in evaluation two−phase thermal transport. It may thus be anticipated that the surface heat flux in natural convective boiling will be expressible as

$$q = \varphi q_c + \gamma q_b \quad (1)$$

where the terms of q_b and q_c correspond to heat flux components of pool boiling and convection respectively. φ and γ are weighting factors to be evaluated by [18]

$$\varphi = (1 + X_{tt}^{-0.5})^{1.78} \quad (2)$$

$$\gamma = 0.9622 - 0.5822\ [tg^{-1}(\frac{Re}{6.18 \times 10^4})] \quad (3)$$

$$here \quad \frac{1}{X_{tt}} = (\frac{x}{1-x})^{0.9} (\frac{\rho_f}{\rho_g})^{0.5} (\frac{\mu_g}{\mu_f})^{0.1} \quad at\ y = 0.5H \quad (4)$$

$$Re = Re_f \cdot \varphi^{1.25} \quad at\ y = 0.5H \quad (5)$$

Eqs.(2)–(4) is local correlation. They should be integrated to obtain height–average values. For simplication, the average values here are taken at the half height.

q_b may be taken from pool boiling correlation for R–113 in [19]

$$q_b^{0.3} = AF\triangle T \quad (6)$$

$$where \quad A = 2.38 \quad (7)$$

$$F = 0.7 + 2\frac{P}{P_c}(4 + \frac{1}{1 - P/P_c}) \quad (8)$$

The determination of q_c is discussed below.

Convective Component. In order to account properly for the influence of vapor–liquid slip velocity and the other different factors, such as void fraction, channel height, and so on, Re should be based on the liquid phase. With the hydraulic diameter equal to 2D, consequently, the two–phase Re can be defined as

$$Re_f = \frac{G(1-x)(2D)}{\mu_f} \quad (9)$$

Inserting this two–phase Reynolds number in the standard McAdams correlation for convective heat transfer [20], h_c is found to be

$$h_c(y) = 0.023\frac{k_f}{2D}[\frac{G(2D)}{\mu_f}]^{0.8} (1-x)^{0.8} Pr_f^{0.4} \quad (10)$$

It is thought that the resistance of convective heat transfer lies mainly in the liquid layer near the wall, so prandtl number and thermal conductivity above are given by thermophysical properties of liquid phase.

The height–averaged convective heat transfer coefficient is found to equal

$$h_c = 0.0128[\frac{G(2D)}{\mu_f}]^{0.8} (\frac{k_f}{2D})\ Pr_f^{0.4}\ \frac{H}{x_e}[1 - (1 - x_e)^{1.8}] \quad (11)$$

Although the gap size has no obvious effects on boiling flow pattern for these channels with larger spacings, its effect on heat transfer must be considered when gap sizes vary from 3mm to 10mm. Incorporating the inflluence of gap sizes in terms of the relationship between the bubble departure diameter and gap size [20], $\sigma / g(\rho_f - \rho_g)D^2$, to determine q_c, the convective component of heat flux is

$$q_c = h_c \cdot \triangle T \cdot K[\frac{\sigma}{g(\rho_f - \rho_g)D^2}]^n \quad (12)$$

with the coefficients of K, n determined empirically from a set of our experimental data of R–113 as

K = 0.031, n = 0.43

II **Channel with smaller gap sizes, D < 3mm.** In above reference, no models for boiling heat transfer in narrow channels are in good agreement with our experimental data because they are under the different heating conditions, i.e., symmetric andasymmetric heating. In Monde's [7] and Fujita's [11] mondels, the residence time (the period of the passing bubble), bubble length, and initial thickness of liquid film required for the determination of the heat flux are difficult to measure accurately. In addition, channel heights, gap sizes, void fraction, and flow along the channel are not taken into account in the meantime in their correlations. Heat transfer mechanism for the channels with smaller gap sizes is very different from that with larger gap sizes, because large squeezed coalesced bubbles are generated within the channels.

Our experimental results of boiling visualization photographs and observation in narrow channels made it possible to develop a new physical model for boiling heat transfer. The model is based on the feature that vapor bubbles flatten out in the process of their growth in the narrow channels, forming large coalesced bubbles between the heating walls and being separated from the walls by a thin film of liquid, and they rise. Since each elementary portion of the wall surface is periodically washed by either a liquid film or a liquid plug, which fills the channel gap, the local heat flux periodically varies in time.

Our model for boiling heat transfer in the channel with smaller gap sizes is based on the assumptions as following: in the period of the ebullience T_o, the heated surface is first cooled by evaporation on the thin liquid film for the time duration from t = 0 to T_v during which it is covered by the coalesced bubbles; liquid flows into the space immediately after the bubbles pass and the heated surface is cooled by liquid convection. Flow along the heated surface is taken into account. Consequently, the time–average heat flux of the heated surface is

$$q = \frac{1}{T_o}\int_o^{T_o} q(\tau)d\tau = \frac{T_v}{T_o}q_v + (1 - \frac{T_v}{T_o})q_c \quad (13)$$

where q_v and q_c stand for heat flux components of evaporation and convection respectively. Wall temperature measurement shows the heat transfer process has its transient nature in our experiments. A transient boundary layer type of formation should be required for q_c. Since q_c is much smaller than q_v from our results, for simplication, the determination of convective component of heat flux q_c is approximately given by equation(12) multiplied by a factor $C_1 = 1 + 2(D/H)$. Eq.(12) has been discussed above.

For the time duration of evaporation from $\tau = 0$ to T_v, the chang in the liquid film thickness due to evaporation is small as compared with its initial thickness——a characteristic of subcritical heat fluxes, then

$$q_v = k_f \cdot \frac{T_w - T_s}{\delta_o} \quad (14)$$

where δ_o is the initial thickness of liquid film.

According to [21], the initial thickness of liquid film can be found by the formula

$$\delta_o = Kv_f(\frac{\rho_f D}{\sigma U_g})^{1/3} \quad (15)$$

here U_g is vapor velocity, the constant K is empirically taken to be 96.8 in term of Monde's experimental data, i.e., K = 96.8

The factor T_v / T_o in equation (13) is very important for the determination of the time–average heat flux. Taking into account the fact that the factor T_v / T_o is the time–averaged void fraction, the value of T_v / T_o is determined from the expression

$$\frac{T_v}{T_o} = \frac{J_g}{U_g} \quad (16)$$

where J_g is vapor reduced velocity

$$J_g = \frac{2}{H_{fg}\rho_g D}\int_o^H q dy = \frac{2qH}{H_{fg}\rho_g D} \quad (17)$$

The curves predicted by Eq.(16) displays the same trend with Fujita's data [11].It is confirmed that the prediction is reasonable.

$$U_g = u_{vj} + j = u_{vj} + \frac{Gx}{\rho_g} + \frac{G(1-x)}{\rho_f} \quad (18)$$

Incorporating equation (12), (14), (15), (16), (17), and (18) into equation (13), the time–average heat flux of the heated surface is obtained.

The void fraction (flow quality), mass flux, and drift velocity in the computational relations (1) and (13) were determined on drift–flux model. It is assumed for simplication that the flow in the channel is one dimensional, the gas phase is saturated, and the pressure of liquid phase is equal to that of gas phase. In addition, neglect the axial conduction, stress, and heat transfer due to the impulse of interface and medium. Based on one–dimensional drift–flux model [22] with modification, the governing equations are

$$\frac{\partial(\rho U)}{\partial y} = 0 \quad (19)$$

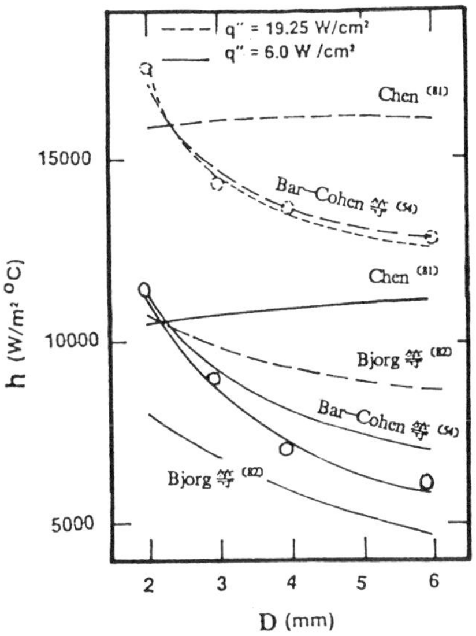

Fig. 6 Comparision among our and other boiling heat transfer correlations, with Bar-cohen's experimental data [10], point: experimental value, line: calculated result

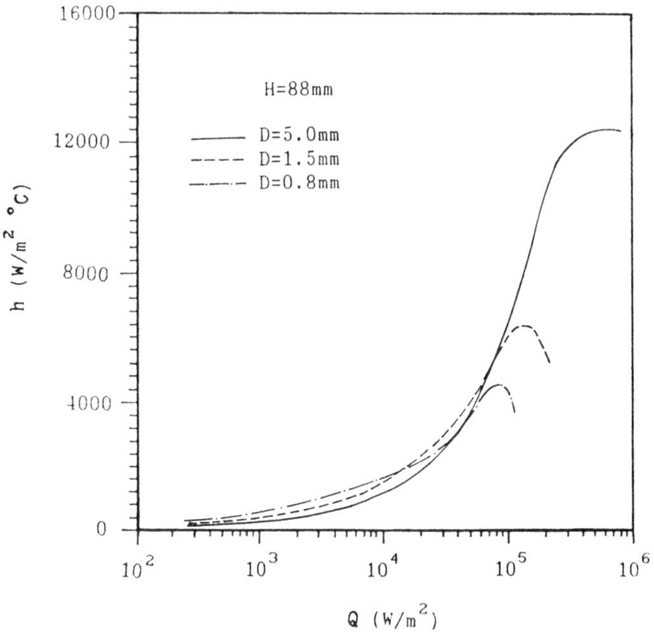

Fig. 7 Relation between boiling heat transfer coefficient and wall heat flux for three typical gap sizes

$$\frac{\partial}{\partial y}[\alpha\rho_g(U + u_{vj}\frac{\rho_f}{\rho})] = \Gamma_g \quad (20)$$

$$\frac{\partial}{\partial y}[\rho U^2 + \frac{\alpha}{1-\alpha}\frac{\rho_g\rho_f}{\rho}u_{vj}^2] = -\frac{\partial P'}{\partial y} + (\rho_s - \rho)g - f_w \quad (21)$$

However, this set of equations is mathmatically "not closed", some closure laws are needed for the practical modeling. The formulus of the drift velocity u_{vj} was adopted from [23] which is applicable to small gap sizes. The Martinelli / Nelson correlation for the in–channel viscous drag is thought as available to determine the mass flux through the narrow channel. The details about this can be seen in [24].

Modelling Results and Comparison with Experimental Data

A computer program was developed to solve the equations of the model. The results given by this program are shown hereafter.

Comparison with experimental data. The calculated results of the influence of heat flux on wall superheat, from the relations (1) and (13), are compared respectively in Fig5a,b against the experimental data at three typical gap sizes. It is found from figure 5a that the experimental data for the larger gap size, D = 5mm, are predicted well by the relation(1), while the agreement becomes less satisfactory for smaller gap sizes, D = 0.8, 1.5mm. On the contrary in Fig.5b, the calculated results from the relation (13) are in fairly good agreement with the experimental data for the smaller gap sizes; for the larger gap size, however, there exist differences between the calculated and experimental results. This reason is due to the fact that boiling characteristics of the coalesced bubble region in the narrow channel with the smaller gap sizes are very different from that of the isolated bubble region with the

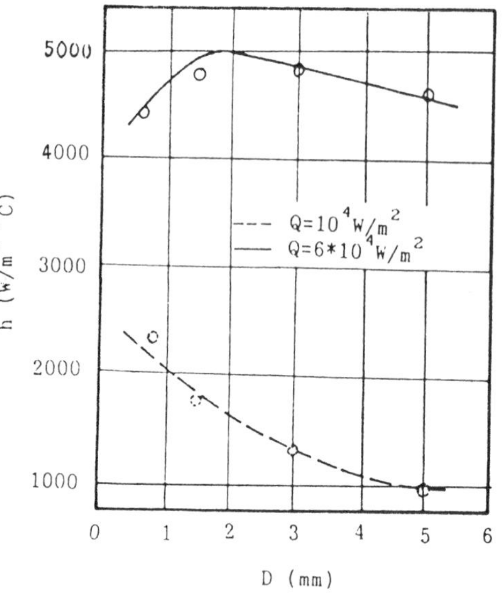

Fig. 8 Variation of boiling heat transfer coefficient with gap size, point: experimental value, line: calculated result

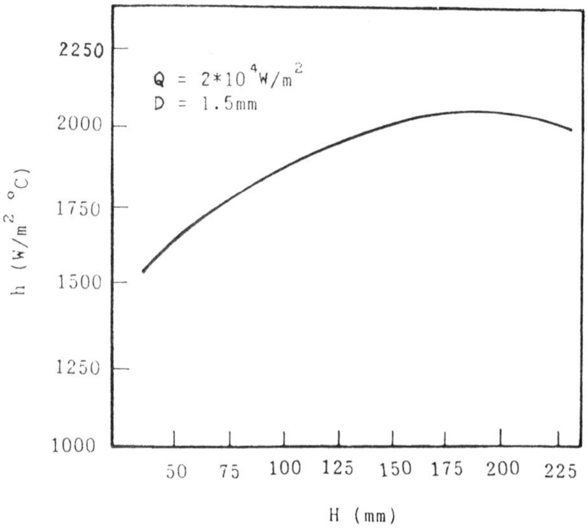

Fig. 9 Influence of gap size and heat flux on wall superheat

larger gap sizes. Nucleate pool boiling plays an important role in the heat transfer in the isolated bubble region, but evaporation on the thin liquid film is more important in the coalesced bubble region. It may be necessary that the different models are established for these two flow patterns of the boiling.

A further comparison among the proposed correlations (1) and (13), the previous correlations, with the referring experimental data of Bar–Cohen and Schweitzer [12], is shown in Fig.6 where the dependence of two–phase heat transfer coefficient on the channel gap size is displayed for two values of heat flux. Obviously, the proposed correlations agree better with the data than the previous correlations, although Bar–Cohen and Schweitzer formulation [12] clearly displays the correct trend and yields nearly their measured values. This means that the present analysis is not only limited to our tests and can be widely used for natural convective boiling in narrow channels.

Heat transfer. Fig.7 illustrates the relations between two–phase heat transfer coefficient and wall heat flux for three typical gap sizes. With a given channel, the heat transfer coefficient increases with wall heat flux by power law. When the heat flux reaches a given value, the heat transfer coefficient begins to decrease rapidly. This may be related with the phenomenon [24] that mass velocity in the channel increases with heat flux first, for the increase of the buoyancy force, then reaches it maximum, finally decreases past the maximum point and void fraction increases. The dry patches form and boiling crisis may happen.

The variation of heat transfer coefficient with gap size can be seen in Fig.8. For a low heat flux, the smaller gap sizes, the larger the boiling heat transfer coefficient, since natural convection and bubble motion enhance boiling heat transfer. The narrow channels may enhance heat transfer by five times of simple pool boiling. With decreasing gap sizes at a high heat flux, the heat transfer increases first, then reduces. Thus, there exists an optimum gap size at which the boiling heat transfer coefficient reaches its maximum.

For a given heat flux, the heat transfer coefficient increases with the channel height; when an assumed total heat load is reached, on the contrary, the heat transfer coefficient begins to decrease with the increase in the channel height, as indicated in Fig.9. There is also an optimun channel height. Combining this result with that in Fig.8, there does exist the optimum channel size in which natural convective boiling reaches its maximum capacity of heat transfer. The choice of the optimum channel size depends on the assumed boiling conditions. This is very valueable in many engineering fields where the enhancement of thermal performance and minimization of packaging sizes and weight are required.

Conclusions

In this study, an experimental investigation of boiling heat transfer has been made for saturated R–113 in a narrow channel which consists of a pair of flat, either transparent or porcelain, and isoflux plates. The influence of channel sizes and heat fluxes on boiling flow pattern and heat transfer is study. Based on experimental findings, a theoretical study has been made of heat transfer enhancement and a model of heat transfer is described in the narrow channel. The calculated results are in good agreement with the experimental values. There exists an optimum channel size at an assumed heat flux in which boiling heat transfer coefficient reaches its maximum.

Reference

1. Fujita, Y., et al., Fundamental research of the heat transfer in boiling on the horizontal tube bundle, JSME Paper 840–3, 1984.

2. Hahne, E. and Müller, J., boiling on a finned type and a finned tube bundle, Int. J. Heat Mass Transfer 26, 849~859, 1983

3. Ishibashi, E. and Nishikawa, K., Saturated boiling heat transfer in a narrow space, Int. J. Heat Mass Transfer 12, 863~894, 1969.

4. Nakashima, K., Boiling heat transfer outside a horizontal multitube bundle, Trans. JSME 42, 1047~1059, 1978.

5. Kusuda, H., Monde, M., Uehara, H., and Nakaoka, T., Bubble influence on boiling heat transfer in a narrow space, Heat Transfer Japen. Res. 9, 49~60, 1981

6. Monde, M., Characteristics of heat transfer enhancement due to bubbles passing through a narrow vertical channel, J. Heat Transfer 110, 1016~1019, 1988

7. Monde, M. and Mitsutake, Y., Enhancement of heat transfer due to bubbles passing through a narrow vertical rectangular channel, Int. J. of Multiphase Flow 15, No.5, 803~814, 1989.

8. Monde, M., Kusuda, H., Uehara, H. and Nakaoka, T., Boiling heat transfer in a narrow rectangular channel, Proc. 8th Int. Heat Transfer Conf., Vol. 4, 2105~2110, 1986.

9. Monde, M., Mihara, S. and Noma, T., Enhancement of heat transfers due to bubbles passing through a narrow vertical rectangular channel (effect of subcooling on heat transfer), Trans. JSME Ser. B55, 483~489, 1989.

10. Yao, S. C. and Chang, Y., Pool boiling heat transfer in a confined space, Int. J. Heat Mass Transfer 26, 841–848, 1983.

11. Fujita, Y., Ohta, H., Uchida, S. and Nishikawa, K., Nucleate boiling heat transfer and critical heat flux in narrow space between rectangular surfaces, Int. J. Heat Transfer 31, pp.229–239, 1988.

12. Bar–Cohen, A. and Schweitzer, H., Thermosyphon boiling in vertical channels, ASME Journal of Heat Transfer 107, 772~ 778, 1985

13. Xia, C., Hu, W., and Guo, Z., Characteristics of natural convective boiling in narrow channels, First European Thermal Science and Third UK NationalHeat Transfer Conference, Univ. of Birmingham, UK, 1992.

14. Kilmenko, V. V., A Generalized correlation for two–phase forced flow heat transfer–Second assessment, Int. J. Heat Mass Transfer 33(10), 2073–2088,1990.

15. Gungor, K. E. and Winterton, R. H. S., Simplified general correlation forsaturated flow boiling and comparisons of correlations with data, Chem. Eng. Res. Des. 65, March 1987.

16. Chen, J. C. A correlation for boiling heat transfer to saturated fluid in convective flow, ASME Paper No. 63–HT–34, 1963

17. Bjorge, R. W., Hall, G. R. and Rohsenow, W. M., Correlation of forced convection boiling heat transfer data, Int. J. Heat Mass Transfer, vol.25, No.6, 753~757, 1982.

18. Schlünder, E. U., Heat Exchanger Design Handbook, Hemisphere Publishing Corporation, 1983.

19. Kakac, S., Bergles, A. E. and Maginger, F., Heat exchangers thermal Hydraulic Fundamentals and Design, Advanced study Institute Book, Hemisphere Pub., Wash., 1981.

20. Rohsenow, W. M. and Choi, H., Heat, Mass and Momentum Transfer, Prentice– Hall, Englewood Cliffs, N.J, 1961

21. Grigoriev, V.A., et al., Towards the problem on the determination of the liquid film thickness under a bubble during boiling in capillary channels, Trudy MEI (Proceedings of the Moscow Power Engineering Institute), No.200, 52~56, 1974.

22. Reocreux, M., et al., The one–dimensional drift–flux model, BNL–NURGE– 50569, pp.36~40.

23. Ünal, H. C., Determination of the drift velocity and the void fraction for the bubble–and–plug–flow regimes during the flow boiling of water at elevated pressures, Int. J. Heat Mass Transfer 21, 1049 ~1056, 1978.

24. Xia, C., Guo, Z. and Hu, W., Thermal drag in natural convective boiling, the 1989 winter Annual Meeting, ASME Special Publication, HTD–Vol. 90, pp.9~13, 1989

HTD-Vol. 197, Two-Phase Flow and Heat Transfer
ASME 1992

EXPERIMENTAL EVALUATION OF CONVECTIVE BOILING OF REFRIGERANTS HFC-134a AND CFC-12

J. Wattelet, J. M. Saiz Jabardo, J. C. Chato, J. S. Panek, and A. L. Souza
Department of Mechanical and Industrial Engineering
University of Illinois at Urbana—Champaign
Urbana, Illinois

ABSTRACT

Experimental heat transfer data are reported for in-tube evaporation of HFC-134a and CFC-12. Testing was conducted in a horizontal, smooth copper tube which had a 10.21 mm diameter and was 2.43 m long. Heat to the test section was provided by tape heaters wrapped along the tube. Sight glasses were installed at the entrance and exit of the test section for observation of flow regimes. Experiments were conducted for HFC-134a and CFC-12 with test conditions as follows: mass flux, 100-500 kg/m^{2}s; heat flux, 5-30 kW/m^2; inlet quality, 20-60 percent; saturation temperature, 5 deg C. Flow regimes observed varied from stratified and wavy at lower mass fluxes and qualities to annular and misty-annular at higher mass fluxes and qualities. The predominant flow regime observed was annular flow. Heat transfer coefficients for HFC-134a were on average 25% higher than those for CFC-12 at similar test conditions in the annular flow regime. For the test conditions considered, minimal effects of heat flux were found and the data was accurately correlated using a purely convective relation. However, as the heat flux was increased, the transition to dryout began at lower qualities.

NOMENCLATURE

A_s	area of the heat transfer surface [m^2]
Bo	boiling number = q''/Gi_{fg}
G	mass flux [kg/m^{2}s]
h	heat transfer coefficient [W/m^{2}K]
i_{fg}	latent heat of vaporization [J/kg]
q	heat input rate [W]
q"	heat flux [W/m^2]
T	temperature [deg C]
x	quality
μ	dynamic viscosity [Pa•s]
ρ	density [kg/m^3]
χ_{tt}	Lockhart-Martinelli parameter = $\left(\frac{1-x}{x}\right)^{0.9}\left(\frac{\rho_v}{\rho_l}\right)^{0.5}\left(\frac{\mu_l}{\mu_v}\right)^{0.1}$

Subscripts

ave	Average value along the test section
b	Bulk
i	Inlet of the test section
l	Either liquid phase or the liquid phase flowing along in the tube
nb	Nucleate boiling

INTRODUCTION

HFC-134a is a potential replacement for CFC-12, which is used in household refrigerators and automotive air conditioners among other refrigeration applications. Because of the phaseout of CFC-12 required by the Montreal Protocol and the tightening energy efficiency standards imposed by the federal government, the performance of HFC-134a relative to CFC-12 needs to be determined for different components of vapor-compression refrigeration systems. One component in these systems is the evaporator. Due to differences in the physical properties of HFC-134a and CFC-12, the evaporator design may need to be reassessed. The refrigerant-side heat transfer coefficient is an important design parameter needed to make this assessment. Information regarding heat transfer coefficients during convective boiling inside horizontal tubes for HFC-134a and CFC-12 may reliably be obtained through experimental evaluation. This paper will summarize a study conducted to provide this information.

Convective boiling heat transfer for CFC refrigerants and other fluids has been extensively examined during the past few decades. Numerous correlations have been developed regarding heat transfer coefficients. During the 1950s and early 1960s, initial correlations were developed to relate the two-phase to the single-phase heat transfer coefficient, based on the premise that the mechanism of heat transfer in forced convection with evaporation was similar to single-phase forced-convection [1]. By applying the Reynold's analogy that relates the energy transport mechanism to momentum transport in convection, it was shown that the ratio between the two-phase flow and the single-phase liquid heat transfer coefficients could be exclusively correlated by the Lockhart-Martinelli parameter, χ_{tt}. The form of this correlation is as follows:

$$\frac{h_{TP}}{h_l} = f(\chi_{tt}) \qquad (1)$$

On the other hand, it was observed that nucleate boiling could occur simultaneously with evaporation along an extensive liquid-vapor interface. To account for these effects, Shrock and Grossman [2] introduced the boiling number, Bo, into the form of Eq. (1) as follows:

$$\frac{h_{TP}}{h_l} = f(\chi_{tt}, Bo) \qquad (2)$$

Chen [3] proposed a correlation based on the superposition of heat transfer coefficients due to nucleate and convective boiling effects as follows:

$$h_{TP} = Sh_{nb} + Fh_l \qquad (3)$$

where S is a suppression factor and F is a function of χ_{tt}. It can be noted that F may assume similar characteristics as $f(\chi_{tt})$ in Eq. (1).

After numerous studies were conducted on specific fluids, correlations were developed during the late 1970s and 1980s which combined these separate studies into large databases. One of these correlations was developed by Shah [4], by breaking up two-phase flow boiling into nucleate boiling, bubble suppression, and convective boiling regimes. Most of the other generalized correlations can be thought of as modifications of those of Chen and Shrock and Grossman. Included are the correlations of Kandlikar [5] and Gungor-Winterton [6].

Recently, a few experimental studies on evaporation heat transfer coefficients of HFC-134a have been conducted. Eckels and Pate [7] compared average heat transfer coefficients for HFC-134a and CFC-12 in a smooth horizontal tube over a quality range of 0.1-0.9. For similar mass fluxes, they found that heat transfer coefficients for HFC-134a were 35-45% larger than those for CFC-12. The correlations of Shah, Kandlikar, Chaddock-Brunemann [8] and Gungor-Winterton predicted their data within ±25%. No attempt was made to correlate their own data. Hambraeus [9] studied the local and average evaporation heat transfer coefficients of HFC-134a and HCFC-22 in a smooth, horizontal tube. She found that heat transfer coefficients for HFC-134a were significantly higher than those of HCFC-22. The Jung and Radermacher correlation [10] overpredicted her results by 50% and the correlation of Pierre [11] predicted her data within ±40%. Takamatsu, Momoki, and Fujii [12] also studied evaporation heat transfer coefficients for HFC-134a and CFC-12 in a smooth, horizontal tube. They found that HFC-134a local heat transfer coefficients were 25% higher than than those CFC-12 for the same saturation temperature, mass flux, heat flux, and quality.

Extensive experimental research is presently being conducted to determine heat transfer coefficients for alternative refrigerants, providing a significant opportunity to reevaluate existing correlations and improve them. The prediction of heat transfer coefficients can only be improved if the flow regimes are identified, the suppression of nucleate boiling is adequately addressed, the dimensionless parameter(s) that influence the particular flow regime is determined, and the transition to dryout is identified with respect to flow conditions. Some of these topics have been previously addressed in the past. However, additional information is still needed to understand the physical mechanisms involved in convective boiling. Some of the aforementioned needs were considered in the research reported in this paper. The objectives of this research were to

(1) experimentally determine heat transfer coefficients for HFC-134a;
(2) compare these results with experimental baseline data for CFC-12;
(3) investigate flow patterns and transitions during the evaporation process;
(4) examine the effect of mass flux, heat flux, and quality;
(5) correlate the data and compare with existing correlations;
(6) compare the data with recent HFC-134a studies.

EXPERIMENTAL FACILITIES

Refrigerant loop

The test apparatus was designed to examine heat transfer characteristics in a horizontal, single-tube evaporator. An overview of this apparatus will be presented in this section. The reader is referred to references [13, 14] for a more detailed description. Figure 1 is a schematic of the experimental test facility. A variable-speed gear pump was used to circulate the refrigerant around the loop, eliminating the need for a compressor and an expansion device found in vapor-compression refrigeration systems. This allowed testing capabilities in a pure refrigeration environment without the influence of the oil from the compressor. A 9-kW preheater was used to control inlet qualities to the test section. Heat input rates in the preheater were set using a 4-20 mA silicon controlled rectifier (SCR), and measured using a watt transducer. This meter was factory calibrated within ±25 W, and its accuracy verified during preliminary single-phase energy balance testing. Heat was removed from the refrigerant using two parallel condensers cooled by an ethylene glycol-water mixture from a chiller. Flow rate was measured using a positive displacement flow meter. This device was also factory calibrated and is accurate to ±0.0379 liters/min. Absolute pressure was measured at several locations using strain-gage pressure transducers (0-400 kPa). These transducers were calibrated using a dead weight tester and are accurate to ±8 kPa. Fluid temperatures were measured using type T thermocouple probes, calibrated to within ±0.2 deg C using a thermostatic bath and a NIST-traceable thermometer in a temperature range between -25 deg C and 25 deg C.

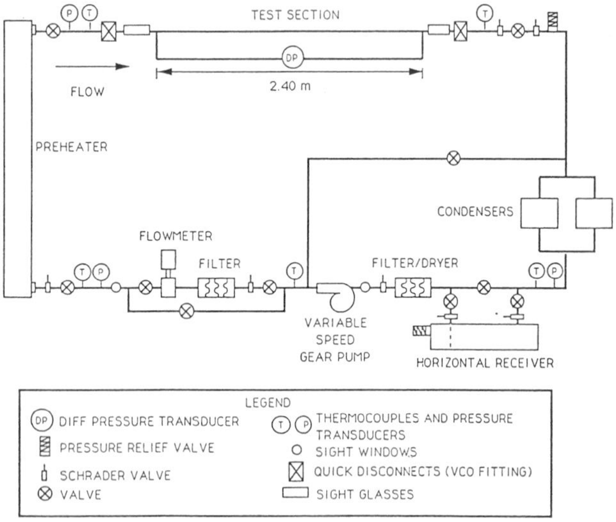

Fig. 1 A schematic diagram of the test loop

Test section

The test section was a 2.43 m long, 10.21 mm i.d. copper tube. Heat input was provided to the test section using SCR-controlled, surface-wrapped heaters. The heat input rate was measured by a watt transducer which was factory calibrated with an accuracy of ±10 W. This calibration was verified during single-phase energy balance testing. To reduce heat gain from the environment, the test section was covered with 0.05 m of foam insulation. Twenty type T thermocouples were soldered in grooves cut longitudinally along the test section to measure the surface temperature. Figure 2 shows the location of these thermocouples. Bulk fluid temperatures were measured at the inlet and outlet of the test section using type T thermocouple probes, calibrated similar to the manner discussed in the previous section. A 0-35 kPa differential pressure transducer was used to measure pressure drop along the test section. This device was calibrated using a dead weight tester and is accurate to within ±0.3 kPa. Sight glasses were installed at the inlet and outlet of the test section to observe flow patterns. Flow visualization was enhanced using a strobe light.

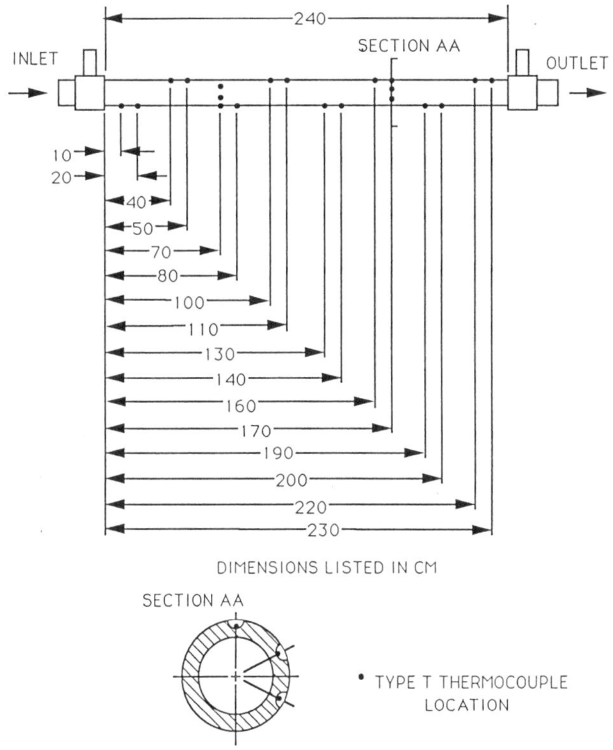

Fig. 2 Location of surface temperature thermocouples along the test section

<u>Data collection</u>

Prior to running, the system loop was evacuated and then charged with refrigerant. Saturation pressure based on thermocouple readings matched saturation pressure based on pressure transducers within 2 kPa. Noncondensibles, such as air, where not found in the system. However, Schrader valves were located in the upper section of the loop for purging purposes if noncondensibles would become a problem.

Data collection was performed using a micro-computer and data acquisition system. The data acquisition system consisted of 5 terminal panels for 40 transducer connections, four data acquisition cards connected to the computer via Nubus slots, and data acquisition software. Testing was conducted at steady-state conditions, which were monitored and controlled by the above-mentioned system. Electric heaters were controlled through 4-20 mA analog output signals sent to the SCRs.

Parameters controlled during tests were mass flux, heat flux, inlet quality, and saturation temperature. Steady-state conditions, reached in approximately 30 minutes to 2 hours, were assumed when the time variation of the saturation temperature was less than 0.1 deg C. The controlled parameters also had to be within the following range of target values: mass flux, ±5%; heat flux, ±5%; inlet quality, ±0.01; saturation temperature, ±0.2 deg C. Once steady-state conditions were achieved, the transducer signals were logged into a data acquisition output file. The channels were scanned once every second for a period of 60 seconds. The values were then averaged and reduced using a spreadsheet macro. Refrigerant transport and thermodynamic properties used during data collection and reduction were obtained from references [15,16,17].

RESULTS

Experimental heat transfer coefficients were determined by Newton's law of cooling using the averaged values of surface and bulk fluid temperatures, surface area of the test section, and heat input rate to the test section as follows:

$$h = \frac{q / A_s}{(T_s - T_b)} \qquad (4)$$

with T_s defined as:

$$T_s = 0.25\, T_{top} + 0.50\, T_{side} + 0.25\, T_{bottom} \qquad (5)$$

where T_{top}, T_{bottom}, and T_{side} are, respectively, the average of the top, bottom, and side temperatures. For testing discussed below, the observed circumferential wall temperature variation was between 0.5 and 2.0 deg C. Since surface temperatures were measured inside grooves located along the external surface of the copper tube, the temperature drop across the tube wall had to be considered. However, its value was found to be negligibly small, and, therefore, it was disregarded. Axial heat conduction along the length of the tube was also neglected. Heat gain from the environment was determined during single-phase energy balance testing.

Uncertainties for the experimental heat transfer coefficients were determined using the method of sequential perturbation outlined by Moffat [18]. Uncertainties in each of the independent variables used to calculate the heat transfer coefficient from Eq. (4) were estimated based on calibration and examination of system-sensor interaction errors, system disturbance errors, and conceptual errors. The experimental uncertainties for the heat transfer coefficients for HFC-134a were between 3.9% and 15.0% and were between 3.0% and 14.8% for CFC-12. For one test of low heat flux (5 kW/m^2) and high mass flux (500 kg/m^{2}s) temperature differences in Eq. (4) were small (under 2 deg C) and resulted in uncertainties of 32.0% for HFC-134a and 28.0% for CFC-12.

Testing was conducted over the range of parameters shown in Table 1. It should be noted that, due to differences in the enthalpy of vaporization between HFC-134a and CFC-12, the resulting outlet qualities of CFC-12 tests were greater than those conducted with HFC-134a.

Table 1. Range of experimental parameters in the evaporation test section

Parameter	**HFC-134a**	**CFC-12**
Temperature (C)	5.0	5.0
Pressure (kPa)	350	363
Mass Flux (kg/m^{2}s)	100, 300, 500	100, 300, 500
Heat Flux (kW/m^2)	5,10,20,30	5,10,20,30
Inlet quality (%)	20, 40, 60	20, 40, 60

Several flow regimes were observed through the sight glasses at the inlet and outlet of the test section. The flow regimes were predominantly annular, with wavy-stratified, wavy-annular, and mist flow regimes also observed. For mass fluxes of 300 and 500 kg/m^{2}s, flow regimes were annular-wavy, annular, and mist. For the 100 kg/m^{2}s cases, the flow regimes were wavy-stratified, annular-wavy, and mist. Dryout occurred during some tests with relatively high qualities, when the annular liquid film on the wall began to completely evaporate. This was detected both by visual observation and a sudden decrease in heat transfer coefficient due to a rise in wall temperature. The thermodynamic quality at this condition can be less than unity, as droplets of liquid are still carried in the vapor core. It should also be noted for this regime that annular flow was intermittently present with the dryout occurring in an oscillatory pattern. In horizontal tubes with uniform heat flux, dryout first occurs in the top region of the tube where the liquid layer is thinner. This was also confirmed by visual observation. The dryout region also varied with the magnitude of the applied heat flux. Dryout began to occur between 55 and 70 percent quality for the 20 and 30 kW/m² cases and between 75 and 85 percent for 5 and 10 kW/m². These results indicate that the dryout region can occur over a large portion of a typical evaporator for higher heat fluxes. Figure 3 presents a rough estimation of the transition between annular flow and the dryout region. This transition is weakly dependent on the mass flux.

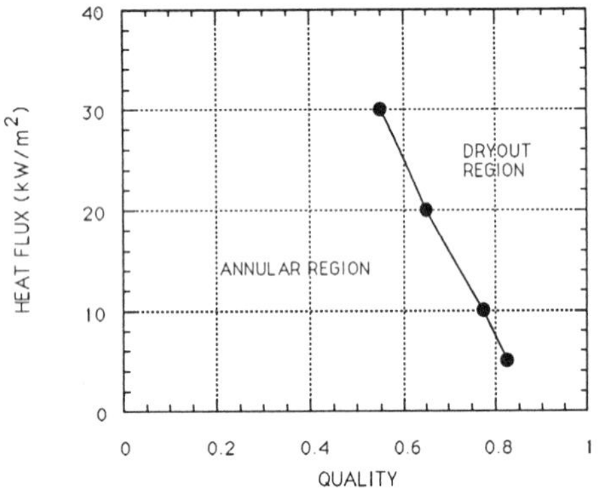

Fig. 3 Transition between annular flow and dryout for varying heat flux for HFC-134a during flow boiling. Mass flux: 300 kg/m^{2}s; saturation temperature: 5 deg C.

The effect of the average test section quality (x_{ave}) can be seen in Fig. 4. In general, the heat transfer coefficient increases with the average quality in the annular flow regime. Intense evaporation at the liquid-vapor interface diminishes the liquid film thickness, reducing the thermal resistance, which is associated with heat conduction across the film. There is no major effect of quality on circumferentially averaged heat transfer coefficients in the wavy-stratified flow regime, that typically occurred for a mass flux of 100 kg/m²s. In this flow regime, top heat transfer coefficients were lower than the bottom ones due to the wetting of the bottom surface.

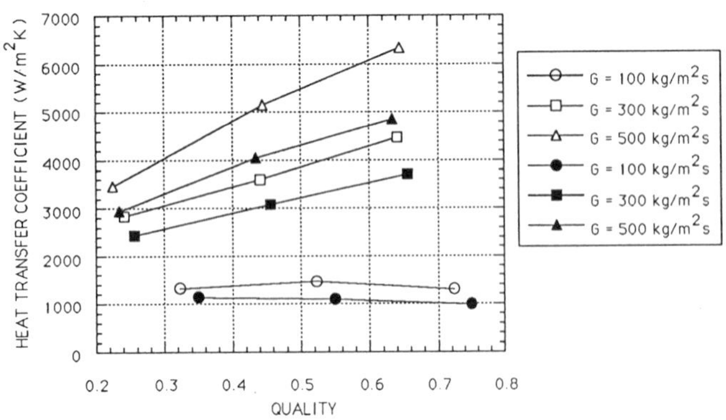

Fig. 4 Heat transfer coefficient versus average quality for HFC-134a (white symbols) and CFC-12 (black symbols) during flow boiling. Heat flux: 5 kW/m^2; saturation temperature: 5 deg C.

The effect of mass flux, G, on the heat transfer coefficient can also be seen in Fig. 5. The heat transfer coefficient increases with the mass flux due to both higher liquid velocities and to transition of flow regimes. At low mass fluxes (G=100 kg/m²s), the flow regime is predominantly wavy-stratified, with only the bottom part of the surface wetted by refrigerant. As mass flux increases, the flow regime changes to an annular pattern. The liquid completely wets the wall, and a vapor core flows in the center of the tube. Heat transfer coefficients are typically much higher in the annular than in the stratified regime due to the complete wetting of the surface of the tube through a thin film. The thinner this film, the lower its thermal resistance, and, as a result, the higher the heat transfer coefficient, as previously noted.

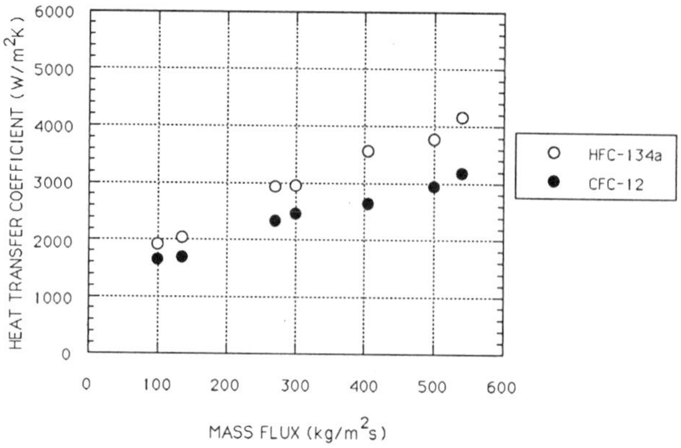

Fig. 5 Heat transfer coefficient versus mass flux for HFC-134a and CFC-12 during flow boiling. Heat flux: 10 kW/m^2; saturation temperature: 5 deg C.

Varying the heat flux in the test section from 5 to 10 kW/m² does not seem to affect the heat transfer coefficient significantly for tests conducted in the annular flow regime, as shown in Fig. 6. Based on this type of variation, it can be inferred that convective boiling was the dominant mode of heat transfer for the test conditions of this study. Nucleate boiling was mostly suppressed due to a deficiency in wall superheat. However, for the 20 and 30 kW/m² cases, with average qualities between 30 and 35 percent, heat flux does appear to have an effect. Nucleate boiling may be present in these cases. For 20 kW/m², at an average quality of 50 percent, the nucleate boiling contribution again appears to be suppressed, probably due to significant surface cooling promoted by the thinning of the film. Overall, out of the 60 tests conducted for the range of parameters shown in Table 1, only 6 appear to present some effect of the heat flux. Further testing in the higher heat flux, lower quality region needs to be conducted to substantiate the contribution of nucleate boiling to the heat transfer coefficient and the point at which it is suppressed.

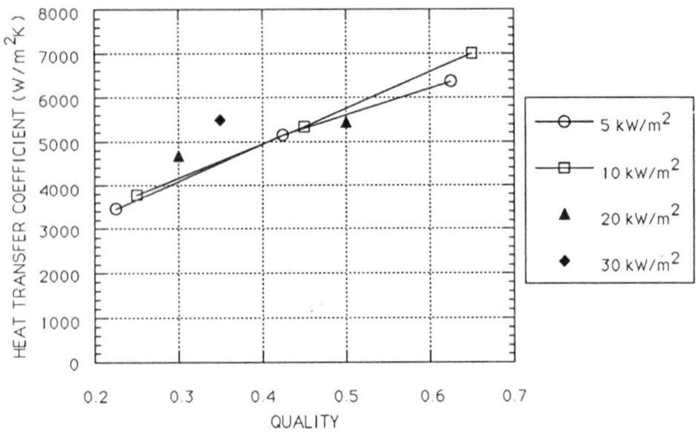

Fig. 6 Effect of heat flux on heat transfer coefficient for HFC-134a during flow boiling. Mass flux: 500 kg/m^{2}s; saturation temperature: 5 deg C.

As can be concluded from examination of Figs. 4 and 5, HFC-134a has a higher heat transfer coefficient than CFC-12 at identical conditions. This can be explained in part by the fact that HFC-134a has a higher liquid thermal conductivity than CFC-12, thus reducing the thermal resistance of the film in the annular flow regime. Heat transfer coefficients for HFC-134a are on average 25% larger than those for CFC-12.

CORRELATION OF RESULTS

Five correlations were selected from the literature to compare with the experimentally obtained heat transfer coefficients. The Chaddock-Noerager correlation was selected because it was based on experimental testing of CFC-12 as the sole evaporating fluid and its method of providing a constant heat flux to the test section. The correlation is of a form similar to Shrock and Grossman, whereby the heat transfer coefficient ratio is a function of the Lockhart-Martinelli parameter and the boiling number.

The correlations of Shah, Gungor-Winterton, and Kandlikar were selected due to their generalized nature, being developed for large databases from several experimenters that use CFC-12 as one of the primary fluids. Shah developed a generalized correlation, based on 780 data points from 19 independent experimental studies, that had a mean deviation of 14 percent for eight different fluids. Results were correlated based on different regimes of boiling. In this correlation, two-phase flow boiling is broken up into a nucleate boiling, a convective boiling, and a bubble suppression regime in which both bubble nucleation and convective effects are significant.

Gungor and Winterton developed a generalized correlation for water, refrigerants, and ethylene glycol, including over 4300 points. The nucleate boiling term was developed by Cooper [19], and it is a function of the reduced pressure and the molecular weight of the fluid among other parameters. This term is multiplied by a suppression factor based on the boiling number, Bo, the Lockhart-Martinelli parameter, χ_{tt}, and liquid Reynolds number, Re_l, in a similar procedure adopted by Chen. The convective term is referenced to the single-phase heat transfer coefficient, h_l, the boiling number, and the Lockhart-Martinelli parameter. The mean deviation for this correlation in the saturated boiling region was 21.4% with respect to the data base.

Kandlikar presented a generalized correlation based on 5246 data points from twenty-four experimental investigations using ten fluids. This correlation is presented in terms of the convection number, Co, Froude number, Fr_l, and several empirically determined constants for the convective term. The nucleate boiling term consists of the boiling number, several empirically determined constants, and a fluid specific term, F_{fl}. This correlation had a mean deviation of 15.9 percent for water, and 18.8% for all refrigerant data. Since the fluid parameter for HFC-134a was not originally provided by Kandlikar, a value of 1.63 was adopted based on suggestion by Eckels and Pate. This value was later on confirmed by Kandlikar [20].

Jung and Radermacher [10] recently developed a correlation based on both pure and mixed refrigerants using a modified form of the Chen correlation. The nucleate boiling term is based on the pool boiling heat transfer coefficient obtained by Stephan and Abdelsalam [21]. This term is multiplied by a suppression factor which is based on the Lockhart-Martinelli parameter and the boiling number. Based on their experimental studies, nucleate boiling was found to be suppressed for qualities above 20% [10]. The convective boiling term is based on the Lockhart Martinelli parameter, with the single-phase heat transfer coefficient based on the liquid flowing alone in the tube. The correlation had mean deviations of 7.2% for pure refrigerants and 9.6% for mixed refrigerants.

Table 2 compares the results obtained from the correlations mentioned above with the experimental heat transfer coefficients obtained in this research. All of the correlations predict a majority of the data within ±20%. The mean deviations of the results as predicted by the correlations with respect to the experimentally obtained ones are as follows: Jung-Radermacher, 7.3%; Chaddock-Noerager, 7.8%; Kandlikar, 10.0%; Shah, 10.8%; and Gungor-Winterton, 16.0%. Figure 7 compares the experimentally measured heat transfer coefficients with those predicted by the Jung-Radermacher's correlation, the best among the ones considered in this study.

Table 2 Comparison of the correlations

Correlation	Mean deviation*	Nucleate boiling contribution	Convective boiling contribution
Jung-Radermacher	7.3%	5%	95%
Chaddock-Noerager	7.8%	3%	97%
Kandlikar	10.0%	27%	73%
Shah	10.8%	**	**
Gungor-Winterton	16.0%	20%	80%
This study	8.4%	0%	100%

* Mean deviation is based on the absolute value of the error between the predicted points and the experimentally determined points.

** Shah does not break up his correlation into nucleate boiling and convective boiling terms. It should be noted that of the 60 experimental points, only 6 are found to be in the bubble suppression regime with the others in the convective boiling regime.

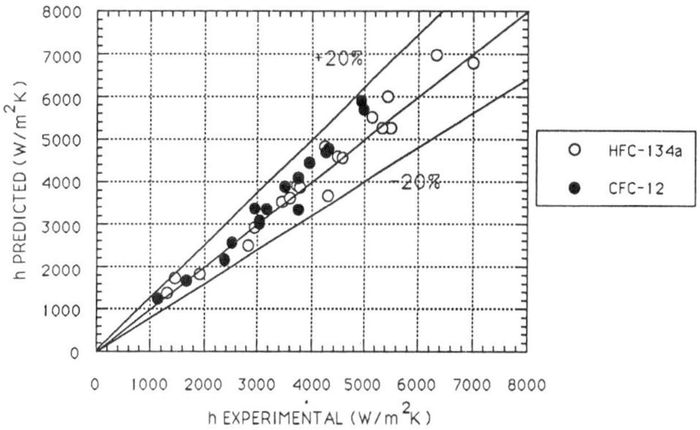

Fig. 7 Predicted heat transfer coefficients (Jung et al.) versus experimental heat transfer coefficients for HFC-134a and CFC-12 during flow boiling.

Although the correlations predict the data well for the annular and stratified flow regimes, they fail to represent adequately the dryout region. It should be noted that the correlations substantially overpredict the experimental values. Figure 3 shows clearly the effect of the heat flux over the transition to dryout. For higher heat fluxes the dryout occurs at lower qualities. Thus, given their inadequacy to predict heat transfer coefficients in the dryout region, those correlations should only be used after a check has been made to verify the flow regime and should not be blindly applied across an arbitrary change of quality. This underlines the need for an accurate prediction of the transition to dryout.

The correlations do an excellent job of predicting the experimental data in the annular flow regime. Upon further examination of Table 2, however, it can be noted that the contributions of the nucleate and convective boiling terms can differ substantially. Thus, the average nucleate boiling contributions for Jung-Radermacher and Kandlikar correlations are, respectively, 5 and 27% based on the experimental conditions.

As previously discussed, there is not a substantial effect of heat flux on the data reported in this study. With this in mind, an attempt was made to correlate the data based on the form of Eq. (1), in which the heat transfer coefficient ratio is only a function of the Lockhart-Martinelli parameter. The resulting equation was as follows:

$$\frac{h_{TP}}{h_l} = 3.37\left(\frac{1}{\chi_{\tau\tau}}\right)^{0.686} \qquad (6)$$

where the coefficient and exponents were determined from a linear least squares curve fit of the data. Figures 8 and 9 are plots of this correlation compared with the experimental data points. The mean deviation of the correlation from the experimental data is 8.4%. For the convective boiling data only, the mean deviation is 7.0%.

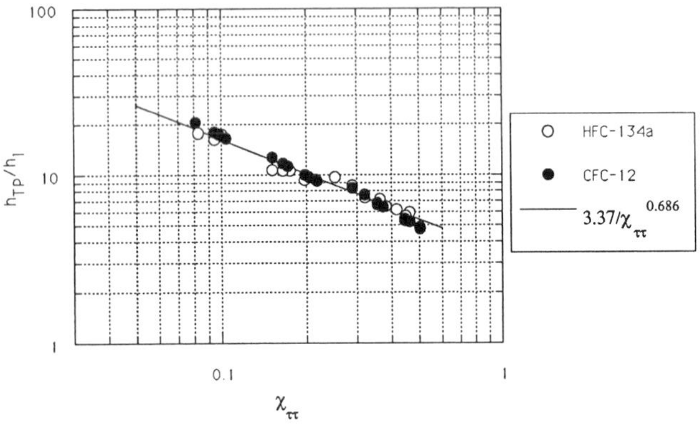

Fig. 8 Two-phase heat transfer multiplier versus $\chi_{\tau\tau}$ for HFC-134a and CFC-12.

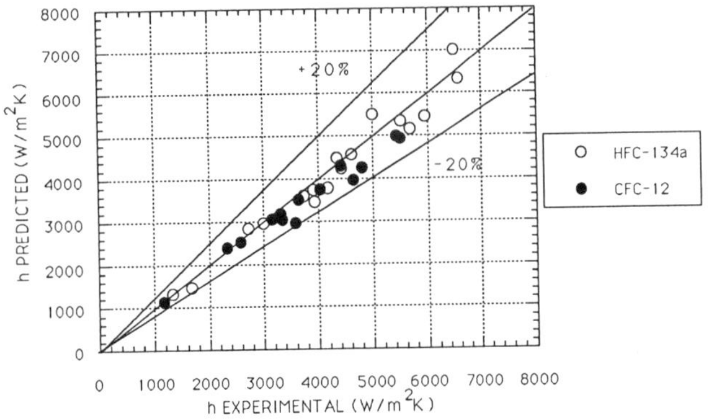

Fig. 9 Predicted heat transfer coefficients from Eq. (6) versus experimental heat transfer coefficients for HFC-134a and CFC-12 during flow boiling.

The excellent correlation of the experimental results by taking into account exclusively convective effects confirms previous predictions of a weak effect of the heat flux. This physical trend was further confirmed by the excellent results obtained from the correlation by Jung and Radermacher, which, as pointed out, presented an insignificant contribution of the nucleate boiling term (5%).

The proposed correlation was able to predict the heat transfer coefficients for the relatively few cases in the wavy-stratified regime within ±20%. Future testing will be conducted in this region to enhance the level of available information. The correlation of data in this case may require the inclusion of the Froude number to account for gravitational effects, such as in the correlations of Kandlikar, Shah, and Gungor-Winterton.

One of the main conclusions regarding Eq.(6) is that it is fluid independent. In other words, the Martinelli parameter is the significant dimensionless group under convective boiling conditions. Another conclusion is that pressure effects are implicitly correlated. The Martinelli parameter accounts for those effects through the property parameter

$$\Gamma = \left(\frac{\rho_v}{\rho_l}\right)^{0.5}\left(\frac{\mu_l}{\mu_v}\right)^{0.1} \qquad (7)$$

which has been proved to be an exclusive function of the reduced pressure by Kenning and Cooper [22], and independent of the particular refrigerant.

COMPARISON WITH OTHER HFC-134a DATA

Another way of examining the experimental data is by direct comparison with other studies using HFC-134a as the test fluid. This is not easily done since none of the recent HFC-134a studies propose correlations of their data. In addition, the range of test parameters for each of the studies is also different. For example, Eckels and Pate determine average heat transfer coefficients for a fixed quality range of 0.1-0.9, whereas the present study determines heat transfer coefficients for small quality changes so that the different flow patterns during the evaporation process can be noted. In addition, Eckels and Pate apply heat to the test section using a counterflow water annulus, whereas the present study applies heat to the test section using electric heaters.

Despite these differences in test conditions, a comparison was attempted by numerically integrating the proposed correlation, Eq. (6), over the quality range used by Eckels and Pate, assuming a linear variation of quality with tube length. Figure 10 shows the comparison between the experimental average heat transfer coefficients from Eckels and Pate for HFC-134a and CFC-12 and predicted average heat transfer coefficients using Eq. (6). The mean deviation was 9.3% for HFC-134a and 6.6% for CFC-12.

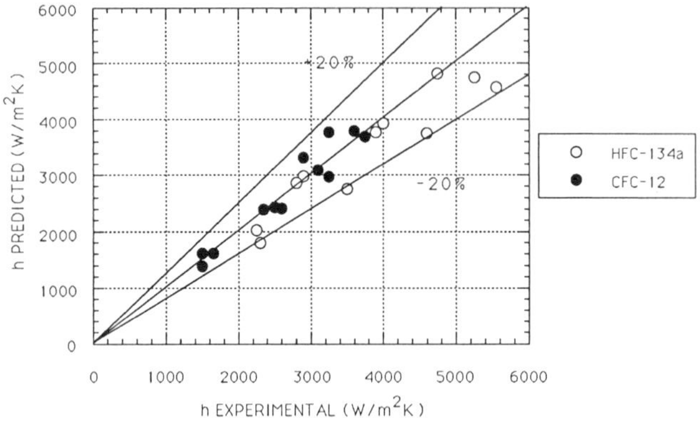

Fig. 10 Predicted heat transfer coefficients from Eq. (6) versus experimental heat transfer coefficients for HFC-134a and CFC-12 from Eckels and Pate [7].

Although the comparison seems quite good, convective boiling probably was not exclusively occurring over the entire quality range. Because of the large quality change in a relatively short (3.67 m) test section for the Eckels and Pate data, the average heat flux in their test section was on the order of 20-30 kW/m². Coupling these high heat fluxes with low inlet qualities, the potential for nucleate boiling exists in the low quality region, which would increase the local heat transfer coefficient. In the high quality region, higher heat fluxes may have induced the dryout regime which would lower the local heat transfer coefficient. Overall, the average heat transfer coefficient for the entire quality range was similar to the average heat transfer coefficient determined as if the entire quality range were in the convective boiling regime only. However, because the convective boiling correlation does predict the data well over a fairly wide range of mass fluxes, heat fluxes, and saturation temperatures, it can be assumed that the effects of nucleate boiling and partial dryout were not substantial for their particular test conditions.

CONCLUSIONS

An experimental investigation for the evaluation of heat transfer coefficients for HFC-134a was conducted and a comparison was made with CFC-12. The flow regimes were predominantly annular, with wavy-stratified, wavy-annular, and mist flow regimes also observed. Heat transfer coefficients for HFC-134a were found to be on average 25% greater than those for CFC-12 in the wavy-stratified and annular flow regimes. As quality and mass flux were increased for tests conducted in the annular flow regime, heat transfer coefficients for both refrigerants increased. For tests conducted in the wavy-stratified flow regime, the heat transfer coefficients remained approximately constant as quality was increased. Heat transfer coefficients did not show a dependence on heat flux, except for tests conducted at low qualities (x_i = 20%) and high heat fluxes (20-30 kW/m^2). For these cases, the effect of heat flux and, hence, nucleate boiling was small compared to convective boiling effects and was suppressed for higher qualities (x_i = 40%,60%). Dryout conditions were observed for several tests at high qualities, especially those conducted at elevated heat fluxes. Based on visual observation at the sight glasses and examination of temperature profiles on the surface of the copper tube, it was possible to estimate that dryout conditions appeared at qualities between 55 and 70 percent for tests with heat fluxes of 20 to 30 kW/m^2 and at qualities of 75 to 85 percent for tests with heat fluxes of 5 and 10 kW/m^2.

Correlations of Jung and Radermacher, Chaddock-Noerager, Kandlikar, Shah, and Gungor-Winterton compared well with the experimental heat transfer coefficients, with the first one predicting the data most accurately. However, the correlations did not predict the heat transfer coefficients for tests including dryout conditions. Although all these correlations predicted the data well, the effect of nucleate boiling on the heat transfer coefficients was overpredicted by the correlations of Kandlikar and Gungor-Winterton. Experimental heat transfer coefficients were accurately correlated based on convective boiling effects alone, hence refuting the possibility of strong effects of nucleate boiling in the range of parameters considered in this study.

ACKNOWLEDGMENTS

The support by the Air Conditioning and Refrigeration Center of the University of Illinois is gratefully acknowledged by the authors. They would also like to extend their appreciation to DuPont Company for providing the refrigerants used in the investigation.

REFERENCES

1. Chaddock, J. B. and Noerager, J. A., "Evaporation of Refrigerant 12 in a Horizontal Tube with Constant Wall Heat Flux," ASHRAE Transactions, Vol. 72, Part I, 1966, pp. 90-103.
2. Shrock, V. E. and Grossman, L. M., "Forced Convection Boiling in Tubes," Nuclear Science and Engineering, Vol. 12, No. 4, 1962, pp. 474-481.
3. Chen, J. C., "Correlation for Boiling Heat Transfer to Saturated Fluids in Convective Flow," Industrial Engineering Chemistry Process Design Development, Vol. 5, No. 3, 1966, pp. 322-339.
4. Shah, M. M., "A New Correlation for Heat Transfer During Boiling Flow Through Pipes," ASHRAE Transactions, Vol. 82, Part II, 1976, pp. 66-86.
5. Kandlikar, S. G., "A General Correlation for Saturated Two-Phase Flow Boiling Heat Transfer Inside Horizontal and Vertical Tubes," ASME Journal of Heat Transfer, Vol. 112, 1990, pp. 219-228.
6. Gungor, K. E. and Winterton, R. H. S., "A General Correlation for Flow Boiling in Tubes and Annuli." International Journal of Heat and Mass Transfer, Vol. 29, No. 3, 1987, pp. 351-358.
7. Eckels, S. J. and Pate, M. B., "An Experimental Comparison of Evaporation and Condensation Heat Transfer Coefficients for HFC-134a and CFC-12," International Journal of Refrigeration, Vol. 14, No. 2, 1990, pp. 70-77.
8. Chaddock, J. B. and Brunemann, H, "Forced Convection Boiling of Refrigerants in Horizontal Tubes-Phase 3", 1967, HL-113, Duke University School of Engineering, Durham, NC.
9. Hambraeus, K., "Heat Transfer Coefficient During Two-Phase Flow Boiling of HFC-134a," International Journal of Refrigeration, Vol. 14, 1991, pp. 357-362.
10. Jung, D. S. and Radermacher, R., "Horizontal Flow Boiling Heat Transfer Experiments with a Mixture of R22/R114" International Journal of Heat and Mass Transfer, Vol. 32, 1989, pp. 131-145.
11. Pierre, B., "Coefficient of Heat Transfer for Boiling Freon-12 in Horizontal Tubes," Heating and Air Treatment Engineer, Vol. 19, 1956, pp. 302-310.
12. Takamatsu, H., Momoki, S., and Fujii, T., "A Comparison of Evaporation Heat Transfer Coefficients and Pressure Drop in a Horizontal Smooth Tube for HFC-134a and CFC-12", 1991, XVIII International Congress of Refrigeration.
13. Wattelet, J. P., "Design, Building, and Baseline Testing of an Apparatus used to Measure Evaporation Characteristics of Ozone-Safe Refrigerants," Master of Science Thesis. Department of Mechanical and Industrial Engineering, University of Illinois at Urbana-Champaign, 1990.
14. Panek, J. S., "Evaporation Heat Transfer and Pressure Drop in Ozone-Safe Refrigerants and Refrigerant-Oil Mixtures," Master of Science Thesis. Department of Mechanical and Industrial Engineering, University of Illinois at Urbana-Champaign, 1991.
15. Wilson, D. P. and Basu, R. S., "Thermodynamic Properties of New Stratospherically Safe Working Fluid-Refrigerant 134a," ASHRAE Transactions, Vol. 94, Part II, 1988, pp. 2095-2118.
16. McLinden, M. O., "Measurement and Formulation of the Thermodynamic Properties of Refrigerants 134a and 123," ASHRAE Transactions, Vol.95, Part II, 1989, pp. 185-196.
17. Jung, D. S. and Radermacher, R., "Transport Properties and Surface Tension of Pure and Mixed Refrigerants," ASHRAE Transactions, Vol. 97, Part I, 1991, pp. 90-99.
18. Moffat, R. J., "Describing the Uncertainties in Experimental Results," Experimental Thermal and Fluid Science, Vol. 1, 1988, pp. 3-17.
19. Cooper, M. G., "Saturation Nucleate Pool Boiling. A Simple Correlation," Transactions of the 1st United Kingdom National Conference on Heat Transfer, Vol. 2, 1984, pp. 785-793.
20. Kandlikar, S. G., "Flow Boiling Maps for Water, R-22 and R-134a in the Saturated Region," Proceedings of the 9th International Heat Transfer Conference, Vol. 2, 1990, pp. 15-20.
21. Stephan, K. and Abdelsalam, M., "Heat Transfer Correlation for Natural Convection Boiling," International Journal of Heat and Mass Transfer, Vol. 23, No. 1, 1980, pp. 73-87.
22. Kenning, D. B. R. and Cooper, M. G., "Saturated Flow Boiling of Water in Vertical Tubes," International Journal of Heat and Mass Transfer, Vol. 32, No.3, 1989, pp. 445-458.

HTD-Vol. 197, Two-Phase Flow and Heat Transfer
ASME 1992

NUCLEATE BOILING CHARACTERISTICS OF A SMALL ENHANCED TUBE BUNDLE IN A POOL OF R-113

S. B. Memory, S. V. Chilman, and P. J. Marto
Department of Mechanical Engineering
Naval Postgraduate School
Monterey, California

ABSTRACT

Heat transfer measurements were made during nucleate boiling of R-113 from a bundle of 15 electrically-heated, copper TURBO-B tubes arranged in an equilateral triangular pitch. The bundle was designed to simulate a portion of a refrigeration system flooded-tube evaporator. The nominal outside diameter of the tubes was 15.9 mm, and the tube pitch was 19.1 mm. Five of the tubes that were oriented in a vertical array on the centerline of the bundle were each instrumented with six wall thermocouples to obtain an average wall temperature and a resultant average heat transfer coefficient. All tests were performed at atmospheric pressure. The majority of the data were obtained with increasing heat flux to study the onset of nucleate boiling and the influence of surface 'history' upon boiling heat transfer.

Data taken during increasing heat flux showed that the incipient boiling point was dependent upon the number of tubes in operation; activating lower tubes in the bundle decreased the incipient boiling heat flux and wall superheat of the upper tubes. In the boiling region, data showed that at low heat fluxes ($\approx$ 1 kW/m^2), the average heat-transfer coefficient for a TURBO-B tube bundle was 4.6 times that obtained for a smooth tube bundle (under similar conditions) and 60% greater than that obtained for a single TURBO-B tube; this 'bundle effect' is similar to that found for a smooth tube bundle. At high heat fluxes (100 kW/m^2), the average heat-transfer coefficient for the bundle was 3.6 times the smooth tube bundle and there was still a significant 'bundle effect' of 22%, which is contrary to the results from a smooth tube bundle (where all 'bundle effect' was eliminated at high heat fluxes).

INTRODUCTION

In recent years significant progress has been made in understanding nucleate boiling heat transfer on the shell side of tube bundles. However, most of the work to date has involved smooth tubes and not enhanced tubes (Leong and Cornwell (1979); Cornwell et al. (1980); Cornwell and Schuller (1982); Cornwell and Scoones (1988); Cornwell (1989)). These works observed the two-phase flow patterns in a smooth tube bundle and deduced that sliding bubbles from lower tubes on upper tubes, and liquid forced convection, can account for significant heat transfer in the top part of the bundle. Two-phase convection effects in a smooth tube bundle have also been studied by Wege and Jensen (1984), Hwang and Yao (1986), and Jensen and Hsu (1988).

The influence of tube position within a bundle of smooth tubes using R-11, R-12, R-22 and R-113 has been studied extensively by Wallner (1974), Fujita et al. (1986) Chan and Shoukri (1987), Rebrov et al (1989) and Marto and Anderson (1992). These investigators, using both square and staggered tube arrangements with a tube pitch-to-diameter ratio of between 1.2 and 2.0, verified that the influence of the lower tubes in a bundle can significantly increase the heat transfer from the upper tubes at low heat fluxes due to two-phase convection effects. At high heat fluxes (typically > 50 kW/m^2) in the fully developed nucleate boiling region, the data for all the tubes come together onto a single curve, representative of a single smooth tube i.e. there is no 'bundle effect'. Furthermore, these works seemed to verify that this behavior at low and high heat fluxes can be predicted by combining pool boiling predictions with liquid forced convection predictions. Similar results (showing an effect of lower tubes on upper tubes at low heat fluxes and little or no effect at high heat fluxes) have been obtained for finned tube bundles by Hahne and Müller (1983), Yilmaz and Palen (1984) and Müller (1986).

There has been a significant amount of work on enhanced (i.e. non-finned) tube bundles; the reader is referred to Jensen (1988) and Thome (1990) for good reviews in this area. Arai et al (1977) found that the 'bundle effect' for a Thermoexcel tube bundle was smaller than that found for a smooth or finned tube bundle. Furthermore, any 'bundle effect' was eliminated at lower heat fluxes (typically 30 kW/m^2) and data from all tubes agreed closely with single Thermoexcel tube results; they attributed this to early domination of convection effects by nucleation effects. For porous coated surfaces, Czikk et al (1970) and Fujita et al (1986) found no 'bundle effect' over a wide range of heat fluxes (1-100 kW/m^2) and bundle data agreed closely with single tube data.

From the above, it is clear that the two-phase interactions that occur in tube bundles during boiling can be very complex and change with heat flux level, fluid properties, tube surface and tube bundle layout. Therefore, it is very difficult to use information from one type of bundle and fluid combination, and apply it to another situation. Instead, a complete range of experimental data are needed in the open literature that covers various fluids, tube surfaces, bundle layouts, heat fluxes and inlet qualities, so that theoretical models can be formulated and appropriately evaluated. This is particularly important in the refrigeration industry where new, alternative refrigerants and refrigerant-oil mixtures are being proposed.

For natural convection in bundles, there is limited published information and the majority of this is for air (Masters (1972), Sparrow and Niethammer (1981)). In addition, limited information is available regarding the onset of nucleate boiling in a bundle, bundle hysteresis effects and the influence of past history upon the incipient boiling condition. Marto and Anderson (1992) report these effects for a smooth tube bundle using pure R-113.

The present paper is a continuation of the work started by Marto and Anderson (1992). Its purpose is to establish additional baseline nucleate boiling data of R-113 from a small bundle of enhanced tubes (TURBO-B). It is part of a program to establish a comprehensive boiling heat transfer database for refrigerants and refrigerant/oil mixtures (e.g. R-113, R-114, R-114/oil (Memory et al (1992)), R-124, R-124/oil) from single tubes and small tube bundles. Although R-113 is not a common refrigerant, it is widely used in the two-phase literature and is convenient for fundamental studies.

This paper emphasizes the influence of tube position in the bundle upon the incipient boiling point and any hysteresis effects common in nucleate boiling processes. It also reports any 'bundle effect' found in the boiling region. The data obtained are compared with the smooth tube data of Marto and Anderson (1992) and are expected to serve as reference data for comparison with other refrigerants and refrigerant-oil mixtures from bundles of smooth as well as enhanced tubes.

EXPERIMENTAL APPARATUS

The basic experimental apparatus used during this investigation is shown in Figure 1 and is identical to the apparatus used by Marto and Anderson (1992). It consisted of an evaporator and condenser arranged to provide reflux operation. The condenser included four instrumented test tubes and five auxiliary coils, all cooled by a refrigerated mixture of water and ethylene glycol. It was designed to permit independent condensation studies while using the evaporator as a source of vapor. During the present investigation, the condenser was used simply to maintain the system pressure at atmospheric conditions.

The evaporator was fabricated from stainless steel plate and formed into a short cylinder. Electrically heated tubes were cantilever-mounted from the back wall of the evaporator to permit easy viewing along the axis of the tubes through the lower of two glass windows mounted on the front. Figure 2 is a schematic sectional view of the evaporator that shows four sets of heated tubes. Two auxiliary heaters, each capable of 4 kW, were installed on each side of the test bundle to maintain the liquid pool at saturated conditions and to provide system pressure control. Five simulation heaters, each also capable of 4

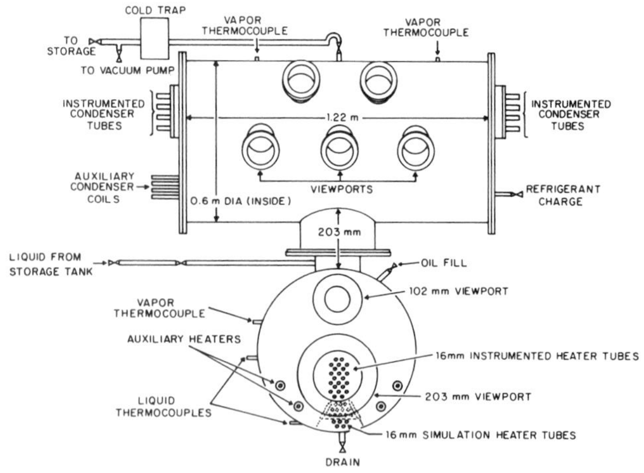

Figure 1: Experimental Apparatus

kW, were mounted below the test bundle in order to simulate additional tube rows and to provide inlet vapor quality into the bottom of the test bundle. The test bundle consisted of fifteen heated enhanced tubes: ten active tubes (marked 'A') which contained 1 kW heaters, and five instrumented tubes (marked 'I') which, in addition to 1 kW heaters, contained wall thermocouples. The instrumented tubes were thus located along the centerline of a symmetrical, staggered tube bundle.

All the enhanced tubes were made from commercially-available TURB0-B copper tubing. The surface of the TURBO-B tube is made from a finned tube which is then cross-grooved and rolled to provide a grid of rectangular flattened blocks that are wider than the original fins (see Figure 3). The tubes are supplied in a variety of nominal diameters and, depending on the rolling process, size of the flattened blocks, thus providing differing gap widths to suit a particular application. The present tubes had a diameter to the base of the enhancement of 14.2 mm. The thickness of the enhancement was 0.85 mm giving a nominal outside diameter of 15.9 mm. They were arranged in an equilateral triangular pitch (i.e., centerline-to-

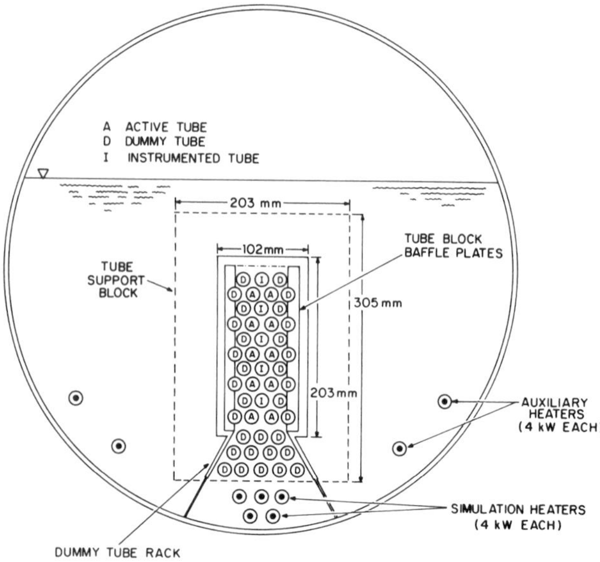

Figure 2: Sectional View of Evaporator

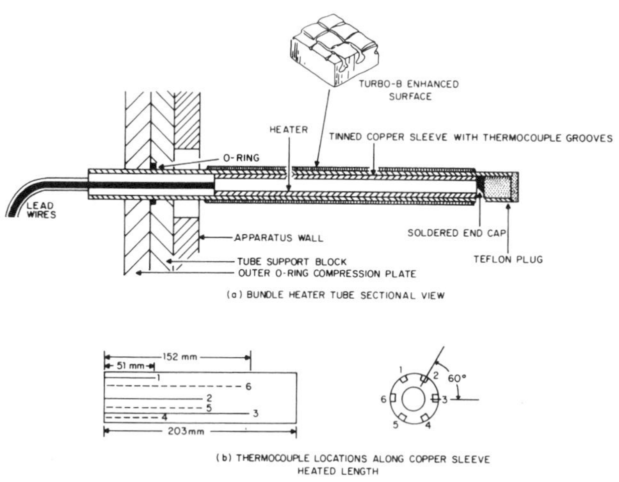

Figure 3: Instrumented Tube Construction

centerline spacing) of 19.1 mm, giving a pitch to outside diameter ratio of 1.20. The heated length of each tube was 190 mm[1]. Each set of auxiliary tubes, simulation tubes and test bundle tubes could be heated independently using three separate rheostat controllers.

The bundle also contained a number of unheated dummy smooth tubes (marked 'D') that were used to guide the two-phase mixture through the bundle. Two vertical baffle plates were installed to restrict side circulation into and out of the bundle. An open space was left, however, on the lower part of each side of the dummy tube bundle, adjacent to the simulation heaters, to permit side entry of liquid into the bottom of the simulation bundle. Thus, liquid circulation was vertically upward over the test tubes with no net horizontal component. The liquid-vapor mixture at the top of the bundle split symmetrically, with the liquid returning down the outside of each of the baffle plates. Vapor from the evaporator flowed upward through a riser section and was distributed axially and circumferentially to the top of the condenser by a vapor shroud. The condensate collected in the bottom of this shroud and returned to the evaporator by gravity. The two-phase mixing and the condensate return flow were observed through the top window in the evaporator.

In measuring boiling heat transfer coefficients, great care must be exercised with the cartridge heater and temperature measuring instrumentation to ensure good accuracy. Various installation techniques have been reviewed by Jung and Bergles (1989). Based upon extensive pool boiling data with R-113, they concluded that the heat transfer coefficient of a single tube in pool boiling is not sensitive to variations in the cartridge heater heat flux provided that enough thermocouples are used to measure an average wall temperature. During this investigation, the instrumented test tubes were fabricated in a similar way to those used by Hahne and Müller (1983), Wanniarachchi et al. (1986) and Marto and Anderson (1992). Figure 3 is a cross-sectional sketch of an instrumented tube, showing the tube construction and the location of the wall thermocouples. To smooth out any non-uniformities in heat flux caused by the continuous cartridge heater and to provide a convenient method to install wall thermocouples, a copper sleeve was used inside the test tube into which the cartridge heater was inserted with a tight mechanical fit. Six 1 mm square slots, spaced 60 degrees apart at varying longitudinal distances from one end (as shown in Figure 3b), were milled into the outside wall of the sleeve to create channels in which the wall thermocouples were soldered. The outer surface of the sleeve and the inner surface of the test tube (with a clearance of 0.1 mm) were then carefully soldered together. Full details of the construction and soldering process used for these tubes are provided by Marto and Anderson (1992).

Each test tube was installed into the tube support block using an O-ring. An outer compression plate was then bolted to the support block to seal each tube. Once all the evaporator tubes were installed, the support block was bolted into the rear wall of the evaporator unit. During this assembly, care was taken to insure that each tube was properly aligned in the bundle and that they were all pushing against the inside face of the lower front viewing window.

MEASUREMENTS AND PROCEDURES

Vapor temperatures were measured by two thermocouples at the top of the condenser and one thermocouple near the top of the evaporator. Liquid temperatures were measured by three thermocouples, two located close to the free surface of the liquid and the third located at the bottom of the pool close to the dummy tube rack. During operation, the top two thermocouples were located in a frothy, two-phase mixture and were considered to be well representative of the saturation temperature at the free surface. All measurements were taken when all three liquid thermocouples indicated the same temperature i.e. no subcooling in the pool.

The average outer wall temperature of the instrumented test tubes was obtained by averaging the six wall thermocouples in the copper sleeve and correcting for the small radial temperature drop due to conduction across the copper wall. The temperature drop across the 0.05 mm thick solder joint between the copper sleeve and the test tube was neglected. The voltage and current to each of the test tubes were measured using sensors. For a given tube, the average heat flux was calculated by dividing the electrical power (after it was corrected for small axial losses from each end of the test tube) by the tube surface area (based on the diameter to the base of the enhancement of 14.2 mm and an active heating length of 190 mm). If the surface area were computed using the nominal outside diameter of 15.9 mm, a 12% increase in area would result. During all the tests, the pressure was kept at atmospheric and the liquid level was kept approximately 100 mm above the top row of tubes. In order to calculate the local saturation temperature for each tube in the bundle, a hydrostatic pressure correction was made between the tube location and the free surface of the liquid (any pressure drop effects due to the two-phase flow in the bundle were neglected). All the data were obtained and reduced with a computer-controlled data acquisition system.

Prior to filling with R-113, the system was evacuated and leak checked. Once the system was felt to be vacuum tight, R-113 was added to the evaporator by drawing it into the evaporator under vacuum. Prior to operating the evaporator, the ethylene glycol/water coolant for the condenser was cooled and maintained at a temperature between -10 to -15 °C using an 8-ton refrigeration system. Circulating pumps were then turned

[1] The heaters used were Watlow Firerod heaters which were continuously wound with an 8 inch nominal diameter and a 7.5 inch heated length.

on to get a desired flow through the condenser tubes and cooling coils.

During this investigation, nucleate boiling data were obtained following two different surface aging procedures[2]. With the first procedure, (surface aging C), the evaporator power was secured overnight. The following morning prior to taking data, the pool temperature was slowly brought up to saturation conditions using the auxiliary heaters; operation of the tubes then commenced with increasing heat flux in pre-determined steps. In this way, boiling incipience and bundle start up problems could be investigated. The second procedure (surface aging D) consisted of boiling at 100 kW/m^2 for 30 minutes followed by immediate operation with decreasing heat flux in pre-determined steps. Once the required heat flux in the evaporator had been fixed, the coolant flow through the condenser was adjusted to maintain the saturation temperature corresponding to 1 atmosphere. It should be noted that the liquid recirculation pattern that existed in the evaporator vessel (and, therefore, across the test bundle) prior to the commencement of each experiment may have been affected by the different aging procedures. However, once system operation was begun, data were not taken until 10-15 minutes after start up. Therefore, most of the recirculation variations would have disappeared prior to actually taking data. Nevertheless, due to possible variations in the recirculation pattern, the uncertainty of the first one or two data points taken at very low heat fluxes, following aging procedure C, may be larger than the uncertainties stated later in the paper.

Seven independent tests (listed in Table 1) were conducted for increasing and decreasing heat flux. Test numbers 1 to 5 progressively activated the 5 instrumented enhanced tubes within the bundle to see the effect of lower heated tubes on the performance of the upper tubes. Test number 6, in addition, looked at the effect of activating the remaining 10 enhanced tubes around these instrumented tubes. All these tests were essentially carried out with zero quality entering the bottom of the bundle. Test number 7 allowed for a given inlet quality by activating the simulation heaters below the bundle.

Table 1: Tests Conducted on Tube Bundle

Test No.	Description
1	Top instrumented enhanced tube activated only
2	Top 2 instrumented enhanced tubes activated only
3	Top 3 instrumented enhanced tubes activated only
4	Top 4 instrumented enhanced tubes activated only
5	All 5 instrumented enhanced tubes activated
6	Whole bundle of 15 enhanced tubes activated
7	Whole bundle of 15 enhanced tubes plus simulation tubes activated

2 Marto and Anderson (1992) used four different aging procedures, A, B, C and D. Although they briefly discussed the results of procedures A and B, all the data presented by them were for procedures C and D.

RESULTS AND DISCUSSION

All data were obtained with R-113 at a pressure of 1 atm in the range of heat fluxes between 1 and 100 kW/m^2 (a typical naval submerged evaporator operates at a heat flux between 15-20 kW/m^2). During increasing heat flux runs, the onset of nucleate boiling was observed through the lower viewing window. This 'point' was defined as the applied heat flux where first nucleation was observed on the instrumented enhanced tubes. When heat flux is plotted versus the wall superheat (defined as the mean wall temperature minus the saturation temperature), the 'incipient boiling' condition is indicated by a dramatic change in slope since the heat transfer mechanism changes from single-phase convection to two-phase convection with the activation, growth and departure of vapor bubbles.

The uncertainty in the experimental data was estimated using a propagation of error analysis. The uncertainty in the wall superheat was dominated by the uncertainty in the wall temperature measurements. At each operating point, the values of the six wall thermocouples were recorded and compared to examine variations in wall temperature caused either by non-uniformities in the cartridge heater coils or by the test tube soldering and assembly procedure. This problem has been explored in more detail by Wanniarachchi et al. (1986). The maximum variation of the six measured wall temperatures was 1.8 °C at the maximum heat flux ($\approx$ 100 kW/m^2) and 0.4 °C at the minimum heat flux ($\approx$ 1 kW/m^2). This variation appeared to be random and independent of thermocouple orientation and was probably caused by the tube soldering process. The uncertainty in the saturation temperature was estimated to be 0.1 °C. These variations in wall temperature and saturation temperature created an estimated uncertainty in the wall superheat of ± 0.9 °C at high heat fluxes and ± 0.2 °C at low heat fluxes. For the heat flux, the uncertainty depends on the current and voltage measurements (which relatively decrease with increasing values) as well as the tube surface area measurement (which remains constant). The corresponding uncertainty in the measured heat flux was estimated to be 5% at low heat flux decreasing to 1.5% at high heat flux.

Throughout this investigation, the instrumented tubes located along the centerline of the tube bundle were numbered consecutively from the top downward as tubes 1, 2, 3, 4 and 5 respectively. It was mentioned earlier that when using aging procedure C, the pool was slowly brought up to saturation conditions using the auxiliary heaters; these auxiliary heaters had to remain on throughout certain tests (with low numbers of activated tubes in the bundle) in order to maintain saturation conditions in the pool. It was found that the value at which the power to these auxiliary heaters was set significantly affected the data in the natural convection region. Figure 4 shows data taken for tube 1 operating alone (test 1) with each of the auxiliary heaters set at 1 kW and 3 kW respectively. Two clear trends are visible. Firstly, with the auxiliary heaters set at 3 kW, the wall superheat is significantly reduced, thereby increasing the heat-transfer coefficient. This is presumably due to the greater buoyancy-induced circulation from the auxiliary heaters enhancing the flow velocity across this tube. Also evident is an earlier point of incipience (indicated by an arrow) at the higher auxiliary heater setting. This could have been due to microscopic bubbles from the auxiliary heaters finding their way through the bundle and impinging on tube 1. This may have the same effect as increasing the amount of dissolved gas; Murphy and Bergles

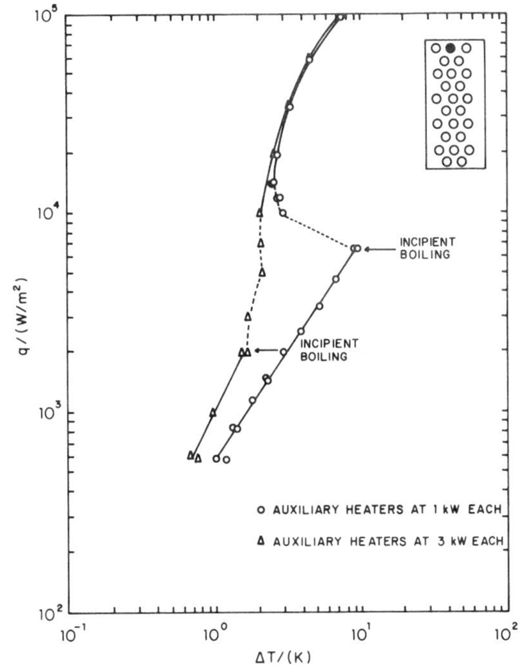

Figure 4: Effect of Auxiliary Heaters on Tube 1 Operating Alone With Increasing Heat Flux (Aging Procedure C)

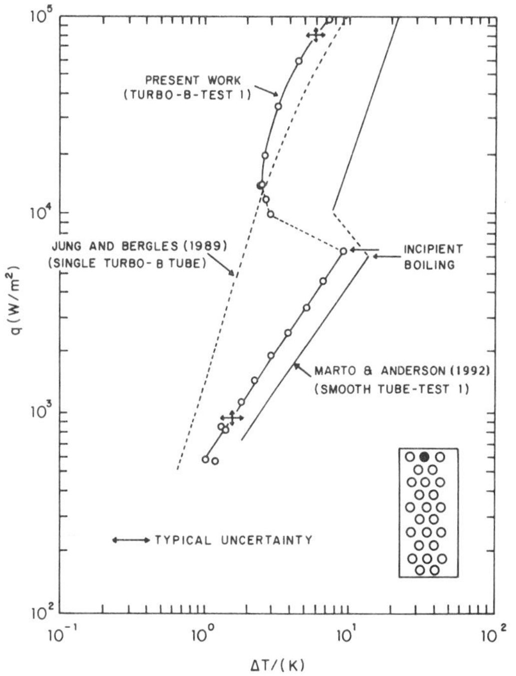

Figure 5: Tube 1 Operating Alone With Increasing Heat Flux (Aging Procedure C)

(1971) show that increasing the dissolved gas can significantly lower the wall superheat at which incipience occurs. As a consequence of these effects, all subsequent tests were conducted with the auxiliary heaters set at 1 kW to eliminate any uncertainty due to their activation.

Figure 5 shows the data for tube 1 operating alone as a single tube (test 1), during increasing heat flux (aging procedure C). Typical uncertainty bands for both q and ΔT are included at low and high heat fluxes; it can be seen that the uncertainty in ΔT is the more significant, especially at low heat flux. Also shown are the data of Marto and Anderson (1992) taken on the same apparatus for a smooth tube operating alone. In the natural convection region, both sets of data have the same slope, but the TURBO-B data indicate a significant increase in the heat-transfer coefficient. This is probably due to the auxiliary heater settings, which were not carefully controlled in the experiments of Marto and Anderson (1992) and have been shown to significantly affect this region (see Figure 4). The point of incipience occurs at a similar value of heat flux (≈ 7 kW/m^2) for both types of tube and both sets of data demonstrate a significant temperature overshoot of about 7 °C. In the boiling region, the TURBO-B tube yields a heat-transfer coefficient between 3 and 4 times that of the smooth tube, the enhancement decreasing as heat flux increases.

The increasing heat flux data of Jung and Bergles (1989) for pool boiling of R-113 at 1 atmosphere from a single short (100 mm heated length), 19.1 mm nominal outside diameter (18 mm diameter to base of enhancement) TURBO-B copper tube are also included in Figure 5 for comparison. Their data show no incipient point and no temperature overshoot, due to a different surface aging procedure. Their tube was pre-boiled at a heat flux of 30 kW/m^2 for 30 minutes and then turned off until the thermocouples recorded the saturation temperature of the pool (length of time not given), i.e. there was no subcooling in contrast to procedure C where the pool was allowed to sit overnight and reach room temperature. Clearly, then, in addition to variations in local flow conditions, the incipient boiling characteristics depend upon the active site distribution on the boiling surface and this can be very different depending upon the immediate past history of the heating surface and the temperature of the surrounding pool. In the fully developed nucleate boiling region, both sets of data are in reasonably good agreement; the small differences seen may be because the bundle geometry in the present investigation, with the inactive, closely-spaced neighboring tubes and the vertical baffles, generated a different liquid circulation pattern over tube 1, than occurred over the single tube used by Jung and Bergles (1989).

Figure 6 shows the results for tubes 1 and 2 operating simultaneously (test 2), during increasing heat flux. The natural convection data of tube 1, compared to the data in Figure 5 when operating as a single tube, shows that there appears to be little influence of the lower heated tube (tube 2) upon the heat transfer of the upper tube (tube 1). However, the natural convection heat transfer coefficient for the lower tube is less than the upper tube. This behavior is exactly the same as that reported by Marto and Anderson (1992) for a smooth tube bundle and may be due to expansion of the flow after it leaves the tightly packed bundle, i.e. differences in the velocity and temperature fields in the wake region of a heated tube within the bundle to those of a heated tube at the top of the bundle.

Figure 6 also shows that the incipient boiling heat flux is reduced to about 5 kW/m^2 for both tubes, although tube 2 shows an irregular nucleation process until higher heat fluxes are reached. This irregular behavior has been observed by Wanniarachchi et al. (1987) during boiling of R-114 from a single enhanced tube containing numerous nucleation sites. They attributed this behavior to incomplete nucleation along the boiling surface due to non-uniform heat flux or non-uniform cavity openings. Tube 1 displays a slightly smaller

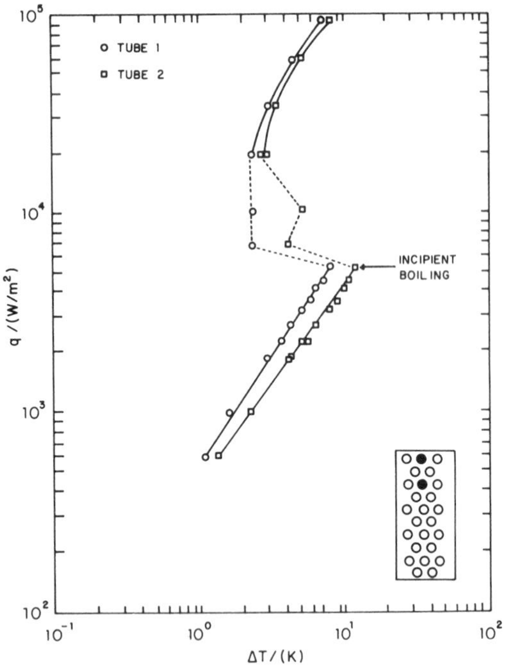

Figure 6: Tubes 1 and 2 Operating Simultaneously With Increasing Heat Flux (Aging Procedure C)

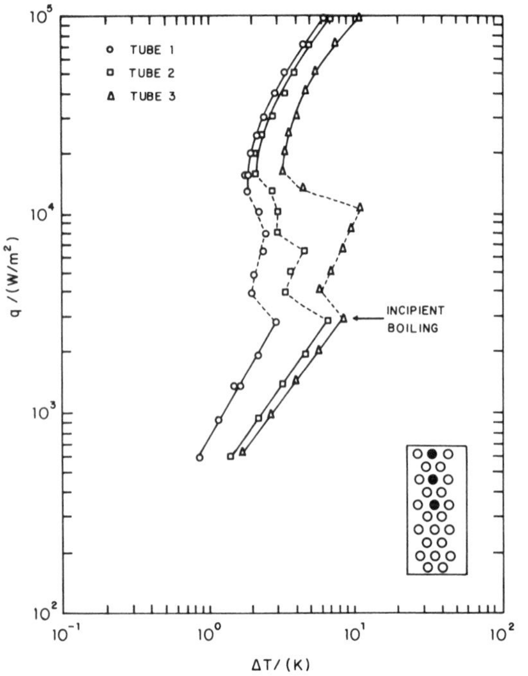

Figure 7: Tubes 1, 2 and 3 Operating Simultaneously With Increasing Heat Flux (Aging Procedure C)

characteristic temperature overshoot (about 6 °C) than shown in Figure 5. Once fully boiling, tube 2 agrees very closely with tube 1 but with a slightly lower heat-transfer coefficient.

Figure 7 shows similar data for three heated tubes (test 3). With tube 3 in operation, there is no apparent influence on the natural convection data of tube 2, confirming the above-mentioned results of Figure 6 that a lower heated tube does not seem to influence the natural convection behavior of an upper tube; this is also similar to that reported by Marto and Anderson (1992) for a smooth tube bundle. However, a lower heated tube does seem to influence the start of incipience, such that all three tubes partially nucleate at a lower heat flux of 3 kW/m². This also suggests that it is the lowest activated tube in the bundle that starts to nucleate first and as a lower tube starts to nucleate, it 'triggers' the tubes above, probably due to the increased activity within the pool caused by bubbles sliding over the upper tubes. This is supported by the fact that at a heat flux of 6 kW/m², as 'more' of tube 2 nucleates (thereby affecting the top tube (tube 1) further), the lower tube (tube 3) remains unaffected. In the fully developed nucleate boiling region, all three tubes exhibit similar behavior, with a substantial decrease in heat transfer as one moves down the bundle. Similar results to the above-described phenomena were obtained when bundles of both 4 and 5 tubes were heated simultaneously (tests 4 and 5).

Figure 8 shows the behavior of tubes 1 through 5 when all 15 active tubes are heated as a bundle (test 6). In the natural convection region, tube 1 still exhibits a lower wall superheat than the other heated tubes. The main effect of having heated tubes around the five instrumented tubes is to reduce the heat flux at which incipience first occurs. The transition from natural convection to full nucleation for all five instrumented tubes occurs over a range of heat flux from 3 kW/m² to 16 kW/m² with the top tube reaching full nucleation first, followed by the other tubes in order down the bundle; however, this transition becomes more 'erratic' as one moves down the bundle i.e. lower tubes partially nucleate as they are affected by the nucleation of other tubes. In the fully developed boiling region, there is a small effect of tube position.

Figure 9 shows the additional effect of including the simulation heaters below the bundle (test 7). The added inlet quality causes both the point of incipience and the temperature overshoot to further decrease for all five instrumented tubes; the top tube

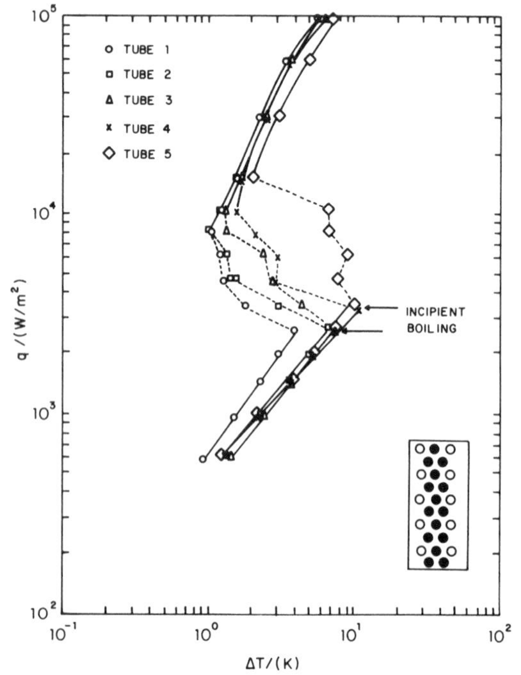

Figure 8: Tube Bundle Operating With No Simulation Heaters With Increasing Heat Flux (Aging Procedure C)

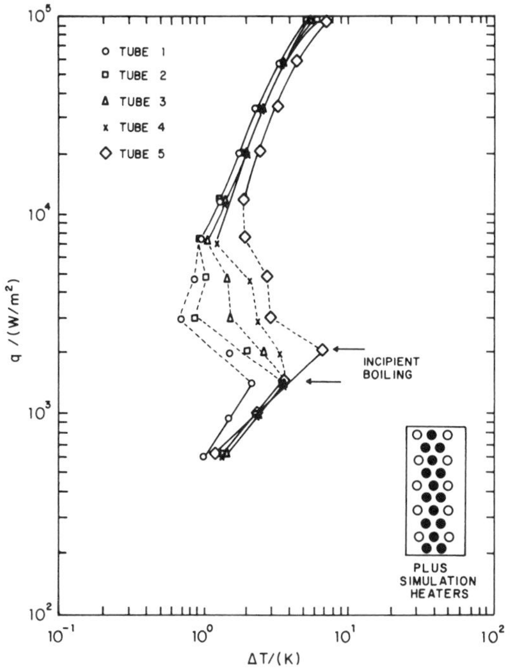

Figure 9: Tube Bundle Operating With Simulation Heaters With Increasing Heat Flux (Aging Procedure C)

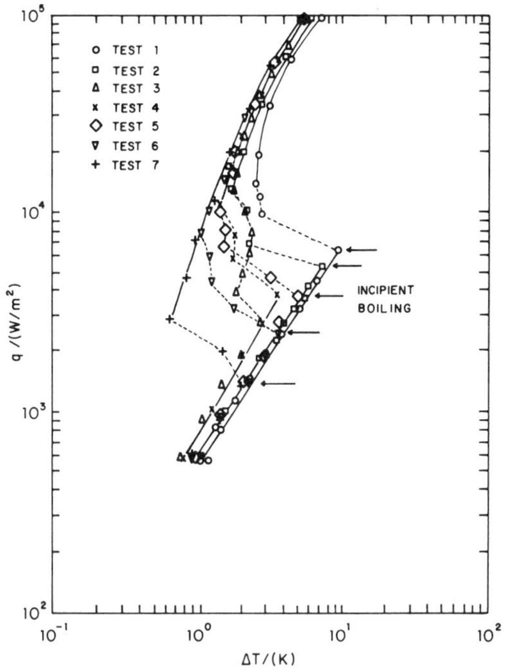

Figure 10: Tube 1 Operating for All 7 Tests With Increasing Heat Flux (Aging Procedure C)

now starts nucleating at a heat flux of 1.5 kW/m^2 and has a temperature overshoot of only 1.5 °C. Again, Figure 9 shows that transition to full nucleation occurs in order down the bundle.

Due to the way in which the tubes were individually manufactured, making too many direct comparisons between data taken on different tubes could lead to error. Of perhaps more interest is to look at what happens to the top tube as succesive tubes below it are activated. Figure 10 shows this data for tube 1 during all seven tests. In the natural convection region, it can be seen that the successive activation of lower tubes slightly enhances the performance of the top tube, as discussed above. There is also slight enhancement evident in the boiling region, larger than that reported by Marto and Anderson (1992) for a smooth tube bundle, indicating that the influence of bubbles from a lower tube impacting on an upper tube is an important mechanism to enhance the heat transfer process with these type of enhanced surfaces. This is discussed in more detail below.

In the region between natural convection and full nucleation, Figure 10 shows that the heat flux at which the incipient point occurs decreases from a value of 7 kW/m^2 for test 1 to 1.5 kW/m^2 for test 7. Furthermore, the temperature overshoot decreases from 7 °C for test 1 to 1.5 °C for test 7. These results indicate that lower tubes help to trigger 'earlier' nucleation on upper tubes. From a theoretical consideration of boiling incipience from an isolated surface (Hsu (1962), Han and Griffith (1965), Howell and Siegel (1967)), one may expect a lower incipient heat flux for an enhanced tube due to the high density of relatively large, active cavities present on such a surface. This has been experimentally verified for single tubes by Memory and Marto (1992). However, comparison of the present work with that of Marto and Anderson (1992) illustrates that incipience occurs at a similar heat flux for both an enhanced and smooth tube bundle for all but test 7 (where all evidence of temperature overshoot was eliminated for the smooth tube bundle). This is an indication that the flow induced by the closely-spaced tubes in the bundle is more important in the initiation of boiling than the cavity size. It further suggests that the presence of two-phase flow at the entry of the bundle (with a quality greater than zero) can certainly diminish, and even eliminate, boiling incipience problems.

Figure 11 shows the data for tube 1 operating separately as a single tube (test 1) for decreasing heat flux (aging procedure D). Uncertainty bands for ΔT are again shown, indicating that care must be taken when interpreting results at low heat flux. The data for tube 1 during decreasing heat flux, when compared to the data for tube 1 during increasing heat flux (Figure 5), show very good agreement at high heat fluxes, within the uncertainties of the measurements, and a definite hysteresis pattern at low heat fluxes. Also shown on Figure 11 are the smooth tube data of Marto and Anderson (1992) for tube 1 operating separately as a single tube (taken on the same apparatus) and the data of Jung and Bergles (1989) for a short TURBO-B tube, both for decreasing heat flux. It can be seen that the two sets of TURBO-B data are in good agreement and both provide enhancements in the heat-transfer coefficient of between 3 and 4 for the whole range of heat flux.

Figure 12 shows the data for five instrumented tubes in operation during decreasing heat flux (test 5). From a comparison with Figure 11, it is clear that the wall superheat of tube 1 has been significantly reduced by the presence of the lower tubes at all heat fluxes, thereby improving heat-transfer performance and indicating a 'bundle effect'. Furthermore, the data for tube 5 is almost exactly the same as for tube 1 from test 1 in Figure 11 i.e. when operating alone. This is due to the fact that tubes 1 and 5 were the lowest activated tubes in the bundle for tests 1 and 5 respectively and consequently had no bubbles impinging on them from heated tubes below.

Figure 13 shows similar data for the whole bundle in operation plus the simulation heaters (test 7). If one compares Figures 12

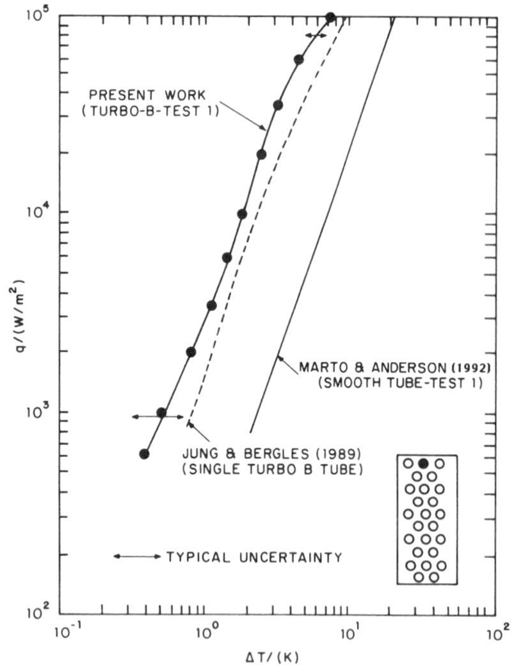

Figure 11: Tube 1 Operating Alone With Decreasing Heat Flux (Aging Procedure D)

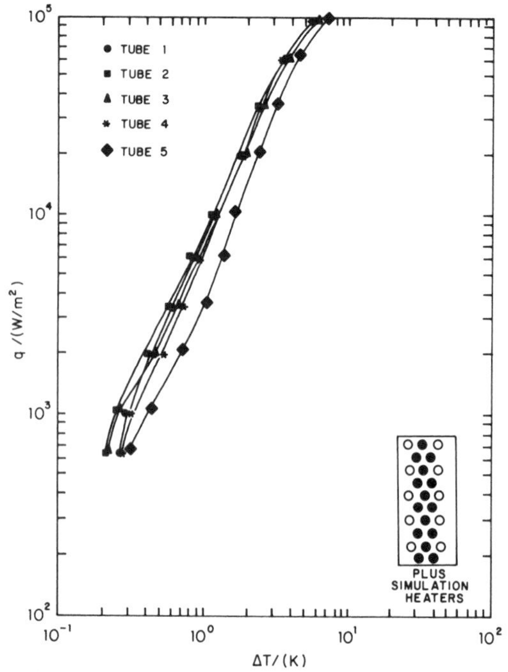

Figure 13: Tube Bundle Operating With Simulation Heaters With Decreasing Heat Flux (Aging Procedure D)

and 13, it is evident that the data with the simulation heaters on show a very slight improvement in the performance of all tubes at all heat fluxes, and especially tube 5 which is now more 'in-line' with the other tubes due to bubble impingement from below; at high heat fluxes, there is no significant change. Figure 14 shows the data for tube 1 only for all 7 tests during decreasing heat flux. It is clear that the performance of tube 1 is enhanced as more tubes within the bundle are activated, indicating a 'bundle effect' at all heat fluxes. When the data in Figure 14 are compared to those in Figure 10, it is clear that the decreasing heat flux data agree very well with the increasing heat flux data taken in the fully developed boiling region.

Marto and Anderson (1992) found similar effects for a smooth tube bundle and attributed them to a number of mechanisms as postulated by Cornwell (1989). The first of these is due to local liquid forced convection which is more important at low heat fluxes, especially for smooth and finned tubes. A second is due

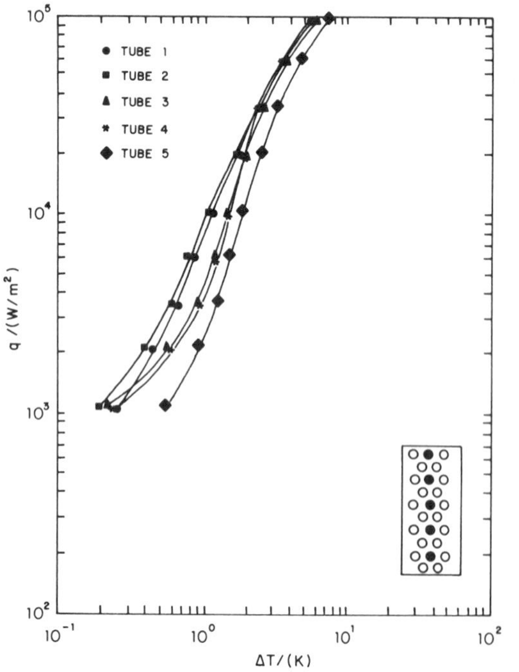

Figure 12: Tubes 1, 2, 3, 4 and 5 Operating Simultaneously With Decreasing Heat Flux (Aging Procedure D)

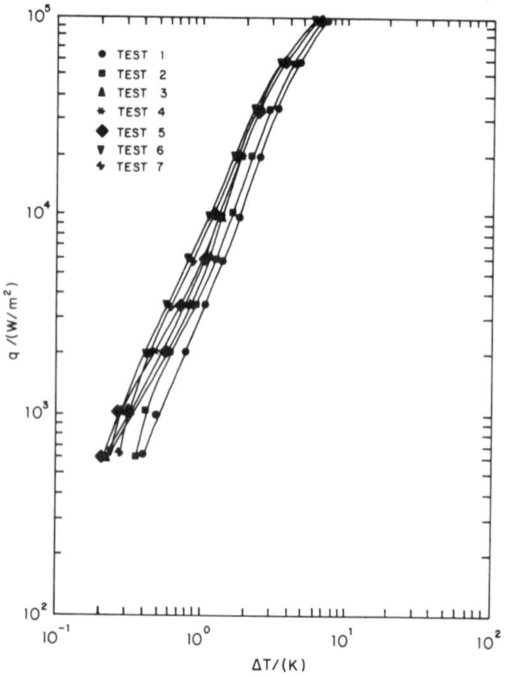

Figure 14: Tube 1 Operating for All 7 Tests With Decreasing Heat Flux (Aging Procedure D)

to bubbles from lower tubes impinging on the upper tubes causing additional turbulence in the superheated liquid boundary-layer. In addition, these bubbles can 'trap' an evaporating microlayer between the heated surface and sliding bubble, creating very large heat-transfer coefficients. A third mechanism is due to bubble nucleation on the surface itself. Obviously, these latter two mechanisms will become more important as the number of bubbles from below increases. At lower heat fluxes, therefore, one may expect these two terms to be more important for an enhanced surface than for a smooth surface.

In addition to these three mechanisms, Marto and Anderson (1992) suggest that secondary nucleation could further increase heat transfer when large sliding bubbles are present. Mesler and Mailen (1977) found that when a bubble bursts through a thin liquid film (such as an evaporating microlayer), many new bubbles grow from that location due to entrained vapor nucleii created from the bursting process. Due to the higher nucleation site density found with an enhanced surface, then this mechanism, if present, would again be more effective for such a surface when compared with a smooth surface.

Table 2 compares the average TURBO-B bundle heat-transfer coefficient (averaged over all 5 instrumented tubes) determined from Figure 13 (test 7) with smooth tube values from Anderson (1989) (taken under similar conditions) at three nominal heat fluxes. It can be seen that a TURBO-B tube bundle enhances the overall heat-transfer coefficient by a factor of 4.6 at low heat flux, dropping to 3.6 at high heat flux (it must be remembered that there is greater uncertainty in the data at low heat flux). Also shown in Table 2 are the values of the heat-transfer coefficient for tube 1 operating alone (test 1). The enhancement in relation to a smooth tube is the same at low heat fluxes, but drops to 3 at high heat fluxes.

Dividing the average heat-transfer coefficient from test 7 with that from test 1 gives a 'bundle effect' provided by bundle operation compared with single tube operation; the 'bundle effect' for test 1 and test 7 are also listed in Table 2. It can be seen that at low heat fluxes, the TURBO-B surface displays a 'bundle effect' of 1.61. This is very similar to the smooth and Thermoexcel tube 'bundle effect' found respectively by Anderson (1989) and Arai et al (1977). However, unlike Anderson (1989) and Arai et al (1977), who found that bundle operation gave no performance enhancement for heat fluxes greater that about 30 kW/m^2, the present data for TURBO-B surfaces still exhibit a 'bundle effect' of 1.22 even at the highest heat flux of 100 kW/m^2. The present data are also contrary to the porous coated bundle data of Czikk et al (1970) and Fujita et al (1986) who essentially obtained a boiling factor of 1 over the whole range of heat flux tested. It may be that some of the heat-transfer mechanisms mentioned above become more important with the TURBO-B type of enhanced surface.

The present results confirm that a 'bundle effect' must be used in the design of flooded evaporators and that the use of single tube results will be conservative. This 'bundle effect' will, however, depend upon the type of boiling surface being used, the tube bundle layout, the operating heat flux and, of course, the type of liquid being boiled.

Table 2: Comparison of TURBO-B and Smooth Tube Bundle

	Heat-Transfer Coefficient/(W/m^{2}K)		
Heat Flux/ (W/m^2)	1.5 x 10^3	1 x 10^4	1 x 10^5
Smooth Bundle			
Test 1$^{(a)}$	488	1511	4658
Test 7$^{(a)}$	790	1782	4670
'Bundle Effect'$^{(a)}$	1.62	1.18	1.00
TURBO-B Bundle			
Test 1	2263	5279	13770
Test 7	3654	8010	16770
'Bundle Effect'	1.61	1.52	1.22
Enhancement (Test 1)$^{(b)}$	4.6	3.5	3.0
Enhancement (Test 7)$^{(b)}$	4.6	4.5	3.6

(a)Data of Anderson (1989) for a smooth tube bundle.
(b)Enhancement defined as ratio of heat-transfer coefficient of the TURBO-B surface to the smooth tube value.

CONCLUSIONS

Nucleate boiling data of R-113 at atmospheric pressure were obtained using a small bundle of TURBO-B copper tubes. The data were obtained for both increasing and decreasing heat flux. Based upon the results pertaining to this particular bundle and apparatus, the following conclusions may be made:

1. The effect of auxiliary heaters (to aid experimentation) can be significant in the natural convection region and must be carefully accounted for.

2. In the natural convection region, heated lower tubes do not have much influence on heat transfer from upper tubes.

3. The presence of heated lower tubes in a bundle reduces the incipient boiling point and the accompanying temperature overshoot for upper tubes.

4. During nucleate boiling, the presence of heated lower tubes enhances the heat-transfer coefficient of upper tubes at all heat fluxes, from 60% at low heat flux to 22% at high heat flux. This leads to a 'bundle effect' that should be included in the design of flooded evaporators that depends on heat flux and the number of tubes in the bundle.

5. When compared to a smooth tube bundle under similar conditions, enhancements of 4.6 and 3.6 were obtained at low and high heat fluxes respectively.

ACKNOWLEDGEMENTS

This work was funded by the David Taylor Research Center, Annapolis, MD. The authors would like to thank Mr. Petur Thors of Wolverine Tube Co. for supplying the TURBO-B tubes and Dr. John Thome for his useful comments.

REFERENCES

Anderson, C.L., 1989, "Nucleate Pool Boiling Performance of Smooth and Finned Tube Bundles in R-113 and R-114/Oil Mixtures", M.Sc Thesis, Naval Postgraduate School, Monterey.

Arai, N., Fukushima, T., Arai, A., Nakajima, T., Fujie, K. and Nakayama, Y., 1977, "Heat Transfer Tubes Enhancing Boiling and Condensation in Heat Exchangers of a Refrigerating Machine", ASHRAE Transactions, vol 83, pt 2, pp 58-70.

Chan, A.M.C. and Shoukri, M., 1987, "Boiling Characteristics of Small Multitube Bundles" Journal of Heat Transfer, Vol. 109, pp. 753-760.

Cornwell, K., 1989, "The Influence of Bubbly Flow on Boiling from a Tube in a Bundle", Proceedings of Eurotherm Seminar No.8, Advances in Pool Boiling Heat Transfer, May 11-12, Paderborn, Germany, pp. 177-183.

Cornwell, K. and Schuller, R.B., 1982, "A Study of Boiling Outside a Tube Bundle Using High Speed Photography" Int. Journal Heat and Mass Transfer, Vol. 25, pp. 683-690.

Cornwell, K, and Scoones, D.J., 1988, "Analysis of Low Quality Boiling on Plain and Low-Finned Tube Bundles", Proceedings 2nd UK Heat Transfer Conf., Vol. 1, pp. 21-32.

Cornwell, K., Duffin, N.W. and Schuller, R.B., 1980, "An Experimental Study of the Effects of Fluid Flow on Boiling Within a Kettle Reboiler Tube Bundle", ASME Paper 80 HT-45, National Heat Transfer Conf., Orlando, Florida.

Czikk, A. M., Gottzmann, C.F., Ragi, E. G., Withers, J.G. and Habdas, E.P., 1970, "Performance of Advanced Heat Transfer Tubes in Refrigerant-Flooded Liquid Coolers", ASHRAE Trans., Vol 76, pp. 96-109.

Fujita, Y., Ohta, H., Hidaka, S. and Nishikawa, K., 1986, "Nucleate Boiling Heat Transfer on Horizontal Tubes in Bundles", Proceedings of 8th Int. Heat Transfer Conf., San Francisco, Vol. 5, pp. 2131-2136.

Hahne, E. and Müller, J., 1983, "Boiling on a Finned Tube and a Finned Tube Bundle", Int. Journal Heat and Mass Transfer, Vol. 26, pp. 849-859.

Han, C. Y. and Griffith, P., 1965, "The Mechanism of Heat Transfer in Nucleate Pool Boiling, Part I, Bubble Initiation, Growth and Departure", Int. J. Ht. Mass Transfer, Vol. 8, No. 6, pp 887-904.

Howell, J. R. and Siegel, R., 1967, "Activation, Growth and Detachment of Boiling Bubbles in Water from Artificial Nucleation Sites of Known Geometry and Size", NASA TN-D-4101.

Hsu, Y.Y., 1962, "On the Size Range of Active Nucleation Cavities on a Heating Surface", J. Ht. Transfer, Vol 84C, No. 3, pp. 207-216.

Hwang, T.H. and Yao, S.C., 1986, "Crossflow Boiling Heat Transfer in Tube Bundles", Int. Communication Heat and Mass Transfer, Vol. 13, pp. 493-502.

Jensen, M. K., 1988, "Boiling on the Shellside of Horizontal Tube Bundles", Two-Phase Flow Heat Exchangers, Kluwer Academic Publishers, S Kakac et al (eds.), pp. 707-746

Jensen, M.K. and Hsu, J.T., 1988, "A Parametric Study of Boiling Heat Transfer in a Horizontal Tube Bundle", Journal of Heat Transfer, Vol. 110, pp. 976-981.

Jung, C.J. and Bergles, A.E., 1989, "Evaluation of Commercial Enhanced Tubes in Pool Boiling", Report DOE/ID/12772-1, Rensselaer Polytechnic Institute, Troy, N.Y.

Leong, L.S. and Cornwell, K., 1979, "Heat Transfer Coefficients in a Reboiler Tube Bundle", The Chemical Engineer, UK, April, pp. 219-221.

Marto, P.J. and Anderson, C. L., 1992, "Nucleate Boiling Characteristics of R-113 in a Small Tube Bundle", Journal of Heat Transfer, vol. 114, (forthcoming).

Marsters, G.F., 1972, "Arrays of Heated Horizontal Cylinders in Natural Convection", Int. Journal Heat and Mass Transfer, Vol. 15, pp. 921-933

Memory, S.B. and Marto, P.J. , 1992, "The Influence of Oil on Boiling Hysteresis of R-114 From Enhanced Surfaces", presented at "Pool and External Flow Boiling", Santa Barbara.

Memory, S. B., Akcasayar, N., Eraydin, H. and Marto, P. J., 1992, "Nucleate Pool Boiling of R-114 and R-114/Oil Mixtures From Smooth and Commercial Enhanced Surfaces: Part II - Small Tube Bundles", to be submitted to the Journal of Heat Transfer.

Mesler, R. and Mailen, G., 1977, "Nucleate Boiling in Thin Liquid Films", AIChE Journal, Vol. 23, p. 954.

Müller, J., 1986, "Boiling Heat Transfer on Finned Tube Bundles: The Effect of Tube Position and Intertube Spacing", Proceedings of 8th Int. Heat Transfer Conf., San Francisco, Vol. 5, pp. 2111-2116.

Murphy, R.W. and Bergles, A. E., 1971, "Subcooled Flow Boiling of Fluorocarbons", MIT Report No. DSR 71903-72.

Rebrov, P.N., Bukin, V.G. and Danilova, G.N., 1989, "A Correlation for Local Coefficients of Heat Transfer in Boiling of R-12 and R-22 Refrigerants on Multirow Bundles of Smooth Tubes", Heat Transfer - Sov. Res., Vol. 21, No. 4, pp. 543-548.

Sparrow, E.M. and Niethammer, J.E., 1981, "Effect of Vertical Separation Distance and Cylinder-to-Cylinder Temperature Imbalance on Natural Convection for a Pair of Horizontal Cylinders", Journal of Heat Transfer, Vol. 103, pp. 638-644.

Thome, J. R., 1990, "Enhanced Boiling Heat Transfer", Hemisphere Publishing Corporation, Ch. 10, pp. 254-260.

Wallner, R., 1974, "Heat Transfer in Flooded Shell and Tube Evaporators", Proceedings 5th Int. Heat Transfer Conf., Tokyo, Vol. 5, pp. 214-217.

Wanniarachchi, A.S., Sawyer, L.M. and Marto, P.J., 1987, "Effect of Oil on Pool Boiling Performance of R-114 from Enhanced Surfaces", Proceedings 2nd ASME-JSME Thermal Engineering Joint Conference, Honolulu, Hawaii, Vol. 1, pp. 531-537.

Wanniarachchi, A.S., Marto, P.J. and Reilly, J.T., 1986, "The Effect of Oil Contamination on the Nucleate Pool Boiling Performance of R-114 from a Porous Coated Surface", ASHRAE Trans., Vol. 92, Pt. 2, pp. 525-538.

Wege, M.E. and Jensen, M.K., 1984, "Boiling Heat Transfer from a Horizontal Tube in an Upward Flowing Two-Phase Crossflow", Journal of Heat Transfer, Vol. 106, pp. 849-855.

Yilmaz, S. and Palen, J.W., 1984, "Performance of Finned Tube Reboilers in Hydrocarbon Service", ASME Paper No. 84-HT-91.

HTD-Vol. 197, Two-Phase Flow and Heat Transfer
ASME 1992

BOILING BEHAVIOR OF AQUEOUS MIXTURES AT ATMOSPHERIC AND SUBATMOSPHERIC PRESSURES

Wade R. McGillis and Van P. Carey
Department of Mechanical Engineering
University of California
Berkeley, California

John S. Fitch and William R. Hamburgen
Western Research Laboratory
Digital Equipment Corporation
Palo Alto, California

Abstract

Water/methanol and water/2-propanol binary mixture boiling behavior is investigated at atmospheric and subatmospheric pressures. Phase equilibrium, heat transfer characteristics, and the critical heat flux (CHF) condition are determined for saturated pool boiling from a discrete heat source while varying the mixture concentrations. The heat source is an upward-facing copper surface submerged in a laterally confined, finite pool. A nonideal aqueous-mixture model provides good agreement with phase equilibrium experimental data. A single-component nucleate boiling correlation is presented to provide reasonable prediction of the low pressure data of this investigation. The mixture wall superheat is increased from that of an ideal mixture and predicted by a model based on phase equilibrium. Small additions of alcohol to water increase the CHF condition above that of pure water. Higher concentrations of alcohol begin decreasing the CHF condition to that of the pure alcohol.

Nomenclature

A, A_0	binary mixture superheat constants
A_{wa}, A_{aw}	activity coefficient constants
c_1, c_2	constants
$c_{b,i}$	pure component i boiling constant
CHF	critical heat flux
$f_{i,v}$	fugacity of component i in the vapor
$f_{i,l}$	fugacity of component i in the liquid
g	gravity
$\bar{g}_i^E$	excess Gibbs energy
h_{lv}	heat of vaporization
m,n	pure fluid boiling curve constants
P	total pressure
P_i	pure component i saturation pressure
P_c	critical pressure
P_r	reduced pressure, P/P_c
Pr	liquid Prandtl number
q''	wall heat flux
$q''_{m,Z}$	maximum Zuber wall heat flux
R	gas constant
R_c	surface active cavity size
T_f	liquid fluid temperature
T_s	liquid saturation or bubble-point temperature
T_w	wall or surface temperature
T_c	critical temperature
T_r	reduced temperature, T/T_c
ΔT	$T_w - T_f$, superheat
ΔT_{ideal}	ideal mixture superheat
x_i	component i liquid mole fraction
y_i	component i vapor mole fraction

Greek Symbols

γ_i	activity coefficient of component i
ϕ_i	fugacity coefficient of component i
ρ_l	liquid density
ρ_v	vapor density
σ	liquid surface tension

subscripts

a	alcohol
w	water
ONB	onset of nucleate boiling

Introduction

The goal of practical boiling research is to predict boiling performance. Boiling performance consists of the superheat required for incipience, boiling surface temperature fluctuations, the fully developed nucleate boiling heat transfer coefficient, and the maximum or critical heat flux condition. In applications where it is desirable to keep the temperature of a boiling surface low while removing the greatest amount of heat, reducing the saturation pressure may be a useful solution. However, reducing the saturation pressure may have an undesirable effect on other aspects of boiling.

Because the primary application of this research was to remove high heat fluxes and maintain low surface temperatures, a major objective of this investigation was to ascertain boiling characteristics at low pressure. A reduction in the pressure causes a corresponding decrease in the saturation or boiling temperature. Consequently, a given superheat level is achieved with a lower surface temperature. This approach is particularly useful when water, having a relatively high saturation temperature at atmospheric pressure, is used as the boiling liquid. Water is a desirable liquid since it has such a high heat of vaporization, is nontoxic, and nonflammable.

Boiling in sealed vessels is a typical application of subatmospheric pressure boiling. Heat pipes, thermosiphons, and some heat pump cycles may rely on subatmospheric pressures to provide low surface temperatures while moving significant quantities of heat. For example, it is often desirable to maintain a low temperature on the heated end of a heat pipe or thermosiphon in spot cooling of electronic components. Heat fluxes from current electronic components are approaching 50 W/cm^2. These fluxes are not easily handled by solid heat sinks. Phase-change heat sinks, which operate with a nearly isothermal interior, are becoming increasingly attractive. Low temperature operation of these heat sinks may be prescribed by creating a saturated liquid and vapor state in the vessel at subatmospheric pressures. Therefore the boiling occurs in the heated end of the vessel at a lower temperature. Knowledge of the boiling characteristics of the small heated surface is necessary to insure that steady and safe operating conditions are maintained.

Table 1: Important fluid characteristics

characteristic	water	methanol	2-propanol
chemical formula	H_20	CH_30H	C_3H_80
molecular weight (g/mol)	18	32	60
water azeotrope	N/A	none	x_p=.6
normal b.p. (°C)	100	64	82
$P_{25°C}$ (kPa)	3	17	6
$h_{lv,25°C}$ (kJ/kg)	2300	1160	700
$\sigma_{25°C}$ (N/m)	.068	.022	.024
$q''_{m,Z}$ (W/cm^2)	130	57	45
toxicity	none	moderate	low
normal m.p. (°C)	0	-72	-67
flash point (°C)	N/A	12.2	11.7

The characteristics of pool boiling of water at low pressure are known to be much different from boiling at atmospheric pressure. Raben et al. (1965), Cole and Shulman (1967), and Van Stralen et al. (1975) investigated saturated nucleate pool boiling of water at subatmospheric pressures off large horizontal surfaces. In their experiments, bubble growth rates, frequencies, and departure diameters for different subatmospheric pressures were investigated. Single component boiling regimes of water and acetone at low pressure in a thermosiphon were examined by Niro and Beretta (1990). Van Stralen (1956) reported experimental CHF data for water as a function of pressure from a 0.02 cm diameter heated platinum wire. Unfortunately, a result of boiling water at subatmospheric pressures is low-frequency bubble departure, creating surface temperature gradients and surface temperature oscillations (McGillis et al. (1991)).

In some electronics cooling applications, it is necessary to keep the boiling behavior steady, with uniform heat removal. Pure water at subatmospheric pressures does not easily provide these attributes, particularly at low fluxes. Under some conditions, the boiling performance can be improved by adding a second liquid to water. This would be classified as boiling of a binary mixture.

Despite the numerous studies of binary mixture boiling processes over the past 50 years, very little is known about the mechanisms responsible for their behavior. State-of-the-art review articles by Shock (1982) and Thome and Shock (1984) provide an extensive overview of prior multicomponent boiling research. There is also a dearth of information in the literature regarding boiling of water with organic liquids at subatmospheric pressures from horizontal surfaces. Bonilla and Perry (1941) investigated combinations of water, ethanol, n-butanol, and acetone from a horizontal chromium plate at pressures ranging from 20 to 130 kPa.

This study considers binary mixtures of water with methanol or 2-propanol at atmospheric and subatmospheric pressures. Table 1 lists important properties of the component liquids. For the reasons stated above, water is a particularly attractive fluid for high heat flux applications, and adding these fully miscible alcohols to water offers further boiling performance improvements. In addition, mixtures of water with methanol or 2-propanol are attractive because their properties are well documented and high purity materials are commercially available.

Experimental Apparatus

Figure 1 shows the experimental test section and system used in this investigation. The copper test section was machined to accommodate two cartridge heaters in the bottom end. The top half of the copper piece was milled to provide a long 1.27 x 1.27 cm square section. Within this section, 0.8 mm holes were drilled to the center to hold thermocouple wires. The copper and thermocouples were then cast in a low viscosity epoxy. The main body of the pool boiling container was made with 2.5 cm I.D. tubing. In order to examine boiling at pressures below 10 kPa, various experimental designs were considered. Nucleate boiling is highly dependent on cavity size, distribution, and wetting properties. The most extraneous nucleation sites to control were at the interface between the copper test section and the epoxy. This interface had to maintain a vacuum seal after thermal cycling. Once the epoxy cured, the top surface of the copper could be prepared. Excess epoxy was milled down flush with the copper surface. The copper/epoxy surface was then finished with #320 emery paper and cleaned with alcohol. The epoxy surface was then bonded to the end of a glass tube to allow observations. The clear tube fit inside an O-ring fitting at the bottom of the condenser so that repeated assembly was simple. The condenser was made of a long section of copper tubing which had radial, helically wound copper fins soldered onto its O.D. Heat was removed from the fins with an air blower.

A thermocouple probe extended down through the inside

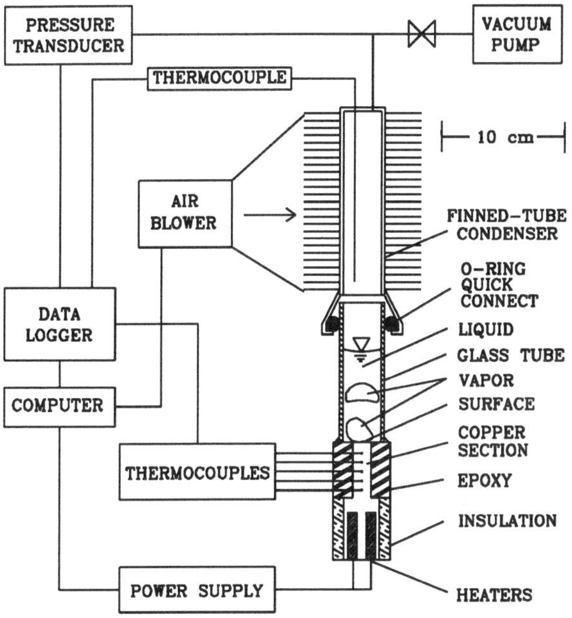

Figure 1: Schematic diagram of test system and test section used in pool boiling experiments.

of the tube and could be positioned vertically to monitor fluid or vapor temperatures. A transducer measured the internal pressure of the thermosiphon. A valve allowed a two-stage vane-type vacuum pump to pull the internal pressure down to very low values. The liquid was boiled during this step to remove gasses from the internal volume. A second deaerated fluid was added to the system via a septa and syringe to keep noncondensible gases out. Once the system was filled, the thermodynamics were verified. A computer and control algorithm varied the blower speed as needed, and proved able to keep the fluid temperature, T_f, within 1°C during continuous boiling. This maintained the internal pressure constant within 0.27 kPa.

A datalogger recorded the temperatures and system pressure, with a sampling rate of up to 1000 Hz. A linear fit of the measured temperature gradient in the copper section was used to calculate the test section heat flux and the heated surface temperature. Experiments and 2-D numerical analyses indicate that the heat losses from the test section were less than 6%, and that the average heat flux at the boiling surface, q'', could be determined within 3%. Experimental uncertainties in the pressure and differential temperature measurements were 0.1 kPa and 0.1°C, respectively.

Steady state for the entire experiment was determined by monitoring the temperature changes with time via the datalogger and computer. When a particular surface was extended to the critical heat flux condition, the final, highest heat flux for which the system reached steady state for nucleate boiling was used as the CHF data point.

Data presented in this paper were measured with a stationary liquid pool height of 7.1 cm. The pressures investigated ranged from about 7 to 101 kPa. Heat flux levels off the 1.27 x 1.27 cm surface were as high as 200 W/cm^2.

Boiling Process in a Thermosiphon

The thermosiphon used in this study could operate in a variety of boiling regimes, from low heat flux, evaporative heat transfer, through nucleate boiling, to critical heat flux or dry out. When there was no heat supplied to the system, and the liquid and vapor were at thermodynamic equilibrium, the pressure in the system corresponded to the saturation pressure at the system temperature. The liquid became superheated with the application of very low heat fluxes. Heat was removed by evaporation at the liquid/vapor free surface and subsequent condensation occurred at the condenser walls.

At higher heat flux levels, the liquid becomes superheated enough to allow bubble growth from cavities on the heated surface. As in any superheated system, the onset of nucleation in a cavity depends on: 1) residual gasses or vapor in the cavity; 2) the size, shape and material of the cavity; 3) the liquid and vapor thermophysical properties; and 4) the amount that the liquid is superheated. If all of the necessary criteria for bubble growth were met, a bubble grew and released from the heated surface.

When a bubble grows and departs from a heated surface, colder fluid replaces the highly superheated fluid. Large bubbles at low pressure cause significant downwash of cool liquid adjacent to the wall and subsequent wall temperature fluctuations. Since bubble growth depends on a sufficient superheating of the surrounding liquid, an appreciable time was required to reheat the liquid in the vicinity of the wall to a superheated state and initiate subsequent bubble growth. This time is termed the waiting time.

The length of time from the beginning of bubble growth to bubble departure depends on how large the bubble must become for release to occur. This interval therefore depended on the rate at which the bubble grew to departure size. The departure bubble size is determined from the net effect of forces acting on the bubble as it grows on the surface. For an upward-facing horizontal heated surface, surface tension holds the bubble down, while buoyancy pulls it up. If the bubble grows rapidly, the inertia associated with the induced liquid flow around the bubble may also tend to pull the bubble up. Bulk liquid motions may also produce lift forces on the bubble causing it to be lifted away.

At higher heat flux levels t_w decreases, and the number of active nucleation sites and bubble departure frequency increases. This results in continuous boiling.

The critical heat flux is not a boiling regime but the maximum heat flux attainable before the system makes a transition to film boiling. For most applications, a transition to film boiling is unacceptable because of the very large wall temperature excursions.

Phase Equilibrium

The thermophysical properties of the fluid mixture must be determined before boiling performance can be predicted. For example, incipience, heat transfer coefficients and critical heat flux are heavily dependent on the fluid properties. Methods for predicting binary mixture boiling phenomena require interpretations of the mixture properties and the mixture phase equilibrium characteristics. Nonequilibrium conditions and the heat and mass transport properties have important effects.

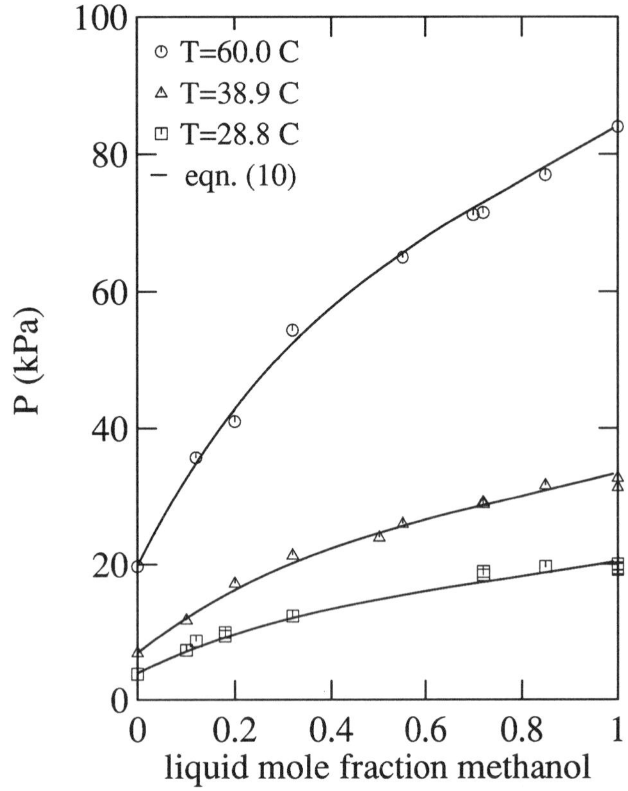

Figure 2: Liquid phase equilibria for water/methanol. Comparison of experimental data with the nonideal mixing model.

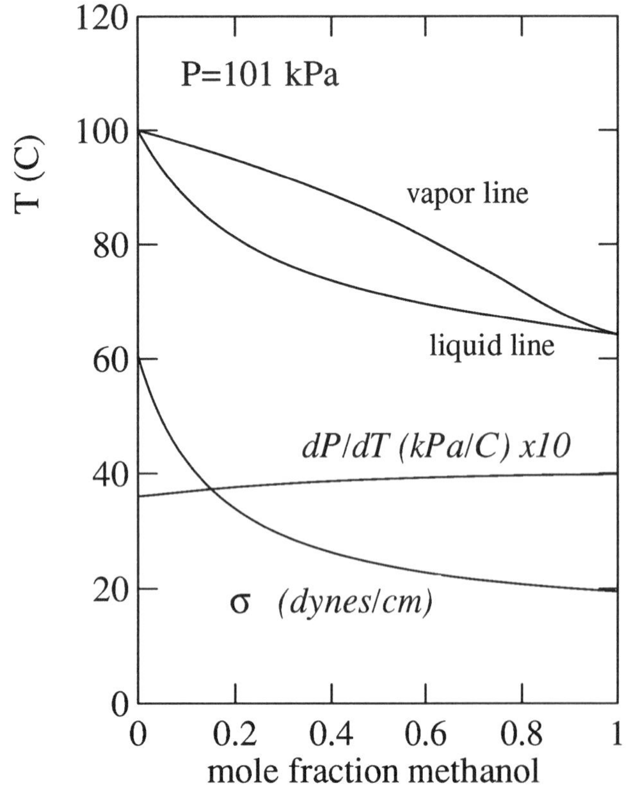

Figure 3: Calculated water/methanol phase equilibria, surface tension, and dP/dT variation at $P = 101\text{kPa}$.

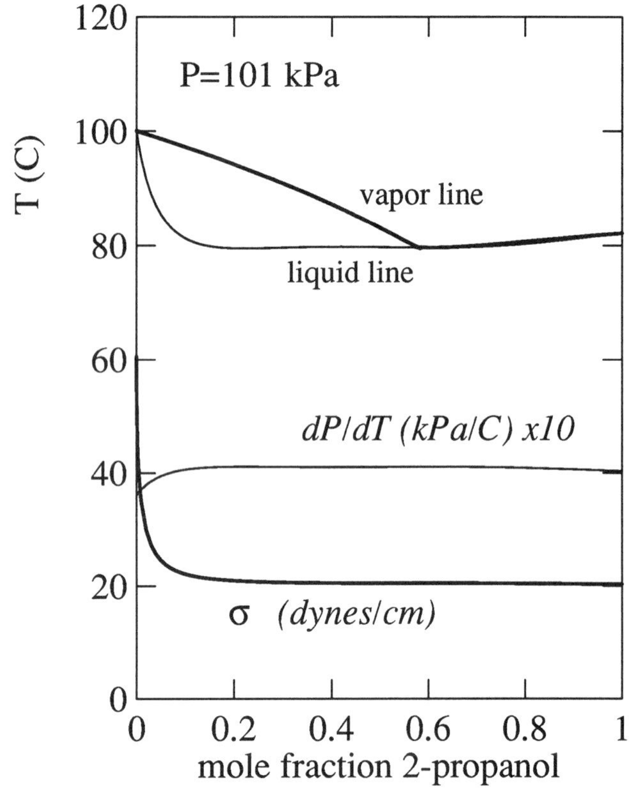

Figure 4: Calculated water/2-propanol phase equilibria, surface tension, and dP/dT variation at $P = 101\text{kPa}$.

The thermophysical properties of a binary mixture can be determined from ideal and nonideal mixture relationships. In these mixtures, the chemical potential and fugacity of component i in the liquid are equal to those of component i in the vapor.

$$f_{i,v} = f_{i,l} \tag{1}$$

The fugacity is an auxiliary function and may be used to simplify the abstract concepts of the chemical potential. The fugacity of component i in the vapor is related to the total pressure by

$$f_{i,v} = \phi_i \, y_i \, P \tag{2}$$

and the fugacity of component i in the liquid is related to the saturation pressure of pure component i by

$$f_{i,l} = \gamma_i \, x_i \, P_i \tag{3}$$

where ϕ_i is the fugacity coefficient, γ_i is the activity coefficient, and y_i and x_i are the the vapor and liquid mole fractions of component i, respectively. For low pressure, $\phi_i = 1$ is a very good approximation. However, γ_i is not unity for nonideal mixtures. The activity coefficient, γ_i, is related to the molar excess Gibbs energy by the relation

$$\bar{g}_i^{\,E} = R\,T \, ln\gamma_i \tag{4}$$

$$\bar{g}_i^{\,E} = 0 \quad at \;\; x_i = 0$$

Prausnitz et al. (1986) describe some possible models to determine activity coefficients for binary mixtures. One popular model is the two-parameter Margules equation

$$\bar{g}_i^E = \{c_1 + 2(c_2 - c_1)\,x_i\}\,x_j^2 \quad (5)$$

For isothermal data at low pressure, the Gibbs-Duhem equation is applicable. In terms of activity coefficients, the Gibbs-Duhem equation is expressed as

$$\sum_i x_i\, d(ln\,\gamma_i) = 0 \quad (6)$$

The Gibbs-Duhem equation provides a thermodynamically consistent relationship between the activity coefficients in the solution. With equations (4), (5), and (6), the activity coefficients of a water/alcohol mixture are determined to be

$$ln\,\gamma_w = \{A_{wa} + 2(A_{aw} - A_{wa})\,x_w\}\,x_a^2 \quad (7)$$

and

$$ln\,\gamma_a = \{A_{aw} + 2(A_{wa} - A_{aw})\,x_a\}\,x_w^2 \quad (8)$$

Substituting (2) and (3) into (1), it follows that for low pressure,

$$y_i\,P = \gamma_i\,x_i\,P_i \quad (9)$$

and using the fact that $y_a + y_w = 1$ and $x_a + x_w = 1$, the saturation pressure of the mixture can be expressed as

$$P = x_a\,\gamma_a\,P_a + (1 - x_a)\,\gamma_w\,P_w \quad (10)$$

Rearranging equation (9), the vapor pressure of the mixture is related to the vapor mole fraction by

$$P = \frac{\gamma_i\,x_i\,P_i}{y_i} \quad (11)$$

Experiments were performed in order to determine whether the nonideal binary mixture behavior agreed with the current predictive methodology shown above. The activity coefficients do not vary significantly over the temperature range investigated so neglecting the temperature dependence is a reasonable approximation. Table 2 lists constants used in equations (7) and (8) that provide good agreement with experimental data of this investigation.

Table 2: Activity coefficient constants

	A_{wa}	A_{aw}
water/methanol mixture	0.6552	0.8106
water/2-propanol mixture	0.9768	2.4153

In these experiments, it was assumed that only small a volume of liquid was needed to occupy the vapor volume. Thus, the liquid composition was assumed not to change throughout the experiments. The saturation pressures for pure fluids and fluid mixtures were measured at different temperatures. Measured pressures are plotted along with equation (10) in Figure 2. Figure 2 shows the liquid portion of the phase equilibrium diagram for water/methanol at different temperatures. It is apparent that the phase equilibria model agrees well with the measured phase equilibria data. The RMS error was calculated to be 1.18 kPa. Equations (10) and (11) provide the determination of the equilibrium concentrations in the vapor and liquid for any system pressure and temperature. Having determined x_i and y_i from equations (10) and (11), the liquid/vapor equilibrium phase diagram can be constructed. The phase diagrams for water/methanol and water/2-propanol are plotted in Figures 3 and 4 at a pressure of 101 kPa.

Onset of Boiling or Boiling Incipience

Bubble growth depends on surface characteristics, fluid properties, and available superheat. Bubble growth is expected to take place from an active cavity (cavity with residual gasses or vapor) when the surrounding liquid reaches the required superheat. The following estimate of superheat required for the onset of nucleation, ΔT_{ONB}, can be derived from static equilibrium.

$$\Delta T_{ONB} = \frac{2\sigma}{R_c\,(dP/dT)} \quad (12)$$

Equation (12) indicates that the superheat required for the onset of nucleation is determined by the liquid surface tension, σ, the active surface cavity size, R_c, and the slope of the vapor pressure curve, dP/dT. The value dP/dT for pure and binary mixtures can be calculated by taking the derivative of equation (10). Fluids with a low surface tension, such as alcohols, will wet a given surface better and have a lower contact angle than higher surface tension fluids such as water. As shown in Figures 3 and 4, the water/alcohol mixture surface tension is reduced from that of pure water, and is extreme in the case of water/2-propanol.

Discussion of Results

Boiling of Pure Fluids

Figure 5 is a comparison of the nucleate boiling data obtained for water at different pressures. A reduction in the pressure for a saturated water system shifted the boiling curve to higher superheat ($T_w - T_f$) levels. However, the decrease in the saturation temperature associated with lower pressures more than compensated for this effect, resulting in lower wall temperatures for a given heat flux.

The shift in the boiling curves may be attributed to a combination of effects. Lower pressures resulted in lower vapor densities and larger bubbles. The lower pressure increased the minimum superheat required for nucleation, resulting in a delayed onset. Also, while greater fluid viscosities result from lower fluid temperatures, the production of small eddies behind the growing and departing bubbles may be reduced which may reduce heat transfer. As can be seen in Figure 5, fluctuations in the wall superheat are more prevalent at 4 kPa than at 6.9 kPa. Fluctuations in wall superheat at 101 kPa are indiscernible.

The general increase of wall superheat with decreasing pressure is also seen in the pure methanol and 2-propanol boiling results shown in Figures 6 and 7. The pure methanol and 2-propanol boiling systems do not have discernible temperature fluctuations caused by intermittent boiling. However, both of these systems have a wall temperature overshoot or excursion at heat fluxes less than 10 W/cm^2. Pure methanol or 2-propanol wet copper surfaces well because of their low contact angle behavior. A fluid that wets well can significantly reduce R_c which, in turn, significantly increases ΔT_{ONB}, causing a large temperature overshoot at incipience.

The three pure fluids have the following noteworthy

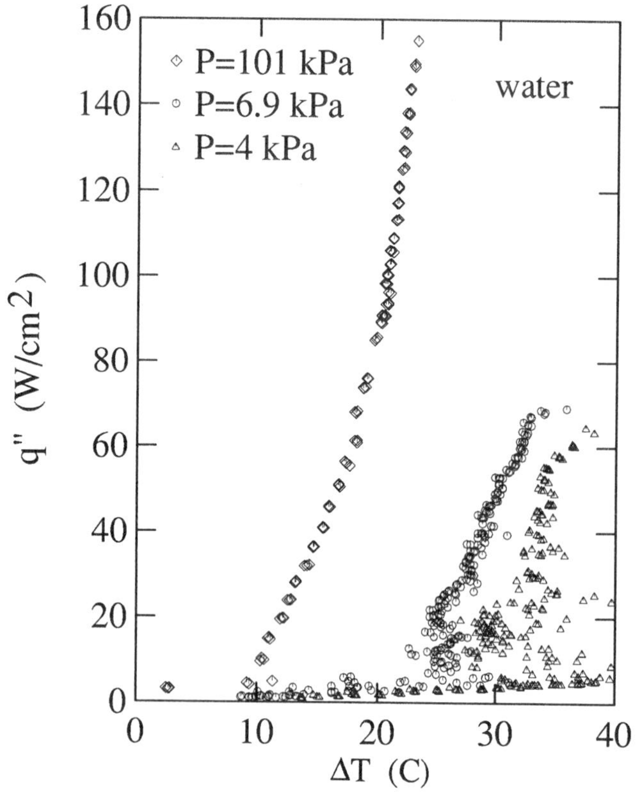

Figure 5: Effect of pressure on the nucleate boiling curves for pure water. CHF reached in each case.

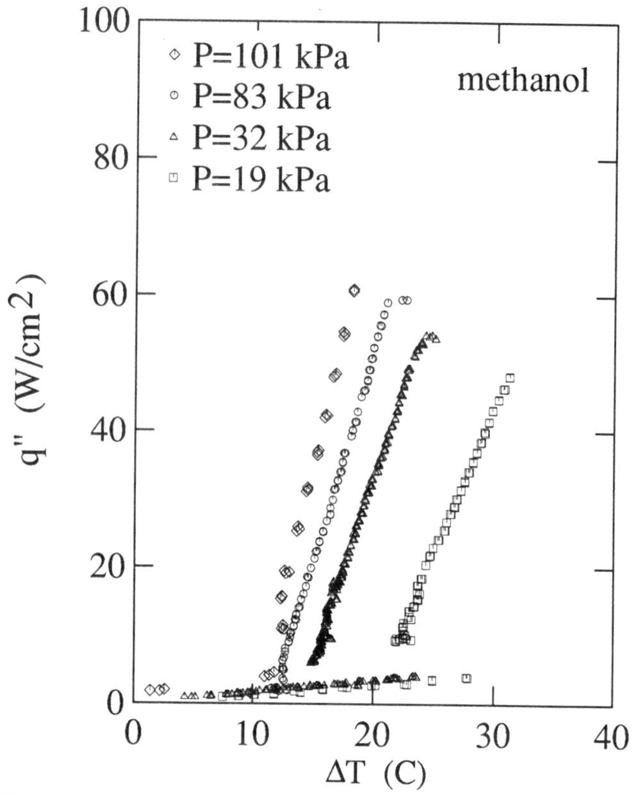

Figure 6: Effect of pressure on the nucleate boiling curves for pure methanol. CHF reached in each case.

behavior at subatmospheric pressure conditions. At subatmospheric pressures, the pure water handled the highest heat flux but demonstrated unsteady boiling behavior at low heat fluxes. The wall temperature of the methanol system was the lowest and provided steady boiling behavior. The 2-propanol system operates with steady boiling behavior but has a relatively small critical heat flux and the highest wall temperature.

There are many pure fluid pool boiling correlations in the literature. Popular ones are outlined in Carey (1992). Some nucleate boiling relations correlated heat transfer with pressure. However, most of these correlations were developed for boiling at pressures equal to or higher than atmospheric. Success at predicting subatmospheric heat transfer characteristics has not been well demonstrated. Correlations based on optimal fits to experimental data provide the greatest accuracy, however, with the least physical basis. Cooper (1982) showed that many of the nucleate boiling correlations based on physical models and dimensional analysis provide similar numerical answers to those correlated to reduced properties. An advantage of reduced property correlations is that they generally provide more simplified expressions, do not require extensive property information, and are less susceptible to calculation error. The following correlation was created to provide a fit to the experimental data of this investigation, for pure fluids.

$$q'' = (c_{b,i} \sqrt{Pr}\, \Delta T)^3 \tag{13}$$

$$c_{b,i} = m\, P_r^{\,n} \tag{14}$$

The constant $c_{b,i}$ is a function of the reduced pressure and varies with fluid type. Most nucleate boiling correlations have a superheat power relation ranging from 3 to 5. In equation (13), the units for q'' are W/m^2, and the units for ΔT are °C.

Values for $c_{b,i}$ are shown in Figure 8. From experimental data of this investigation, the optimal fit for m and n for each fluid is listed in Table 3.

Table 3: Boiling curve constants

i	water	methanol	2-propanol
m	49.4	13.4	15.4
n	0.50	0.45	0.65

When correlation of the pure fluid nucleate boiling behavior at low pressure was attempted without the Prandtl number dependence, accuracy was greatly compromised. In order to keep this relation dependent on reduced properties only, the Prandtl number can be correlated in terms of reduced properties. Cooper (1982) represented fluid properties in reduced property form. The Prandtl number correlation can be constructed and expressed as

$$Pr = P_r^{-0.49}\, T_r^{1.51}\, (1 - T_r)^{-0.64}\, 10^{-3.77} \tag{15}$$

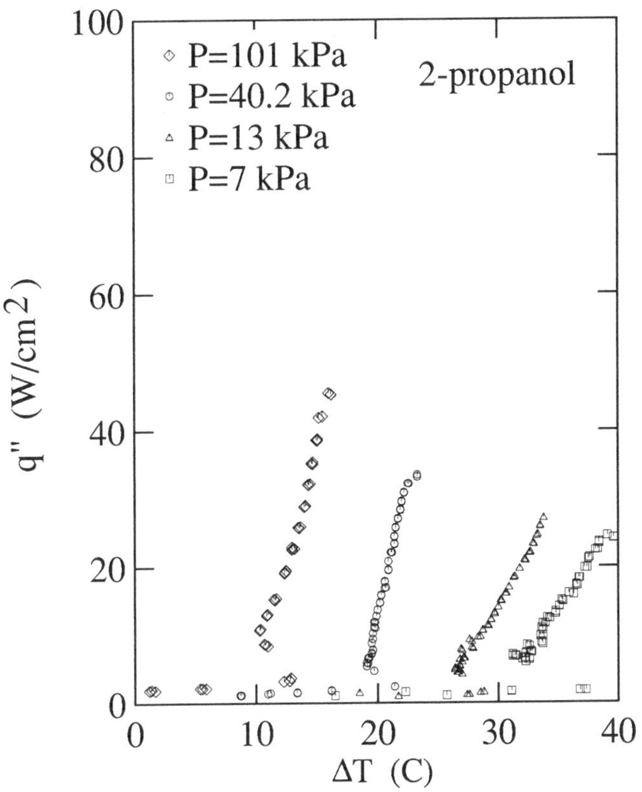

Figure 7: Effect of pressure on the nucleate boiling curves for pure 2-propanol. CHF reached in each case.

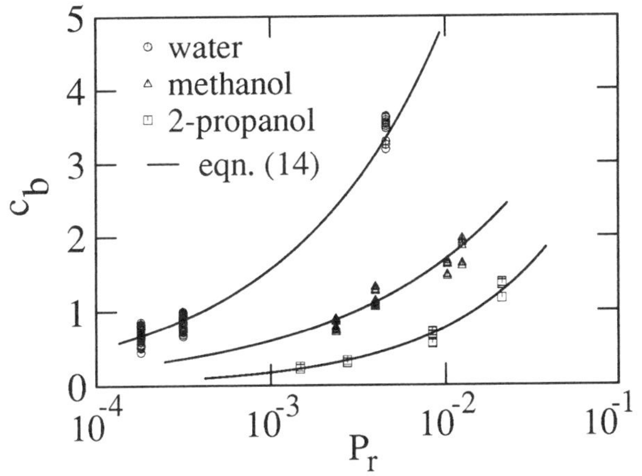

Figure 8: Comparison between measured and calculated pure fluid boiling constant versus reduced pressure.

Boiling of Aqueous Mixtures

Figure 9 compares boiling curves of water/methanol and water/2-propanol mixtures at atmospheric pressure and a constant heat flux. The superheat is increased above that of the pure fluids, particularly when small amounts of alcohol are added to water.

Complex mechanisms govern binary mixture boiling. Binary mixture properties have multiple effects on the onset condition (equation (12)) and therefore on the range of active cavities. The smaller the value of ΔT_{ONB}, the more sites of size R_c will be active. An additional effect on the boiling process is the mass transfer resistance which takes place during the growth of the binary mixture

bubble. As a binary mixture bubble grows from the heated surface, evaporation of the more volatile fluid (in this case alcohol) from the liquid/vapor interface depletes its concentration in the liquid/vapor interface. Diffusion or convection of the more volatile fluid to the interface is not fast enough. Consequently, the departing vapor bubbles are smaller and the amount of heat absorbed due to evaporation is reduced.

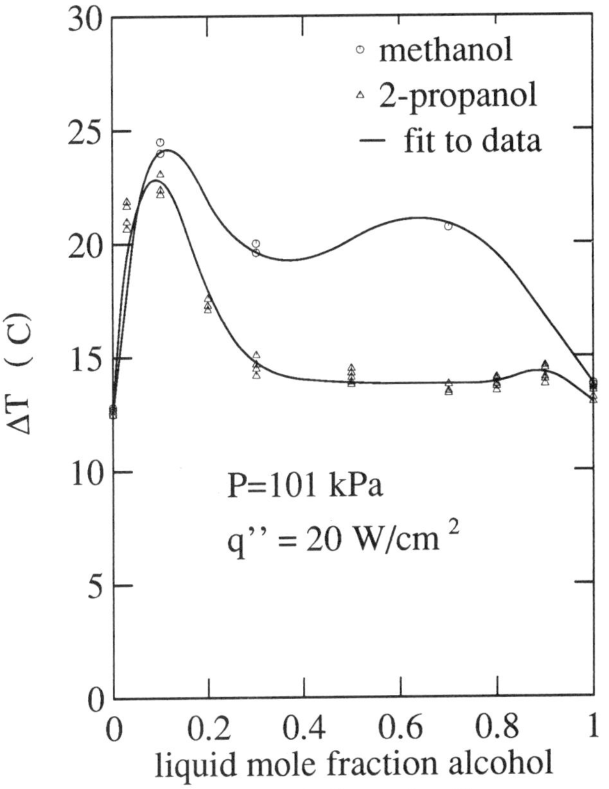

Figure 9: Wall superheat data for boiling methanol/water and water/2-propanol mixtures at atmospheric pressure.

Figures 10 and 11 compare the boiling curves of pure water and alcohol to various mixtures of them. In each figure, the binary mixture boiling curve is plotted against the pure fluids at the same system temperature of 38.9°C. Additions of water to methanol increased the wall superheat from that of pure methanol and reduced the fluctuations in wall temperature from those of pure water. The critical heat flux condition reached a maximum in the vicinity of $x_{methanol}=0.1$.

Additions of water to 2-propanol decreased the wall superheat from that of pure 2-propanol. Additions of 2-propanol to water also reduced the fluctuations in wall temperature from that of pure water. This may be due to the reduction in the bubble departure size. Although the average superheat did not significantly vary through the range of concentrations of 2-propanol in water, the magnitude of the critical heat flux did, reaching a maximum in the vicinity of $x_{2\text{-}propanol}=0.03$.

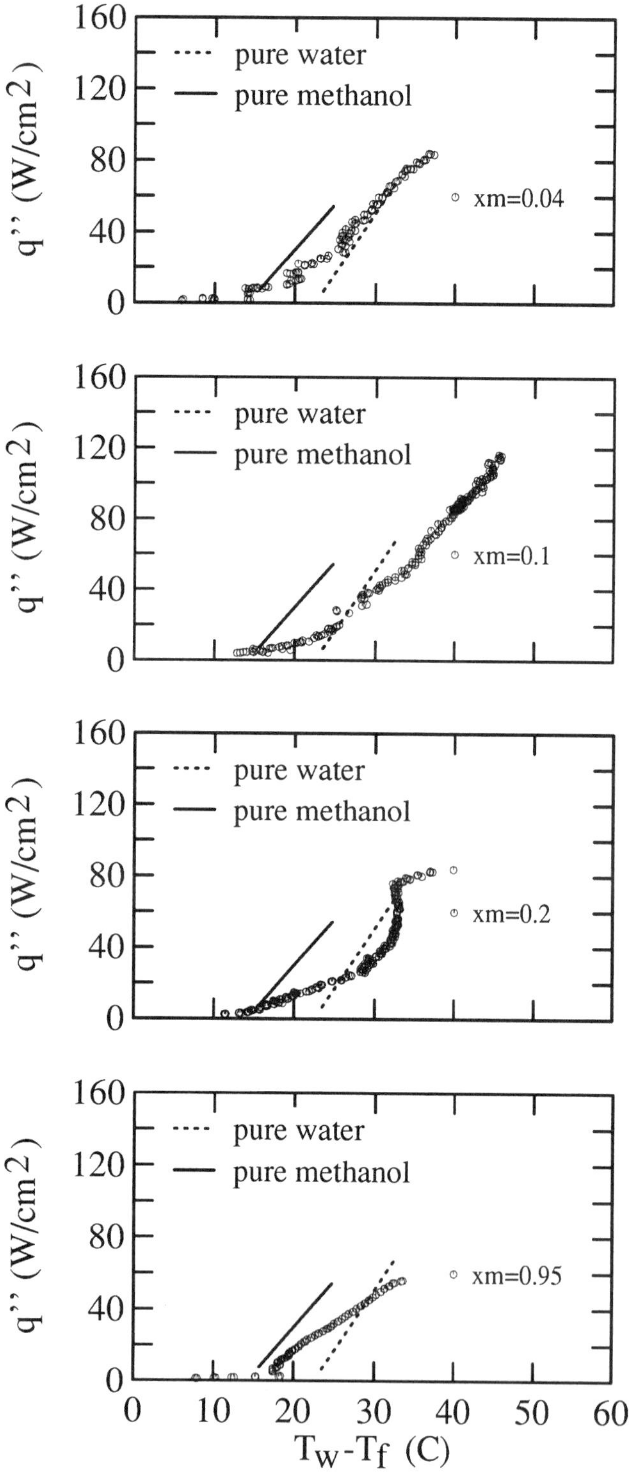

Figure 10: Boiling curves for methanol and water mixtures. Approximate fit to pure fluid data plotted for comparison. CHF reached in each case. $T_f = T_s = 38.9$ °C.

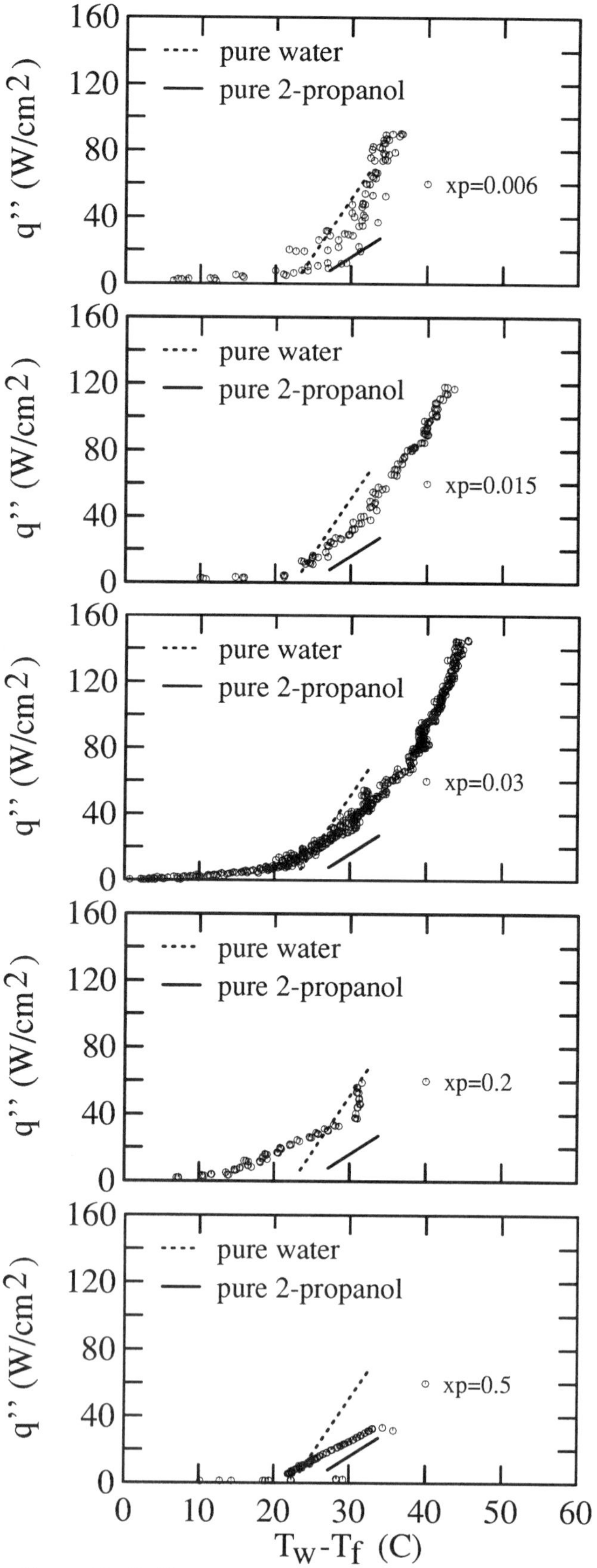

Figure 11: Boiling curves for 2-propanol and water mixtures. Approximate fit to pure fluid data plotted for comparison. CHF reached in each case. $T_f = T_s = 38.9$ °C.

For an ideal mixture, the superheat may be postulated to be a weighted average of the pure fluid superheats.

$$\Delta T_{ideal} = x_a \, \Delta T_a + (1-x_a) \, \Delta T_w \tag{16}$$

The pure component superheats, ΔT_a and ΔT_w, in equation (16) are determined at the system equilibrium pressure and prescribed wall heat flux using equation (13). For a wall heat flux of 20 W/cm^2, Figures 12 and 13 compare the actual superheat of the binary mixture to the ideal superheat determined from equation (16). In each case, the system temperature is kept constant. This implies that the system pressure varies with concentration. By calculating ΔT_{ideal} at the actual system pressure corresponding to the liquid concentration and system temperature, the effect of pressure is eliminated.

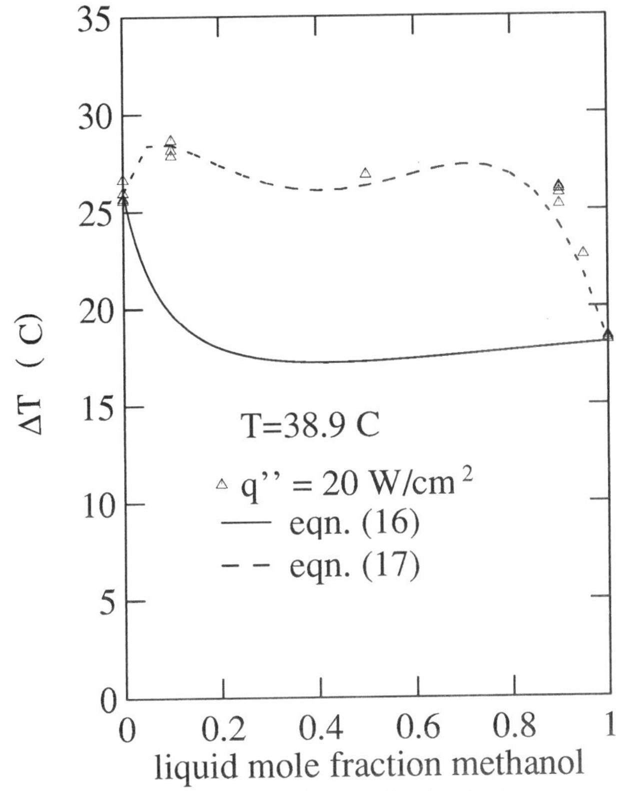

Figure 12: Wall superheat data for water/methanol mixtures. Ideal mixture superheat shown for comparison.

As seen in Figures 10 and 11, the binary mixture superheats at subatmospheric pressure fell mostly between the pure component boiling curves at the same system temperature. However, Figures 12 and 13 show that measured wall superheats are higher than the wall superheats the ideal mixture model predicts at the same system pressure.

A discussion of some binary mixture superheat predictive techniques is given in both Thome and Shock (1984) and Carey (1992). It appears that there is presently no method that can be applied for all mixtures, pressures, and over varying heat flux levels. There also appears to be no attempt to predict superheats for the binary mixtures studied in this investigation at subatmospheric pressures. Stephan and Korner (1969) correlated superheats of organic/organic mixtures and of aqueous/organic mixtures for pressures ranging from 100 to 1000 kPa. They correlated the mixture superheat in terms of the absolute value of the difference between the more volatile component vapor and liquid molar concentrations. The wall superheat was given as

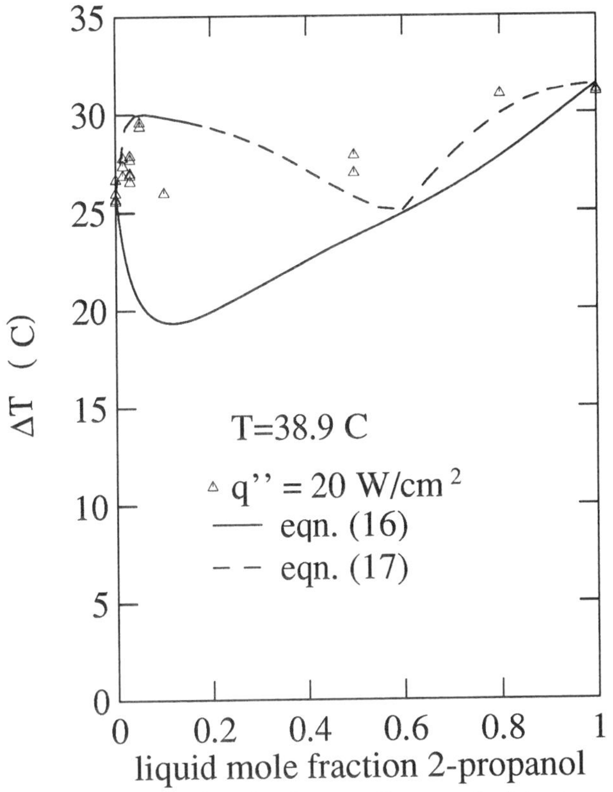

Figure 13: Wall superheat data for water/2-propanol mixtures. Ideal mixture superheat shown for comparison.

$$\Delta T = \Delta T_{ideal}(1 + A\,|y_a - x_a|) \tag{17}$$

where A is a function of fluid type and pressure. They provided a value of A for the water/2-propanol mixture at atmospheric pressure, but not for a water/methanol mixture. They also provided an empirical expression for A to account for a pressure range of 100-1000 kPa. The subatmospheric pressure data of this investigation could not be correlated with the pressure dependence they provided. For a first approximation to the subatmospheric pressure data of this investigation, A is correlated by the relation

$$A = A_0(0.55 + 0.0045P) \tag{18}$$

where P is in kPa and A_0 is the value of A at atmospheric pressure. Table 4 lists values of A_0 that provide reasonably good agreement to data of this investigation. The value of A_0 for the water/2-propanol mixture is the same as that given by Stephan and Korner (1969). A constant value of A_0 for the water/methanol did not provide reasonable accuracy over all concentrations. An equation which is dependent on the liquid mole fraction of methanol is provided for accurate agreement with data.

Table 4: Binary mixture constant	
mixture	A_0
water/methanol	$1+14.8x_a^3-6.2x_a^2+1.2x_a^{.2}$
water/2-propanol	2.04

For a heat flux of 20 W/cm^2 and a constant temperature of 38.9°C, the calculated mixture superheats are plotted in Figures 12 and 13. Equation (17) provides a reasonable prediction of the mixture superheats studied in this investigation at atmospheric and subatmospheric pressures.

Critical Heat Flux

To avoid the risk of film boiling and the accompanying high surface temperatures, the ability to predict the CHF condition is useful. The effect of pressure on the critical heat flux condition for pure fluids is shown in Figures 5, 6, and 7. The maximum heat flux in the nucleate boiling curves is the critical heat flux. For the range of pressures considered, increasing the pressure increased the CHF condition. This is consistent with the CHF behavior observed for a variety of other pool boiling circumstances. A commonly used model which predicts the critical heat flux in saturated pool boiling for a surface of infinite extent was given by Zuber (1959). For low pressures, the Zuber critical heat flux is

$$q''_{m,Z} = 0.131\sqrt{\rho_v}\; h_{lv}\,(g\,\rho_l\,\sigma)^{.25} \qquad (19)$$

Zuber's analysis assumes the critical heat flux is attained when the large vapor jets leaving the surface become Helmholtz unstable. In equation (17), $q''_{m,Z}$ varies with $\sqrt{\rho_v}$, so that with a given increase in pressure, there is a corresponding increase in the vapor density, and the critical heat flux is increased. An increase in the density of the vapor allows more energy removal per unit volume of departing vapor.

The critical heat flux condition of the aqueous mixtures considered in this study is not monotonic with concentration. As seen in Figures 10 and 11, small additions of alcohol to water increased the CHF condition above that of pure water. Higher concentrations of alcohol began decreasing the CHF condition to that of the pure alcohol. The data of this investigation could not be predicted by the basic Zuber model, particularly at low pressures.

Conclusions

This study investigated the pool boiling of water/methanol and water/2-propanol mixtures from a small horizontal surface in a finite pool at atmospheric and subatmospheric pressures. The results support the following conclusions.

1. A binary mixture phase equilibrium model is shown to accurately predict experimental phase equilibrium data.
2. An empirical model is presented for predicting each pure fluid's pool boiling heat transfer coefficient. The model appears to work well for a wide range of pressures, including subatmospheric pressures.
3. This investigation provides heat flux versus wall superheat data for water/methanol and water/2-propanol mixtures at subatmospheric pressures. For a given heat flux and pressure, the mixture wall superheats are shown to increase significantly from calculated ideal mixture wall superheats. A more accurate model, based on weighted phase equilibrium characteristics, is presented.
4. At subatmospheric pressures, the water/alcohol mixtures studied in this investigation provide a means of maintaining low wall temperatures, while reducing the large wall temperature fluctuations characteristic of pure water systems and increasing the CHF condition above that of pure water or of pure alcohol.

References

C. F. Bonilla and C. W. Perry, ''Heat Transmission to Boiling Binary Liquid Mixtures'', *Chemical Engineering Progresses Symposium Series*, Vol. 37, 1941, pp. 685-705.

V. P. Carey, *Basic Elements of Liquid-Vapor Phase-Change Phenomena,* Hemisphere Publishing Corp., New York, 1992.

R. Cole and H. L. Shulman, ''Bubble Departure Diameters at Subatmospheric Pressures'', *Chemical Engineering Progresses Symposium Series*, Vol. 62, No. 64, 1967, pp. 6-16.

M. G. Cooper, ''Correlations for Nucleate Boiling - Formulation using Reduced Properties'', *PhysicoChemical Hydrodynamics*, Vol. 3, No. 2, 1982, pp. 89-111.

W. R. McGillis, V. P. Carey, J. S. Fitch and W. R. Hamburgen, ''Pool Boiling on Small Dissipating Elements at Subatmospheric Pressure'', *Proceedings on Phase Change Heat Transfer*, 1991 ASME National Heat Transfer Conference, Minneapolis, Minnesota, July 1991, pp. 27-36.

A. Niro and G. P. Beretta, ''Boiling Regimes in a Closed Two-Phase Thermosyphon'', *International Journal of Heat and Mass Transfer*, Vol. 33(10), 1990, pp. 2099-2110.

J. M. Prausnitz, R. N. Lichtenthaler, E. Gomes de Azevedo, *Molecular Thermodynamics of Fluid-Phase Equilibria,* Prentice-Hall, New Jersey, 1986, Second Edition.

I. A. Raben, R. T. Beaubouef, and G. E. Commerford, ''A Study of Heat Tranfer in Nucleate Pool Boiling of Water at Low Pressure'', *Chemical Engineering Progresses Symposium Series*, Vol. 61, No. 57, 1965, pp. 249-257.

R. A. W. Shock, *Multiphase Science and Technology,* Hemisphere Publishing Corp., New York, Vol. 1, 1982..

Stephan, K. and Korner, M., ''Calculation of Heat Transfer in Evaporating Binary Liquid Mixtures'', *Chemie-Ingenieur Technik*, Vol. 41, 1969, pp. 409-417.

J. R. Thome, R. A. W. Shock, *Advances in Heat Transfer,* Academic Press, Inc., New York, Vol. 16, 1984..

S. J. D. Van Stralen, ''Heat Transfer to Boiling Binary Liquid Mixtures at Atmospheric and Subatmospheric Pressures'', *Chemical Engineering Sciences*, Vol. 5, 1956, pp. 290-296.

S. J. D. Van Stralen, R. Cole, W.M. Sluyter, and M. S. Sohal, ''Bubble Growth Rates in Nucleate Boiling of Water at Subatmospheric Pressures'', *International Journal of Heat and Mass Transfer*, Vol. 18, 1975, pp. 655-669.

N. Zuber, ''Hydrodynamic Aspects of Boiling Heat Transfer'', AEC Report No. AECU-4439, Physics and Mathematics, AEC, 1959.

HTD-Vol. 197, Two-Phase Flow and Heat Transfer
ASME 1992

CAUSES OF THE APPARENT HEAT TRANSFER DEGRADATION FOR REFRIGERANT MIXTURES

M. A. Kedzierski
National Institute of Standards and Technology
Gaithersburg, Maryland

J. H. Kim
Electric Power Research Institute
Palo Alto, California

D. A. Didion
National Institute of Standards and Technology
Gaithersburg, Maryland

ABSTRACT

This paper presents an investigation into the causes of the apparent heat transfer degradation associated with horizontal-annular flow evaporation of refrigerant mixtures. The apparent heat transfer degradation is the difference between the measured heat transfer coefficient and the heat transfer coefficient that would be obtained from a linear interpolation of the single component values. The degradation is apparent since the linearly interpolated values have no physical basis. For horizontal-annular flow evaporation, most of the heat transfer degradation is a consequence of the use of the locally uniform equilibrium temperature in the measurement and calculation of the heat transfer coefficient. In reality, both circumferential and radial composition gradients can exist within the liquid film which cause temperature distributions that deviate significantly from a uniform saturation temperature. If the actual liquid-vapor interface temperatures (local vapor temperatures) were used in the calculation of the measured heat transfer coefficient for the impose heat flux condition, most of the apparent degradation would not exist. The remainder of the heat transfer degradation is due to nonlinear mixture property effects. Previously published measured heat transfer coefficients for three mixtures were investigated. The focus of the study was to determine the magnitude and the cause of the individual components of the heat transfer degradation of the studied mixtures.

NOMENCLATURE

English symbols

c_p	specific heat (kJ/kg-K)
k	thermal conductivity (W/m-K)
h_i	$h_{2\phi}$ from linear interpolation of single components (W/m^2-K)
h_p	$h_{2\phi}$ predicted using single component correlation (W/m^2-K)
$h_{2\phi}$	two-phase heat transfer coefficient (W/m^2-K)
$\dot{m}$	mass flow rate (kg/s)
P	absolute pressure (Pa)
P_c	critical pressure (Pa)
P_r	reduced pressure, P/P_c
q''	heat flux (W/m^2)
Q	mass flux (kg/m^2-s)
r	coordinate perpendicular to heat transfer surface (m)
T_i	temperature of liquid-vapor interface (K)
T_s	saturated fluid temperature (K)
T_w	inside tube wall temperature (K)
x	mole fraction of more volatile component
x_m	mass fraction of more volatile component
x_q	thermodynamic quality
y	coordinate along heated surface (m)

Greek symbols

$\Delta h_{2\phi}$	$h_i - h_{2\phi}$ (W/m^2-K)
ρ	density (kg/m^3)
ρ_{exp}	experimentally measured density (kg/m^3)
μ	viscosity (kg/m-s)

Subscripts

b	bottom of tube
l	liquid
m	mixture or mass
t	top of tube
v	vapor
1	component number one
2	component number two

INTRODUCTION

Pool boiling of mixtures has been practiced since antiquity. The ancient Greeks made their drinking water by distilling sea water. In the 19th century oil was refined by distillation to make kerosene for lamps. Although the practical application of mixture heat transfer is very old, the experimental and theoretical study of it is relatively new. The study of in-tube flow boiling of refrigerant mixtures is especially recent. During this short period of study, researchers have found that liquid mixtures do not evaporate as efficiently as single component liquids. However, mixtures of refrigerants can be used to enhance the efficiency of refrigeration equipment as compared to single component refrigerants (Mulroy et al., 1988). Unfortunately, the decrease or degradation in the efficiency of the evaporation of mixtures increases the costs of the performance improvements that can be achieved by using mixtures in cycles. This study is an attempt to further the understanding of horizontal flow boiling of mixtures with the hope of generating ideas that might lead to reduction of the heat transfer degradation associated with refrigerant mixtures.

There are two fundamental thermodynamic differences between single component fluids and mixtures which, in turn, cause fundamental differences between the phase change characteristics of single component fluids and mixtures. First, at constant pressure, the mixture temperature rises during evaporation, while the temperature of the single component fluids remains constant. Second, the liquid and vapor compositions are different in the mixture, while they are identical in the single component. These points are demonstrated by the phase equilibrium diagrams for the three binary mixtures investigated here, i.e., R22/R114, R12/R152a, and R13B1/R152a which are shown in Fig.s 1 through 3, respectively.

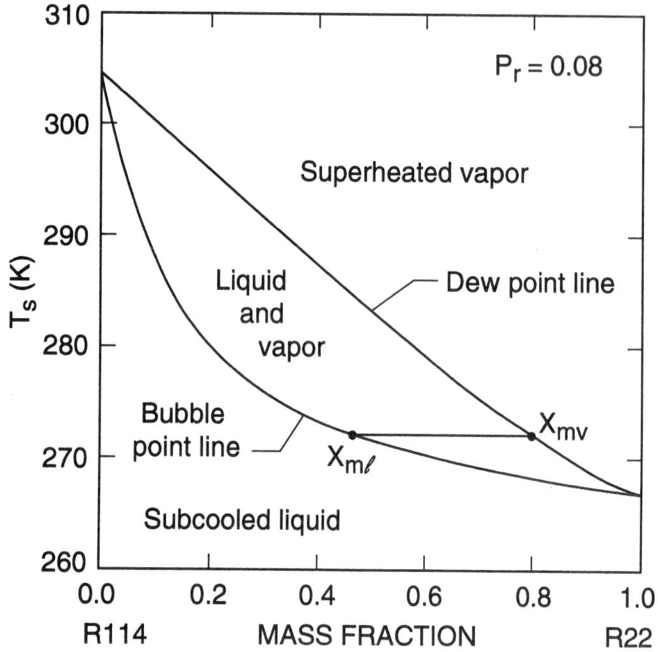

Fig. 1 Phase equilibrium diagram for R22/R114 at P_r = 0.08

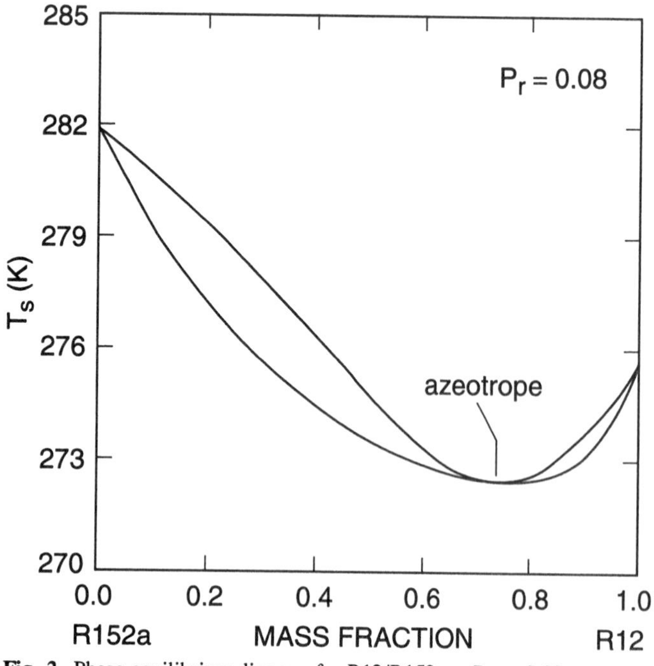

Fig. 2 Phase equilibrium diagram for R12/R152a at P_r = 0.08

Figure 1, the phase equilibrium diagram for the R22/R114 mixture at a reduced pressure (P_r) of 0.08, represents the thermodynamic state of the mixture at equilibrium conditions. The phase equilibrium diagram is a

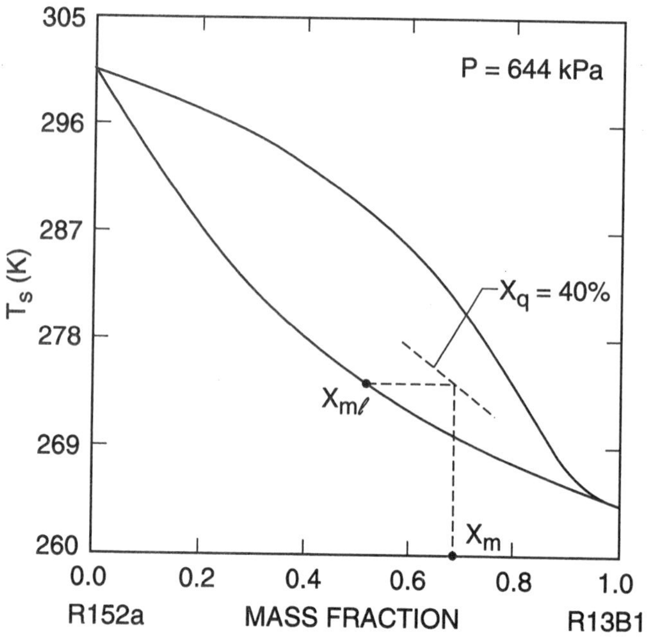

Fig. 3 Phase equilibrium diagram for R13B1/R152a at P = 644 kPa

plot of equilibrium temperatures versus the mass fraction of the more volatile component. The more volatile component is R22, since its equilibrium or saturation temperature is lower than that of R114 at the same pressure. The lower line, the bubble point line, represents the variation of the liquid saturation temperature with composition. The upper line, the dew point line, represents the variation of the saturated vapor temperature with composition. The area between the dew point and bubble point lines represents a two-phase mixture with a liquid of composition x_{ml} and a vapor of composition x_{mv} in coexistence. The distance between the dew point and bubble point lines, i.e., the temperature glide, may loosely be used to determine the potential for the heat transfer degradation of a particular mixture. The heat transfer of a binary mixture, which has a large temperature glide, is likely to be lessened by concentration gradients. Correspondingly, an azeotrope, which is a mixture that has no temperature glide, is likely to exhibit very little heat transfer degradation. Figure 2 shows that an azeotrope exists at x_m = 73.8% R12 as depicted by the intersection of the dew and bubble lines.

Figure 4 illustrates the concentration gradients that are established within a liquid mixture film as a consequence of preferential evaporation of the more volatile component. Notice that there are concentration gradients in two directions: (1) parallel to the uniformly heated surface (dx_{ml}/dy), and (2) perpendicular to (or radially from) the uniformly heated surface (dx_{ml}/dr).

The concentration gradient parallel to the heated surface shown in Fig. 4 is induced by the constant heat flux boundary condition and the varying film thickness. As a mixture evaporates, the bulk fluid is depleted of the more volatile component which reduces the mass fraction of the mixture. The mass fraction coordinate is measured along the wall, increasing in the vertical y-direction. The mass fraction of the thin-film (x_{mlt}) region is less than that of the thick-film (x_{mlb}) region since the thin-film region, having less mass, is further along the two-phase spindle of the equilibrium diagram. Consequently, a film thickness gradient has induced a concentration gradient along the heated surface. Since gravity imposes a nonuniform circumferential film thickness distribution for horizontal annular flow, the above argument describes the mechanism of the circumferential concentration gradients for annular flow within a horizontal tube. The temperature of the liquid-vapor interface at the top of the tube

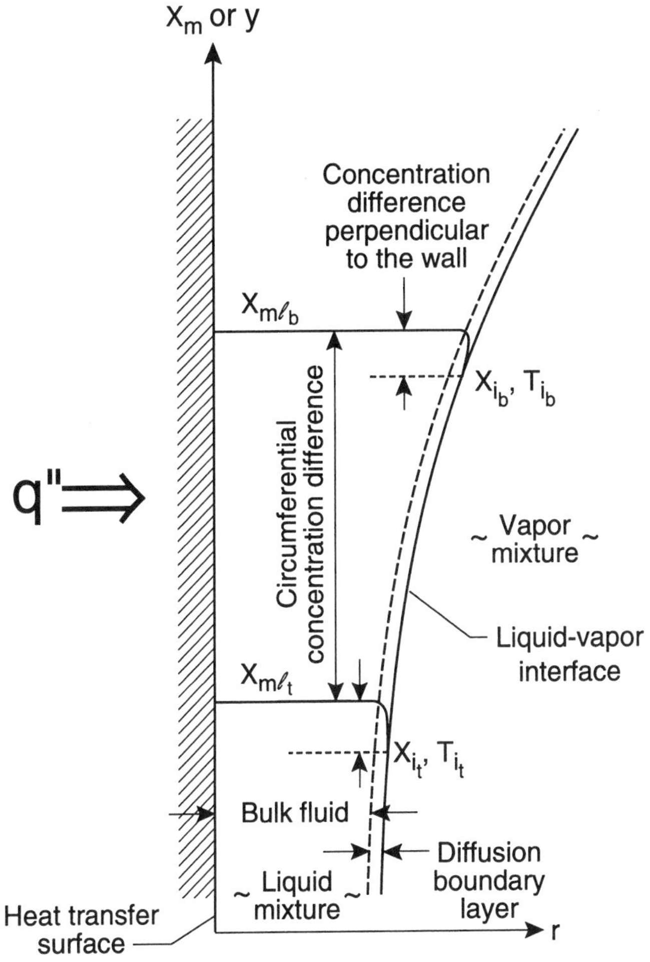

Fig. 4 Concentration gradients within the liquid film

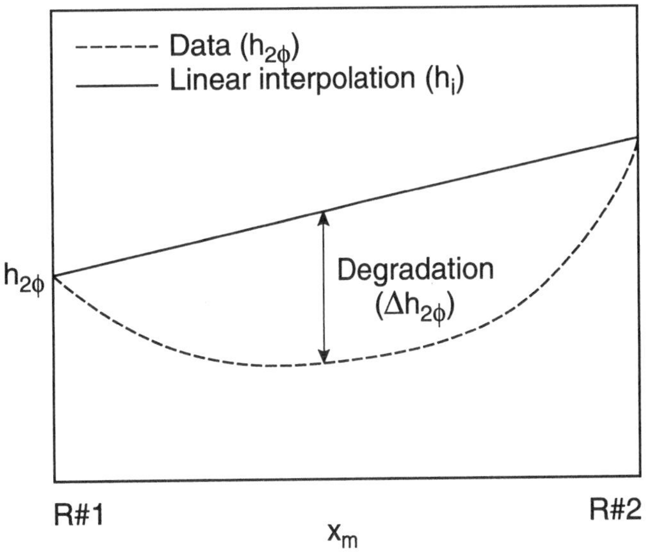

Fig. 5 Typical relationship of mixture flow boiling with respect to composition

(T_{it}) is greater than that at the bottom of the tube (T_{ib}). The magnitude of the liquid-vapor interface temperatures are determined by both the circumferential and the radial concentration gradients.

The concentration gradient perpendicular to the heated surface exists primarily within the thickness of the diffusion boundary layer at the liquid-vapor interface, as shown in Fig. 4. Turbulent mixing prevents the formation of concentration gradients within the bulk liquid. In summary, evaporation depletes the diffusion boundary layer of the more volatile component and the convection confines the concentration gradient to a narrow region within or close to the liquid-vapor interface. This describes the mechanism of the radial concentration gradients for annular flow within a horizontal tube.

Figure 5 demonstrates the variation of the two-phase heat transfer coefficient with respect to composition for an illustrative mixture. The dashed line represents experimental data. Notice that the heat transfer coefficient can be less than that for either pure component. The solid line, h_i, is a linear interpolation between the heat transfer coefficients of the pure components. The straight line has no physical meaning; however, it is used as a reference from which the degradation of the heat transfer coefficient of binary mixtures can be quantified. The heat transfer degradation ($\Delta h_{2\phi}$) is the difference between the interpolated heat transfer coefficient (h_i) and the measured heat transfer coefficient ($h_{2\phi}$) for a given composition.

Several explanations for the heat transfer degradation associated with mixtures have been postulated. Two popular reasons are the loss of available superheat and mass transfer resistance. Also, Ross et al. (1987), Jung and Didion (1989) and Stephan and Preusser (1979) attribute a portion of the degradation to the nonlinear variation of the thermodynamic and transport properties of mixtures with composition. The nonlinear variation of properties with respect to composition may contribute to a nonlinear degradation of the heat transfer coefficient. Stephan and Korner (1969) suggest still another reason for the degradation: mixtures must produce more work to form bubbles than for an equivalent pure fluid.

Loss of available superheat, as described by Shock (1982), is the loss of the heat transfer driving potential due to the increase of the fluid temperature upon evaporation. For a fixed tube-wall temperature, an increase in the fluid temperature results in a reduction or a loss of temperature difference between the wall and the fluid. The data examined here are for a constant, imposed heat flux boundary condition, not a constant wall temperature boundary condition. For the constant heat flux case, the heat transfer driving potential is the heat flux, which dictates the tube-wall temperature. A loss in driving potential cannot be imposed by an increase in fluid temperature upon evaporation since the wall temperature will rise, as the fluid temperature rises, to satisfy the imposed heat flux boundary condition. For the above reasons, the argument for the loss of available superheat does not strictly apply to the constant heat flux boundary condition and, consequently, cannot be investigated in this paper.

Mass transfer, as defined in this paper, is the movement of a liquid component due solely to a concentration gradient. The motion of the liquid is induced by the tendency of the liquid to achieve a uniform equilibrium concentration. The magnitude of the mass flux, due to mass transfer, is insignificant compared to that due to evaporation by heat exchange. Therefore, mass transfer cannot significantly reduce the heat transfer by a movement of fluid which is opposed to the evaporation. However, mass transfer can indirectly effect the heat transfer by determining the magnitude of the concentration gradients. In turn, the concentration gradients establish the temperature distributions which control the heat transfer. In summary, the mass transfer effects the heat transfer coefficient primarily by altering the temperature through concentration gradients and not by the movement of fluid.

Mass transfer resistance is defined in this paper as a resistance to the neutralization of concentration gradients. The mass transfer resistance indirectly causes a degradation in the heat transfer by raising the liquid

temperatures. For clarity, concentration gradients rather than mass transfer resistance is used to discuss the heat transfer degradation.
The additional work of bubble formation cannot be studied in the convective region. All of the data which were examined are in the convective, evaporative flow region. Nucleate boiling is suppressed in the convective region (Chen, 1966). Therefore, the additional resistance due to bubble formation in a mixture cannot be examined using the cited data.

For the above reasons, the focus of the investigation is on the effects of concentration gradients and nonlinear mixture properties on the heat transfer degradation. It cannot be known with any certainty that the above two effects are the only effects that contribute to the heat transfer degradation. However, it is assumed that the heat transfer degradation that is not due to the fluid property effect is due to concentration gradients.

For evaporative flow, it is speculated that most of the heat transfer degradation associated with concentration gradients results from the use of the saturated equilibrium temperature in the calculation of the heat transfer coefficient. If concentration gradients are present in the liquid, the actual liquid-vapor interface temperature (vapor temperature) will be greater than the saturated equilibrium temperature which is obtained from an overall energy balance and the measured pressure. If the actual liquid-vapor interface temperature was used to calculate the measured heat transfer coefficient the heat transfer coefficient would be greater than that calculated from the equilibrium temperature. Consequently, a large portion (possibly all) of the heat transfer degradation associated with concentration gradients can be attributed to the use of the equilibrium temperature in the calculation of the measured heat transfer coefficient.

EXPERIMENTAL WORK INVESTIGATED

Only binary-mixture, horizontal-flow boiling with a constant tube wall heat flux is considered here (Ross, 1985 or Ross et al., 1987, and Jung and Didion, 1989 or Jung et al., 1989). The local two-phase heat transfer coefficient ($h_{2\phi}$) was calculated as:

$$h_{2\phi} = \frac{q''}{T_w - T_s} \qquad (1)$$

where q" is the heat flux at the outside wall, T_w is the inside wall temperature, and T_s is the saturated fluid temperature evaluated at the measured pressure. Local wall temperature measurements were made along the tube length. The tube wall temperature was measured at four circumferential positions, 90 degrees apart, at each axial position along the tube. Four circumferential heat transfer coefficients (top, bottom, right, and left side) were calculated and averaged to obtain the value for the heat transfer coefficient at the given axial position.

Ross et al. (1987) have measured the local flow boiling heat transfer coefficient for various compositions of the R13B1/R152a mixture. Figure 6 presents the measured two-phase heat transfer coefficient ($h_{2\phi}$) versus the mass fraction of the more volatile component R13B1. The data were taken for a saturation temperature, at the exit of the test section, of 270 K, a mass flux of 460 kg/(m^2 s), a thermodynamic quality of 40%, and an incident heat flux of 30 kW/m^2. The solid line is a linear interpolation between the heat transfer coefficients for the single components. The degradation as compared to the single component reference line is greatest (65% of h_i) at x_m = 0.82.

Jung et al. (1989) measured the local flow boiling heat transfer coefficient for various compositions of the R12/R152a mixture and the R22/R114 mixture. His data for a reduced pressure (P_r) of 0.08 at the exit of the test section, a constant imposed heat flux of 17 kW/m^2, a thermodynamic quality of 65%, and a mass flow rate ($\dot{m}$) of 0.023 kg/s are shown in Fig.s 7 and 8. The maximum heat transfer degradation for the R22/R114 mixture is located at an overall mass composition of 0.61. Its magnitude is 2040 W/m^2-K or 37% lower than the linear interpolation between the $h_{2\phi}$ for the pure components.

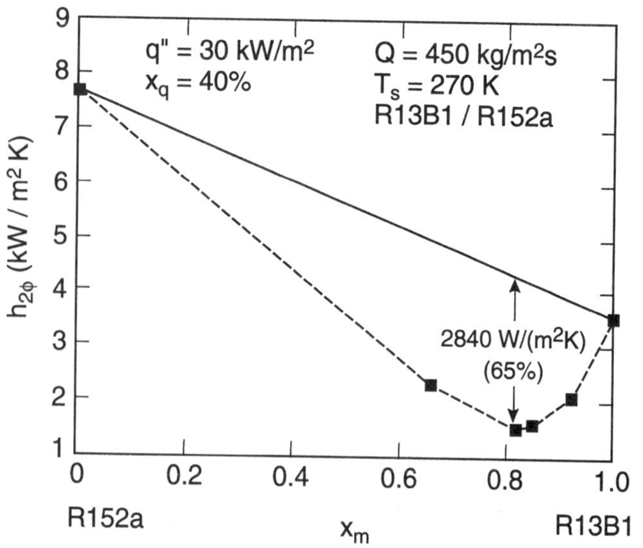

Fig. 6 Measured horizontal flow boiling heat transfer coefficients for the R13B1/R152a mixture (Ross et al. (1987))

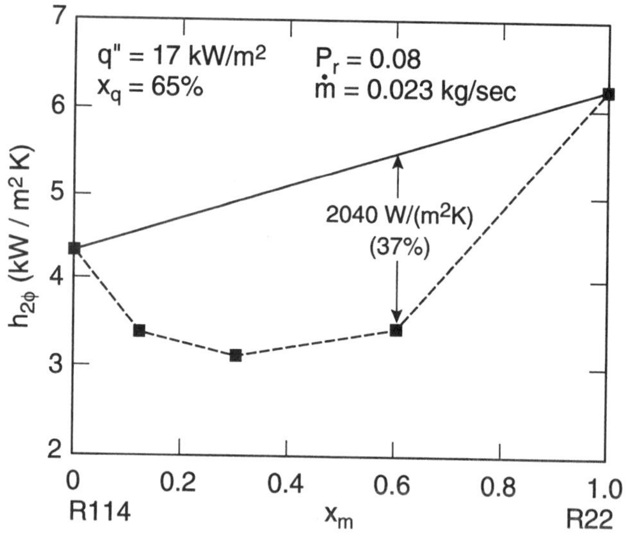

Fig. 7 Measured horizontal flow boiling heat transfer coefficients for the R22/R114 mixture (Jung and Didion, 1989)

The heat transfer degradation for the R22/R114 mixture is large, but not as large as that present for the R13B1/R152a mixture. It is intuitively reasonable to suggest that the large heat transfer degradation associated with the R13B1/R152a data is a result of the coupled effects of: (1) the large difference in mass concentration between the liquid and vapor phases ($x_{mv} - x_{ml}$ = 0.12 at x_m = 0.82) and, (2) the relatively large molecular mass (149 g/mole) of the more volatile component (R13B1). The difference in concentration between the liquid and vapor phases represents the potential for concentration gradients within the liquid. The difference between the vapor and liquid composition ($x_{mv} - x_{ml}$) for the R22/R114 mixture is approximately 0.02 mass fraction greater than that for the R13B1/R152a mixture from about 0.2 to 0.8 liquid mass fraction. Therefore, the potential for mass transfer resistance is slightly greater for the R22/R114 than it is for the R13B1/R152a mixture. However, the molecular mass of R22 is 86 g/mole, and that of R13B1 is 149 g/mole.

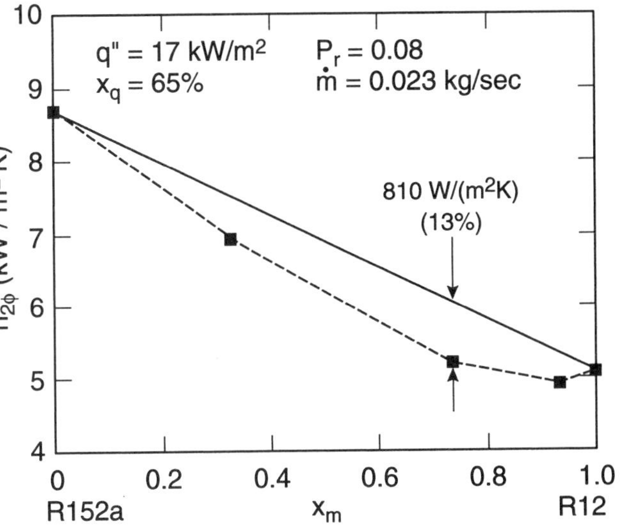

Fig. 8 Measured horizontal flow boiling heat transfer coefficients for the R12/R152a mixture (Jung and Didion, 1989)

In other words, the more volatile component of the R22/R114 mixture is lighter than that of the R13B1/R152a mixture. The speed at which molecules diffuse determines the magnitude of the concentration gradient (McCabe and Smith, 1976). For binary liquids, the rate of diffusion is primarily a function of: (1) the liquid viscosity, (2) the derivative of the log of the activity with respect to the log of the mole fraction of the more volatile component, and (3) the molecular mass of the components (Bird et al., 1960). The viscosity and the activity-composition data for the two mixtures do not differ significantly. If an analogy with vapor diffusion is permitted, then it is reasonable to assume that heavy liquid molecules, like those of R13B1, would diffuse more slowly than the lighter R22 liquid molecules. Consequently, there would be larger concentration gradients present in the R13B1/R152a mixture than in the R22/R114 mixture. Therefore, it is hypothesized that the heat transfer degradation of the R13B1/R152a mixture is larger than that of the R22/R114 mixture because the molecular mass of the more volatile component of the R13B1/152a mixture is greater than that of the R22/R114 mixture.

The phase equilibrium diagram for the R12/R152a mixture, in Fig. 2, shows that the maximum difference between the liquid and vapor compositions for that mixture is approximately 0.1 mole fraction. The fact that the composition difference is small suggests that the potential for heat transfer degradation should be small. The measured two-phase heat transfer coefficient for the R12/R152a mixture, shown in Fig. 8, satisfies the speculation by exhibiting only a 13% degradation in the heat transfer from the linear.

EFFECT OF FLUID PROPERTIES ON $h_{2\phi}$

Mixing Rules

The calculation of the heat transfer coefficient, as given by equation (1), requires relatively few fluid properties. However, its correlation and prediction rely heavily on the estimated or measured fluid properties. For this reason, it is essential that correlations are presented along with the fluid property mixing rules that were used to fit the data. The following analysis demonstrates the effect of the mixing rule on the determination of the two-phase heat transfer coefficient for mixtures in the convection-dominated regime.

The functional form of the Dittus-Boelter (1930) equation ($k_l^{0.6}(c_{pl}/\mu_l)^{0.4})\rho_l^{0.8}$ is frequently used to correlate the convection-dominated region of two-phase flow within a tube. The k_l is the thermal conductivity of the liquid; c_{pl} is the specific heat of the liquid; μ_l is the viscosity of the liquid; and, ρ_l is the density of the liquid. These are the primary fluid properties which are necessary for the correlation and prediction of heat transfer coefficients for two-phase flow boiling.

An estimate of the fluid properties of a mixture can be obtained from the fluid properties of the pure components using mixing rules. Three typical mixing rules are: (1) linear, (2) ideal, and (3) non-ideal. The simplicity of the linear mixing or mass fraction averaging rules is attractive:

$$k_{l_m} = x_{m_1} k_{l_1} + (1 - x_{m_1}) k_{l_2} \tag{2}$$

$$c_{p_{l_m}} = x_{m_1} c_{p_{l_1}} + (1 - x_{m_1}) c_{p_{l_2}} \tag{3}$$

$$\mu_{l_m} = x_{m_1} \mu_{l_1} + (1 - x_{m_1}) \mu_{l_2} \tag{4}$$

Linear mass fraction weighing mixing rules are seldom used to approximate the liquid thermal conductivity and the liquid viscosity. However, the linear mixing rule can be used to closely approximate the specific heat of a mixture.

The ideal mixing rules are slightly more complex than the linear mixing rules, but closely approximate the properties of a mixture where the pure components have similar vapor pressures and come from similar chemical families. The ideal mixing rules assume that there are no mixing effects that enhance or reduce the value of the property due to mixing (Reid et al., 1977).

$$k_{l_m} = \exp[x_{m_1} \ln(k_{l_1}) + (1 - x_{m_1}) \ln(k_{l_2})] \tag{5}$$

$$\mu_{l_m} = \exp[x_{m_1} \ln(\mu_{l_1}) + (1 - x_{m_1}) \ln(\mu_{l_2})] \tag{6}$$

The non-ideal mixing rules chosen for this study have an additional term to account for the effects of mixing:

$$k_{l_m} = x_{m_1} k_{l_1} + (1 - x_{m_1}) k_{l_2} - 0.72 x_{m_1} (1 - x_{m_1}) |k_{l_1} - k_{l_2}| \tag{7}$$

$$\mu_{l_m} = \exp[x \ln(\mu_{l_1}) + (1 - x) \ln(\mu_{l_2})] + 0.85[\rho_{l_{exp}}(\frac{x}{\rho_{l_1}} + \frac{(1-x)}{\rho_{l_2}}) - 1] - 0.085 \tag{8}$$

Equation (7) was obtained from Reid et al. (1977) and equation (8) was obtained from Jung and Didion (1990).

Figure 9 is used to examine the impact of the mixing rule on the prediction of the heat transfer coefficient for mixtures. The figure consists of four graphs of the predicted $h_{2\phi}$ versus the mass fraction (x_m) for the R22/R114 mixture. The uppermost line is the linear interpolation between the heat transfer coefficients of the single components, R22 and R114. The remaining three graphs are of the predicted flow boiling heat transfer coefficient (h_p), using three different mixing rules to estimate the fluid properties used in the correlation. Jung's (1989) flow boiling correlation for single component fluids was used so that only the effect of the mixing rule on the heat transfer coefficient were examined. The predictions can be viewed as a heat transfer coefficient for the mixture if there were no concentration gradients present within the liquid.

Three general characteristics of Fig. 9 are evident. First, note that the predicted heat transfer coefficient is nonlinear with respect to the composition for all of the mixing rules. Second, the apparent heat transfer degradation is the greatest at a mass fraction of 70%, which is different from the mass fraction for the greatest value of $x_{mv} - x_{ml}$. This indicates that the nonlinear property effects are acting to minimize the heat transfer by a mechanism which is different from that of the concentration gradient effects. The consequence of the property effects interacting with the

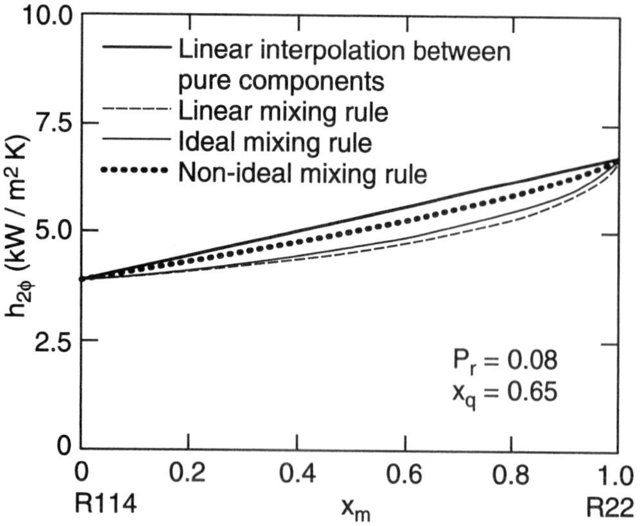

Fig. 9 Impact of the mixing rule on the apparent heat transfer degradation of the R22/R114 mixture

concentration gradient effects is to minimize the heat transfer at a composition which is a compromise between the two effects. Third, the apparent maximum heat transfer degradation becomes less as the mixing rule for the fluid property estimations progress from the linear to the ideal and finally to the non-ideal. In summary, the maximum deviation from the linear interpolation is 18% for the linear mixing rule, 14% for the ideal mixing rule, and 6% for the non-ideal mixing rule. Figure 9 demonstrates the importance of consistency in the use of mixing rules for estimating the fluid properties to be used in the heat transfer coefficient correlations.

The Less Volatile Component

The removal of heat from the wall by convection for two-phase flow within a tube is governed by both the molecular conduction of heat through the liquid film and the transport of the liquid along the tube wall. Hence, the local properties of the liquid film determine the local rate of heat transfer. Figure 3 illustrates that the composition of the liquid increases in the less volatile component as the fluid evaporates. For this reason, care should be taken to evaluate the liquid properties at the liquid composition (x_{ml}) (not the overall composition (x_m)) in both the correlation and prediction of the flow boiling heat transfer coefficient with respect to flow quality (x_q).

Figure 10 is a plot of the predicted heat transfer coefficient, using Jung's (1989) single component model, versus the overall composition. The figure shows that as large as a 15% error in the prediction of the heat transfer coefficient can occur if the fluid properties are evaluated at the overall composition rather than the liquid composition. The heat transfer coefficient for the R12/R152a mixture will be underestimated by using the overall composition to evaluate the liquid properties. Contrary to this, the heat transfer coefficient for the R22/R114 mixture will be overestimated by using the overall composition to evaluate the liquid properties.

In general, the liquid properties of a two-phase mixture more closely resemble those of the less volatile component than that which would be anticipated considering the overall composition. For a R22/R114 mixture, this results in heat transfer coefficients which are always below the linear interpolation of the heat transfer coefficients of the single components. However, the predicted heat transfer coefficients evaluated at x_{ml} for the R12/152a mixture are slightly above the h_i values. The liquid properties of R152a are more beneficial for convection than the liquid properties of R12, resulting in an enhancement with respect to the overall composition. The opposite is true for the R22/R114 mixture where the liquid properties

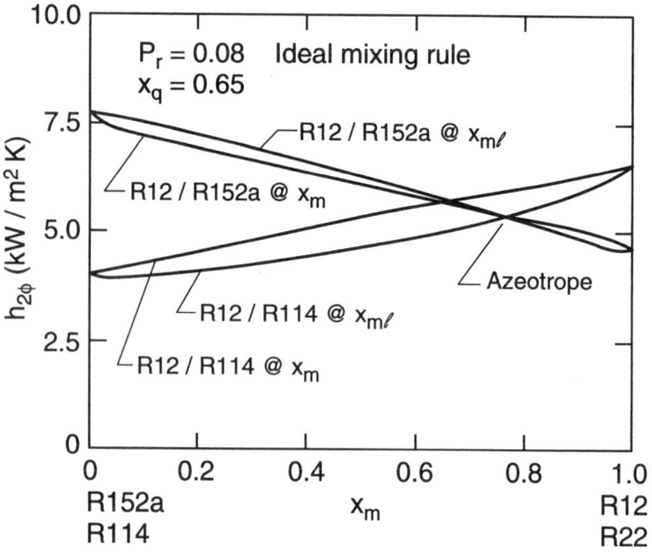

Fig. 10 Effect of evaluating liquid properties at x_m and x_{ml} on $h_{2\phi}$

of the less volatile component do not promote the convection as well as the more volatile component. It may be possible to tailor a mixture which has heat transfer coefficients above the linearly interpolated values by selecting the less volatile component to have the best convective characteristics of all the components.

COMPONENTS OF DEGRADATION

Following is an attempt to isolate and quantify the individual components of the total heat transfer degradation depicted in Fig. 5. The first section concentrates on determining the influences of fluid properties and liquid concentration gradients on the heat transfer coefficient. The last section attempts to isolate the proportions of the $\Delta h_{2\phi}$ that are due to the circumferential and radial concentration gradients.

In order to analyze the influence of fluid properties on the heat transfer coefficient in the absence of concentration gradients, the correlation for two-phase single component horizontal flow boiling by Jung and Didion (1989) was utilized. Figure 11 shows the two-phase heat transfer coefficient for the R12/R152a mixture, as predicted using the Jung and Didion (1989) correlation, versus the mass fraction of the more volatile component. The predicted values were adjusted to facilitate a fair analysis of the nonlinear property effects on the heat transfer. The difference between the prediction and the linear interpolation between the predicted single component heat transfer coefficients was transferred to Fig. 11 as a heat transfer degradation for the measured heat transfer coefficients due to nonlinear property effects. Figure 11 compares the experimental data to the adjusted predicted values for the R12/R152a mixture. The fluid properties degrade the heat transfer coefficient by 6% at $x_m = 0.5$. Concentration gradients are assumed to be responsible for the remaining 7% degradation in the heat transfer coefficient at $x_m = 0.7$.

Figure 12 demonstrates that both fluid properties and concentration gradients reduce the heat transfer coefficient for the R22/R114 mixture. Only a small portion, 14% of the apparent degradation ($\Delta h_{2\phi}$) is due to fluid property effects of mixtures. The majority, 86% of the apparent degradation is due to concentration gradients within the liquid film. Both circumferential and radial concentration gradients contribute to the degradation due to concentration gradients. A closer look into the degradation due to the concentration gradients follows.

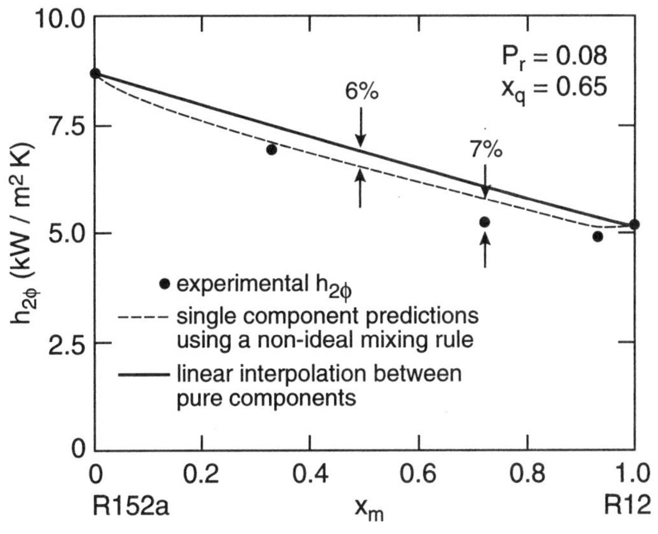

Fig. 11 Comparison of fluid property, and concentration gradient effects on the degradation of R12/R152a heat transfer

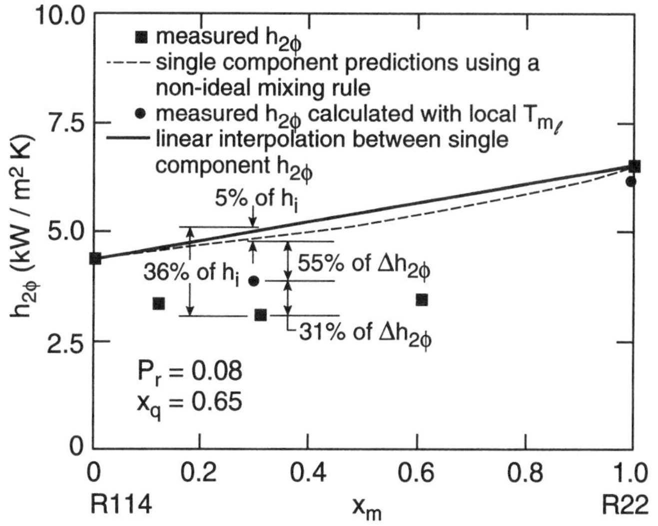

Fig. 12 Comparison of fluid property, radial and circumferential gradient effects on the degradation of R22/R114 heat transfer

Figure 13 is a plot of the heat transfer degradation ($\Delta h_{2\phi}$) versus the difference between the vapor and liquid mass fractions (x_{mv} - x_{ml}). Recall that this difference represents the potential for concentration gradients within the liquid. Figure 13 shows that the heat transfer degradation of the R22/R114 data correlates well with the vapor-liquid composition difference, having a standard deviation of 1.5% from the straight line. Consequently, the heat transfer degradation appears to increase linearly with the difference in the liquid and vapor compositions. The linear rate of increase of the degradation with composition difference, for the R13B1/R152a mixture, is not as well defined by a straight line as it is for the R22/R114 mixture. The evidence for this is that the standard deviation of the $\Delta h_{2\phi}$ from the linear least squares fit for the R13B1/R152a mixture is $\pm$12% which is much greater than that for the R22/R114 mixture. Notice that the heat transfer degradation for the R22/R114 mixture and the R13B1/R152a mixture have approximately the same rate of increase with respect to the increase in the difference in the vapor and liquid compositions (3.4 kW/m^2-K-mass fraction difference). A given increase in the difference between the compositions of the phases results in the same net change in the heat transfer coefficient for both the R22/R114 and the R13B1/R152a mixtures. Also, notice that the intercept of the R22/R114 mixture gives a degradation of 0.3 kW/m^2-K which is the degradation that would be expected in the absence of concentration gradients. Figure 9 confirms this hypothesis since the property effects alone cause nearly the same degradation over x_m = 0.4 to 0.8. This suggests two postulations. First, the heat transfer degradation may be approximated as a sum of the degradation due to approximately constant property effects and that due to concentration gradients which are linearly dependent upon the difference between the vapor and liquid mass fractions. Second, the heat transfer degradation, due to the circumferential and radial concentration gradients, is directly related to the difference between the composition of the liquid and vapor phases.

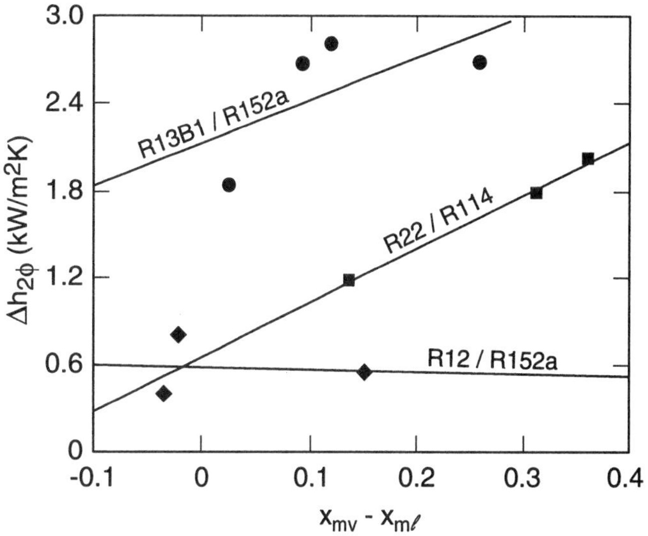

Fig. 13 Effect of vapor-liquid composition difference on the heat transfer degradation

The heat transfer degradation of the R12/R152a mixture appears to have little dependence upon the vapor-liquid mass fraction difference. This suggests that there are no liquid concentration gradients associated with the R12/R152a mixture since the degradation is not a function of the composition difference. Yet, the degradation is still greater than that due to the mixture property effects. The remaining apparent degradation may be a consequence of: (1) the uncertainty in the heat transfer coefficient measurements, and/or (2) the inability of the Jung and Didion (1989) model for the single component heat transfer coefficients to precisely predict the behavior of mixtures without concentration gradients.

Figure 14 is a graph of the difference in vapor and liquid mass fraction versus the liquid mass fraction for the three mixtures. The data symbols correspond to the liquid mass fraction at which the heat transfer coefficients of Fig.s 6 through 8 were measured. The vapor-liquid mass fraction difference for the R22/R114 mixture is the largest of the three mixtures peaking at near 0.4 mass fraction difference. The vapor-liquid mass fraction difference for the R13B1/R152a mixture is nearly the silhouette of the R22/R114 mixture, but is approximately 0.02 lower than that for R22/R114 over most of the liquid mass fraction range.

If it is assumed that the largest x_{mv} - x_{ml} will produce the greatest heat transfer degradation, then Fig. 14 shows that this occurs at a liquid composition of approximately 0.28 for both the R22/R114 and the R13B1/R152a mixture. The vapor-liquid composition difference associated with the maximum degradation lies between the vapor-liquid composition difference for the measured points for the R22/R114 mixture. Therefore, the maximum R22/R114 degradation was probably revealed by the measured heat transfer coefficients, shown in Fig. 7. However, the liquid compositions, for which data were taken for the R13B1/R152a mixture,

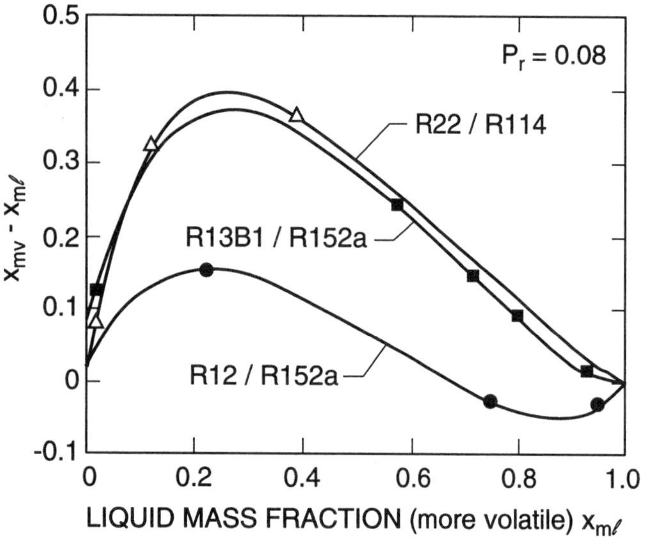

Fig. 14 Concentration difference versus liquid composition

are all above the composition that would demonstrate the largest possible heat transfer degradation. Figure 14 shows that at $x_{ml} = 0.53$, the potential for concentration gradients is 35% below its maximum value. Consequently, if the heat transfer coefficient was measured at a liquid mass fraction of 0.28, the heat transfer degradation may have been substantially greater than that which was measured at a mass fraction of 0.82.

Oddly, the R12/R152a data is inconsistent with the above trends. The maximum, measured heat transfer degradation of the R12/R152a mixture corresponds to the azeotropic composition of the mixture. The degradation associated with the R12/R152a mixture is of the same magnitude as the accuracy of the heat transfer coefficient measurement (± 10%). Consequently, the composition at which the actual maximum heat transfer degradation occurs cannot be known with much confidence.

CIRCUMFERENTIAL AND RADIAL GRADIENTS

Ross et al. (1987) had hypothesized the existence of a circumferential liquid concentration gradient for evaporating mixtures. Jung et al. (1989) validated this speculation by taking concentration samples from the top, the bottom, and the side of the tube. Figure 15 shows that the measured concentration variation for the R22/R114 mixture, at an overall composition of 48% mole R22, is as large as 0.07 mole fraction from the top to the bottom of the tube. Jung's measurement was probably not of the liquid-vapor interface composition. Most likely, the composition that was measured was that of the bulk liquid, as shown in Fig. 4.

When the heat transfer coefficient is calculated, the intention is to define it across the actual temperature difference for which the heat transfer is occurring: the temperature of the heated surface minus the temperature of the liquid-vapor interface ($T_w - T_i$). The temperature of the liquid-vapor interface (i.e, the vapor temperature), in the Ross et al. (1987) and the Jung et al. (1989) studies, was assumed to be circumferentially uniform and equal to the equilibrium temperature. The equilibrium or saturation temperature was determined from the measured vapor pressure, the measured overall composition, the calculated thermodynamic quality and an equation of state (EOS). The saturation temperature will be circumferentially uniform only if the composition of the bulk liquid is uniform around the circumference of the tube. Jung et al. (1989) have shown that a circumferentially uniform liquid composition cannot be guaranteed, which implies that the liquid-vapor interface temperature can also vary. The importance of the circumferential variation is evident in the calculation of the heat transfer coefficient for a particular axial

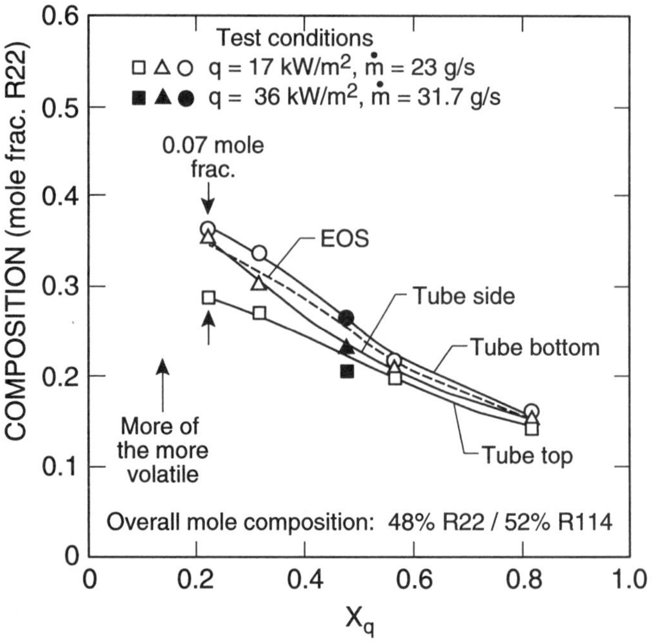

Fig. 15 Measured circumferential liquid film concentration distribution for a 48% mole R22/R114 (Jung et al., 1989)

position, which is obtained by averaging the heat transfer coefficients for the four positions around the circumference of the tube. Thus, the calculation of the heat transfer coefficient using the circumferentially uniform saturation temperature assumption may be inaccurate for some annular flow patterns.

Ideally, both the radial and circumferential concentration gradients should be considered in the evaluation of the liquid-vapor interface temperature for the calculation of the local heat transfer coefficient. Since Jung et al. (1989) circumferentially measured the composition of the bulk liquid, they were able to calculate the circumferential variation of the local saturation temperature for that composition. This temperature still does not represent the temperature of the liquid-vapor interface since: (1) there is a further drop in composition from the bulk of the liquid to the liquid-vapor interface, and (2) the bulk of the liquid will be superheated. The liquid-vapor interface is the only portion of the liquid which is possibly in thermodynamic equilibrium. The liquid-vapor interface temperature is, most likely, equivalent to the equilibrium temperature at the composition of the interface. Therefore, both radial and circumferential concentration gradients influence the temperature distribution of the liquid-vapor interface.

Jung et al. (1989) calculated the local heat transfer coefficient based on the equilibrium temperature obtained from the circumferentially varying composition of the bulk liquid (x_{ml}). The circumferentially averaged local heat transfer coefficient, if calculated using the circumferentially varying saturation temperature, is higher than that calculated using the uniform saturation temperature obtained from the thermodynamic quality and pressure. Figure 12 shows the value of the $h_{2\phi}$ (at $x_m = 0.31$) if it is calculated using the circumferentially varying saturation temperature. The "corrected" heat transfer coefficient is 18% greater than that calculated using the uniform saturation temperature. Consequently, the apparent heat transfer degradation is not as large as the uniform saturation temperature calculation method suggests. In fact, 35% of the apparent degradation due to concentration gradients is due to the use of a circumferentially uniform saturation temperature in the calculation of the heat transfer coefficient. It is speculated that the remaining degradation due to concentration gradients is caused by the radial concentration gradients within the liquid film. In summary, for the heat transfer conditions shown in Fig. 12 at (x_m

= 0.31), 14% of the heat transfer degradation is due to the nonlinear property effects, 31% is due to the circumferential concentration gradients, and 55% is due to the radial concentration gradients. Therefore, it is likely that, if the actual liquid-vapor interface temperature was used to calculate the heat transfer coefficient, the degradation due to nonlinear property effects would be the only degradation present.

Since a limited amount of data are presented, it cannot be assumed that the circumferential averaged heat transfer coefficient calculated using the composition of the bulk liquid will always be higher than that using the uniform equilibrium temperature. However, an argument can be made to establish that, in general, the above trend should be true for most horizontal-annular flow evaporation with a constant heat flux boundary condition. The phenomenon is a result of the nonuniform liquid film thickness around the circumference of the tube, i.e., a thick-film region in the bottom half of the tube and a thin-film region near the top half of the tube. The velocities of the tube-top and tube-bottom liquid films are nearly the same, due to the imposed axial pressure gradient of the vapor phase. Thus, the liquid mass flow rate of the tube-top is less (possibly much less) than that for the bottom of the tube. The liquid stream with the smaller mass flow rate will progress from liquid to vapor sooner than the stream with the larger mass flow rate for the identical heat input (uniform heat flux boundary condition). The temperature and quality of the mixture increases as it evaporates. For this reason, the liquid in the thin-film region will always be at a higher temperature than the liquid in the thick-film region. Also, the liquid near the top of the tube (the thin-film region) will be at a higher temperature than that for a flow with a circumferentially uniform film thickness distribution. For this reason, the heat transfer coefficient at the top of the tube, using the local liquid temperature, will always be higher than that using the uniform equilibrium temperature. Likewise, the heat transfer coefficient at the bottom of the tube calculated from the local liquid temperature will always be marginally lower than that calculated using the uniform equilibrium temperature. But, the top of the tube will have a significantly higher heat transfer coefficient compared to the uniform method and the bottom of the tube will have a marginally lower heat transfer coefficient compared to the same method. Consequently, the average of the top and bottom heat transfer coefficients, using the actual liquid-vapor interface temperature, will tend toward being larger than that calculated from the uniform equilibrium temperature.

CONCLUSIONS

Several precautions must be used when predicting and correlating two-phase heat transfer data for mixtures. First, non-ideal mixing rules which most closely approximate the behavior of the mixtures should be used to correlate the data. The presentation of a new correlation should include the mixing rules that were used to generate it. The liquid composition and not the overall composition of the mixture should be used to evaluate the local fluid properties since it is the liquid film which locally controls the heat transfer.

The two-phase heat transfer coefficient predicted using the linear, the ideal, and the non-ideal mixing rules all resulted in a nonlinear variation of the heat transfer coefficient with the mass fraction. The heat transfer coefficients predicted using the non-ideal mixing rule were closest to the linear interpolation of the heat transfer coefficient between the single component fluids.

The degradation of the heat transfer coefficient by concentration gradient effects appears to be more significant for mixtures of fluids having widely different boiling points. The R22/R114 mixture has a maximum mass concentration difference which is more than double that of the R12/R152a mixture. The heat transfer degradation for the R12/R152a mixture is consequently much smaller than that for the R22/R114 mixture. For equal differences between the vapor and liquid mass fractions, concentration gradient effects appear to be more significant where the more volatile component has a relatively high molecular mass. For example, the R22/R114 and the R13B1/R152a mixtures have approximately the same concentration difference between the liquid and the vapor. But the R13B1 molecule is heavier than the R22 molecule and consequently the R13B1/R152a mixture has a larger heat transfer degradation. In summary, it is speculated that the $x_{mv} - x_{ml}$ is the potential for concentration gradients and the molecular mass of the more volatile component determines the magnitude of the concentration gradient which in turn determines the magnitude of the heat transfer degradation.

The largest measured heat transfer degradation, 2840 W/m^2-K, of all the studied mixtures was for the R13B1/R152a mixture. The difference between the vapor and the liquid mass fraction associated with this data point was 0.13 which was 35% below its maximum value. Consequently, a heat transfer degradation greater than that which was measured may be expected at a mass fraction of 0.53 since this composition corresponds to a vapor-liquid composition difference of 0.28.

An argument can be made to establish that the heat transfer coefficient calculated accounting for the circumferentially varying saturation temperature will be higher than that calculated using a circumferentially uniform saturation temperature for most horizontal-annular flow evaporation. The R22/R114 heat transfer coefficient, for one particular set of conditions, calculated using the varying saturation temperature was 18% greater than that calculated using a uniform saturation temperature.

Typically, the measured flow boiling heat transfer coefficient is calculated using a uniform equilibrium temperature. The apparent heat transfer degradation for this situation can be attributed to three phenomena: (1) fluid property effects, (2) radial liquid concentration gradients, and (3) circumferential concentration gradients. For the R22/R114 mixture, 14% of the heat transfer degradation is due to fluid property effects, 55% is due to radial concentration gradients, and the remaining 31% is attributed to the circumferential concentration gradient. It is postulated that, if the actual liquid-vapor interface temperature were used to calculate the measured heat transfer coefficient it would be greater than that calculated from the equilibrium temperature.

Although concentration gradients within the liquid film tend to degrade the heat transfer, nonlinear fluid property effects may act to enhance the heat transfer. This can occur when the heat transfer characteristics of the less volatile component are more favorable for high heat transfer rates than those of the more volatile component. The heat transfer performance more closely resembles that of the less volatile component. Thus, a mixture could be tailored to minimize the heat transfer degradation by selecting a mixture such that the heat transfer coefficient of the less volatile component is greater than that of the more volatile component.

ACKNOWLEDGEMENTS

This work was funded jointly by NIST and DOE DE-AI01-91CE23808 under project manager Terry G. Statt. Additional funding was provided by EPRI RP 8006-2. The authors would also like to thank the following NIST personnel of the Thermophysics Division for their valuable inputs towards the completion of this work: Dr. M. O. McLinden, Dr. M. R. Moldover, Dr. G. Morrison, and Dr. D. Ripple.

REFERENCES

Bird, R. B., Stewart, W. E., and Lightfoot, E. N., 1960, Transport Phenomena, Wiley, New York.

Chen, J. C., 1966, "Correlation for Boiling Heat Transfer to Saturated Fluids in Convective Flow," I&EC Process Design and Development, Vol. 5, No. 3, July, pp. 322-329.

Dittus, F. W., and Boelter, L. M. K., 1930, Univ. Calif., Publ. Eng., Vol. 2, p. 443.

Jung, D. S., and Didion, D., 1990, "Mixing Rule for Liquid Viscosities of Refrigerant Mixtures," Int. J. Refrig., Vol. 13, July.

Jung, D. S., and Didion, D., 1989, "Horizontal-Flow Boiling Heat Transfer Using Refrigerant Mixtures", EPRI ER-8364, May, EPRI.

Jung, D. S., McLinden, M., and Radermacher R., 1989, "Measurement Techniques for Horizontal Flow Boiling Heat Transfer with Pure and Mixed Refrigerants," Experimental Heat Transfer, Vol. 2, pp. 237-255.

McCabe, W. L., and Smith, J. C., 1976, Unit Operations of Chemical Engineering, 3rd. ed., McGraw-Hill, New York.

Mulroy, W., Kauffeld, M., McLinden, M., and Didion, D., 1988 "Experimental Evaluation of Two Refrigerant Mixtures in a Breadboard Air Conditioner," Proceeding of the IIR Conference, Purdue University, West Lafayette, IN.

Reid, R., C., Prausnitz, J. M., and Sherwood, T. K., 1977, The Properties of Gases and Liquids, 3rd ed., McGraw-Hill, New York.

Ross, H., Radermacher, R., Dimarzo, M. and Didion, D., 1987, "Horizontal Flow Boiling of Pure and Mixed Refrigerants," Int. J. Heat Mass Transfer, Vol. 30, No. 5, pp. 979-992.

Ross, H., 1985, "An Investigation in Horizontal Flow Boiling of Pure and Mixed Refrigerants," Ph.D. Thesis, University of Maryland, College Park, MD.

Shock, R. A. W., 1982, "Boiling in Multicomponent Fluids," Multiphase Science and Technology, Hemisphere Publishing Corp, Vol. 1, pp. 281-386.

Stephan, K., and Korner, 1969, Chem.-Ing.-Tech., Vol. 41, p. 409.

Stephan, K., and Preusser, P., 1979, Ger. Chem. Eng. (Engl. Transl.), Vol. 2, p. 161.

HTD-Vol. 197, Two-Phase Flow and Heat Transfer
ASME 1992

INFLUENCE OF VAPOR BOUNDARY-LAYER SEPARATION ON FORCED CONVECTION FILM CONDENSATION IN DOWNFLOW ON A HORIZONTAL TUBE

S. B. Memory
Department of Mechanical Engineering
Naval Postgraduate School
Monterey, California

J. W. Rose
Department of Mechanical Engineering
Queen Mary and Westfield College
University of London
London, United Kingdom

ABSTRACT

For high velocity forced convection film condensation on a horizontal tube, such as occurs in a low pressure power condenser, vapor boundary-layer separation is thought to play a significant role in the determination of the heat-transfer coefficient. This paper reports tests in which the effects of boundary-layer separation were avoided or suppressed by:

1) making heat flux and wall temperature measurements for the upstream half of the tube only, using a specially manufactured tube, and

2) using a 'Thwaites (1947) flap' to suppress or delay separation.

In addition, dropwise condensation tests were conducted enabling visual determination of the position of separation under various conditions - droplets upstream of separation moved rapidly down the tube surface while downstream drops were virtually stationary.

The experiments were conducted using ethylene glycol at low pressure (1 to 15 kPa), giving vapor velocities up to 140 m/s. The results are compared with laminar condensate flow theory.

NOMENCLATURE

a	constant defined in equation (13)
b	constant defined in equation (14)
d	diameter of condenser tube
F	dimensionless parameter, $(gd\mu h_{fg})/(k\Delta T U_\infty^2)$
g	acceleration due to gravity
G	dimensionless parameter, $(k\Delta T/\mu h_{fg})(\rho\mu/\rho_v\mu_v)^{1/2}$
h_{fg}	latent heat of vaporisation
k	thermal conductivity of condensate
Nu	mean Nusselt number for a whole tube, $(\alpha d/k)$
$Nu_{\pi/2}$	mean Nusselt number for a half-tube, $(\alpha_{\pi/2}d/k)$
P	dimensionless parameter, $(\rho_v h_{fg}\mu/\rho k\Delta T)$
P_i	$(P_\infty - \Delta P_1)$
P_∞	saturation vapor pressure
q	mean heat flux
R	specific ideal gas constant
Re	two-phase Reynolds number, $(\rho U_\infty d/\mu)$
T_i	'corrected' mean condensate surface temperature (eqn. (15))
$T_{sat}(P_\infty)$	saturation temperature at P_∞
U_∞	free-stream vapor velocity
α	mean heat-transfer coefficient for a whole tube
$\alpha_{\pi/2}$	mean heat-transfer coefficient for a half-tube
γ	ratio of principal specific heat capacities
ρ	density of condensate
ρ_v	density of vapor
μ	viscosity of condensate
μ_v	viscosity of vapor
ΔP_1	given by equation (13)
ΔP_2	given by equation (14)
ΔT	mean temperature drop across the condensate film
ξ	suction parameter, $(q/h_{fg})(d/\rho_v\mu_v U_\infty)^{1/2}$
φ_c	angle around tube at which predicted rate of increase of condensate film thickness with angle becomes infinite
φ_s	angle around tube at which vapor boundary-layer separates

INTRODUCTION

For filmwise condensation on a horizontal tube at low vapor velocity, good agreement has been found between heat-transfer measurements and laminar condensate flow theory. Based on the approach of Shekriladze and Gomelauri (1966) and Fujii et al. (1972), Rose (1984) gave the following equations for laminar condensate flow around a horizontal tube:-

$$NuRe^{-1/2} = \frac{0.9 + 0.728F^{1/2}}{\left(1 + 3.44F^{1/2} + F\right)^{1/4}} \tag{1}$$

$$NuRe^{-1/2} = \frac{0.9\left(1+G^{-1}\right)^{1/3} + 0.728F^{1/2}}{\left(1 + 3.44F^{1/2} + F\right)^{1/4}} \tag{2}$$

where F and G are dimensionless parameters given respectively by $(gd\mu h_{fg})/(k\Delta T U_\infty^2)$ and $(k\Delta T/\mu h_{fg})(\rho\mu/\rho_v\mu_v)^{1/2}$. In equations (1) and (2), the effects of vapor boundary-layer separation and pressure variation in the condensate film are neglected. Equation (1) is a very accurate representation of numerical solutions first obtained by Shekriladze and Gomelauri (1966) who used the asymptotic (infinite suction) expression for the shear stress at the vapor/condensate interface. As discussed by Lee and Rose (1982), this underestimated the shear stress and hence the heat transfer. Equation (2) incorporates a correction to account more accurately for the vapor shear stress based on the integral approach of Fujii et al. (1972). It can be seen that equation (2) approaches equation (1) with increasing values of G (increasing condensation rate). Both equations approach the Nusselt (1916) solution at high values of F (low vapor velocity).

Fujii et al. (1979) obtained further solutions (see also Lee and Rose (1982)) which included the effect of vapor boundary-layer separation. The vapor shear stress and vapor boundary-layer separation position were determined by an integral approach due to Truckenbrodt (1956) for uniform suction. Rose (1984) approximated the angle of separation given by this approach by:

$$\varphi_s = 1.76 + 0.164\xi + 0.00869\xi^2 \quad (3)$$

where ξ is the suction parameter $(q/h_{fg})(d/\rho_v\mu_v U_\infty)^{1/2}$. By adopting the method of Truckenbrodt (1956) as modified by Fujii et al (1979), Rose (1984) further obtained:

$$\varphi_s = 2.93 - 1.02\exp(-0.147\xi - 0.0127\xi^2) \quad (4)$$

Mayhew et al (1965), using an approximate analysis due to Prandtl (1935), gave an equation for the angle of vapor boundary-layer separation, again based on uniform suction, as:

$$\varphi_s = \cos^{-1}\left(-(\xi/4.36)^2\right) \quad \text{for } \xi < 4.36 \quad (5a)$$

$$\varphi_s = \pi \quad \text{for } \xi > 4.36 \quad (5b)$$

Equations (3) and (5) predict that the vapor boundary-layer does not separate for $\xi > 6.33$ and $\xi > 4.36$ respectively. Equation (4) predicts that the angle of separation approaches an asymtotic value of 168° with increasing ξ, i.e. that separation is not completely suppressed.

In experiments with condensation of ethylene glycol with vertical vapor downflow over a horizontal tube at high vapor velocities, Memory and Rose (1986) found significant discrepancies between experiment and theory. An abrupt thickening of the condensate film on the lower half of the tube was also noted under certain conditions. This had been reported earlier by Nicol and Wallace (1976) for condensation of steam and attributed to separation of the vapor boundary-layer which gives rise to an abrupt reduction in the tangential shear stress on the condensate film.

Rose (1984) drew attention to a second explanation of the abrupt thickening of the condensate film. He showed that when the pressure term is retained in the momentum equation for the condensate film, under certain conditions theory predicts an infinite rate of increase of condensate film thickness with angle, φ_c, at some location on the lower part of the tube. This occurred at:

$$\varphi_c = \cos^{-1}\left[\frac{-(1 - 21.5\,(F/8P)P^{1.05})}{(1 + 21.5P^{1.05})}\right] \quad (6)$$

for

$$(F/8P) < 1 \quad (7)$$

where P is a dimensionless parameter given by $(\rho_v h_{fg}\mu/\rho k\Delta T)$. This is due to the fact that on the lower half of the tube and in the absence of vapor boundary-layer separation, the pressure gradient in the condensate film opposes the vapor shear stress. Rose obtained the following conservative equation for the mean heat-transfer coefficient:

$$\mathrm{Nu}\mathrm{Re}^{-1/2} = \frac{0.64(1 + 1.81P)^{0.209}(1 + G^{-1})^{1/3} + 0.728F^{1/2}}{(1 + 3.51F^{0.53} + F)^{1/4}} \quad (8)$$

Equation (8) approaches the Nusselt solution at low vapor velocity. At high vapor velocity, the pressure gradient term is included for the upper half of the tube but the contribution to heat transfer from the lower half of the tube is neglected.

In order to investigate uncertainties for the lower half of the tube, two separate experiments have been designed by the present authors to try and throw further light on the effects of separation; both were performed with vertical vapor downflow over a single horizontal tube. Measurements were first made for the upper half of the tube only where separation does not occur. The lower half of the tube was internally well insulated. In the second experiment, a vertical 'Thwaites (1947) flap' was attached to the bottom of the tube to delay the point of separation.

MODIFICATIONS OF THEORY TO PROVIDE RESULTS FOR UPPER HALF OF TUBE ONLY

The integrals which give the mean heat-transfer coefficient leading to equations (1), (2) and (8) have been re-evaluated by Memory (1989). The analysis of Nusselt (1916) leads, for the upper half of a tube only, to:

$$\mathrm{Nu}_{\pi/2}\mathrm{Re}^{-1/2} = 0.866F^{1/4} \quad (9)$$

When using the asymptotic approximation for the interfacial shear stress, the following expression for the mean Nusselt number for the upper half of the tube is obtained:

$$\mathrm{Nu}_{\pi/2}\mathrm{Re}^{-1/2} = \frac{1.273 + 0.866F^{1/2}}{(1 + 3.51F^{0.53} + F)^{1/4}} \quad (10)$$

A correction to equation (10) to account more accurately for the interfacial shear stress (after Fujii et al. (1979)) may be incorporated to give:

$$\mathrm{Nu}_{\pi/2}\mathrm{Re}^{-1/2} = \frac{1.273(1 + G^{-1})^{1/3} + 0.866F^{1/2}}{(1 + 3.51F^{0.53} + F)^{1/4}} \quad (11)$$

When the effects of pressure variation in the condensate film around the tube are included (see Rose (1984)), the following result is obtained:

$$\mathrm{Nu}_{\pi/2}\mathrm{Re}^{-1/2} = \frac{1.273(1 + 1.81P)^{0.209}(1 + G^{-1})^{1/3} + 0.866F^{1/2}}{(1 + 3.51F^{0.53} + F)^{1/4}} \quad (12)$$

Equation (12) is close to equation (11), indicating that inclusion of the pressure gradient term in the momentum equation for the condensate film has little effect when considering a half-tube.

APPARATUS AND PROCEDURE

The apparatus is shown in Figure 1 and is essentially the same as that used by Memory and Rose (1986). Ethylene glycol vapor flowed vertically downwards over a horizontal copper test condenser tube (length = 76.2 mm, OD = 12.7 mm, ID = 8.15 mm) through which cooling water passed. The vapor velocity was calculated from the precisely-measured power input to the boilers with a small correction to account for the thermal loss from the apparatus. The coolant temperature rise was measured using a 10-junction thermopile, the measuring junctions being located after well-insulated mixing boxes. A second measurement of coolant temperature rise was taken from single thermocouples at inlet and exit from the condenser tube. The two values generally agreed to within 0.05 K. Great care was taken to ensure adequate isothermal immersion of the thermocouple leads and to prevent significant thermal conduction between the condenser tube and the body of the test section and the environment.

The vapor temperature was obtained from two thermocouples located in the test section 65 mm upstream of the condenser tube. The vapor pressure could be measured to within ± 50 Pa using a mercury and ethylene glycol manometer fitted with a vernier scale. The highest estimated uncertainty in the experimental value of the heat-transfer coefficient was approximately 10% and occurred at thr lowest vapor pressure of 1 kPa and the lowest coolant flow rate of 0.4 l/min. At the highest vapor pressure of 15 kPa (with a similar flow rate), the estimated uncertainty was always less than 2%.

Strenuous efforts were made to ensure that the apparatus was leak-tight. With the boilers off and the apparatus at ambient temperature under a vacuum of 500 Pa, the pressure rise in a 12 hour period never

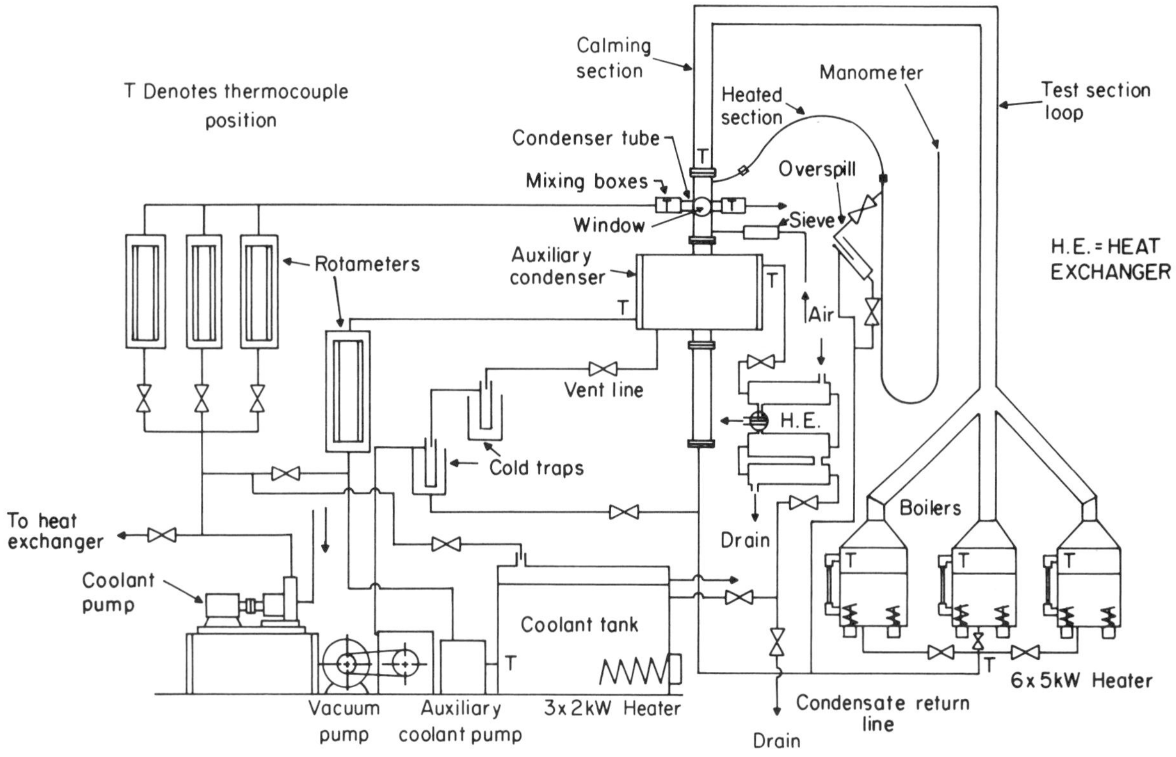

Figure 1: Apparatus

exceeded 400 Pa (3 mmHg). This 'leak-rate' was checked frequently between test runs throughout the course of the investigation. The tube surfaces were also checked frequently through the viewing window during the tests to ensure complete wetting and film condensation.

Two methods were used to control the auxiliary condenser condensation rate so as to maintain steady pressures when operating with different vapor and coolant flow rates. One was to adjust the coolant temperature at inlet to the auxiliary condenser and the other was to admit air into the auxilary condenser. Tests were carried out at pressures of 1, 1.5, 2, 3, 4, 5, 7, 10 and 15 kPa for the half-tube and 1, 1.5, 2, 3, 5 and 8 kPa for the whole tube. At each pressure, different ranges of vapor velocity could be obtained by increasing or decreasing the power input to the boilers. At low pressures (< 2 kPa), the maximum vapor velocity was limited by the condensing capacity of the auxiliary condenser, i.e. further increase in boiler power led to an increase in pressure. At higher pressures, the vapor velocity was limited only by the maximum power input to the boilers.

The tubes tested are shown in Figure 2. For the half-tube (Figure 2a), copper was machined from the inside surface of the lower half of the tube to leave a wall thickness of only 0.5 mm. A half cylinder of PTFE ('teflon') was then inserted as shown. When calculating the heat-transfer rate to the upper half of the tube, a small 'fin-type' correction was made for heat transfer to and along the thin lower-side copper layer.

The second tube (Figure 2b) had a vertical strip of metal (Thwaites (1947) flap) attached to the bottom of the tube parallel to the axis of the tube. The top of the strip fitted tightly into a slot 1.5 mm deep x 1.5 mm wide cut into the underside of the tube. Tests were conducted using both a copper and a stainless-steel strip to assess the fin contribution to the heat transfer due to the strip.

The wall temperature in each case was measured by thermocouples embedded in the tube wall at a diameter of 9.5 mm (3 for the half-tube and 6 for the tube with flap). These thermocouples were spaced

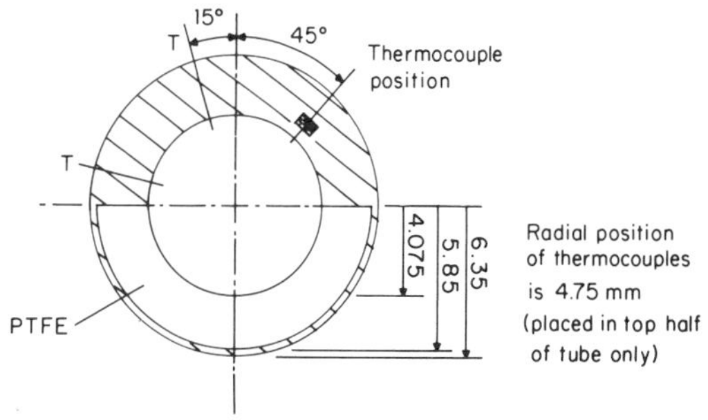

Figure 2a: Half-Tube

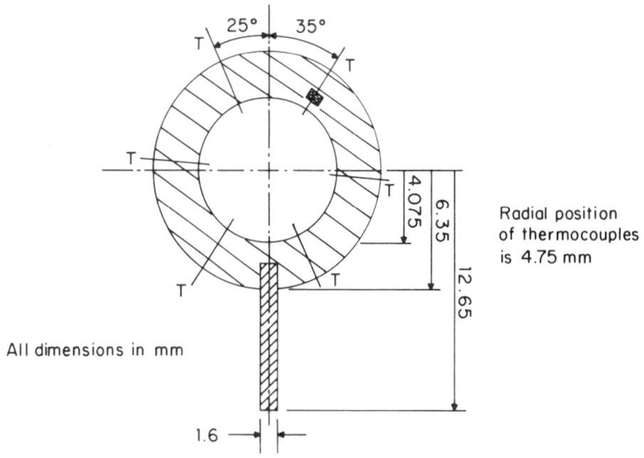

Figure 2b: Tube Fitted With Thwaites Flap

evenly around the circumference of the tube, but offset from the top of the tube by 15° and 25° respectively for the half-tube and tube with flap as shown. The thermocouple junctions were located midway along the length of the tubes in longitudinally machined grooves (1.6 mm x 1.6 mm) enclosed by tightly-fitting soldered copper strips. The tubes were finished by copper plating and machining down to the required outside diameter. The wall outer surface temperature was taken as an arithmetic mean of the wall thermocouples, with a small correction to account for their depth below the surface.

DROPWISE CONDENSATION VISUAL INVESTIGATION

During the course of the investigation, dropwise condensation occurred 'fortuitously' due to contamination in the apparatus. Figures 3a-e show photographs of dropwise condensation during vapor downflow. During dropwise condensation, the position of vapor boundary-layer separation is indicated by the fact that below the separation point, the drops are almost stationary while higher on the tube, the droplets move rapidly down the surface under the influence of vapor shear stress. Four coolant flow rates (condensation rates) ranging from 1.2 to 10 l/min (shown in the Figures as (i)-(iv)) were used at each of five vapor velocities between 25 to 93 m/s.

The photographs in Figure 3 exhibit an abrupt increase in droplet size at some angle around the tube at low condensation rate indicating the position of vapor boundary-layer separation. The angle of vapor boundary-layer separation from the top of the tube can be measured from the photographs. Table 1 gives the measured angles for each vapor velocity and condensation rate. It was not possible to determine the angle precisely and for each case a range is given. It can be seen that the angle of separation increases with increase in condensation rate or 'suction' as expected. A small increase in the angle of separation with increase in vapor velocity for a constant condensation rate was also noted.

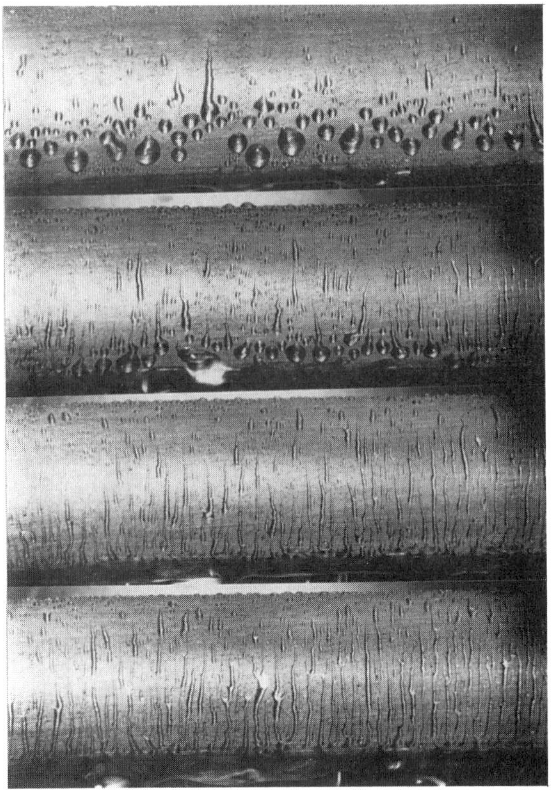

2(i)

2(ii)

2(iii)

2(iv)

Figure 3b: Dropwise Condensation Photographs (U_∞ = 45 m/s)

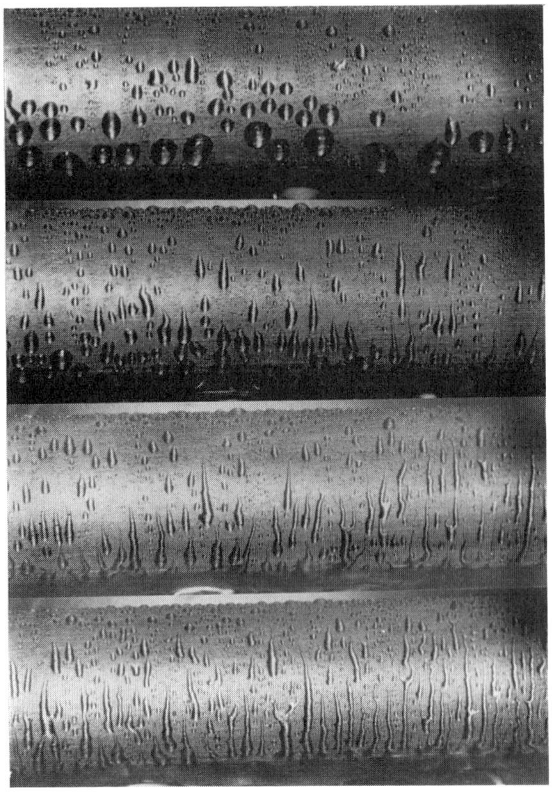

1(i)

1(ii)

1(iii)

1(iv)

Figure 3a: Dropwise Condensation Photographs (U_∞ = 25 m/s)

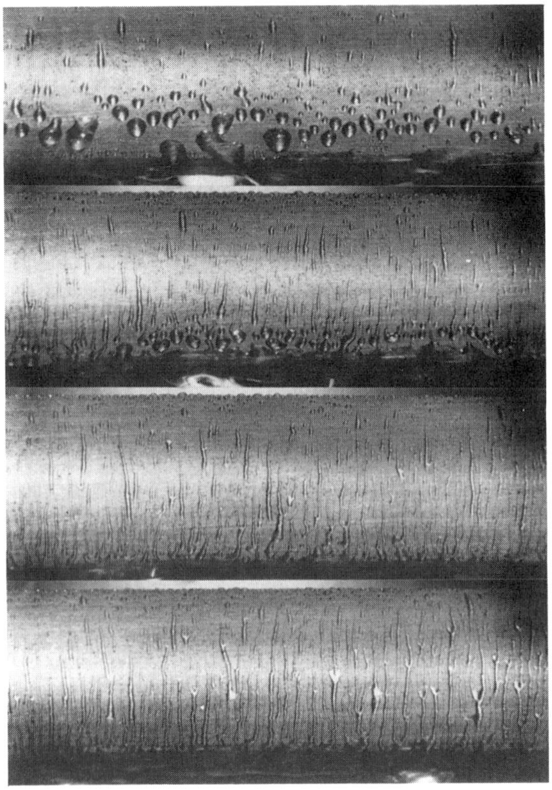

3(i)

3(ii)

3(iii)

3(iv)

Figure 3c: Dropwise Condensation Photographs (U_∞ = 52 m/s)

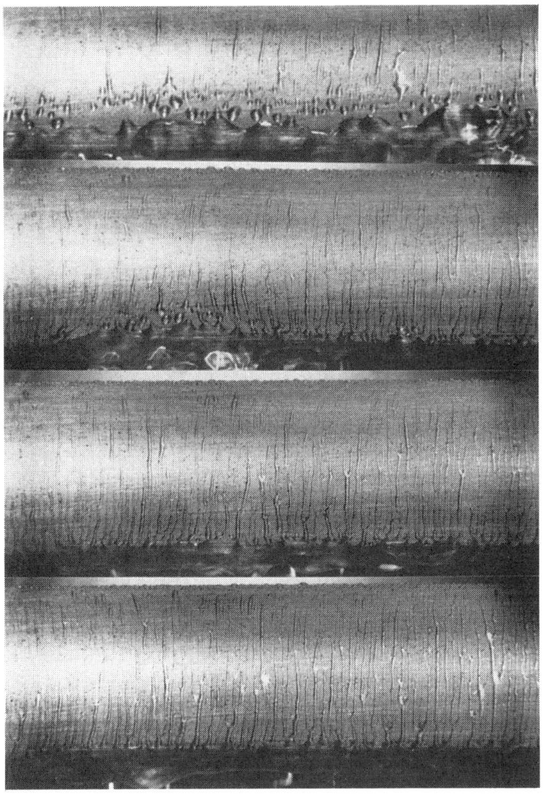

4(i)

4(ii)

4(iii)

4(iv)

Figure 3d: Dropwise Condensation Photographs (U_∞ = 69 m/s)

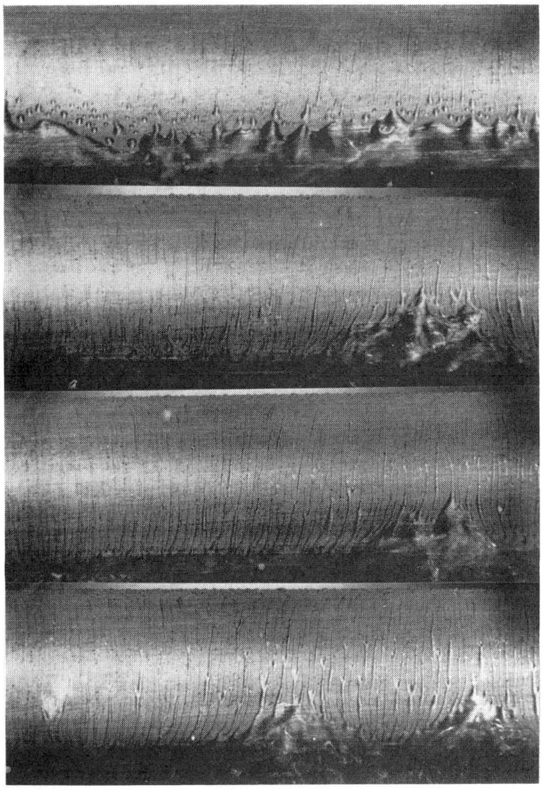

5(i)

5(ii)

5(iii)

5(iv)

Figure 3e: Dropwise Condensation Photographs (U_∞ = 93 m/s)

For those cases where, in addition to the photographs, complete heat-transfer data were taken, the suction parameter, ξ, has been determined and included in Table 1. The calculated values of ξ from equations (3)-(5) have also been included in the Table. Figure 4 gives a comparison between the predicted and experimental values of ξ given in Table 1. It can be seen that the limited number of experimental points are generally in fair agreement with theory.

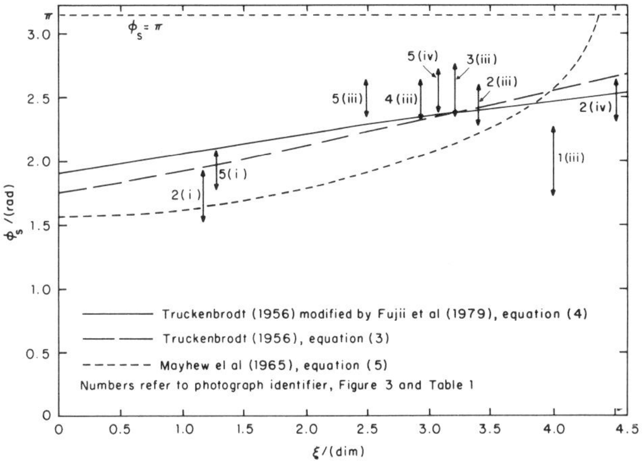

Figure 4: Comparison of Experimentally Determined Angle of Vapor Boundary-Layer Separation With Theory

RESULTS AND DISCUSSION

Whole Tube

It was suggested by Memory and Rose (1986) that corrections to the condensate surface temperature to account for the effects of pressure variation around the tube and interphase mass-transfer resistance should be made under conditions of high vapor velocity and low pressure respectively.

Separation of the vapor boundary-layer leads to a mean pressure over the tube surface which is lower than the free-stream pressure. Following Butterworth (1977), this pressure difference can be estimated by:

$$\Delta P_1 = a\,\rho_v U_\infty^2 \qquad (13)$$

where a is a constant of order unity. This leads to a condensate surface temperature lower than that of the vapor. Suppression of separation by use of the "Thwaites flap", i.e. $\Delta P_1 = 0$, might therefore be expected to increase the mean vapor-side heat-transfer coefficient.

In addition, Memory and Rose (1986) indicated that interphase mass-transfer resistance can be significant for vapors with low values of γ under conditions of low pressure. Following Rahbar and Rose (1984), the pressure drop at the vapor/liquid interface due to interphase mass-transfer can be estimated by :

$$\Delta P_2 = \frac{b\,q(\gamma+1)\left(R\,T_{sat}\,(P_i)\right)^{1/2}}{4\,h_{fg}(\gamma-1)} \qquad (14)$$

where b is a constant of order unity and $P_i = (P_\infty - \Delta P_1)$. Equations (13) and (14) were used to evaluate a 'corrected' mean condensate surface temperature:

$$T_i = T_{sat}\,(P_\infty - \Delta P_1 - \Delta P_2) \qquad (15)$$

Using equation (15), a corrected value of the temperature drop across the condensate film could be evaluated. It was found by Memory and Rose (1986) that values of a = 0.7 and b = 0.8 gave good agreement with theory (equations (1) and (2)).

Table 1: Comparison of Experimental and Theoretical
Values of Vapor Boundary-Layer Separation Angle

Photo No. (Fig.3)	Coolant Flowrate (l/m)	Vapor Velocity (m/s)	Experimental Range of φ_s (deg)	Calculated ξ (dim)	Calculated φ_s (deg)		
					Eqn. (3)	Eqn. (4)	Eqn. (5)
1(i)	1.2	25	80 - 98	-	-	-	-
1(ii)	4.0	25	96 - 106	-	-	-	-
1(iii)	6.0	25	100 - 130	4.00	146	142	147
1(iv)	10.0	25	119 - 149	-	-	-	-
2(i)	1.2	45	87 - 110	1.16	112	119	94
2(ii)	4.0	45	113 - 133	-	-	-	-
2(iii)	6.0	45	131 - 149	3.40	139	138	127
2(iv)	10.0	45	133 - 151	4.50	153	144	180
3(i)	1.2	52	96- 112	-	-	-	-
3(ii)	4.0	52	122 - 137	-	-	-	-
3(iii)	6.0	52	135 - 158	3.20	136	136	123
3(iv)	10.0	52	137 - 158	-	-	-	-
4(i)	1.2	69	104 - 119	-	-	-	-
4(ii)	4.0	69	124 - 149	-	-	-	-
4(iii)	6.0	69	133 - 151	2.93	133	134	117
4(iv)	10.0	69	135 - 155	-	-	-	-
5(i)	1.2	93	103- 120	1.26	113	120	95
5(ii)	4.0	93	131 - 151	-	-	-	-
5(iii)	6.0	93	135 - 151	2.50	127	131	109
5(iv)	10.0	93	137 - 155	3.07	135	135	120

For the tube fitted with the Thwaites flap, it was expected that vapor boundary-layer separation would be delayed or suppressed and that maintenance of downward shear stress over a larger proportion of the surface would lead to higher heat-transfer coefficients. Both corrections (as discussed above) with a = 0.7 and b = 0.8 were applied to the data. Figure 5 shows the dependence of heat flux on mean vapor-to-wall temperature difference for the tube fitted with and without the copper flap under similar experimental conditions of vapor pressure and vapor velocity. It is clear that the Thwaites flap has little discernible effect on the mean measured heat transfer rate, suggesting that vapor boundary-layer separation, rather than condensate film instability, was the cause. However, below a vapor velocity of around 100 m/s, visual observation did indicate that the abrupt thickening of the condensate film (similar to that mentioned by Nicol and Wallace (1976) and Fujii et al. (1979)) moved back around the tube for increases in condensation rate. The fact that suppression of boundary-layer separation apparently has little effect on the heat transfer may in part be due to the smaller contribution from the lower (and thicker) part of the condensate film and also to the fact that vapor boundary-layer separation, with abrupt thickening of the condensate film, may lead to turbulence and enhanced heat transfer for the lower part of the tube.

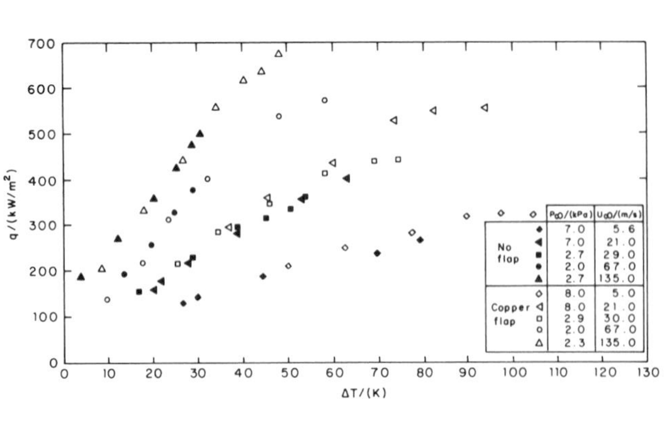

Figure 5: q vs ΔT For Tube Fitted With and Withour Copper Flap

Above 100 m/s, it was noted that the position of the abrupt thickening of the condensate film was insensitive to condensation rate and further increases in vapor velocity. This is consistent with shock-induced boundary-layer separation (see Shapiro (1953)). For the present tube and duct dimensions, and for the vapor pressures used, it was estimated (see Memory (1989)) that local sonic conditions would occur at about the position of the abrupt thickening of the condensate film for vapor velocities in the range 100 to 110 m/s.

Half-Tube

For the half-tube, the calculation for determining the mean heat-transfer coefficient is essentially the same as for the tube with the flap except that the correction for the overall pressure drop (equation (13)) is not applicable since separation never occurs on the top half of the tube. Consequently, only the interphase mass-transfer correction was applied to the half-tube data with $P_i = P_\infty$ and b = 0.8 in equation (14).

The dependence of heat flux on mean vapor-to-wall temperature difference for the half-tube is shown in Figure 6 for a relatively high vapor pressure (7 kPa). At high vapor velocity, better agreement with half-tube theory is obtained with the corrected experimental data, which was also the case for the whole tube (Memory and Rose (1986)).

The experimental data in Figure 6 suggest points of inflexion at the higher values of ΔT, after which the heat flux increases more rapidly than indicated by laminar condensate flow theory. It is also seen that the inflexion occurs at lower values of ΔT for higher vapor velocities. These facts are consistent with onset of turbulence in the condensate film at high vapor shear stress.

Figure 7 compares the half-tube data (for three vapor pressures) with theory on a dimensionless basis. Equation (12), based on the analysis of Rose (1984), is given for the extreme combinations of G and P obtained from the data. The agreement between experiment and theory is good and can be attributed to there being no uncertainties associated with the lower half of the tube. It may be noted that, for low pressures, agreement with theory was significantly worse when the interphase mass transfer correction was omitted.

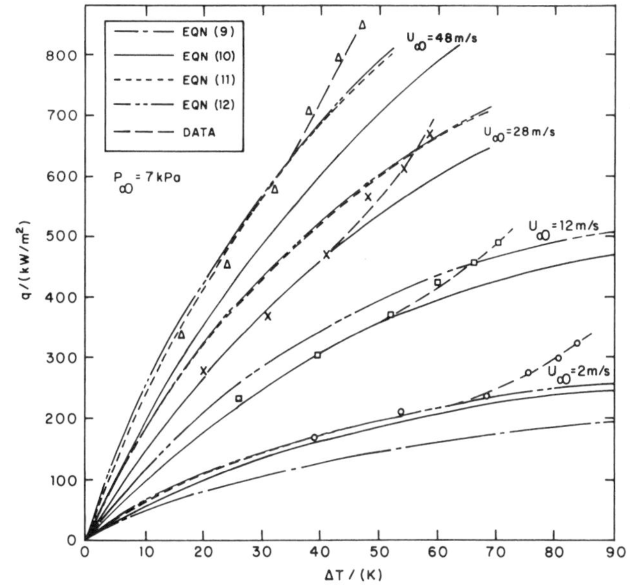

Figure 6: q vs ΔT For Half-Tube at High Vapor Pressure

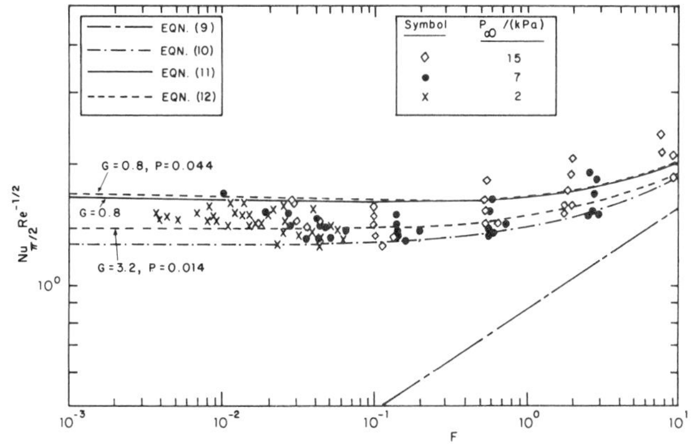

Figure 7: Comparison of Corrected Half-Tube Data With Theory

The data in Figure 7 indicate a slight upturn in the value of $NuRe^{-1/2}$ for decreasing F (increasing vapor velocity), similar to that reported by Memory and Rose (1986) for a whole tube. The minimum value of $NuRe^{-1/2}$ for a whole tube was found to occur at a value of F around 0.1 and was attributed by Memory and Rose to onset of turbulence in the condensate film. Figure 7 indicates that the minimum occurs at a similar value of F for the half-tube. The fact that the upturn is seen for the half-tube is further evidence of onset of turbulence in the condensate film.

CONCLUSIONS

Evidence of vapor boundary-layer separation has been seen in photographs of dropwise condensation. These indicate that vapor boundary-layer separation occurs at locations broadly in agreement with theory.

For the tube fitted with the Thwaites flap, visual observation suggests that vapor boundary-layer separation can be suppressed under conditions of strong suction for vapor velocities below about 100 m/s. However, no significant increase in the measured heat-transfer was noted. This may be due to the presence of turbulence in the condensate film beyond the point of separation.

The half-tube data are in good agreement with laminar condensate film theory when evaluated for the upper half of a tube only, with significantly less scatter than the data for a whole tube. Discrepancies between experiment and theory for a whole tube are therefore attributable to uncertainties (vapor boundary-layer separation, turbulence, film instability) relating to the lower half of the tube.

For vapor velocities above 100 m/s, both visual observation and calculations, based on vapor flow over a cylinder within a confined duct, indicate that the separation of the vapor boundary-layer may be shock-induced.

REFERENCES

Butterworth, D. (1977), "Developments in the Design of Shell and Tube Condensers", ASME Paper No. 77-WA/HT-24.

Fujii, T., Uehara, H. and Kurata, C. (1972), "Laminar Filmwise Condensation of Flowing Vapour on a Horizontal Cylinder", Int. J. Heat Mass Transfer, Vol. 5, pp. 235-246.

Fujii, T., Honda, H. and Oda, K. (1979), "Condensation of Steam on a Horizontal Tube - The Influence of Oncoming Vapor Velocity and Thermal Condition at the Tube Wall", Condensation Heat Transfer, 18th Nat. Heat Transfer Conf., San Diego, California, pp. 35-43.

Lee, W. C. (1982), "Filmwise Condensation on a Horizontal Tube in the Presence of Forced Convection and Non-Condensing Gas", PhD Thesis, University of London, UK.

Lee, W. C. and Rose, J. W. (1982), "Film Condensation on a Horizontal Tube - Effect of Vapour Velocity", Proc. 7th Int. Heat Transfer Conf, Munich, Vol. 5, pp. 101-106.

Mayhew, Y. R., Griffiths, D. J. and Phillips, J. W. (1965), "Effect of Vapor Drag on Laminar Film Condensation on a Vertical Surface", Proc. Inst. Mech. Eng., Vol. 180, pp. 280-289.

Memory, S. B. (1989), "Forced Convection Film Condensation on a Horizontal Tube at High Vapour Velocity", PhD Thesis, University of London, UK.

Memory, S. B. and Rose, J. W. (1986), "Filmwise Condensation of Ethylene Glycol on a Horizontal Tube at High Vapour Velocity", Proc. 8th Int. Heat Transfer Conf., San Francisco, Vol. 4, pp. 1607-1614.

Nicol, A. A. and Wallace, D. J. (1976), "Condensation With Appreciable Vapor Velocity and Variable Wall Temperature", Symp. on Steam Turbine Condensers, NEL Report No. 619, pp. 27-38.

Nusselt, W. (1916), "Die Oberflachenkondensation des Wasserdampfes", Z. des Vereines Deutscher-Ing., Vol. 60, pp. 569-575.

Prandtl, L. (1935), "The Mechanics of Viscous Fluids", Aerodynamic Theory (W. F. Durand, ed.), Vol. III, pp. 34-208.

Rahbar, S. and Rose, J. W. (1984), "New Measurements for Forced Convection Film Condensation", 1st UK Nat. Heat Transfer Conf., Leeds, Vol. 1, pp. 619-632.

Rose, J. W. (1984), "Effect of Pressure Gradient on Forced Convection Film Condensation on a Horizontal Tube", Int. J. Heat Mass Transfer, Vol. 27, pp. 39-47.

Shapiro, A. H. (1953), "The Dynamics and Thermodynamics of Compressible Fluid Flow", Ronald Press Co., New York.

Shekriladze I. G. and Gomelauri, V. I. (1966), "Theoretical Study of Laminar Film Condensation of Flowing Vapour", Int. J. Heat Mass Transfer, Vol. 9, pp. 581-591.

Thwaites, B. (1947), "The Production of Lift Independently of Incidence - The Thwaites Flap", R. & M. No. 2611, pp. 625-644.

Truckenbrodt, E. (1956), "Ein einfaches Naherungsverfahren zum Berechnen der Laminaren Riebungsicht mit Absangung", Forschung Ing. Wes., Vol. 32, pp. 147-157.

LOCAL HEAT TRANSFER COEFFICIENTS FOR FORCED CONVECTION CONDENSATION OF STEAM IN A VERTICAL TUBE IN THE PRESENCE OF AIR

M. Siddique, M. W. Golay, and M. S. Kazimi
Department of Nuclear Engineering
Massachusetts Institute of Technology
Cambridge, Massachusetts

ABSTRACT

An experimental investigation has been conducted to determine the local condensation heat transfer coefficient of steam in the presence of air flowing downward inside a 46 mm internal diameter vertical tube. The air/steam mixture flow rate was measured with a calibrated vortex flow meter before it entered the 2.54 m long test condenser. Cooling water flow rate in an annulus around the tube was measured with a calibrated rotameter. Temperatures of the cooling water, the air/steam mixture, and the tube inside and outside surfaces, were measured at 0.3 m intervals in the test condenser. Inlet and exit pressures and temperatures of the air/steam mixture and of the cooling water were also measured. The local heat flux was obtained from the slope of the coolant axial temperature profile and the coolant mass flow rate. The ranges of the test variables were: mixture inlet temperatures 100, 120, and 140°C, inlet air mass fraction 10 to 35%, mixture inlet Reynolds number of about 5000 to 22,700. The local heat transfer coefficient varied from 100 W/m^{2}K to about 15,000 W/m^{2}K. The local Nusselt number increased with the mixture Reynolds number and decreased with the Jakob number and the air mass fraction.

NOMENCLATURE

A	= Tube Inside Surface Area
A_f	= Cross-sectional Area of Tube
C	= Constant
C_p	= Heat Capacity
D	= Diffusion Coefficient
d	= Tube Inner Diameter
h	= Condensation Heat Transfer Coefficient
H_{fg}	= Latent Heat
h_m	= Mass Transfer Coefficient
H	= Enthalpy
Ja	= Jakob Number
k	= Thermal Conductivity
L	= Length
M	= Molecular Weight
$\dot{m}$	= Mass Flowrate
$\dot{m}''$	= Interfacial Mass Flux
MIT	= Massachusetts Institute of Technology
Nu	= Nusselt Number
P	= Pressure
Pr	= Prandtl Number
Q	= Heat Transferred
q''	= Heat Flux
Re	= Reynolds Number
R^2	= Coefficient of Determination
Sc	= Schmidt Number
T	=Temperature
u, v	= Velocity Components
UCB	= University of California at Berkeley
V	= Average Mixture Velocity
W	= Air Mass Fraction
x	= Axial Position
x, y	= Axial and Lateral Coordinates
D	= Change
r	= Density

Subscripts

a	= Air
b	= Bulk Air/Steam mixture
c	= Cooling Water
cond	= Condensate
in	= Inlet to Tube
mix	= Mixture
s	= Steam
tot	= Total
w	= Tube Inside Wall

NOTE: Mixture Properties are Unsubscripted

INTRODUCTION AND EXAMPLE APPLICATION

It has been well established that the presence of noncondensable gases in vapors can greatly inhibit the condensation process. The mass transfer resistance to condensation results from a buildup of noncondensable gas concentration at the liquid/gas interface leading to a decrease in the corresponding vapor partial pressure and thus the interface temperature at which condensation occurs. In an application where this is important, the proposed advanced passive boiling water reactor design (Simplified Boiling Water Reactor, SBWR), utilizes as a main component of the Passive Containment Cooling System (PCCS) the Isolation Condenser (IC). The function of the IC is to provide the ultimate heat sink for the removal of the Reactor Coolant System sensible heat and core decay heat. In performing this function, the IC must have the capability to remove sufficient energy from the reactor containment in order to prevent the containment from exceeding its design pressure shortly following design basis events and to significantly reduce containment pressure in the longer run. After a typical postulated loss of coolant accident, the steam/air mixture from the reactor containment/pressure vessel may flow to the IC which will then reject decay heat to a pool of water. The steam condenses in downward flow through a bundle of tubes and the condensate drains back to the reactor pressure vessel.

During forced in-tube condensation of a vapor in the presence of a noncondensable gas, the condensed liquid flows as an annular film adjacent to the cooled tube wall and the vapor/gas mixture flows through the tube core. The bulk motion of the core sweeps away the noncondensable gas leading to a lower buildup of the gas concentration at the interface and at the same time the shear stress exerted by the gas phase leads to a thinning of the condensate film. Both of these factors lead to an increase in the heat transfer coefficient compared to the free convection case. In the IC the heat transfer coefficients vary greatly along the length of the tube. The decrease in the rates of heat and mass transfer with distance down the tube is mainly caused by the progressively increasing air mass fraction and the decreasing flow velocity, resulting from the progressive dehumidification of the mixture. Since the rate of heat transfer is strongly coupled to the hydrodynamic characteristics of the PCCS, a detailed knowledge of the variation of local heat transfer coefficients is necessary in order to predict the overall performance of the PCCS and to optimize the design of the IC.

Many analytical studies have been conducted to investigate the effect of noncondensable gases on steam condensation for both stagnant and forced convective situations (e.g., Refs. 1 - 7). However, most of the theoretical analyses are confined to laminar flow situations for either flow over a plate or condensation outside a horizontal tube. Thus, they are not directly applicable to the non-similar in-tube condensation case where both the flow velocity and air concentration change along the tube length. Only recently Wang and Tu (1988) presented a numerical solution using the heat and mass transfer analogy together with correction factors for the suction effect to show the detrimental effect of noncondensable gases on vapor condensation. On the experimental side much of the data reported pertains to stagnant atmospheric conditions (e.g., Refs. 9 - 12). Here, also, most of the studies are also confined to either an external plate flow geometry or condensation outside a horizontal tube. For the few rare forced convection in-tube condensation studies (e.g., Refs. 13 - 16), the reported results are only for overall or mean values of the heat transfer coefficient. Furthermore, the test conditions are usually restrictive as they are not representative of the operating conditions of the IC. Because of the strong dependance of the local gas concentration and flow conditions upon the local heat transfer coefficient, the need exists for localized.

The main objective of the investigation described in this paper was to measure local heat transfer coefficients for steam condensing in the presence of air inside a tube such as could be typical for an IC. Emphasis was placed upon obtaining data spanning the range of inlet air concentrations, operating temperature, and steam flow rate, which would simulate the operating conditions of the IC in a typical Loss of Coolant Accident (LOCA). A heat transfer correlation was formulated from the collected data, which permits a localized description of the system in terms of the effects of air concentration, the dynamic gas motion, and the thermal driving force of the wall subcooling.

DESCRIPTION OF THE TEST FACILITY

The experimental apparatus consisted of an open cooling water circuit and an open air/steam loop, as shown in the flow diagram of Figure 1. Steam was generated in a 5.0 m high by 0.45 m inside diameter, cylindrical stainless steel vessel, by boiling water using four immersion type sheathed electrical heaters. The heaters could be individually controlled (on or off) and were nominally rated at 7 kW each. A Variac was wired to one of the heaters for finer control of the power level. A high precision wattmeter indicated the power input to the boiler. Compressed air was supplied to the base of the steam generating vessel via a pressure regulating valve, a calibrated rotameter, and a flow control valve, respectively.

This vessel also served as a mixing chamber, where the air while rising up attained a thermal equilibrium with the steam as well as formed a homogeneous mixture with it. The air/steam mixture left the pressure vessel from the top through an isolation valve fitted to the side of this vessel. Downstream from the boiler the mixture flow rate was measured with an accurately calibrated vortex flow meter. The mixture temperature and pressure were also measured before the inlet to the test section which consisted of a single vertical stainless steel (SS) tube inside which condensation would occur. The condenser tube dimensions are: 50.8 mm outside diameter (OD), 46.0 mm inside diameter (ID), and 2.54 m effective length. A 62.7 mm inside diameter concentric jacket

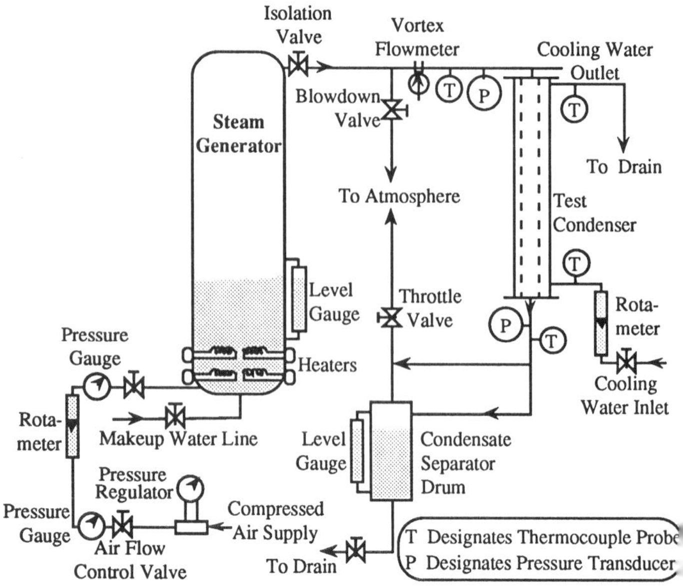

Fig. 1 Schematic of Experimental Apparatus

pipe surrounded the test condenser. The air/steam mixture flowed down through the tube while cooling water flowed counter-currently up through the annulus. A sliding joint obtained through the use of an O-ring at the lower end of the condenser, allowed relative motion between the condenser tube and the jacket pipe, and thus prevented thermal stresses from building up. Condensed liquid leaving the test condenser was separated in the condensate separator/collector drum. This drum was vented to the atmosphere via a throttle valve on the line carrying the residual air/steam mixture from the end of the test condenser. The steam generating vessel, the test condenser, and all connecting pipes were thoroughly insulated with fiber-glass to minimize heat losses to the environment.

Cooling water was supplied to the test section in a once-through mode via a calibrated rotameter and a flow control valve, and was dumped into a drain after use.

As shown in Figure 2, the respective temperatures of the cooling water, the tube centerline gases, and the tube inside and outside wall, were measured at nine stations spaced 0.3 m apart along the length of the condenser. Figure 3 shows the details of the test section instrumentation. All temperature measurements were made with J-type (iron/constantan) thermocouples which were calibrated after installation. Provision was also made for taking gas mixture samples at each measuring station. This was achieved by horizontally inserting a 3.2 mm OD SS tubing through the jacket wall and the condenser tube wall, respectively, using compression fittings. A 1.6 mm OD SS sheathed rigid thermocouple probe was then inserted through the same tubing to measure the centerline gas mixture temperature. The clearance between the temperature probe and the tubing was sufficient to allow the mixture to flow past when the sampling valve was opened. Unfortunately the steam in the mixture would condense and block the sample line to the gas analyzer. The sample line was wrapped with electrical resistance heating wire and insulated to prevent this condensation, but both of these measures failed as these lines were very thin and long. No meaningful data could be collected by this technique. Regrettably all gas sampling had to be abandoned.

The outer wall temperatures were measured by means of 0.8 mm OD SS sheathed thermocouple probes, embedded in longitudinally machined grooves 0.9 mm wide x 0.4 mm deep x 12.7 mm long, brazed with silver solder. The inner wall temperatures were measured by means of 0.8 mm OD SS sheathed thermocouple probes, embedded in drilled holes filled with silver solder. The holes which enter the tube wall at an angle were drilled deep enough to just encounter the inner tube surface without piercing through. This located the physical junction of the thermocouple at less than 0.2 mm from the inner surface of the tube wall. However, no correction was made to take into account

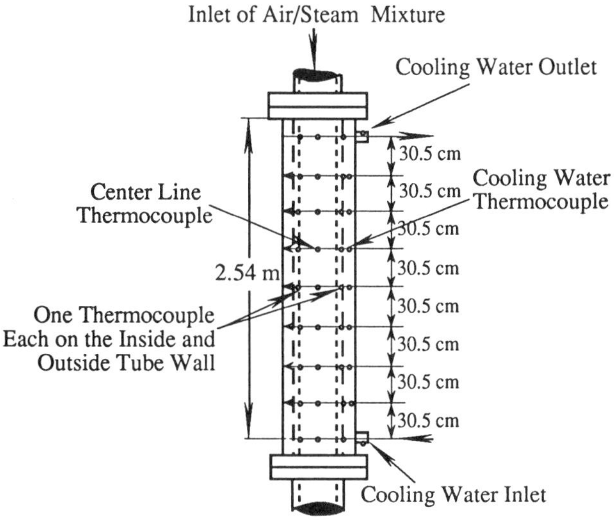

Fig. 2 Temperature Points along Length of Condenser

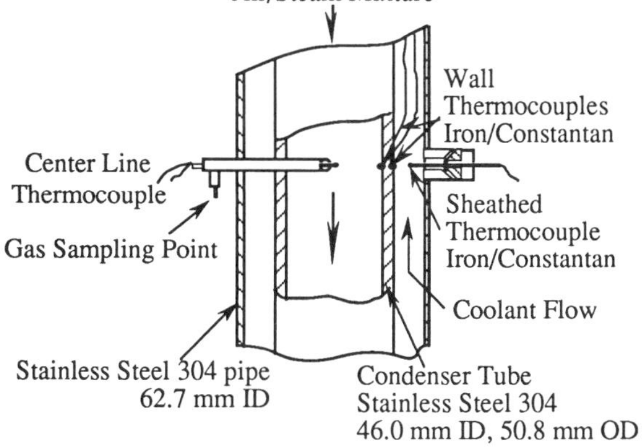

Fig. 3 Detail of the Test Section Instrumentation

the difference between the actual wall temperature and the measured wall temperature. It was estimated that for the values of the heat fluxes employed in these experiments, the measured wall temperature was in error by about 0.5°C in the worst case. All the wall thermocouple wires were taken out through the coolant jacket walls, via compression fittings located at the ends of the test condenser.

The cooling water temperatures were measured by means of 1.6 mm OD SS sheathed rigid thermocouple probe inserted through the cooling jacket wall to the mid point of the annular gap. The coolant flow rate had to be kept low enough to provide a large coolant temperature change per unit length of the annulus so that accurate local heat flux measurements could be made. However, this led to errors in measuring the coolant bulk temperature under low coolant velocity conditions due to inadequate mixing. Under laminar flow conditions the probes were unable to sense the correct bulk temperature due to the presence of steep temperature gradients near the tube wall. The problem was solved by bubbling small amounts of air (less than 1% of the coolant mass flow rate) through the annulus along with the cooling water, which enhanced turbulence and mixing. Thus accurate bulk coolant temperature measurements in the annulus were made possible. The coolant inlet and exit temperatures were also measured downstream from, well insulated 61.0 cm long, mixing lengths of 38.0 mm OD pipe filled with wire mesh.

Two high precision calibrated pressure transducers measured the pressure of the air/steam mixture at the entry and exit of the test section. The pressure measurements were accurate to 0.1 kPa.

The test section inlet air mass flow rate was taken from the measured air flow rate into the boiler. The steam mass flow rate was separately calculated based on the measured heater power input to the boiler, corrected for heat losses to the surroundings, and the heat gained by the air in attaining thermal equilibrium with the steam. Boiler heat loss calibration was carried out at all the three operating temperatures employed in this study. The heat loss was determined by bringing the boiler to the desired temperature. Then using the Variac the heater power was adjusted to find the heater input power necessary to compensate for the losses and thus maintain the boiler temperature at a constant level. In these tests the room temperature was the same as when the actual runs were made.

The inlet mixture flow rate, as measured by the vortex flow meter, was checked against the flow rate obtained by summing the individual steam and air flow rates. The values so obtained agreed closely with those measured by the vortex meter, the largest discrepancy being around 3% less than the vortex flow meter reading.

In preliminary shake down tests using pure steam the total condensate flow rate was determined by noting the rise in the level of the condensate separator tank over a measured time interval. Knowing the inside diameter of the tank and the condensate temperature, the condensate flow rate thus determined agreed closely with the measured steam flow rate as well as with the heat balance on the coolant side. This shows that the heat losses from the outer jacket were negligible, as well as confirms the reliability of the steam flow measurements.

The air fraction in the inlet stream was calculated by the individual mass flow rates of air and steam. This was checked against the value obtained by using the inlet pressure and temperature measurements (assuming saturation conditions) and the Gibbs-Dalton ideal gas mixture equation:

$$W = \frac{P_{tot} - P_s(T_{in})}{P_{tot} - \left(1 - \frac{M_s}{M_a}\right) P_s(T_{in})} \tag{1}$$

Excellent agreement was found between the air mass fraction determined by the two methods. Similar observations have also been made by Lee and Rose (1984).

A data acquisition (DA) system collected thermocouple, pressure transducer, and vortex flow meter data from 40 input channels. The DA system consisted of a scanner and a data processor linked to a personal computer. It was programmed to scan at a speed of 40 channels per second. The processor recorded data every minute and performed average and standard deviation calculations. The attainment of a steady state was indicated by the stabilization of the input readings and very low values of the standard deviations.

EXPERIMENTAL METHOD

Before the start of an experimental run, the boiler was isolated from the rest of the system. The water level in the boiler was adjusted and heaters were then turned on. All instrumentation was checked for proper operation and all zero readings were recorded. Depending on the water inventory, it took from about 1 to 1-1/2 hours for the boiler to reach the set temperature which was from 2 to 4°C above the desired mixture inlet temperature. While the boiler was heating up the coolant flow rate to the test condenser was adjusted to the desired value. Once the boiler had reached the set temperature, the boiler power was adjusted to the desired steam production rate corrected for the heat losses by using the Variac.

Depending on the inlet air fraction desired, the air flow rate for the set steam flow rate was calculated. Air at room temperature was supplied to the base of the boiler at the required rate. At the same time the boiler isolation valve was fully opened. The throttle valve on the vent line of the condensate separator drum was then manually adjusted until the system pressure stabilized. All data were taken after the facility reached a steady state condition where the boiler and test condenser pressure and temperature readings had stabilized to a nearly constant value for about 15 minutes. Then for a two minute period, data were recorded for each channel. Coolant and air flow rates were recorded separately.

After the run was completed the next air fraction setting was obtained by increasing the air flow, as calculated, and stabilizing the system pressure using the throttle valve on the vent line. Before the start of each successive run the water level in the boiler and the condensate separator drum was checked through the sight glass level. The separator drum was emptied by draining off the condensate. When the boiler inventory was reduced to the threshold level the tests were terminated.

DATA ANALYSIS

In order to calculate local heat transfer coefficients, the local air/steam mixture bulk temperature, local innerwall temperature, and the local heat flux must be known.

The local inner wall temperature was measured directly and the local heat flux was obtained from a knowledge of the coolant temperature profile. From a steady state energy balance on the coolant we get:

$$q''(x)dA = \dot{m}_c C_{p,c} dT_c \,, \tag{2}$$

or

$$q''(x) = \frac{\dot{m}_c C_{p,c}}{pd} \frac{dT_c}{dL}(x) \,. \tag{3}$$

The slope of the coolant temperature profile, i.e., dTc/dL was determined from a least squares polynomial fit of the coolant temperature as a function of condenser length. The temperature drop across the tube wall was to be used to obtain a second estimate of the local heat flux, but a large scatter in the tube outer wall temperature measurements (possibly due to fluctuations in the cooling water flow field) rendered this option useless.

From a sectionwise steady state heat balance the sectional condensate flow rate was obtained. The heat balance neglected the gas phase sensible heat transfer, considering only the latent heat transfer as:

$$\Delta Q = \Delta \dot{m}_{cond} H_{fg} \,. \tag{4}$$

This is a good assumption for air/steam mixtures corroborated by many theoretical and experimental studies (e.g, Asano et al., 1979 and Siddique et al., 1989). Here H_{fg} was calculated at the average wall temperature of the section. The local condensate flow rate at a particular location is the sum of the incremental condensate flow rates up to that point. The local steam flow rate was determined by subtracting the local condensate flow rate from the known steam flow rate at the inlet to the test section. As the air flow rate is constant, knowing the local steam flow also fixes the local air fraction as:

$$W(x) = \frac{\dot{m}_{a(in)}}{\dot{m}_{a(in)} + \dot{m}_s(x)} \,. \tag{5}$$

Pressure drop measurements indicated that for most cases the drop was very small. The largest pressure drop measured was of the order of 6.9 kPa for a total inlet pressure of 500 kPa. For these reasons the pressure drop through the test condenser was neglected in calculating the local steam partial pressure. Using the total inlet pressure and the local air mass fraction one may calculate the local steam partial pressure employing the Gibbs-Dalton ideal gas mixture equation. The latter was used along with the Steam Tables to obtain the local bulk steam temperature, assuming saturation conditions. This value compared well with the measured centerline temperature readings, the latter readings being higher by up to a maximum of 4 K. The local condensing heat transfer coefficient was obtained as:

$$h(x) = \frac{q''(x)}{(T_b - T_w)} \,. \tag{6}$$

The local mixture density, viscosity, and thermal conductivity were calculated at the bulk mixture temperature. The density was evaluated on the basis of ideal gas mixtures, the viscosity by the method of Wilke (1950), and the thermal conductivity as explained in Reed et al. (1977). The local mixture velocity was calculated as:

$$V(x) = \frac{\dot{m}_{a(in)} + \dot{m}_s(x)}{\rho A_f} \,. \tag{7}$$

The Reynolds number and the Nusselt number were based on mixture properties and the tube inside diameter.

EXPERIMENTAL ERROR

The maximum uncertainty associated with the heat transfer coefficient obtained from the data was calculated to be ±17.3%, as shown in more detail in the Appendix.

RESULTS AND DISCUSSION

Experiments were performed for air/steam mixture inlet temperatures of 100, 120, and 140°C, respectively. At each inlet temperature setting, the steam flow rate was varied by using heater power levels of 6, 13, and 20 kW. In addition, for each inlet temperature and power level setting, six runs were performed varying the inlet air mass fraction from 10 to 35%. Pertinent results and observations are discussed in this section. Complete sets of reduced data are given in Siddique et al. (1992).

For a fixed steam inlet flow rate Figure 4 shows the axial variation of the mixture Reynolds number at different values of the inlet air mass fraction. Here all the three runs are performed for an inlet temperature of 120°C with inlet air mass fractions of 0.113, 0.155, and 0.224, respectively.

The cooling water flow rate is nearly the same in each of these runs. We observe that for all three runs the Reynolds number decreases down the length of the condenser. This is due to the fact that removal of steam by condensation decreases the total mixture mass flow rate and thus the local Reynolds number decreases. The greater decrease in the Reynolds number for the low air mass fraction run is due to the low air flow rate associated with it. For the same three runs Figure 5 shows the variation of the heat flux along the length of the condenser tube. The heat flux at the inlet decreases as the inlet air mass fraction increases. Lower down in the tube when the condensation is nearly complete, the three curves come close together.

Figure 6 shows the effect of increasing the inlet mixture temperature on the heat flux profile. The three runs are for nearly the same steam inlet flow rate, inlet air fraction, and cooling water flow rate. We

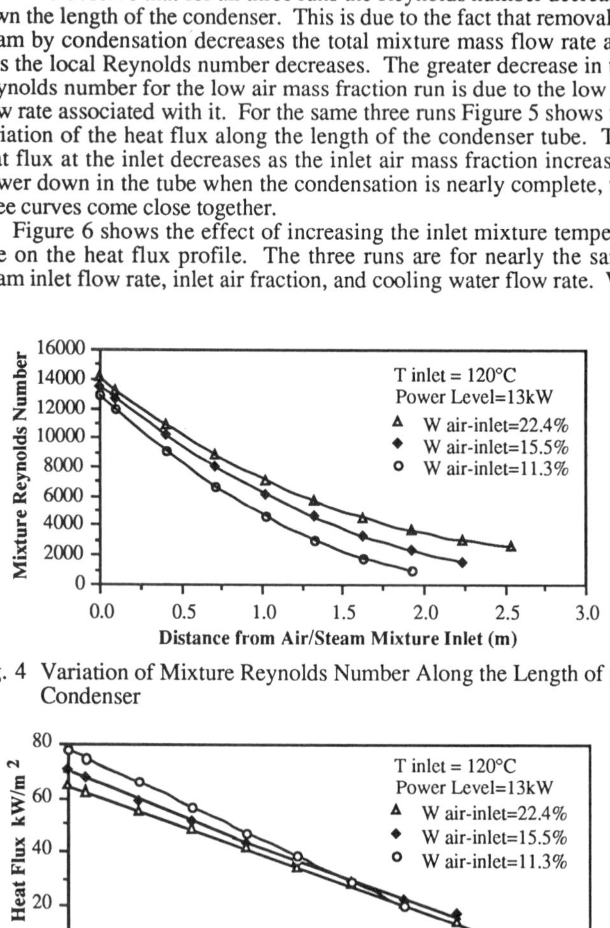

Fig. 4 Variation of Mixture Reynolds Number Along the Length of the Condenser

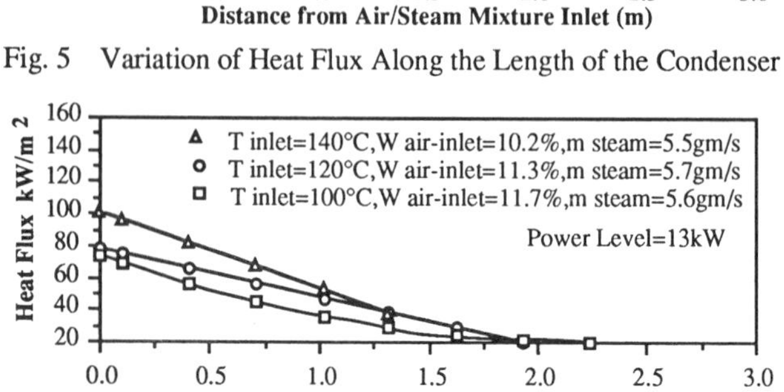

Fig. 5 Variation of Heat Flux Along the Length of the Condenser

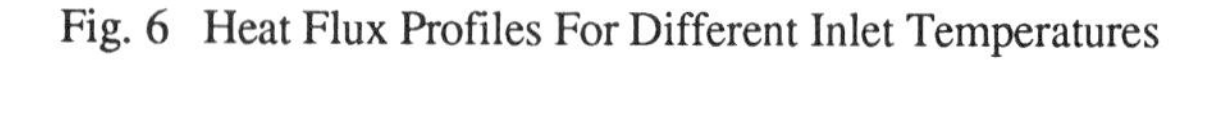

Fig. 6 Heat Flux Profiles For Different Inlet Temperatures

observe that as the inlet temperature increases, the average heat flux increases with a consequent decrease in the condensing length. This should be expected as greater wall subcooling increases the thermal driving force for heat transfer, and the higher pressure associated with higher inlet temperature leads to a greater number of molecular collisions helping in the diffusive transport of energy.

For a fixed mixture inlet temperature of 120°C Figure 7 illustrates the axial variation of the wall heat flux at three different values of the mixture inlet Reynolds number. As the variation in the inlet air mass fractions is very small (values of 0.08, 0.099, and 0.113, respectively, were used) what we see is basically the effect of the mixture velocity upon the heat transfer. The cooling water flow rate is nearly the same in each of these runs. We observe that as the mixture inlet Reynolds number increases, the average heat flux over the length of the condenser increases. This identifies the mixture Reynolds number as one of the important parameters in the convection process.

In pure vapor condensation the condensate film provides the only heat transfer resistance, whereas if small amounts of a noncondensable gas are also present, then the main resistance to heat transfer lies in the gas/vapor boundary layer. This has been shown in many studies (e.g., Sparrow et al., 1967, Denny et al., 1971, and Kim and Corradini, 1990). Under forced convection conditions the resistance of the condensate film is further diminished. As it gains momentum from the gas phase, the film becomes thinner, its velocity increases , and turbulence is initiated at much lower values of Re_{film}. Critical film Reynolds number values for transition from laminar to turbulent have been suggested to be as low as 70 (Rohsenow et al., 1956). Turbulence promotes rippling and waviness in the film which further reduces the film resistance by inducing convection heat transfer augmenting the conduction in the film. Moreover the heat conduction in the film is enhanced because of a spectrum of film thicknesses analogous to the improvement in heat transfer experienced in dropwise condensation. In other words the loss in heat transfer through thickened regions is more than compensated by the increase through the thinned regions (Drummond and Rigg, 1976). Barry and Corradini (1988) suggested that the film waviness causes mixing and turbulence in the gas/vapor boundary layer as well. In many experiments dealing with forced convection of pure vapors inside tubes, heat transfer coefficients up to 50% higher than the Nusselt value have been reported (Tepe and Mueller, 1947) (Nusselt analysis assumes that condensate film is in laminar flow). Corradini (1984) modelled forced convection condensation of steam in the presence of air on a flat plate using the heat and mass transfer analogy corrected for high mass transfer rates for the air/steam boundary layer and simple Nusselt analysis for the condensate film. He found that including the film heat transfer resistance did not significantly effect the overall heat transfer even though the Nusselt film analysis gives a conservative estimate of the film resistance. Based on the above discussion we can conclude that in forced convection the gas/vapor boundary layer provides the controlling resistance. Toward the end of the condensing tube the gas phase Reynolds number decreases and the film thickness increases. Intuitively one might reason that the aforementioned conclusion may not remain valid in this region but we must not forget that the noncondensable gas concentration has also increased. The gas is the major resistance to heat transfer in the gas/vapor boundary layer.

Model for Nondimensional Correlation

The governing conservation equations of the air/vapor boundary layer were used as a basis for defining the appropriate nondimensional groups for correlating the data. In the IC, as most of the heat transfer

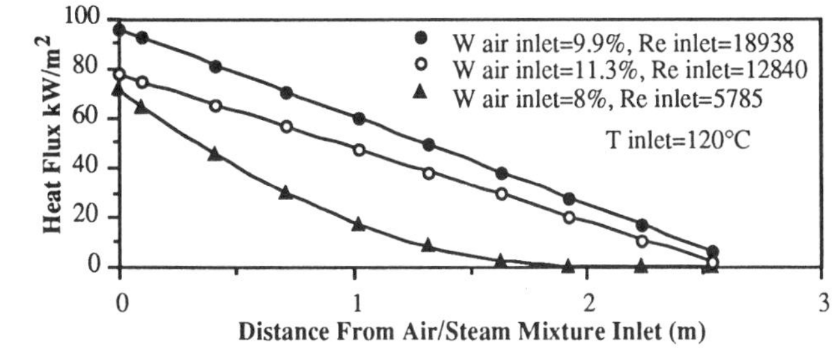

Fig. 7 Heat Flux Profiles For Different Inlet Reynolds Numbers

takes place in the first 0.5 to 0.6 m of the condenser tube it may be important to consider the developing flow effects in the heat-mass transfer model. In a related analytical study by Siddique et al. (1992), it was found that for the range of inlet mixture Reynolds numbers employed in these experiments, the developing flow effects are secondary and can be neglected. From the dimensionless momentum equation we can readily define the Reynolds number as the parameter by which the effective momentum layer thickness of geometrically similar systems can be related. The energy equation can be written in terms of the enthalpy as:

$$u\frac{\partial H}{\partial x} + v\frac{\partial H}{\partial y} = \frac{k}{\rho}\frac{\partial^2 T}{\partial y^2} . \tag{8}$$

Eq. 8 was normalized by defining the following dimensionless variables:

$$x^* = \frac{x}{d} \text{ and } y^* = \frac{y}{d} , \tag{9}$$

$$u^* = \frac{u}{V} \text{ and } v^* = \frac{v}{V} , \tag{10}$$

$$T^* = \frac{(T - T_w)}{(T_b - T_w)} \text{ and } H^* = \frac{(H - H_w)}{(H_b - H_w)} . \tag{11}$$

Substituting Eqs. 9 to 11 in the enthalpy equation of the air/steam boundary layer we obtain the result:

$$u^*\frac{\partial H^*}{\partial x^*} + v^*\frac{\partial H^*}{\partial y^*} = \frac{1}{PrRe}\frac{C_p(T_b - T_w)}{H_{fg}}\frac{\partial^2 T^*}{\partial y^{*2}} . \tag{12}$$

Here the group $C_p(T_b - T_w)/H_{fg}$, known as the Jakob number, is the ratio of the sensible energy to the phase change energy while Pr relates momentum diffusion to the diffusive sensible heat transfer which, as mentioned earlier, is not important under the present conditions. We know that the diffusive mass transfer resistance in the air/steam boundary layer is controlling. The presence of air is the actual cause of the existence of both temperature and concentration gradients. The concentration equation can be written as:

$$u\frac{\partial W}{\partial x} + v\frac{\partial W}{\partial y} = D\frac{\partial^2 W}{\partial y^2} . \tag{13}$$

Eq. 13 can be normalized by defining a nondimensional concentration

$$W^* = \frac{(W - W_w)}{(W_b - W_w)} . \tag{14}$$

Substituting Eqs. 9, 10 and 14 in the concentration equation of the air/steam boundary layer we obtain the result:

$$u^*\frac{\partial W^*}{\partial x^*} + v^*\frac{\partial W^*}{\partial y^*} = \frac{1}{ScRe}\frac{\partial^2 W^*}{\partial y^{*2}} . \tag{15}$$

The nondimensionalized concentration equation indicates that the solution to the concentration profile will depend upon the Reynolds and Schmidt numbers. In other words the mass transfer Sherwood number is a function of the Reynolds and Schmidt numbers. For our purposes we can define the Nusselt number and relate it to the other variables of this analysis as follows:

At any axial location,

$$q''(x) = h_{tot}(T_b - T_w) \sim \dot{m}''_{cond} H_{fg}. \tag{16}$$

A mass balance on the interface yields the following equations:

$$\dot{m}''_s = \left(-\rho D\frac{\partial W_s}{\partial y}\right)_w + W_{s,w}(\dot{m}''_{tot})_w , \tag{17}$$

and

$$\dot{m}''_a = \left(-\rho D\frac{\partial W_a}{\partial y}\right)_w + W_{a,w}(\dot{m}''_{tot})_w , \tag{18}$$

As the condensate surface is impermeable to air, Eq. 18 can be simplified as

$$\left(\rho D\frac{\partial W_a}{\partial y}\right)_w = W_{a,w}(\dot{m}''_{tot})_w . \tag{19}$$

Also $$W_a + W_s = 1, \tag{20}$$

therefore, $$\frac{\partial W_a}{\partial y} = -\frac{\partial W_s}{\partial y}. \tag{21}$$

Solving for $\dot{m}''_{tot}$ from Eq. 19 and substituting in Eq. 17 together with Eq. 21 we get

$$\dot{m}''_s = \left(-\rho D\frac{\partial W_s}{\partial y}\right)_w + W_{s,w}\left(-\frac{\rho D}{W_{a,w}}\frac{\partial W_s}{\partial y}\right)_w , \tag{22}$$

Simplifying Eq. 22 we get

$$\dot{m}''_s = \left(-\rho D\frac{\partial W_s}{\partial y}\right)_w \Big/ (1 - W_{v,w}) = h_m\frac{(W_{s,b} - W_{s,w})}{(1 - W_{s,w})}. \tag{23}$$

Here h_m is the mass transfer coefficient. Defining the Nusselt number as $Nu = h_{tot}\, d/k$, Eq. 16 can now be written as

$$Nu\,(x) = h_m\frac{(W_{a,w} - W_{a,b})}{W_{a,w}} H_{fg}\frac{d}{k(T_b - T_w)} . \tag{24}$$

By using the definition of the Sherwood number, $Sh = (h_m\, d/\rho D)$, Eq. 24 can be written as

$$Nu\,(x) = Sh\left(\frac{W_{a,w} - W_{a,b}}{W_{a,w}}\right)\frac{Pr}{Sc}\left(\frac{H_{fg}}{C_p(T_b - T_w)}\right) . \tag{25}$$

Therefore, we can say that $Nu = f(Re, Sc, (W_{a,w} - W_{a,b})/W_{a,w}, Ja)$. As only air/steam mixtures were used in this part of work, the dependence of Nu on Sc was not included. The data from this investigation are correlated by a relationship of the form

$$Nu\,(x) = C\,Re^a\left(\frac{W_{a,w} - W_{a,b}}{W_{a,w}}\right)^b Ja^c . \tag{26}$$

The above expression was linearized by employing a logarithmic transformation. A multiple linear regression analysis was then used to find the coefficients. The final correlation is of the form

$$Nu\,(x) = 6.123\,Re^{0.223}\left(\frac{W_{a,w} - W_{a,b}}{W_{a,w}}\right)^{1.144} Ja^{-1.253} , \tag{27}$$

which applies in the range of the experiments:

$0.1 < W_{air} < 0.95,$
$445 < Re < 22{,}700,$
and $0.004 < Ja < 0.07.$

Figure 8 shows the comparison of the Nu numbers as calculated by Eq. 27 with the experimental data. The adjusted R^2 value for the fit of the data to the correlation is 0.885, signifying that the correlation model explains 88.5% of the variability in the data. The standard deviation of the correlation predicted Nusselt numbers from the experimentally obtained Nusselt numbers was calculated to be 0.40 or 40%.

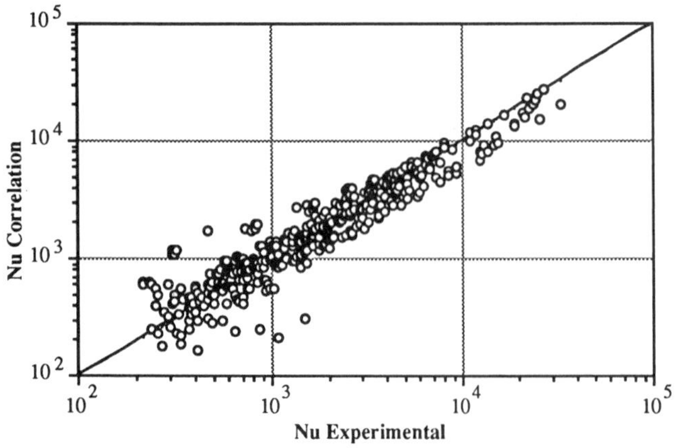

Fig. 8 Comparison of Air-Steam Experimental Nusselt Numbers with Correlation Nusselt Numbers

COMPARISON WITH UCB CORRELATION

The only directly comparable work available on this subject is that of Vierow (1990) at UC Berkeley. Her study was made using a 22.0 mm ID vertical tube, natural circulation air/steam system. The operating conditions of inlet temperature and steam flow rate were comparable to ours, except that the maximum inlet air mass fraction was 0.14. The local heat transfer coefficients have been correlated as follows:

$$h(x) = 0.005\,Re_{cond}\,(x)^{0.45} W_{air}\,(x)^{-1.1}\, h_{nusselt}\,(x) , \tag{28}$$

where $h_{nusselt}$ is defined as the heat transfer coefficient for pure steam calculated using the Nusselt theory.

As the two correlations are based on completely different theoretical premises, a direct comparison between Eqs. 27 and 28 is possible only if the local values of the Reynolds number and air mass fraction of the air/steam mixture are known, and the corresponding value of the condensate film Reynolds number at that particular axial location is also known. From the data that we have obtained, the local values of the condensate film Reynolds number were also computed and thus a comparison with the predictions of the UCB correlation was made possible. Figure 9 shows the comparison between the UCB correlation and MIT correlation predicted local heat transfer coefficients. The reported standard deviation band of 30% of the the heat transfer coefficient associated with the UCB correlation is also drawn. We observe that starting at a heat transfer coefficient value of 700 W/m^{2}K, the UCB correlation progressively underpredicts the MIT correlation predicted values. The maximum difference observed is as great as a factor of ten. High heat transfer coefficients are associated with relatively high mixture Reynolds numbers and/or low air mass fractions. Below heat transfer coefficient values of 700 W/m^{2}K, most of the MIT correlation predicted values fall below the predictions of the UCB correlation. Figure 10 shows the heat

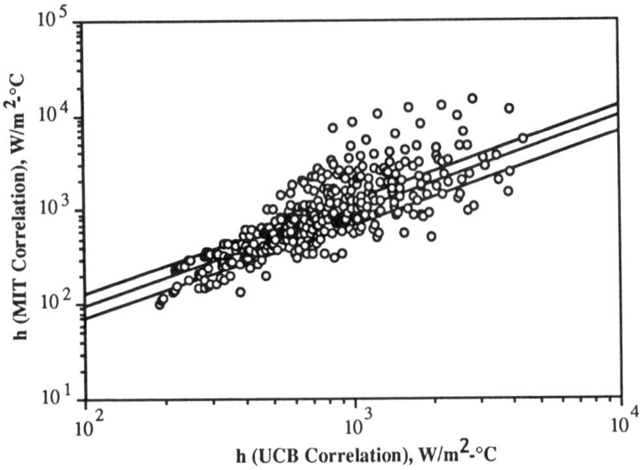

Fig. 9 Comparison of the Predictions of MIT and UCB Correlations

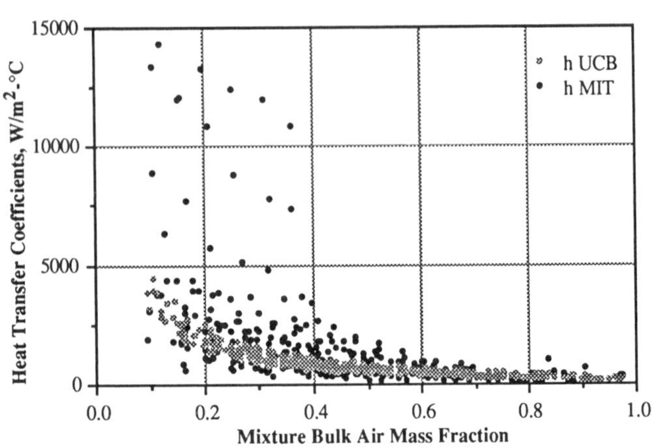

Fig. 10 Heat Transfer Coefficients Versus the Air Bulk Mass Fraction

transfer coefficients as predicted by the UCB correlation and those experimentally obtained plotted against the mixture bulk air mass fraction. We observe that the UCB correlation predicts a monotonous decrease in the heat transfer coefficients as the air mass fraction (W_{air}) increases. The drop in the heat transfer coefficient value is gradual up to a value of $W_{air} = 0.5$, beyond which the heat transfer coefficient assumes an asymptotic value. On the other hand in the region $W_{air} = 0.1$ to 0.35, for any fixed value of W_{air}, the experimentally obtained values of the heat transfer coefficient have very high as well as very low values. Above $W_{air} = 0.35$ the experimental heat transfer coefficients gradually decrease and assume a low asymptotic value. The reason for this behavior becomes clear if we bear in mind that these tests were conducted with inlet W_{air} values of 0.1 to 0.35 and three different inlet temperatures and steam flow rates. Therefore in the region $W_{air} = 0.1$ to 0.35, at any fixed value of W_{air} we will have different values of the heat transfer coefficient depending on the local values of Re_{mix} and bulk mixture temperature. This behavior is not observed above $W_{air} = 0.35$ because, first of all our test matrix did not have inlet values of $W_{air} > 0.35$; and secondly, at values of W_{air} of about 0.30 to 0.35 the film Reynolds number in the first 10.0 cm of the test condenser is still less than 20. Such data points were not used for this comparison as they were outside the range of validity of the UCB correlation. Figures 11a and 11b show the plots of the heat transfer coefficients as predicted by the UCB correlation and those experimentally obtained as a function of the mixture Reynolds number for parametric values of the bulk air mass fraction. In each of these plots the average value of W_{air} as stipulated is actually a band of air mass fraction values covering ±2% of the stated average

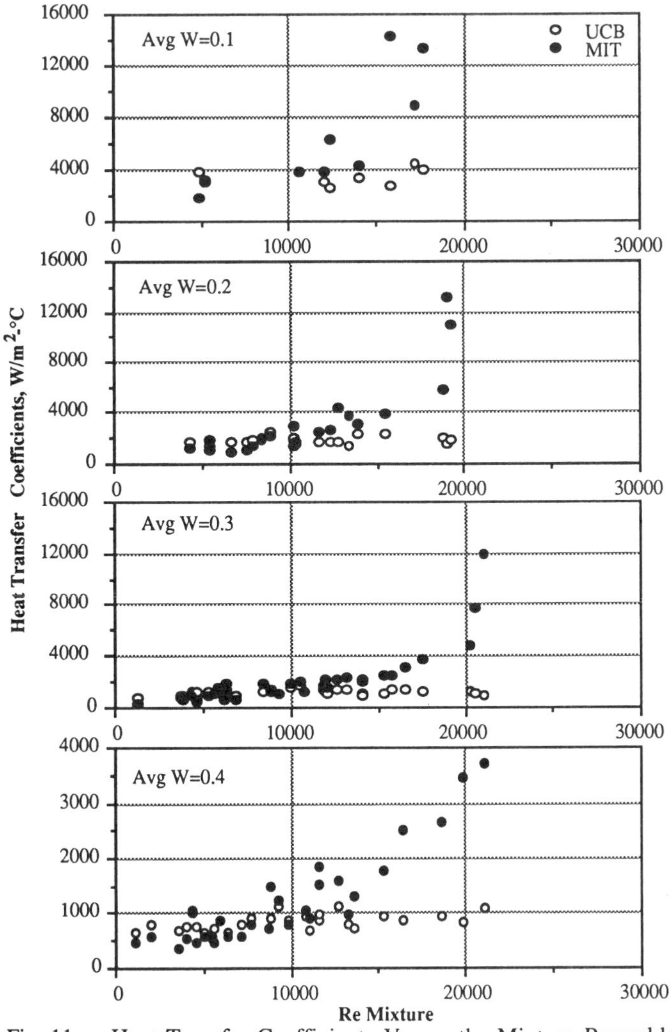

Fig. 11a Heat Transfer Coefficients Versus the Mixture Reynolds Number

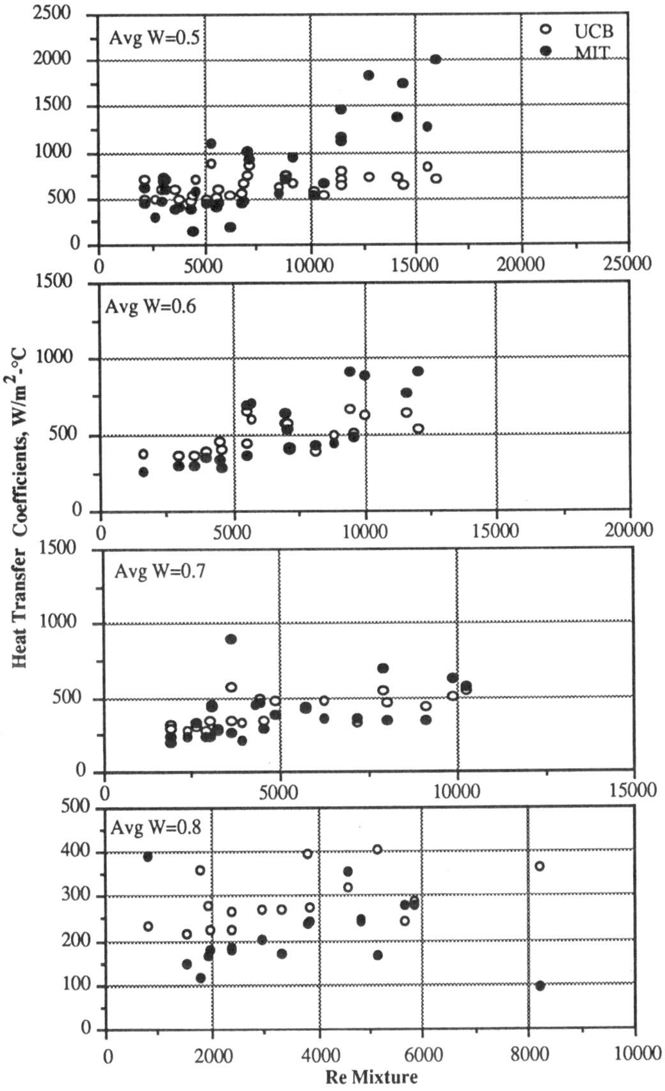

Fig. 11b Heat Transfer Coefficients Versus the Mixture Reynolds Number

value. We observe that for any single value of W_{air} between 0.1 to 0.5 the experimental heat transfer coefficient increases as the mixture Reynolds number increases. On the other hand the UCB correlation predicted heat transfer coefficients remain insensitive to the increase in the mixture Reynolds number. Above $W_{air} = 0.5$, the experimental and the UCB correlation predicted heat transfer coefficients do not have such a huge discrepancy, since in this region the heat transfer rates are low because of relatively low mixture Reynolds number and high air mass fractions.

CONCLUSIONS

Local heat transfer coefficients were obtained for forced convection condensation of steam/air mixtures inside a tube such as could be typical for an IC,under conditions simulating a LOCA. The governing conservation equations of the air/vapor boundary layer were used as a basis for identifying the appropriate nondimensional groups for correlating the data. The model developed correlates the data reasonably well and can be used to predict the local heat transfer coefficient inside tubes for the range of conditions utilized in these experiments. Comparison of the results obtained to the predictions of the UCB correlation showed that the UCB correlation is conservative at high values of mixture Reynolds number and low bulk air mass fractions, and nonconservative at low values of mixture Reynolds number and high bulk air mass fractions, with respect to our correlation.

REFERENCES

Al-Diwani, H.K. and Rose, J.W., "Free Convection Film Condensation of Steam in the Presence of Noncondensing Gases," *Int. J. Heat Mass Transfer*, Vol. 16, 1973, pp. 1359-1369.

Asano, K., Nakano, Y. and Inaba, M., "Forced Convection Film Condensation of Vapor in the Presence of Noncondensing Gas on a Small Vertical Flat Plate," *J. Chem. Engr. of Japan*, Vol. 12, 1979, pp. 196-202.

Barry, J.J., and Corradini, M.L., "Film Condensation in the Presence of Interfacial Waves," *ASME Proceedings*, National Heat Transfer Conference, HD-96, Vol. 2, 1988, pp. 523-529.

Borishanskiy, V.M., et al., "Effect of Uncondensable Gas Content on Heat Transfer in Steam Condensation in a Vertical Tube," *Heat Transfer Soviet Research*, Vol. 9, No. 2, March-April 1977, pp. 35-42.

Borishanskiy, V.M., et al., "Shell Side Coefficient of Heat Transfer from Steam Contaminated with Noncondensable Gases," Heat Transfer Soviet Research, Vol. 14, No. 3, May-June 1982.

Corradini, M.L., "Turbulent Condensation on a Cold Wall in the Presence of a Noncondensable Gas," Nuclear Technology, Vol. 64. 1984, pp. 186-195.

Denny, V.E., Mills, A.F. and Jusionis, V.J., "Laminar Film Condensation From a Steam-Air Mixture Undergoing Forced Flow Down a Vertical Surface," *J. Heat Transfer*, Vol. 93, 1971, pp. 297-304.

Drummond, G. and Rigg, D.R., "Waves in Condensate Films," NEL Report No. 619, Aug. 1976, pp.19-26.

Ivashchenko, N.I., et al., "Heat Transfer With Steam Condensation From a Steam-Gas Mixture," *Heat Transfer Soviet Research*, Vol. 21, No. 1, Jan-Feb 1989. pp. 42-47.

Kim, M.H. and Corradini, M.L., "Modeling of Condensation Heat Transfer in a Reactor Containment," *Nuclear Eng & Design*, Vol. 118, 1990, pp. 193-212.

Kroger, D.G. and Rohsenow, W.M., "Condensation Heat Transfer in the Presence of a Noncondensable Gas," *Int. J. Heat Mass Transfer*, Vol. 11, 1968, pp. 15-23.

Lee, W.C. and Rose, J.W., "Forced Convection Film Condensation on a Horizontal Tube With and Without Noncondensing Gases," *Int. J. Heat Mass Transfer*, Vol. 27, 1984, pp. 519 -528.

Minkowycz, W.J. and Sparrow, E.M., "Condensation Heat Transfer in the Presence of Noncondensables, Interfacial Resistance, Superheating, Variable Properties and Diffusion," *Int. J. Heat Mass Transfer*, Vol. 9, 1966, pp. 1125-1144.

Reed, R.C., Prausnitz, J.M. and Sherwood, T.K., *The Properties of Gases and Liquids*, Mc Graw Hill, New York, 1977.

Rohsenow, W.M., Webber, J.H. and Ling, A.T., "Effect of Vapor Velocity on Laminar and Turbulent Film Condensation," *Trans. ASME*, Vol. 78, No. 8, 1956, pp. 1637-1643.

Rose, J.W., "Condensation of a Vapour in the Presence of a Noncondensing Gas," *Int. J. Heat Mass Transfer*, Vol. 12, 1969, pp. 233-237.

Siddique, M., Golay, M.W. and Kazimi, M.S., "The Effect of Hydrogen on Forced Convection Steam Condensation," *AIChE Symposium Series*, Vol. 85, No. 269, 1989, pp. 211-216.

Siddique, M., Golay, M.W. and Kazimi, M.S., "The Effects of Noncondensable Gases on Steam Condensation Under Forced Convection Conditions," MIT Report No. MIT-ANP-TR- 010, March 1992.

Sparrow, E.M. and Lin, S.H., "Condensation Heat Transfer in the Presence of a Noncondensable Gas," *J. of Heat Transfer*, Vol. 9, Aug. 1964, pp. 430-436.

Sparrow, E.M., Minkowycz, W.J. and Saddy, M., "Forced Convection Condensation in the Presence of Noncondensables and Interfacial Resistance," *Int. J. Heat Mass Transfer*, Vol. 10, 1967, pp. 1829-1945.

Tepe, J.B. and Mueller, A.C., "Condensation and Subcooling Inside an Inclined Tube," *Chem. Eng. Progress*, Vol. 43, 1947, p. 267-271.

Uchida, H., Oyama, A. and Togo, Y., "Evaluation of Post-Incident Cooling Systems of Light Water Reactors," *Proc. 3rd Int. Conf. on the Peaceful Uses of Atomic Energy*, Vol. 13, 1964, pp. 93-103.

Vierow, K.M., "Behavior of Steam-Air Systems Condensing in Cocurrent Vertical Downflow," MS Thesis, Dept. of Nuclear Eng. University of California at Berkeley, 1990.

Votta, Jr., F. and Walker, C.A., "Condensation of Vapor in the Presence of Noncondensing Gas," *AIChE Journal*, Vol. 4, 1958, pp. 413-417.

Wang, C.Y. and Tu, C.J., "Effects of Noncondensable Gas on Laminar Film Condensation in a Vertical Tube", *Int. J. Heat Mass Transfer*, 1988, Vol 31, pp. 2339-2345.

Wilke, C.R., "A Viscosity Equation for Gas Mixtures," *J. Chem. Phys.*, Vol. 18, 1950, pp. 517-519.

Appendix

EXPERIMENTAL ERROR ANALYSIS

The local heat transfer coefficient was defined as:

$$h(x) = \frac{\dot{m}_c C_p}{\pi d (T_b - T_w)} \frac{dT_c}{dL}(x)\,, \tag{A.1}$$

Therefore:

$$\frac{\partial h}{\partial \dot{m}_c} = \frac{C_p}{\pi d (T_b - T_w)} \frac{dT_c}{dL}(x)\,, \tag{A.2}$$

$$\frac{\partial h}{\partial (T_b - T_w)} = -\frac{\dot{m}_c C_p}{\pi d (T_b - T_w)^2} \frac{dT_c}{dL}(x)\,, \tag{A.3}$$

and

$$\frac{\partial h}{\partial (dT_c/dL)} = \frac{\dot{m}_c C_p}{\pi d (T_b - T_w)}\,. \tag{A.4}$$

The total error in h is given by:

$$\sigma_h^2 = \left(\frac{\partial h}{\partial \dot{m}_c}\sigma_{\dot{m}_c}\right)^2 + \left(\frac{\partial h}{\partial (dT_c/dL)}\sigma_{(dT_c/dL)}\right)^2 + \left(\frac{\partial h}{\partial (T_b - T_w)}\sigma_{(T_b-T_w)}\right)^2 \tag{A.5}$$

Substituting Eqs A.2 to A.4 in Eq. A.5 and dividing both sides of Eq. A.5 by h^2 obtained from Eq. A.1 we get:

$$\frac{\sigma_h}{h} = \sqrt{\left(\frac{\sigma_{\dot{m}_c}}{\dot{m}_c}\right)^2 + \left(\frac{\sigma_{(dT_c/dL)}}{dT_c/dL}\right)^2 + \left(\frac{\sigma_{(T_b-T_w)}}{(T_b - T_w)}\right)^2} \tag{A.6}$$

The standard deviation of the thermocouples was typically less than 0.3 K, with the fluctuations more pronounced in the outer surface temperatures of the tube wall.

The slope of the coolant axial temperature profile, i.e., dT_c/dL was determined from a least squares polynomial fit of the coolant temperature as a function of condenser length. The data was fit with polynomials up to the fourth order. For almost all the runs the adjusted R^2 value for the fit of the data to the polynomial was greater than 0.98. Thus the error associated with the curve fitting was very small. As seen from Eq. A.6, it should be noted however, that the error associated with the heat transfer coefficient increases when the coolant temperature change per unit length of the annulus (dT_c/dL) is small. This happens toward the end of the condensing tube where the noncondensable gas concentration is very high and consequently the heat transfer is severely degraded.

In the absence of an independent method for determining the heat transfer coefficient, it is assumed here that the deviation in the value of the coolant axial temperature gradient was within ±15% of the value obtained from the curve fit. Although this assumption might be too conservative for most part of the condensing length, in reality it only reflects the error associated with the value of dT_c/dL toward the end of the condensing tube.

The minimum wall subcooling was about 5°C for the data used in the regression analysis. A few data points with a value of wall subcooling less than 5°C were not used in the regression analysis as the high uncertainty associated with these data points, would have unrealistically biased the total error in the value of the experimentally determined heat transfer coefficient to a high value. The flow meter accuracy specified by the manufacturer was within ±1% of the full scale reading which was verified during the flow meter calibration. The minimum flow rate used was 65% of the full scale reading. Therefore:

$$\sigma_{\dot{m}_c} = 0.01\,\dot{m}_c(\text{Full Scale}) \tag{A.7}$$

$$\dot{m}_c = 0.65\,\dot{m}_c(\text{Full Scale}) \tag{A.8}$$

$$\sigma_{T_b} = \sigma_{T_w} = 0.3 \tag{A.9}$$

$$(T_b - T_w) = 5.0 \tag{A.10}$$

$$\sigma_{(T_b - T_w)} = \left(\sigma^2_{T_b} + \sigma^2_{T_w}\right)^{1/2} \quad (A.11)$$

$$\sigma_{(dT_c/dL)} = 0.15(dT_c/dL) \quad (A.12)$$

Substituting values from Eqs A.7 to A.12 in Eq. A6 and simplifying we get:

$$\left[\frac{\sigma_h}{h}\right]_{max} = 0.173$$

Therefore the maximum uncertainty associated with the heat transfer coefficient obtained from the data is ±17.3%

Accuracy of the Correlations

The standard deviation of the correlation predicted Nusselt numbers from the experimentally obtained Nusselt numbers was assessed as follows:

Define, $$Nu_{error} = \frac{(Nu_{correlation} - Nu_{experimental})}{Nu_{correlation}} \quad (A.13)$$

and $$\sigma_{Nu} = \left[\frac{\Sigma(Nu_{error})^2}{n - 1}\right]^{1/2} . \quad (A.14)$$

Using the above equations the standard deviation of the correlation predicted Nusselt numbers (S_{Nu}) was calculated.

HTD-Vol. 197, Two-Phase Flow and Heat Transfer
ASME 1992

NEW MEASUREMENTS FOR CONDENSATION OF MERCURY: IMPLICATIONS FOR INTERPHASE MASS TRANSFER

A. K. Kosasie and J. W. Rose
Department of Mechanical Engineering
Queen Mary and Westfield College
University of London
London, United Kingdom

ABSTRACT

Measurements of the vapor-condensate interface resistance by Niknejad and Rose (1981) for condensation of mercury, where care was taken to minimise non-condensing gas effects, showed general agreement with interphase mass transfer theory and indicated that the condensation coefficient for mercury exceeds 0.9. However, at the higher condensation rates, and at the higher temperatures, discrepancies between experiment and theory suggested that non-condensing gas effects may still have been significant. A new apparatus has been built with a vacuum integrity better than that of Niknejad and Rose (1981) by a factor of at least 1000. As in the earlier case, the temperature of the vertical plane condensing surface and the heat flux, or condensation rate, have been measured with very high accuracy. Preliminary results for high vapor temperatures and condensation rates, where the data are most sensitive to effects of non-condensing gases, indicate significantly lower vapor-to-surface temperature differences than those obtained formerly under the same conditions. Details of the new apparatus and experimental procedure are given. The new results are reported and compared with the earlier data and with theory.

NOMENCLATURE

m	condensation mass flux
P_V	vapor pressure
$P_{sat}(T)$	saturation pressure at temperature T
Q	heat flux
R	specific ideal-gas constant
T	thermodynamic temperature
T_S	temperature at condensate surface
T_V	vapor temperature
ξ	defined by equation (1)
ΔT	vapor-to-surface temperature difference.

INTRODUCTION

For condensation of metals, particularly at low pressures, the temperature drop at the vapor-liquid interface is significant. As discussed by Niknejad and Rose (1981), several related approximate theories of interphase mass transfer give similar results. Most theories incorporate the so-called "condensation coefficient" i.e. the fraction of those vapor molecules striking the liquid surface that remain in the liquid phase. While values as low as 0.005 have been reported for this parameter, it is now generally accepted as being close to unity, at least for monatomic molecules.

The various theoretical approaches give results which, in the low condensation rate limit, are expressible in the form

$$m = \xi(P_V - P_{sat}(T_S))/(RT_S)^{1/2} \qquad (1)$$

where ξ, which can be regarded as a dimensionless "interface mass-transfer coefficient", is a constant which incorporates the condensation coefficient. If the condensation coefficient is unity, then Schrage's (1953) "simple" and more elaborate theories give values for ξ of 0.798 and 0.627 respectively. A more recent result of Labuntsov and Kryukov (1979) gives ξ = 0.664. The experimental data of Niknejad and Rose (1981), for condensation of mercury vapor, gave values of ξ which, when extrapolated to zero condensation rate, indicated a value close to 0.7 for eight vapor temperatures in the range 378 K to 493 K. For condensation of potassium, the data of Ishiguro et al. (1986) indicate, for condensation rates aproaching zero, values of ξ ranging from around 0.9 at a vapor temperature of 553 K to around 0.6 at 633 K.

It is interesting to note that theory indicates that the dimensionless mass-transfer coefficient *increases* with increasing condensation rate and with decreasing vapor pressure. This is illustrated in Fig. 1 for condensation of saturated mercury vapor. These trends were found in the experimental investigations of Niknejad and Rose (1981) for mercury and Ishiguro et al. (1986) for potassium.

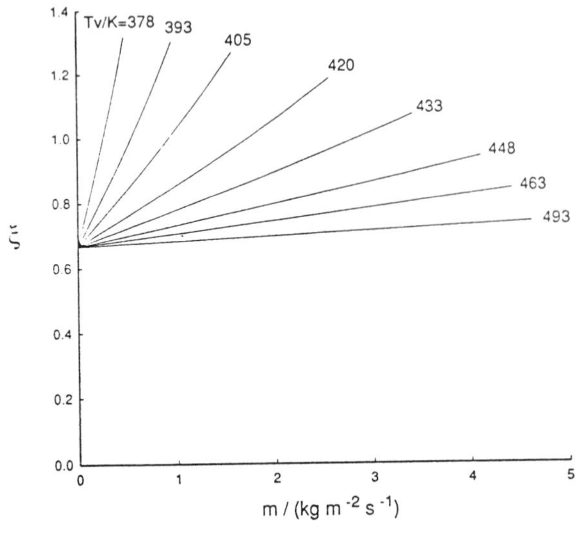

Figure 1. Dependence of ξ on m, for different vapour temperatures, given by the theory of Labuntsov & Kryukov (1979).

In the case of the mercury data of Niknejad and Rose (1981), measurements at the highest vapor temperatures indicated values of ξ *falling* (to values approaching 0.4) with increasing condensation rate, where theory indicates only that the *rate of increase* of ξ with condensation rate falls with increasing temperature and that ξ should be virtually constant for the highest temperature used by Niknejad and Rose (1981) (see Fig. 1).

A new apparatus has been built with the aim of obviating any possible error due to non-condensing gas. A few preliminary runs have been carried out to date at the higher temperatures and condensation rates used by Niknejad and Rose (1981), where the results are most sensitive to effects of non-condensing gases. The results, reported here, strongly suggest that the earlier data were affected by the presence of small amounts of non-condensing gas, despite the strenuous efforts taken to avoid this.

APPARATUS

The apparatus, in the form of a stainless-steel closed loop, comprising boiler, test section, butterfly valve, auxiliary condenser, cold trap/ionization gauge and vacuum pump is shown in Fig. 2. The test section is shown in detail in Fig. 3.

Mercury vapor flowed downwards over the vertical plane surface of the test condenser block which was cooled on the rear side by water. The auxiliary condenser and butterfly valve were used to prevent mercury vapor migrating to the vacuum pump, and in particular (by opening the butterfly valve immediately before taking a set of readings) to eliminate trace amounts of non-condensing gases which might have remained despite thorough outgassing and which would otherwise accumulate near the condensing surface. The condensate from both condensers was returned by gravity to the boiler. Small bore, stainless-steel tubes for thermocouples were welded in the boiler and test condenser sections as indicated in Figs. 2 and 3. The test condenser surface could be viewed through a window and the whole apparatus was thermally well insulated.

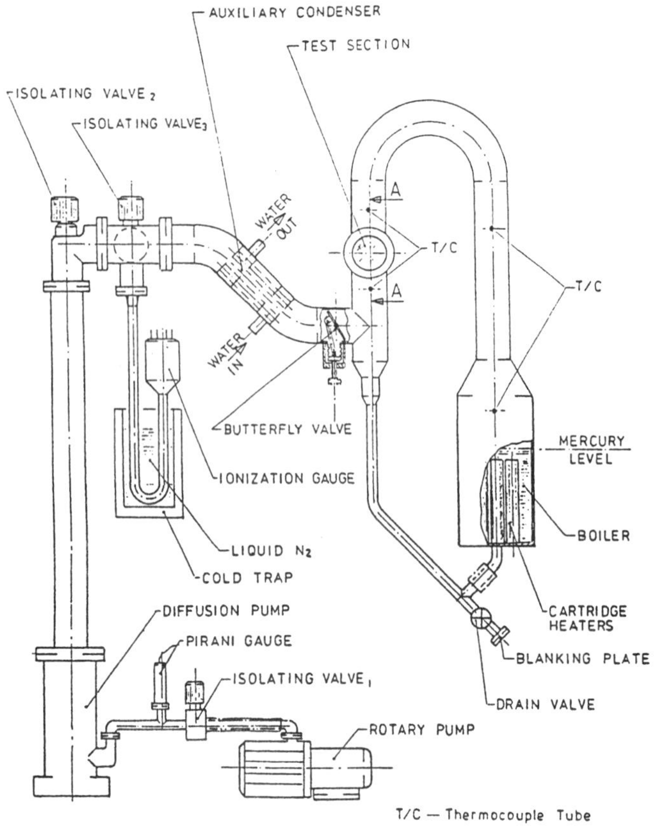

Figure 2 Apparatus

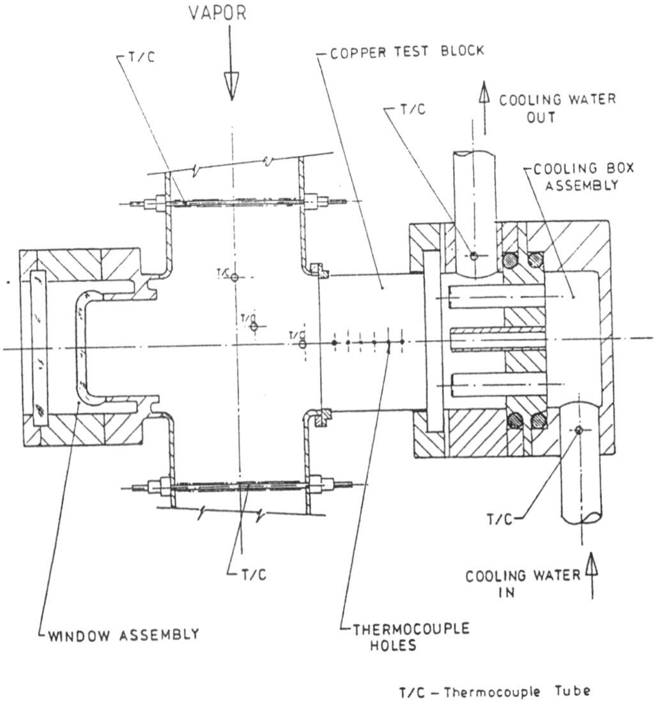

Figure 3 Test Section

Fine-wire thermocouples, accurately located and spaced through the high-purity copper condenser block, served to measure the surface temperature by extrapolation, and the heat flux from the temperature gradient. Six thermocouple holes (0.4 mm diameter) were drilled (from both sides) through the block and ran from side to side parallel to the condensing surface in central horizontal plane of the condenser block. The positions of the thermocouple holes were precisely determined using a travelling microscope. The thermocouples were butt-welded, and located so that their junctions lay in central vertical plane of the block, and the leads ran out to either side along isotherms (to avoid error due to conduction along the leads).

The condenser block was attached to the test section by a stainless steel ring which was first vacuum-brazed to the condenser block and the assembly was then welded to the test loop. To prevent attack of the copper by mercury and to ensure film condensation, the condensing surface of the block was nickel-plated to a nominal thickness of 0.3 mm. The nickel plating thickness was carefully measured before the block was assembled with the condenser loop.

The thermocouple wires (0.2 mm diameter, nichrome-constantan), used to measure the temperatures of the condenser block and vapor, were from the same reels. Before use, the thermocouples were silicone-varnished and annealed at 570 K. The emfs were measured and recorded using a data logging system with an accuracy 2 μV. Calibration was made against a platinum resistance thermometer using a high-precision, constant-temperature bath. The calibration accuracy was better than 0.01 K and calibration points were fitted by a polynomial from which the standard deviation of the points was 0.01 K.

TEST PROCEDURE

Before being filled, under vacuum, the apparatus was out-gassed (by encasing the loop in a portable oven) continuously for 6 days at a temperature of around 540 K. The filling procedure was similar to that given in Niknejad (1979).

As reported by Niknejad and Rose (1981), a period of 42 hours of operation was needed to achieve a complete wetting of the nickel surface. In the work reported here, before making measurements, the apparatus was run continuously at a vapor temperature of 445 K for 2 days with a coolant flow rate that gave (as indicated by subsequent measurements) a condensation rate of about 1.1 kg m^{-2} s^{-1}. In all subsequent tests, and during shut-down periods, the surface remained completely wetted.

Throughout all tests and during shut-down periods the apparatus was vented via the vacuum pump and the partial pressure of non-condensing gases in the apparatus (as indicated by the ionization gauge) was around 2 x 10^{-8} Torr.

Measurements were made (while venting continuously via the vacuum pump) at several vapor temperatures. To eliminate possible traces of non-condensing gases prior to taking a set of measurements, the butterfly valve was partially opened for about 10 minutes, and both the auxiliary condenser and test condenser coolant flow rates adjusted so that the condensation rate in the auxiliary condenser was much higher than that on the test condenser surface. When taking tests the butterfly valve was in the closed position (the lower end of the closing plate having a small clearance slot to allow the condensate to drain back into the loop) and a fixed coolant flow rate was used in the auxiliary condenser. Under these conditions the condensation rate in the auxiliary condenser was around 5% of that on the test surface.

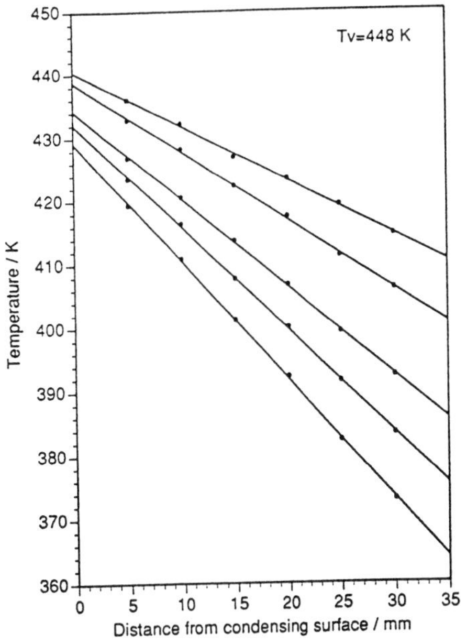

Figure 4. Specimen temperature distributions in the test condenser block

For each vapor temperature, a range of coolant (test condenser) flow rates was used. After a change in coolant flow rate the power input to the boiler heaters was adjusted to maintain the desired vapor temperature. The recorded quantities (when the apparatus was operating under steady conditions) were:- condenser block, vapor, boiler and coolant temperatures; the coolant flow rates (test and auxiliary condensers) and the pressure (ionization gauge) at exit from the 'cold trap'.

The heat flux and surface temperature were determined from the temperature distributions in the copper condensing block (see Fig. 4). For the purpose of the calculations discussed below, the vapor temperature used was that indicated by the thermocouple nearest the test condenser surface.

RESULTS AND DISCUSSION

Fig. 5 compares the vapor-to-surface temperature difference *versus* heat flux measurements of Niknejad and Rose (1981) with theory. The theoretical lines were determined using the result of Labuntsov and Kryukov (1979) (with condensation coefficient = 1) and assuming a saturated vapor, to obtain the interface temperature difference and the equation of Fujii and Uehara (1972) for calculating the temperature difference across the condensate layer in forced convection vapor downflow. The interface temperature difference dominates in all cases, ranging from around 75% of ΔT at the highest vapor temperatures and lowest heat flux, to around 95% of ΔT at the lowest vapor temperature and highest heat flux.

Details of these theoretical results are given in (Niknejad and Rose (1981)). As may be seen from Fig. 5, the temperature differences measured by Niknejad and Rose significantly exceed the calculated values.

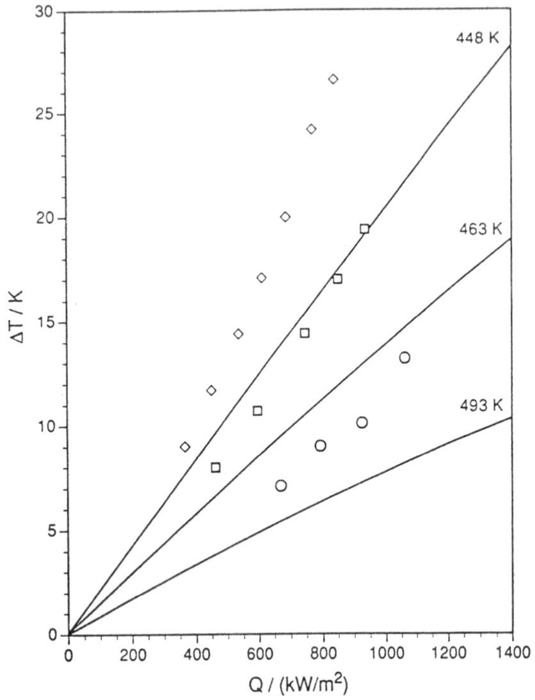

Figure 5. Variation of vapor-to-condensing surface temperature difference with heat flux.

lines - given by the theory of Labuntsov & Kryukov (1979) and the Fujii equation as described in the Appendix of Niknejad & Rose (1981).
◇ - Niknejad 448 K
□ - Niknejad 463 K
○ - Niknejad 493 K

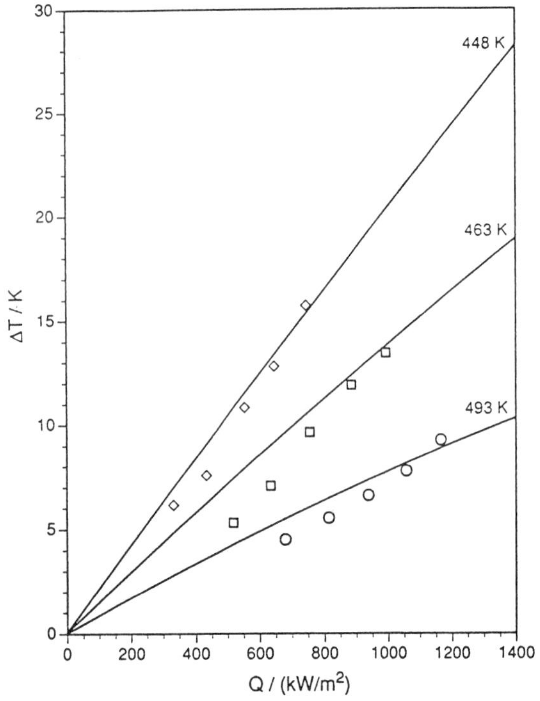

Figure 6. Variation of vapor-to-condensing surface temperature difference with heat flux.

lines - given by the theory of Labuntsov & Kryukov (1979) and the Fujii equation as described in the Appendix of Niknejad & Rose (1981).
◇ - present 448 K
□ - present 463 K
○ - present 493 K

The present provisional measurements are compared with the same theoretical results in Fig. 6. Agreement with theory is evidently better than for the earlier data. This is consistent with the suggestion that non-condensing gases affected the measurements of Niknejad and Rose (1981) and that non-condensing gas effects have been significantly reduced or eliminated in the present apparatus.

The values of ξ obtained by using equation 1 with the present data were somewhat scattered in the approximate range 0.6 to 0.9. Detailed comparison with theory is deferred pending acquisition of data for lower vapor temperatures and consideration of possible effects of non-saturation vapor conditions, stagnation temperature rise on the vapor thermocouple probes and radiation error in the vapor temperature measurement.

Finally, it should be noted that the conclusion of Niknejad and Rose (1981), that the condensation coefficient exceeds 0.9, was based on extrapolation of their data to zero condensation rate, where non-condensing gas effects would be negligible. The present provisional results suggest higher values and lend support to the view that the condensation coefficient for mercury is, in fact, unity.

REFERENCES

1. Fujii T and Uehara H (1972). "Laminar filmwise condensation on a vertical surface." Int. J. Heat Mass Transfer, Vol.15, pp. 217-233.
2. Ishiguro R and Sugiyama K (1986). "An experimental study on intensive condensation of potassium." Heat Transfer 1986 ed. C.L. Tien, Hemisphere, pp. 1635-1640.
3. Labuntsov D A and Kryukov A P (1979). "Analysis of intensive evaporation and condensation." Int. J. Heat Mass Transfer, Vol. 22, pp. 989-1002.
4. Niknejad J (1979). "An investigation of heat transfer during filmwise and dropwise condensation of mercury." PhD thesis, Univ. London.
5. Niknejad J and Rose J W (1981). "Interphase matter transfer : an experimental study of condensation of mercury." Proc. R. Soc. London A, Vol. 378, pp. 305-327.
6. Schrage R W (1953). "A theoretical study of interface mass transfer." New York: Columbia University Press.

HTD-Vol. 197, Two-Phase Flow and Heat Transfer
ASME 1992

DIFFUSION LAYER THEORY FOR TURBULENT VAPOR CONDENSATION WITH NONCONDENSABLE GASES

P. F. Peterson, V. E. Schrock, and T. Kageyama
Department of Nuclear Engineering
University of California
Berkeley, California

ABSTRACT

In turbulent condensation with noncondensable gas, a thin noncondensable layer accumulates and generates a diffusional resistance to condensation and sensible heat transfer. By expressing the driving potential for mass transfer as a difference in saturation temperatures and using appropriate thermodynamic relationships an effective "condensation" thermal conductivity is derived here. With this formulation, experimental results for vertical tubes and plates demonstrate that condensation obeys the heat and mass transfer analogy, when condensation and sensible heat transfer are considered simultaneously. The sum of the condensation and sensible heat transfer coefficients becomes infinite at small gas concentrations, and approaches the sensible heat transfer coefficient at large concentrations. The "condensation" thermal conductivity is easily applied to engineering analysis, and the theory further demonstrates that condensation on large vertical surfaces is independent of the surface height.

NOMENCLATURE

c molar density
c_p constant pressure specific heat
d diameter
D mass diffusivity
Gr Grashof number, Eq. (20)
h heat transfer coefficient
h_{fg} latent heat of vaporization
k thermal conductivity
L vertical length of surface
M molecular weight
Nu Nusselt number, $h_s L/k_s$
P absolute pressure
q'' heat flux
R universal gas constant
Pr Prandtl number, $c_p\mu/k_s$
Ra Rayleigh number, GrSc or GrPr
Re Reynolds number, $\rho v d/\mu$
Sc Schmidt number, $\mu/\rho D$
Sh Sherwood number, $h_c L/k_c$
T absolute temperature
v velocity
$\tilde{v}$ molar average velocity
v_{fg} difference between vapor and liquid specific volumes
x mole fraction
x_{ave} average mole fraction, Eq. (5)
y coordinate normal to interface
δ diffusion layer thickness
ϕ gas/vapor log mean concentration ratio, Eq. (11)

Subscripts

b bulk
c condensation
g noncondensable gas species
i liquid/vapor interface
s sensible heat
t total
v vapor species
w wall, film, and external resistance
wb wet bulb
∞ bulk cooling medium

Superscripts

s saturation

INTRODUCTION

Noncondensable gases are well known for degrading condensation heat transfer in a variety of condensing geometries. Predicting this degradation is crucial to the design of new systems for safer, passive containment cooling for next-generation nuclear power reactors. In condensation the liquid is practically impermeable to the noncondensable species, so noncondensable accumulates next to the liquid/vapor interface. A balance occurs between the bulk convection of noncondensable gas toward the surface and the diffusion of noncondensable gas from the interface. The balance between convection and diffusion results in a logarithmic gas concentration distribution near the interface. Colburn and Hougen (1934) first proposed the theory that condensation mass transport is controlled by diffusion across a thin layer, or film, with the driving potential across the film given by the difference in the bulk and interface gas partial pressures, divided by the log mean gas partial pressure ratio. As the ambient gas concentration approaches zero the resistance to condensation becomes negligible, the limit one expects.

This work examines two significant classes of

condensation problems. For forced convection, extensive experiments were performed inside a vertical tube, providing detailed measurements of multi-dimensional noncondensable gas distributions. For natural convection, large vertical surfaces were studied. Using the proper formulation of the condensation and sensible heat transfer coefficients, standard heat transfer correlations were found to accurately predict the experimental heat flux measurements. Interestingly, the role of sensible heat transfer was found to be substantial at higher gas concentration, a result credited to mist formation in the saturated vapor/gas mixture. With experiments and analysis Mori and Hijikata (1973) found augmentation by a factor of two due to mist formation.

On vertical surfaces, concentration and temperature gradients develop next to the surface, creating a natural convection boundary layer and augmenting the condensation and sensible heat transfer rate. In the case of laminar boundary layers, Sparrow and Lin (1964) and others have developed analytical treatments, as summarized by Peterson (1992). However, for typical surfaces over 0.25m in height, buoyancy forces are normally sufficient to generate turbulent flow. Traditionally, for vertical surfaces, nuclear applications have used an empirical curve fit of heat transfer coefficient data versus the gas to steam weight ratio measured by Uchida et al. (1965). Recently Kataoka et al. (1991) reconfirmed these results in a 4.2m high apparatus, showing that the condensation heat transfer coefficient follows that measured by Uchida et al. closely when plotted versus the steam to gas weight ratio. Dehbi et al. (1991) performed additional experiments for condensation on the outside of a vertical tube with both air and helium present.

Akers et al. (1960) applied the film model to predict condensation rates on vertical tubes and small vertical surfaces at lower Rayleigh numbers. Corradini (1984) applied a film model to predict the experimental data of Uchida et al. and Tagami (1965) for vertical surfaces. Subsequently Kim and Corradini (1990) improved the Corradini model by adopting the mass transfer coefficient recommended by Bird et al. (1960) for film problems, with the recommended correction factor for wall suction. Kim and Corradini report extensive investigations of film waviness effects and two-dimensional modeling.

Film models of the Colburn-Hougen type can be cumbersome in practice, requiring extensive iterations to match the condensation mass flux to heat transport through the condensate film and external thermal resistances, as outlined by Collier (1981). The complexity of film models has prompted researchers like Henderson and Marchello (1969) to correlate condensation data as the ratio of the experimental heat transfer coefficient, defined as $q_t''/(T_b^s - T_w)$, to the Nusselt solution for pure vapor. This simplified method was employed successfully by Vierow and Schrock (1991) modeling condensation in a natural circulations loop.

Physically, however, condensation is best described by the resistances of the condensate film and noncondensable/vapor mixture placed in series. In the present paper a condensation heat transfer coefficient is defined based on the difference between the interface temperature and bulk saturation (dew-point) temperature, as done by Corradini (1984, 1990). Then to maintain simplicity, a "condensation" thermal conductivity is derived from thermodynamics and the fundamental solution for mass transport in diffusion layers with noncondensables. The resulting condensation thermal conductivity varies from zero at large gas concentrations to infinity at zero gas concentration. When combined with the sensible heat transfer contribution, for fully developed flow excellent prediction of experimental data is obtained with standard heat transfer correlations, over the full range of gas concentrations. Care should be taken in entrance regions, however, as temperature inversions and other phenomena cause large departures from fully-developed behavior (Vierow and Schrock, 1991).

The significant advantage of the condensation thermal conductivity formulation is a follows. Because the condensation heat transfer coefficient is defined based on temperatures, in application it can be combined in parallel with the sensible heat transfer coefficient and in series with the film and wall resistances to determine the overall resistance to heat transfer. Further, using this formulation to reduce experimental data the significant influence of sensible heat transfer and mist formation has been demonstrated.

DIFFUSION LAYER THEORY DEVELOPEMENT

Figure 1 shows typical noncondensable gas mole fraction and temperature profiles for free convection condensation on vertical surfaces, where the gas-species molecular weight is greater than the vapor-species molecular weight. Noncondensable gas accumulates at the liquid/vapor interface, reducing the interface saturation temperature T_i^s below the bulk mixture saturation temperature T_b^s. The difference between the interface temperature T_i^s and the cooling medium temperature T_∞ governs the total heat flux q_t'' through the wall. The heat flux through the film and wall must equal the sum of the flux of latent heat q_c'' and the sensible heat q_s'' through the vapor to the liquid/vapor interface,

$$h_w(T_i^s - T_\infty) = q_t'' = q_c'' + q_s'' = -h_{fg}\, c\, M_v\, \tilde{v}_i + k_v \left(\frac{\partial T}{\partial y}\right)_i \qquad (1)$$

where h_w is an effective heat transfer coefficient combining the condensate film, wall, and external thermal resistances, h_{fg} the latent heat, c the total molar density, M_v the molecular weight of the vapor species, k_v the vapor/gas mixture thermal conductivity, and y the coordinate normal to the surface. For higher temperatures, as may occur with hydrogen combustion, an additional term for radiation is required. The average molar velocity away from the interface, $\tilde{v}_i$, is related to the noncondensable gas mole fraction x_g by Fick's law,

$$c\, v_{gi} = c x_{gi}\, \tilde{v}_i - cD \frac{\partial x_g}{\partial y} \qquad (2)$$

where D is the mass diffusion coefficient. The interface is

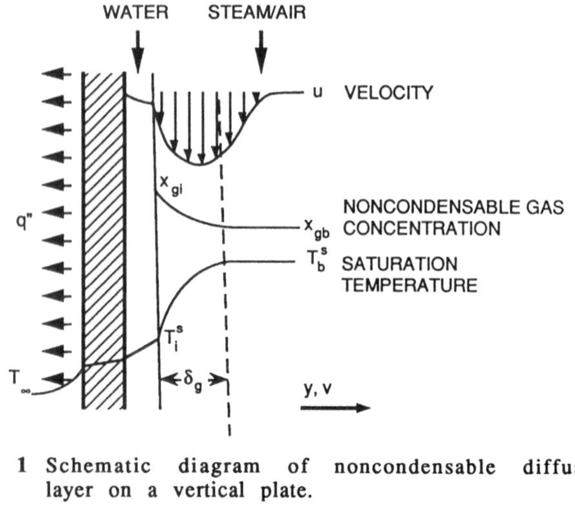

Fig. 1 Schematic diagram of noncondensable diffusion layer on a vertical plate.

impermeable to the noncondensable gas, so the absolute gas species velocity at the interface equals zero, $v_{gi}=0$, and the condensation velocity is

$$\tilde{v}_i = \left(D \frac{1}{x_g}\frac{\partial x_g}{\partial y}\right)_i = \left(D \frac{\partial}{\partial y} \ln(x_g)\right)_i \tag{3}$$

The gradient of $\ln(x_g)$ at the interface can be related to the interface and bulk gas concentrations, x_{gi} and x_{gb}, by considering an effective thickness of the diffusion layer δ_g, shown in Fig. 1, where the thickness δ_g is defined through the relationship

$$\tilde{v}_i = \frac{D}{\delta_g} (\ln(x_{gb}) - \ln(x_{gi})) \tag{4}$$

It is useful to define a log mean mole fraction as

$$x_{ave} = \frac{x_b - x_i}{\ln(x_b/x_i)} \tag{5}$$

Under this definition, $x_b<x_{ave}<x_i$, and the condensation velocity can be written as

$$\tilde{v}_i = \frac{D}{x_{g,ave}\delta_g} (x_{gb} - x_{gi}) \tag{6}$$

Assuming ideal gas behavior, the mole fractions can be expressed as

$$\tilde{v}_i = \frac{D}{P_t x_{g,ave}\delta_g} (P_{vi} - P_{vb}) \tag{7}$$

where P_{vi} and P_{vb} are the partial pressures of the vapor at the interface and in the bulk fluid respectively, and P_t is the total pressure.

The difference of partial pressures in Eq. (7) is not convenient for heat transfer calculations. By expressing the difference in terms of saturation temperatures, then the wall, film, sensible, and condensation heat transfer coefficients can be combined. A modified Clausius-Clapeyron equation, where v_{fg} is an appropriate mean value in the boundary layer, provides a relationship between saturation pressures of Eq. (7) and the saturation temperatures at the interface and in the bulk, T_i^s and T_b^s. Using $\Delta P/\Delta T^s = h_{fg}/T v_{fg}$ and the approximation $v_{fg} = R T_{ave}/M_v x_{v,ave} P_t$, where the log mean vapor concentration $x_{v,ave}$ is defined by Eq. (5), the molar condensation velocity is

$$\tilde{v}_i = \frac{D\, h_{fg} M_v x_{v,ave}}{R\, T_{ave}^2 x_{g,ave}\delta_g} (T_i^s - T_b^s) \tag{8}$$

where M_v is the molecular weight of the vapor species, R the universal gas constant, and $T_{ave}=(T_i^s + T_b^s)/2$ the average temperature in the diffusion layer. For water vapor Eq. (8) gives velocity values within 3% of Eq. (7).

The Sherwood number relates the effective diffusion layer thickness δ_g to the characteristic system dimension L. Combining Eqs. (1), (3), and (8) and replacing the molar density c using the ideal gas law gives

$$Sh = \frac{L}{\delta_g} = \left(\frac{q_c''}{T_i^s - T_b^s}\right) L\, \phi \left(\frac{R^2 T_{ave}^3}{h_{fg}^2 P_t M_v^2 D}\right) \tag{9}$$

The first term on the right side is readily recognized as a condensation heat transfer coefficient. The saturation temperatures determine the driving potential for the condensation heat transfer coefficient h_c, such that

$$h_c = \frac{q_c''}{T_i^s - T_b^s} \tag{10}$$

The third term of Eq. (9) is the gas/vapor log mean concentration ratio, given by

$$\phi = \frac{x_{g,ave}}{x_{v,ave}} = -\frac{\ln((1-x_{gb})/(1-x_{gi}))}{\ln(x_{gb}/x_{gi})} \tag{11}$$

Combined, the third and last terms of Eq. (9) have the inverse units of thermal conductivity, and can be viewed as an inverse *effective condensation thermal conductivity*, $1/k_c$. Because the mass diffusion coefficient varies inversely with pressure and with approximately the square of temperature (Bird et al., 1960), the effective condensation thermal conductivity can be written

$$k_c = \frac{1}{\phi\, T_{ave}}\left(\frac{h_{fg}^2 P_0 M_v^2 D_0}{R^2 T_0^2}\right) \tag{12}$$

where diffusion coefficient D is given in terms of the value D_0 at a reference temperature T_0 and pressure P_0.

The condensation thermal conductivity k_c increases inversely as the gas/vapor log mean concentration ratio ϕ decreases. As required, the condensation thermal conductivity rapidly becomes infinite when the gas concentration reaches zero, and approaches zero as the gas concentration approaches unity. Furthermore, the condensation thermal conductivity is only weakly dependent upon pressure (as is the sensible thermal conductivity for gases), though it decreases as the average temperature T_{ave} increases. The Sherwood number for condensation now takes a simple form,

$$Sh = \frac{h_c L}{k_c} \tag{13}$$

Eq. (1) for the heat flux can now be rewritten as

$$h_w(T_i^s - T_\infty) = q_c'' + q_s'' = h_c\left(T_b^s - T_i^s\right) + h_s\left(T_b - T_i^s\right) \tag{14}$$

where h_c is the condensing heat transfer coefficient given by Eq. (10), and h_s the sensible heat transfer coefficient. Note that the driving potential for the condensing heat transfer coefficient depends on the bulk *saturation* temperature T_b^s, as this saturation temperature gives the driving potential for mass transfer. The driving potential for sensible heat transfer is the *actual* bulk temperature T_b, which allows for superheated vapor conditions.

When the bulk mixture is saturated, the condensation and sensible heat transfer coefficients can be combined into a total heat transfer coefficient, $h_t=h_c+h_s$. The total heat transfer coefficient, which can be measured experimentally, takes the correct limiting behavior with gas concentration. For small gas concentrations, $\phi<<1$, the condensation heat transfer coefficient dominates over the sensible heat transfer coefficient, which remains approximately constant with gas concentration. For large gas concentrations $\phi>>1$ and $h_t \rightarrow h_s$, as condensation becomes negligible.

Because the condensation mass transfer has been expressed in terms of a temperature difference, in engineering applications the heat transfer resistances can be summed in Eq. (14) to eliminate the interface temperature, giving the total heat flux as

$$q_t'' = \frac{h_c\left(T_b^s - T_\infty\right) + h_s\left(T_b - T_\infty\right)}{1 + \frac{h_c + h_s}{h_w}} \qquad (15)$$

With an appropriate correlation for the Nusselt number and initial guess for the interface gas concentration, this formulation can be applied iteratively, converging to the correct heat flux and gas/vapor log mean concentration ratio ϕ. This process is significantly simpler than iteratively matching the mass flux predicted by a mass transfer correlation to the sum of sensible heat transport to the interface and conduction away through the liquid film.

CONDENSATION IN VERTICAL TUBES

Turbulent heat transfer in tubes is commonly correlated in the form

$$Nu = C\,Re^{0.8}Pr^{0.35} \qquad (16)$$

where $Re=\rho v d/\mu$ is the Reynolds number, $Pr=\mu c_p/k$ the Prandtl number, v the mass-average bulk velocity, d the pipe diameter, and μ the viscosity. The scaling coefficient is commonly given the value C=0.023.

In the experiments performed here, the gas/vapor mixture was saturated, so that $T_b=T_b^s$. Using Eq. (16), the condensation and sensible heat transfer coefficients can be expressed as $h_c=C_c(k_c/L)Re^{0.8}Sc^{0.35}$ and $h_s=C_s(k_s/L)Re^{0.8}Pr^{0.35}$. Inserting these into Eq. (14), a total Nusselt number can be defined in terms of experimentally measurable quantities,

$$Nu_t = \frac{q_t'' d/\left(T_b^s - T_i^s\right)}{k_c + \frac{C_s}{C_c}\left(\frac{Pr}{Sc}\right)^{0.35} k_s} = C_c Re^{0.8}Sc^{0.35} \qquad (18)$$

where the Schmidt number is $Sc = \mu/\rho D$ and separate scaling constants C_c and C_s are provided for the condensation and sensible heat transfer coefficients.

The local Nusselt number predicted by Eq. (18) was compared with values measured in an apparatus equipped with a traversing miniature wet-bulb probe. The local gas concentration comes from the total pressure and the saturation pressure determined from the wet-bulb temperature, $x_g=1-P_{wb}^s/P_t$, as described by Peterson and Tien (1987). With the apparatus and probe, shown in Fig. 2, radial and axial gas concentration measurements could be made throughout the condensing test section.

The test section consisted of a 41.3-mm-ID, 4.76-mm-wall pyrex glass tube with an inlet honeycomb flow straightening section. Water circulated in the annulus between the inner and outer glass tubes to cool the 1.0m-long condenser section. Measurements were made in locations at least L/D>10 down the tube, to insure that the local heat transfer and turbulence was close to fully developed. The wall heat flux was determined from temperature measurements on both sides of the glass wall. Because temperature difference across the glass was large (>30°C), error in the heat flux measurement was under 0.5%. The transducer used to measure the total pressure was calibrated by measuring the temperature of pure, saturated steam with the wet-bulb probe. This direct calibration for total pressure allowed gas concentrations to be measured with the wet-bulb probe with an accuracy of ±0.002.

The steam flow rate was determined within 0.5% by volumetric measurements of the condensate flow rate. The gas flow rate was determined within 1.0% using volumetric measurements, capturing the gas in a submerged vessel over a measured time period. The bulk gas concentration was estimated from a mass balance accounting for the upstream condensation rate. Transport properties were evaluated at the arithmetic mean of the bulk and interface temperatures and gas concentrations, using the mixture property values summarized in the appendix.

In the tests, the dominant source of experimental error was the evaluation of the gas/vapor log mean concentration ratio ϕ. For turbulent flow with smaller gas concentrations, the difference between the bulk and interface concentrations becomes small. The ratio ϕ then becomes very sensitive to absolute errors in the interface temperature measurement. The interface temperature was estimated from wall temperature measurements made with a thermocouple pressed against the inside wall. Due to conduction along the leads, the measured temperature fell between the actual wall and interface temperatures. Measurements made with pure steam, where the interface temperature equals the saturation temperature, showed that the interface temperature was consistently 0.4 to 0.7°C higher than the measured wall temperature. Thus the measured wall temperature was increased by 0.3 to 0.7°C, depending on heat flux, giving the interface temperature with an estimated accuracy of ±0.3°C. As shown in Table 1 for typical turbulent flow measurements, the resulting error in the measured Nusselt number was large for small bulk gas concentrations, but the error dropped quickly as the bulk gas concentration increased. The accuracy was better for laminar flow, because the larger condensation resistance resulted in larger differences between the interface and bulk concentrations. This behavior is

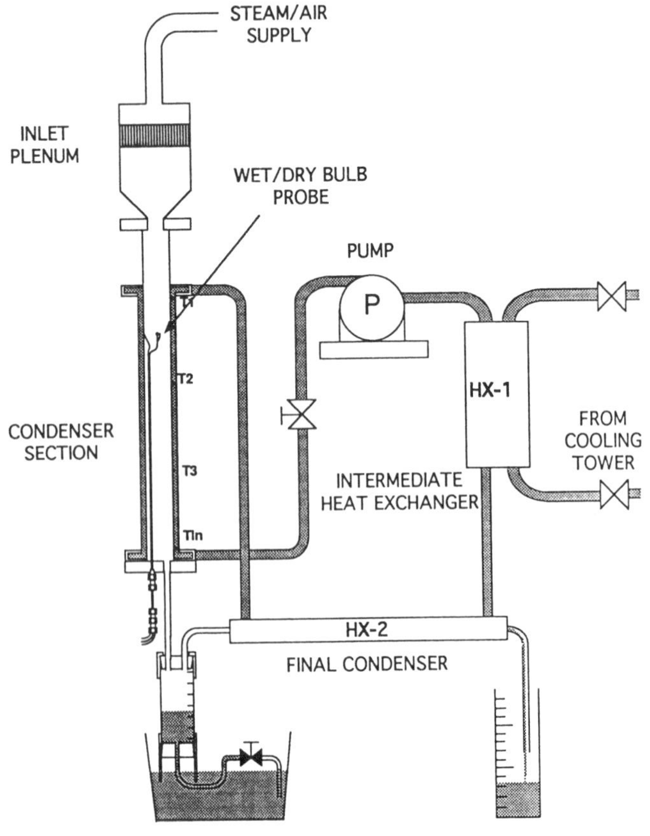

Fig. 2 Schematic of experimental apparatus for condensation in vertical tubes.

reflected in the experimental data, which scatters widely for smaller gas concentrations in turbulent flow, but converges nicely otherwise.

Figure 3 compares experimentally measured Nusselt numbers with the product $Re^{0.8}Sc^{0.35}$. Relatively wide scatter is observed for turbulent flow ($Re^{0.8}Sc^{0.35}>370$) with bulk gas mole fractions less than 0.2, as expected due to the uncertainty in measuring the interface temperature. For $x_{gb}>0.2$ and Re>10,000 excellent agreement is obtained using $C_c=1.2 \times 0.023=0.0276$ and $C_s/C_c=7.0$, giving a standard deviation of 4.7%. For Re>2,000 and $x_{gb}>0.2$, the agreement still gives a standard deviation of 13.2%, showing that the correlation can be applied in the transition region. The factor 1.2 multiplying the standard Dittus-Boelter coefficient is attributed to the effects of surface roughness due to film waviness, and to suction due to condensation. The augmentation of the sensible heat transfer factor C_s by an additional factor of 7.0 is attributed to mist formation as discussed in more detail for vertical plates. Also interesting is the fact that the Nusselt number approaches a constant value of 8.7 for laminar vapor flow with standard deviation 12%, in contrast to the constant value 4.364 obtained analytically for laminar constant heat flux heat transfer in tubes (here the high thermal resistance of the glass resulted in a relatively uniform condensation rate, mimicking constant heat flux behavior). This form of the Dittus-Boelter correlation has also been found to work well with helium as the noncondensable gas, and in tubes of metallic construction, as will be reported in a separate paper.

Table 1 - Typical percentage change in Nusselt number from a ± 0.3°C change in interface temperature for turbulent flow.

x_{gb}	0.033	0.068	0.223	0.309
$\Delta T_i = 0.3$	+79.7	+25.1	+11.3	+4.3
$\Delta T_i = -0.3$	-26.7	-15.4	-9.0	-3.9

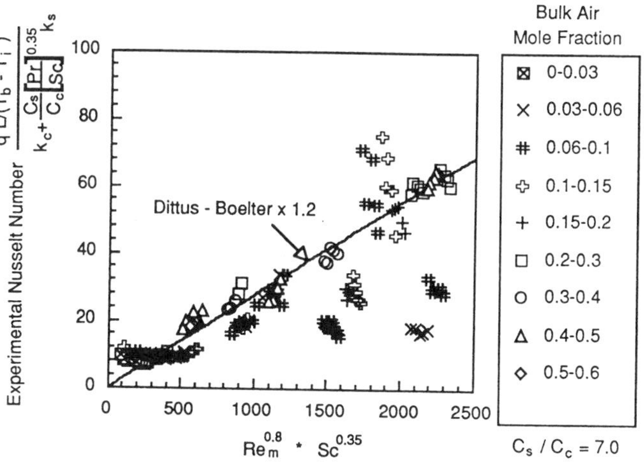

Fig. 3 Experimentally measured Nusselt number for condensation in vertical tubes. Primary data scatter occurs at low gas concentrations.

CONDENSATION ON VERTICAL SURFACES

With natural convection condensation on vertical surfaces, both temperature and concentration gradients contribute to the density difference between the fluid at the surface and the ambient fluid. Assuming ideal gas behavior, the Rayleigh number for mass transfer under natural convection can be expressed as

$$Ra_c = GrSc = \frac{g L^3}{\mu D} (\rho_i - \rho_b)$$

$$= \frac{g L^3 \rho_{vo}}{\mu D} \left\{ \frac{T_0}{T_i} \left[1 + x_{gi}\left(\frac{M_g}{M_v} - 1\right)\right] - \frac{T_0}{T_b} \left[1 + x_{gb}\left(\frac{M_g}{M_v}\right) - 1\right] \right\} \quad (19)$$

where ρ_{vo} is the density of pure vapor at the total pressure P_t and a reference temperature T_0, and M_g and M_v the molecular weights of the gas and vapor species. The average mixture viscosity μ is evaluated at the arithmetic mean of the interface and bulk temperatures and concentrations, using Wilke's method as discussed in the appendix.

For the range of Rayleigh number values $10^9<Ra<10^{13}$, which is typical for condensation on surfaces greater than 0.5m in height, for heat transfer Warner and Arpaci (1968) give

$$Nu = C\,(Ra_s)^{1/3} = C\,(GrPr)^{1/3} \quad (20)$$

where they recommend C=0.10. In this high Rayleigh number, turbulent regime, the Nusselt number increases linearly with the surface height L. As shown later the mass transfer can be treated analogously to heat transfer (using Sh=Nu and $Ra_c=Ra_sSc/Pr$), and thus the condensation heat transfer coefficient is independent of the surface height in this turbulent regime, a result of importance in extrapolating smaller scale experiments to full scale containment modelling.

Kataoka et al. (1991) performed extensive condensation experiments on the 4.2m high wall of a water-wall simulator, obtaining Rayleigh numbers ranging from $9 \times 10^{10}<Ra_c<1 \times 10^{12}$. They measured the initial gas pressure, total pressure, and bulk and wall temperatures. The heat flux through the wall was measured with thermocouples on either side of the wall. Table 2 of the appendix summarizes their experimental measurements. Here the total Nusselt number calculated from the data of Kataoka et al. is compared with the Rayleigh number to the 1/3 power, where the Nusselt number is

$$Nu_t = \frac{q_t'' L / \left(T_b^s - T_i^s\right)}{k_c + \frac{C_s}{C_c}\left(\frac{Pr}{Sc}\right)^{0.33} k_s} = C_c(Ra_c)^{0.33} = C_c(GrSc)^{0.33} \quad (21)$$

and separate scaling constants C_c and C_s are provided for the condensation and sensible heat transfer coefficients. Here the effect of the condensate film resistance is neglected, as calculations for laminar films show it to be less than less than 5% of the total resistance. The bulk gas concentration x_{gb} was calculated from the total pressure, bulk temperature, and initial pressure, assuming ideal gas behavior. The interface concentration was calculated from the saturation pressure based on the wall temperature and total pressure.

Figure 4 shows a comparison of the total Nusselt number (Eq. 21) and condensation Rayleigh number from the data of Kataoka et al.. Using the standard scaling coefficient $C_c=0.10$ for condensation and a larger value $C_s=0.7$ for sensible heat transfer, the ratio of the experimental total heat transfer coefficient to the predicted value scatters by a maximum of 7%, with a standard deviation of 4.0%. Figure 5 shows the relative contributions of condensation and sensible heat transfer as the average gas concentration changes. As the gas concentration becomes small, $\phi \rightarrow 0$, the condensation heat transfer coefficient becomes large. For gas mole average ratios $\phi > 6$ (bulk mole fraction $x_{gb} > 0.8$) sensible heat transfer begins to contribute more than half of the total heat transfer.

The scaling coefficient for condensation, $C_c = 0.10$, matches the value found for heat transfer (Warner and Arpaci, 1968). However the sensible heat scaling coefficient is larger, $C_s/C_c = 7.0$, is almost an order of magnitude greater than that found for dry sensible heat transfer. This effect is attributed to mist formation due to the subcooling of the saturated steam/gas mixture induced by sensible heat transfer, which substantially increases the effective specific heat in the term $Pr/Sc = \rho c_p D/k_s$. Mist formation has been observed experimentally in this type of system by Mori and Hijikata (1973). Dehbi et al. (1991) visually observed gas mixture layer movement in their experiments, noting a drifting motion toward the wall that would imply the presence of mist. Applying an integral boundary layer technique, they showed that mist formation augmented the total heat transfer by up to a factor of 2, implying a larger factor for the sensible heat transfer contribution. No data is available, but it would be expected that C_s/C_c would approach unity as the gas-vapor mixture becomes superheated.

Uchida et al. (1965) performed a series of experiments on a 0.3m high by 0.14m wide plate. Tagami (1965) performed experiments with vertical cylinders 0.3 and 0.9m in height and 0.64m in diameter inside a 3.3m ID, 6m high, $42 m^3$ volume containment. The surfaces were cooled to maintain at a constant temperature of 322K. The experiments covered a Rayleigh number range from $1 \times 10^7 < Ra_c < 6 \times 10^{10}$. The vessels were filled with an initial quantity of air, nitrogen, or argon, and then steam was introduced. By increasing the steam flow rate in increments, the total pressure was increased, reducing the bulk gas concentration. The inlet and outlet cooling water, condensing surface, and vapor bulk temperatures were measured to determine the heat flux and overall heat transfer coefficient, but unfortunately only graphical results for the heat transfer coefficient versus bulk gas mass ratio were presented, so the vapor temperature and (for Uchida et al.) total pressure must be inferred by assuming saturated conditions and ideal gas behavior, with an initial gas pressure of 1.0bar.

Figure 6 compares Nusselt numbers versus Rayleigh numbers from data measured by Uchida et al. (1965) and Tagami (1965). Agreement is good, except for the data of Tagami for the 0.9m high cylinder. The ratio of the experimental total heat transfer coefficient to the predicted value scatters by a maximum of 46%, with a standard deviation of 23%. The transition to turbulence is known to occur between $10^8 < Ra < 10^{10}$, and the correlation form $Ra^{1/3}$ is often used down to 10^8. Here this form was found to work well for $Ra > 10^7$.

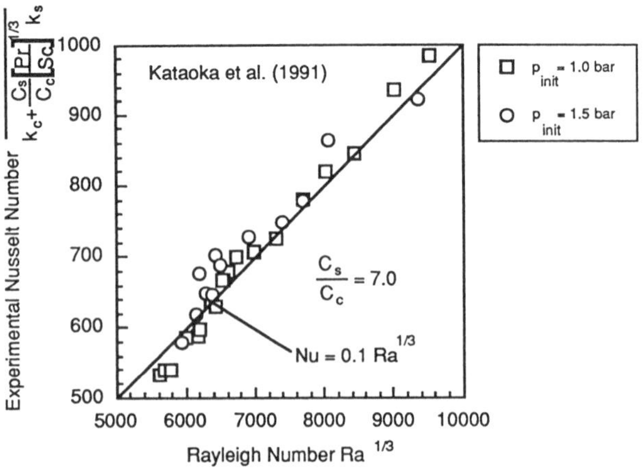

Fig. 4 Comparison of calculated total heat transfer coefficient with data of Kataoka et al. (1991).

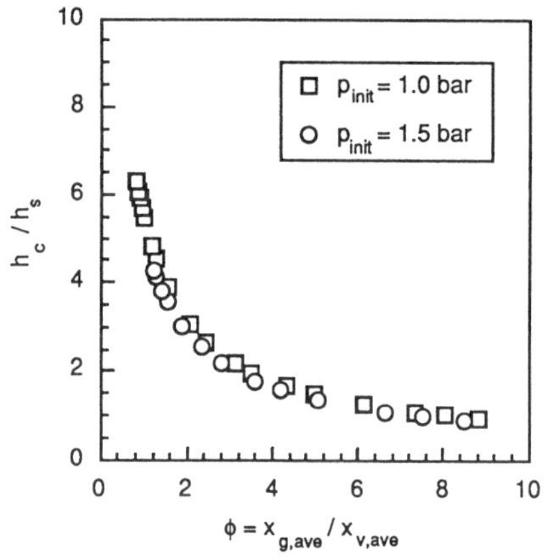

Fig. 5 Relative contributions of condensation and sensible heat transfer as a function of average gas concentration.

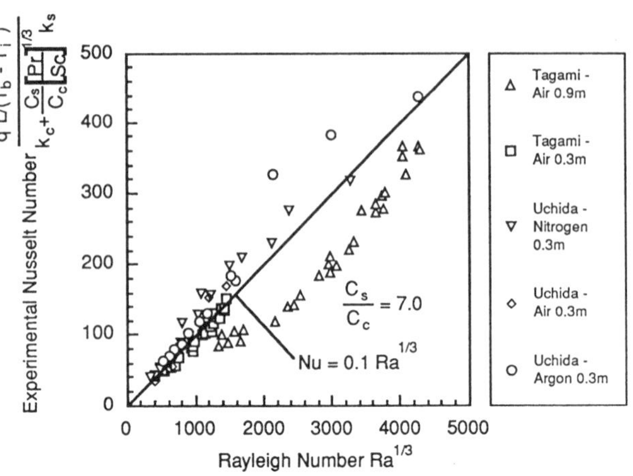

Fig. 6 Comparison of calculated total heat transfer coefficient with data of Uchida et al. (1965) and Tagami (1965).

CONCLUSIONS

Beginning with the fundamental mass transfer equations and boundary conditions for condensation with noncondensable gases, an effective "condensation thermal conductivity," Eq. (12), was derived. With this simple parameter, combined sensible and condensation heat transfer can be predicted using the standard forms for heat transfer correlations.

For turbulent forced convection heat transfer (Re>2000) from saturated vapor/gas mixtures in vertical tubes, the total heat transfer coefficient can be calculated using Eq. (18) with C_c=0.0276 and C_s/C_c=7.0. For laminar flow (Re<2000) the Nusselt number is constant and equals 8.7. For turbulent natural convection heat transfer ($Ra>10^7$) from saturated vapor/gas mixtures onto vertical flat plates and cylinders, the total heat transfer coefficient can be calculated using Eq. (21) with C_c=0.10 and C_s/C_c=7.0. As the vapor/gas mixture becomes superheated, C_s/C_c can be expected to drop toward unity as mist formation is reduced. For conservative design, C_s/C_c=1.0 should be used for all cases.

ACKNOWLEDGEMENTS

The support of the National Science Foundation under Presidential Young Investigator grant CTS-9057258 is sincerely appreciated.

REFERENCES

Akers, W.W., Davis, S.H., and Crawford, J.E., 1960, "Condensation of a Vapour in the Presence of Non-Condensing Gas," *Chem. Engng. Prog. Symp. Series* No. 30, Vol 56, pp. 139-144.

Bird, R.B., Stewart, W.E., and Lightfoot, E.N., 1960, *Transport Phenomena*, John Wiley and Sons, New York, pp. 511, 658-671.

Colburn, A.P., and Hougen, O.A., 1934, "Design of Cooler Condensers for Mixtures of Vapours With Non-Condensing Gases," *Ind. Engng. Chem.*, Vol. 26, No. 11, pp. 1178-1182.

Collier, J.G., 1981, *Convective Boiling and Condensation*, 2nd Ed., McGraw-Hill, New York, pp. 323-328.

Corradini, M.L., 1984, "Turbulent Condensation on A Cold Wall in the Presence of a Noncondensable Gas," *Nuclear Technology*, Vol. 64, pp. 186-195.

Dehbi, A.A., Golay, M.W., and Kazimi, M.S., 1991, "The Effects of Noncondensable Gases on Steam Condensation Under Turbulent Natural Convection Conditions," Report No. MIT-ANP-TR-004, Massachusetts Institute of Technology.

Henderson, C.L., and Marchello, J.M., 1969, "Film Condensation in the Presence of a Noncondensable Gas," *Journal of Heat Transfer*, Vol. 91, pp. 447-450.

Kataoka, Y., Fukui, T., Hatamiya, S, Nakao, T., Naitoh, M., and Sumida, I, 1991, "Experimental Study on Convection Heat Transfer Along a Vertical Flat Plate Between Different Temperature Pools," National Heat Transfer Conference, Minneapolis, Minn., ANS Proceedings, Vol. 5, pp. 99-106.

Kim, M.H., and Corradini, M.L., 1990, "Modeling of Condensation Heat Transfer," *Nuclear Engineering and Design*, Vol. 118, pp. 193-212.

Mori, Y. and Hijikata, K., 1973, "Free Convective Condensation Heat Transfer with Noncondensable Gas on a Vertical Surface," *International Journal of Heat and Mass Transfer*, Vol. 16, pp. 2229-2240.

Peterson, P.F. and Tien, C.L., 1987, "A Miniature Wet-Bulb Technique for Measuring Gas Concentrations in Condensing or Evaporating Systems," *Experimental Heat Transfer*, Vol. 1, pp. 1-15.

Peterson, P.F., 1992, "Species-Controlled Condensation in Enclosures," *Annual Review of Heat Transfer*, Ed. C.L. Tien, Hemisphere, New York, pp. 247-295.

Reid, R.C., Prausnitz, J.M., and Poling, B.E., 1988, *The Properties of Gases and Liquids*, McGraw-Hill, New York, pp. 121, 407, 531.

Sparrow, E.M., and Lin, S.H., 1964, "Condensation Heat Transfer in the Presence of Noncondensable Gas," *Journal of Heat Transfer*, vol. 86, pp. 430-436.

Tagami, T., 1965, "Interim Report on Safety Assessments and Facilities Establishment Project for June 1965," No. 1, Japanese Atomic Energy Research Agency, unpublished work (See Corradini, 1984).

Uchida, H., Oyama, A., and Togo, Y., 1965, "Evaluation of Post-Incident Cooling Systems of Light-Water Power Reactors," Proceedings of the Third International Conference on the Peaceful Uses of Atomic Energy, Geneva, Aug. 31-Sept. 9 1964, Vol. 13, United Nations, New York, pp. 93-104.

Vierow, K.M., and Schrock, V.E., 1991, "Condensation in a Natural Circulation Loop with Noncondensable Gases Part I-Heat Transfer," Proc. Intl. Conf. on Multiphase Flows, Tsukuba, Japan, pp. 183-186, September.

Warner, C.Y., and Arpaci, V.S., "An Experimental Investigation of Turbulent Natural Convection in Air at Low Pressure along a Vertical Heated Flat Plate, *International Journal of Heat and Mass Transfer*, Vol. 11, pg. 397, 1968.

Appendix - Data and Property Values

This section summarizes the experimental measurements of Kataoka et al (1991) and the property values used to reduce the experimental data.

Table 2 - Experimental measurements for vertical surface (Kataoka, 1991).

Vapor Temp (°C)	Wall Temp. (°C)	Heat Flux (kW/m^2)	Total Pres. (bar)
P_{init}=1.0bar			
59.90	41.50	1.12	1.35
62.40	43.90	1.20	1.38
65.00	46.20	1.28	1.41
68.90	50.30	1.52	1.48
72.90	55.70	1.60	1.57
76.70	60.20	1.70	1.63
81.30	66.20	1.89	1.74
84.40	69.90	1.94	1.81
90.60	77.60	2.20	1.98
94.90	83.20	2.30	2.12
101.10	92.60	1.95	2.35
106.40	99.00	1.95	2.56
109.30	101.70	2.25	2.73
114.30	107.00	2.48	3.00
116.60	108.90	2.93	3.15
117.30	109.10	3.28	3.22
120.10	111.00	3.88	3.38
122.80	112.60	4.94	3.57
126.30	115.00	6.00	3.80
P_{init}=1.5bar			
67.40	54.90	0.83	1.98
69.30	57.00	1.02	2.04
72.40	60.80	0.94	2.09
78.70	68.00	1.06	2.23
83.80	73.60	1.13	2.35
87.90	78.10	1.29	2.45
94.80	86.00	1.34	2.66
99.50	90.90	1.55	2.86
106.20	97.80	1.81	3.17
112.00	104.30	1.98	3.48
115.30	107.20	2.45	3.67
121.90	111.00	3.79	4.08
124.50	112.50	4.76	4.30

Due to the multiple species and relatively large temperature differences, the evaluation of transport properties is important in problems of condensation with noncondensables. Here property values were evaluated at the arithmetic mean of the interface and bulk temperatures and gas mole fractions. The schemes recommended by Reid et al. (1988) for evaluating mixture properties, as listed below, were applied here.

Gas/Vapor Viscosity (Wilke's method)

$$\mu_{water} = (0.1483\sqrt{T(K)} - 1.620) \times 10^{-5} \ \text{kg/m·s} \quad (A.1)$$

$$\mu_{air} = (0.1641\sqrt{T(K)} - 1.000) \times 10^{-5} \ \text{kg/m·s} \quad (A.2)$$

$$\mu_{N_2} = (0.3773\sqrt{T(K)} - 4.972) \times 10^{-5} \ \text{kg/m·s} \quad (A.3)$$

$$\mu_{Ar} = (0.2177\sqrt{T(K)} - 1.5089) \times 10^{-5} \ \text{kg/m·s} \quad (A.4)$$

$$\mu_{He} = (0.1543\sqrt{T(K)} - 0.6472) \times 10^{-5} \ \text{kg/m·s} \quad (A.5)$$

$$\mu_{mixture} = \frac{x_g\mu_g}{x_g + (1-x_g)\phi_v} + \frac{(1-x_g)\mu_v}{x_g\phi_g + (1-x_g)} \quad (A.6)$$

$$\phi_v = \frac{\left[1 + \left(\frac{\mu_g}{\mu_v}\right)^{1/2}\left(\frac{M_v}{M_g}\right)^{1/4}\right]^2}{\left[8\left(1 + \frac{M_g}{M_v}\right)\right]^{1/2}} \quad (A.7)$$

$$\phi_g = \frac{\left[1 + \left(\frac{\mu_v}{\mu_g}\right)^{1/2}\left(\frac{M_g}{M_v}\right)^{1/4}\right]^2}{\left[8\left(1 + \frac{M_v}{M_g}\right)\right]^{1/2}} \quad (A.8)$$

Gas/Vapor Thermal Conductivity (Mason and Saxena modification)

$$k_{water} = 0.00308\sqrt{T(K)} - 0.03548 \quad \text{W/m°C} \quad (A.9)$$

$$k_{air} = 0.00277\sqrt{T(K)} - 0.02166 \quad \text{W/m°C} \quad (A.10)$$

$$k_{N_2} = 0.00267\sqrt{T(K)} - 0.02005 \quad \text{W/m°C} \quad (A.11)$$

$$k_{Ar} = 0.00105\sqrt{T(K)} - 0.00184 \quad \text{W/m°C} \quad (A.12)$$

$$k_{He} = 0.01056\sqrt{T(K)} - 0.03295 \quad \text{W/m°C} \quad (A.13)$$

$$k_{mixture} = \frac{x_g k_g}{x_g + (1-x_g)\phi_v} + \frac{(1-x_g)k_v}{x_g\phi_g + (1-x_g)} \quad (A.14)$$

where ϕ_v and ϕ_g come from Eqs. (A.7) and (A.8).

Gas/Vapor Specific Heat:

$$c_{p,water} = 2.494 - 0.00114(T(K)) \quad \text{kJ/kg°C} \quad (A.15)$$

$$c_{p,air} = 0.9808 + 0.00008(T(K)) \quad \text{kJ/kg°C} \quad (A.16)$$

$$c_{p,Ar} = 0.52 \quad \text{kJ/kg°C} \quad (A.17)$$

$$c_{p,He} = 5.2 \quad \text{kJ/kg°C} \quad (A.18)$$

$$c_{p,mixture} = x_g\, c_{p,g} + (1-x_g)\, c_{p,v} \quad (A.19)$$

Mass diffusivity D_0 (p_0=101.3kPa, T_0=298K):

$D_{water/air} = 0.256 \times 10^{-4}$ m²/s $D_{water/N_2} = 0.264 \times 10^{-4}$ m²/s

$D_{water/Ar} = 0.257 \times 10^{-4}$ m²/s $D_{water/He} = 0.838 \times 10^{-4}$ m²/s

Latent heat of vaporization (water):

$h_{fg} = 3184.68 - 2.486(T(K))$ kJ/kg

Water Vapor Density (ideal gas, p_0=101.3kPa, T_0=298K):

$\rho_{vo} = 0.750$ kg/m³

Water Saturation Pressure:

$\ln(P^s(\text{bar})) = 13.48243 - 5028.91/T(K)$

HTD-Vol. 197, Two-Phase Flow and Heat Transfer
ASME 1992

CONDENSING ECONOMIZERS: THERMAL PERFORMANCE AND PARTICULATE REMOVAL EFFICIENCIES

Thomas A. Butcher
Brookhaven National Lab.
Upton, New York

Noh Park
Stony Brook Scientific, Ltd.
Norristown, Pennsylvania

Wai Lin Litzke
Brookhaven National Laboratory
Upton, New York

ABSTRACT

Condensing economizers can be used to increase the thermal efficiency of boilers and furnaces. This project has involved a study of these specifically for application to coal-water mixture fuels although the results can be extended to other fuels. Experimental studies to evaluate thermal performance and removal of particulates across indirect contact economizers have been performed. The test arrangement incorporates oil firing with the injection of flyash into the flue gas to simulate coal combustion products. Water sprays into the combustion products are used to achieve variable flue gas moisture content and a variable amount of condensation in the economizers. The economizers are tubular with flue gas on the outside of the tubes. Tube surfaces are plastic coated to prevent corrosion.

The gas temperature and condensation profiles through the economizers have been predicted and overall predicted performance has been compared with test results. Mechanisms for particle removal are discussed and predicted removal efficiencies as a function of particle diameter are presented. It is shown that inertial impaction is the dominant mechanism and particle removal efficiencies up to 89% have been realized.

NOMENCLATURE

English Letter

A_1, A_2, A_3,	=	constants dependent upon units
A_f	=	$1/(2Ku)$
C_c	=	Slip correction factor
D_p	=	Brownian diffusivity (Stokes-Einstein relation) $kTC_c/3\pi\mu_g d_p$
d_p	=	Particle diameter
d_w	=	heat exchanger tube diameter
h_{fg}	=	water vapor latent heat of vaporization
h_i	=	heat transfer coefficient between film and cooling fluid. Includes inside convective coefficient and thermal resistance of the tube and Teflon coating
h_o	=	outside convection coefficient
K_m	=	water vapor mass transfer coefficient
Ku	=	α-3/4-α^2/4-1/2lnα = Kuwabara No. for Kuwabara flow field
k	=	Boltzmann constant
P	=	total pressure (roughly atmospheric)
P_c	=	critical pressure
P_e	=	Peclet No = $d_w U_\infty / D_p$
P_g	=	atmospheric pressure
P_{wb}	=	water vapor partial pressure in the bulk flue gas
Stk	=	Stokes No. = $\rho_p\, d^2_p U_\infty/(g\mu_g d_w)$
Stke	=	effective Stokes No. = Stk·non-Stokes particle drag correction factor
P_{wc}	=	water vapor partial pressure (saturation) at the condensate temperature
T	=	absolute temperature
T_b	=	flue gas bulk temperature
T_c	=	condensate film temperature
T_f	=	cooling fluid temperature
U_∞:	=	free stream gas velocity

Greek Letter

α	=	volume fraction of heat exchanger tubes
μ_g	=	gas viscosity
ρ_p	=	particle density

INTRODUCTION

Condensing economizers are used to improve the thermal efficiency of small boilers and furnaces. In these systems, the flue gas temperature is reduced below the dew point of the water vapor. Latent heat is recovered and system efficiency increases markedly. Some smaller heating equipment is available with the condensing heat exchanger integral. In other cases a condensing economizer may be added onto an existing boiler or furnace. This is exclusively the case with commercial and industrial scale equipment and both direct and indirect contact condensing economizers are commercially available. With condensing systems flue gas exit temperatures of about 38 C and boiler or furnace efficiencies (based on gas analysis only) of 95% are not uncommon [1,2].

To date, condensing heat exchangers have been applied primarily to gas-fired equipment. However, there have also been some applications to oil- and wood-fired boilers. Relative to other fuels, the condensate from natural gas is less corrosive. In addition, gas combustion products have higher moisture contents which leads to increased water vapor dew points and increased energy efficiency improvements with condensation. Dew points with 30% excess air for gas, oil, and coal firing are 56, 48, and 41 C, respectively (10% moisture in coal as fired).

An important factor in the application of condensing economizers is the utility of heat at the low temperatures needed for condensation. In the residential sector most of the available products are for gas-fired warm air furnaces. In warm air heating systems the return air temperature to the furnace is generally less than 27 C. Hydronic heating systems, in contrast, have return water temperatures well above 49 C and generally above the dew point of the water vapor in the flue gas. Application of condensing economizers to hydronic systems requires oversized radiators in the heated spaces to reduce the return water temperature. Circulating water to preheat building make-up air has also been used. In the commercial and light industrial sectors preheating of make-up or domestic water is a common application [1,2]. In many applications the flue gas temperature after the economizer is slightly higher than the dew point. These "near-condensing" systems must be built to withstand corrosion resulting from transient or local condensation. Corrosion of condensing heat exchanger surfaces has been an area of considerable emphasis and stainless steels, plastics and glass have been used [3,4].

In addition to removing water vapor, condensing economizers remove particulates from the flue gas. In one field test of a boiler plant fired with #6 fuel oil the particulate removal efficiency across the condensing economizer was found to be 70% [5]. Particle removal is dependent primarily upon the following unit mechanisms:

1. Inertial Impaction
2. Interception
3. Diffusion
4. Particle growth due to nucleation and condensation of water vapor
5. Thermophoresis
6. Stefan Flow

In any practical heat exchanger these are affected by factors including: flue gas temperature and moisture content, heat exchanger surface temperature, excess air, total surface area, gas side pressure drop, and heat exchanger configuration.

In some configurations water sprays may be used, either upstream of the condensing heat exchanger in a presaturator section or in the heat exchanger itself. The sprays can enhance the particulate removal by impacting with the particles or simply increasing the amount of water condensed in the heat exchanger. There has been a considerable amount of work done on the use of saturator/condenser combinations for particle growth and collection.

This study is aimed at the application of condensing economizers to coal water slurry-fired systems with an emphasis on enhancing the particulate removal. The work in this program is specifically aimed at residential, commercial, and light industrial applications where there would be the most likely need for the low temperature heat source.

Figure 1 shows a general sketch of a condensing heat exchanger fit to a small coal slurry-fired boiler. At the exit from the conventional heat exchanger section there is a saturator section, where water sprays are used to increase the flue gas moisture content at the expense of temperature. Figure 2 shows the effect of water spray rate in the saturator on the flue gas temperature and flue gas water dew point, assuming coal slurry firing and an inlet gas temperature of 232 C. Saturation of the flue gas would be expected to increase condensation rates and particulate removal due to some of the mechanisms discussed above. The heat available in the economizer is degraded by presaturation and the possible benefits of sprays in particle removal must be weighed against this.

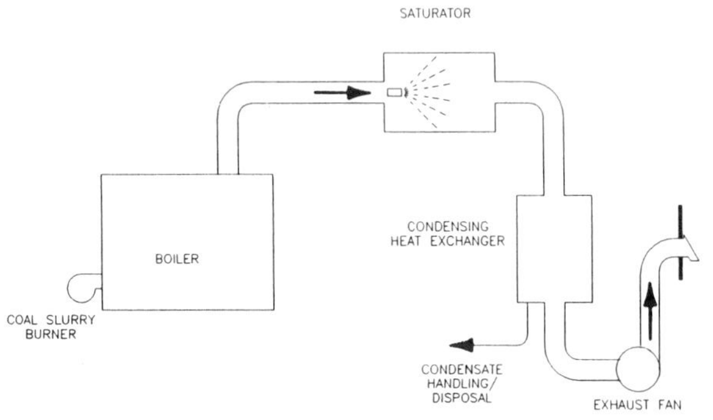

Figure 1. General Condensing Heat Exchanger Concept

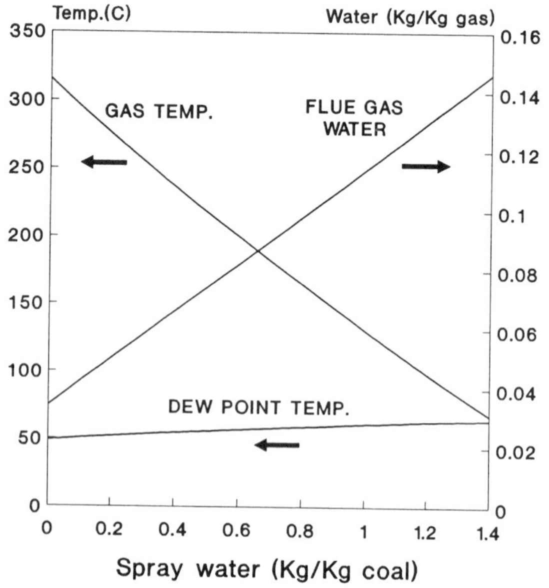

Figure 2. Saturator Spray Water Effects

Following the saturator the flue gas passes through the condensing economizer and is then exhausted through a "wet stack". In many residential applications plastic exhaust vents are used and, in some cases, these may exhaust horizontally through building walls.

This paper presents experimental results on thermal performance and particulate removal efficiency using a condensing economizer. These results are then compared with predictions.

EXPERIMENTAL

A sketch of the experimental arrangement used is provided in Figure 3. All measurements and testing have been performed firing No. 2 fuel oil and the experiments were set up to simulate coal-fired conditions with respect to the concentration of fly ash found in the flue gas. Fly ash is introduced into the combustion chamber with a spray assembly that injects an ash/water slurry. Flue gas moisture content is adjusted by injecting water into the flue gas upstream of the economizer. EPA Method 5 procedures are used for particulate sampling from the flue gas upstream and downstream of the economizer. With simultaneous sampling at both points the removal efficiencies under different operating conditions can be obtained. Details of the equipment setup and the procedures used are summarized below.

Heating System

The boiler is a sectional cast iron hot water boiler and has a capacity rating of 586,000 W. All of the tests were done at a firing rate of 82,040 W or less firing No. 2 fuel oil with a pressure atomized, retention head burner. The boiler unit was modified such that the combustion area, including most of the top section, is lined with refractory boards. At such low firing rates maintaining desired flue gas temperature requires that only three of the boiler heat exchanger passages remain open; the other passages are blocked with refractory material. The amount of heat exchanger surface which is effective can be adjusted to trim boiler exit temperature.

Combustion gases leaving the boiler are directed through the condensing economizer using a draft induced, side-wall power venter system. The piping consists of a series of 13 cm diameter flue piping upstream of the economizer and 10 cm diameter flue piping downstream.

A bypass system with dampers was set up to control the volume of flue gas entering the economizer. Flue gas velocities were measured at the inlet and outlet of the condenser using a pitot tube.

Condensing Economizers

Both air and water-cooled condensing economizers have been used in tests. The air cooled unit is a cross-current heat exchanger, constructed of aluminum and the passages and tube side that come in contact with the flue gas are Teflon coated to resist corrosion. Figure 4 provides an illustration of this air-cooled heat exchanger as well as the specifications relevant to thermal performance. The heat exchanger consists of two sections, one stacked on top of the other; each section contains 8 rows with each row containing 9 tubes across. The tubes are 2.5 cm in diameter and are arranged in a staggered order. A 1/3 hp blower is connected to the economizer to draw cooling air through the tubes at an average bulk velocity of 2.16 m/s.

The water cooled heat exchanger, illustrated in Figure 5, is much longer that the air cooled exchanger but has a smaller cross section. The total surface area in the water cooled unit is considerably less. Gas velocities in the water cooled exchanger are about 5 times greater than in the air cooled unit and this should improve particle removal due to inertial impaction.

Water and Ash/Slurry Injection System

Fly ash was obtained from the precipiator hopper at a pulverized coal-fired utility boiler. This material is then mixed with water to make a slurry which is injected directly into the combustion chamber using an air siphon nozzle. The system is mounted at the back end of the furnace, which is the end opposite that of the burner.

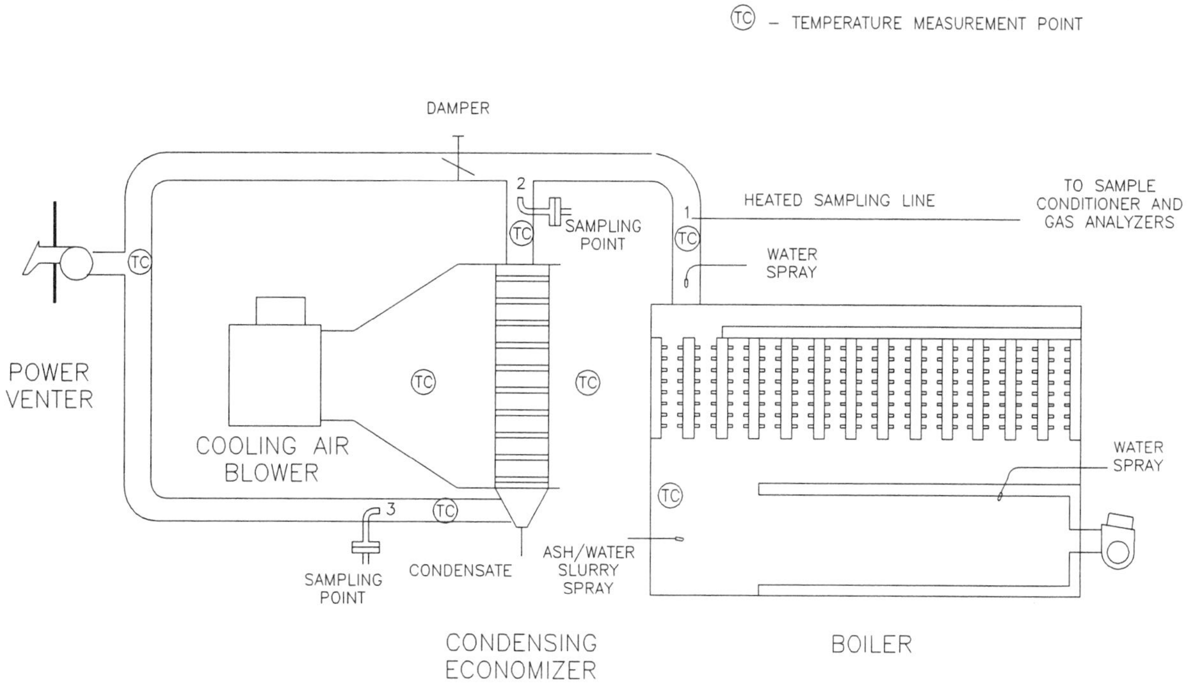

Figure 3. Experimental Arrangement

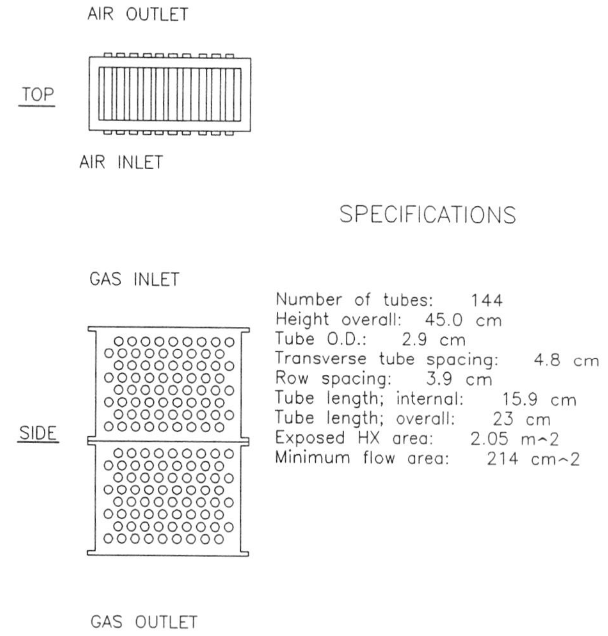

Figure 4. Air-Cooled Condensing Heat Exchanger

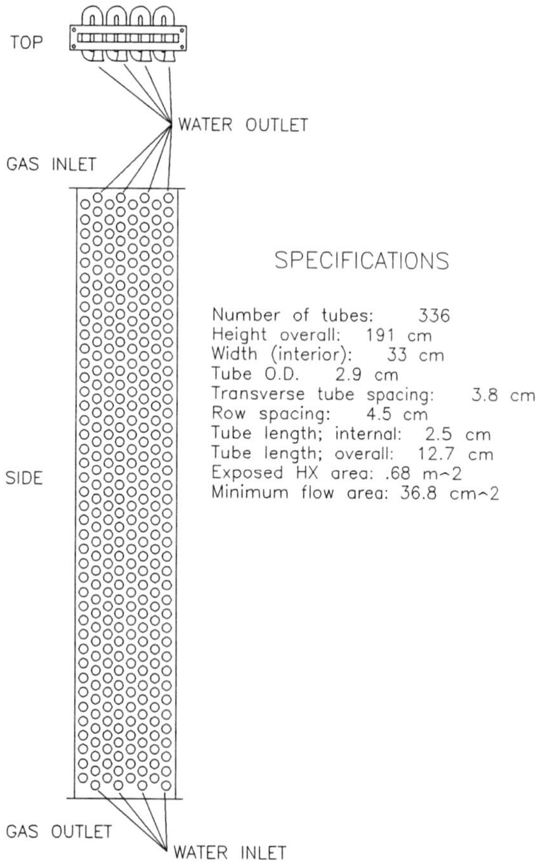

Figure 5. Water-Cooled Condensing Heat Exchanger

To modify the flue gas water vapor content, water can also be introduced directly into the combustion chamber or into the flue pipe, upstream of the economizer, using the same type of spray system. The locations of the spray points used have been selected based on numerous tests in which the water injection rate and the flue gas moisture content were measured. In some configurations the flue gas water content was found to be less than predicted, indicating that not all of the water was evaporating but was instead collecting on duct walls or tops of heat exchanger surfaces. In the configurations used for particulate tests, a good mass balance for the spray water was realized.

Sampling Equipment and Locations

The basic sample components and data acquisition equipment consist of the following:

a. Thermocouples/ Data Logger
b. Sample Gas Conditioner
c. Gas Analyzers
d. Particulate Sampling Trains - modified EPA Method 5

Thermocouples were installed to measure the flue gas temperature changes from the boiler exit to the vent termination and also economizer cooling water or air at the inlet and outlet. Multiple point temperature measurements were made at the cooling fluid outlet in both cases to monitor differences in the parallel flow paths. In the case of the water cooled economizer, the water flow rate was monitored continuously.

Refer to Figure 3 for the locations of three sampling points in the case of the air cooled economizer. The oxygen content of the flue gas is measured at sample point 1, which is located at the boiler outlet, and is analyzed using a paramagnetic oxygen analyzer.

Sampling points for particulate collection are located upstream and downstream of the economizer, sampling points 2 and 3 respectively. These sampling points are also used to collect water vapor to determine the moisture content of the flue gas. EPA Methods 4 and 5, [6] with modifications to the sampling train, were used as reference procedures for collecting the data. In most cases, the measurements upstream and downstream of the condensing heat exchanger were made simultaneously using two sampling trains.

HEAT TRANSFER DESIGN

The method developed by Colburn and Hougen [7,8,9] has been used to calculate the latent and sensible heat transfer rates, the water vapor condensation rates and the temperature profiles through the heat exchangers. This general method for evaluating the condensation of a single vapor from a non-condensible gas has been specifically applied to flue gas condensing economizers by Razgatis, et al. [10].

A heat balance is prepared where heat flow from the flue gas to the condensate film is equal to the heat flow from the film to the cooling fluid. This heat balance, together with an expression for the relationship between temperature and partial pressure for the water vapor provide relations which can be solved for the unknown - the condensate film temperature. Sensible and latent heat flows follow directly. The approach involves several assumptions including: uniform tube temperature along the length of each tube, negligible thermal resistance of the condensate film, partial pressure of water vapor above the condensate film equal to the equilibrium pressure at the condensate temperature.

The heat balance can be written as:

Sensible Heat Transferred From Flue Gas to the Tube Surface	+	Latent Heat Transferred From Flue Gas the Tube Surface	=	Heat to the Cooling Fluid

or:

$$h_0 \cdot (T_b - T_c) + h_{fg} \cdot K_m \cdot P_g \cdot \ln\left(\frac{P - P_{wc}}{P - P_{wb}}\right) = h_i \cdot (T_c - T_f) \quad (1)$$

For given flue gas and cooling fluid temperatures and flue gas moisture content, the above equation contains two unknowns, P_{wc} and T_c. A relation between these is used to complete the system of equations and Antoine's formula [11] has been used.

The outside film coefficient (h_0) is dependent upon gas properties, flow rates, and row number. Correlations by Kays and London [12] have been used (see also [10]). The water vapor mass transfer coefficient has been calculated by analogy with the outside convection coefficient. The inside film coefficient has been calculated using relations from Kays and London [12]. With the air-cooled case this is complicated by a transition Reynolds number (around 3300 in our experimental arrangement) and a small value of tube length/diameter ratio (6.3). With both the air and the water-cooled exchangers, the length of the tube is significantly longer than the width of the flue gas passage. The part of the tube which extends beyond the duct walls (not in contact with the flue gas), contributes significantly to the effective internal heat transfer resistance. This is particularly important in the case of the air-cooled exchanger, where forced convection from the inside of these tube ends and natural convection from the outside of these ends add 50-70% to the heat transfer rate.

Calculations were done on each section of the heat exchanger assuming constant properties across each section. For the results shown in this paper, each section included two rows of tubes. Calculations with more and less rows in each section did not indicate a significant difference in the overall heat exchanger performance.

The thermal performance of the heat exchangers is defined by the rates of sensible and latent heat transfer, which affect the temperature and moisture content of the exiting flue gas. The performance of both of the heat exchangers included in this program has been predicted using the methodology described above. In this section the predicted thermal performance of the exchangers is presented for two specific cases. The first case assumes that a coal water slurry fuel is being fired with 65% concentration of a Pennsylvania bituminous coal. Flue gas enters the condensing heat exchanger at 177 C and there is no water spray in the saturator. The second case assumes the same fuel but uses a water spray to cool the flue gas down to 85 C before entering the heat exchanger.

Figure 6 shows the predicted performance of the air-cooled heat exchanger in the first case. Included are the temperature profile of the gas as it passes through the exchanger, the flue gas mole fraction of water vapor, and the temperature of the condensate film on the outer surface of the tubes ("Cond. Temp."). Also presented are the rates of heat transfer (sensible, latent and total) on a per-row basis. Overall, in this case, the heat transfer rate is 4,340 W in the exchanger. At the assumed firing rate, 82,100 W, the efficiency improvement across the exchanger is 5.8%. Nearly all of this is sensible heat transfer, however, and the rate of moisture condensation is very small. Examination of the water fraction curve in Figure 6 shows that the surface falls below the dew point, and water begins condensing only at the very end of the heat exchanger.

Figure 7 shows similar curves for the second case for the air-cooled exchanger, where the flue gas has been cooled in the saturator before entering the condensing heat exchanger. Here latent heat transfer is much more important and most of the heat exchanger rows are wet. Overall, the heat transfer rate is considerably less than in the first case, only 3,020 W. This provides an efficiency increase of 3.6%.

The performance of the water-cooled heat exchanger in Case 1 is shown in Figure 8. For all row numbers greater than 13 the surface is below the dew point and latent heat contributes significantly to the total heat transfer rate. Clearly this is doing a much better job of both cooling the gas and removing water vapor than the air cooled case. Overall the heat transfer rate is 7,330 W and the efficiency improvement is 8.9%.

The water-cooled heat exchanger, Case 2, is presented in Figure 9. Again, the addition of water to the flue gas increases the importance of latent heat in the exchanger and all of the surfaces are wet. As in the air-cooled case the addition of water to the flue gas reduces the total heat transfer rate but the effect is not nearly as great. A total of 6,150 W is transferred in this case for an efficiency improvement of 7.4%

PARTICULATE REMOVAL PREDICTION

In this section the mechanisms which could be expected to influence particulate removal across a condensing economizer are discussed and estimates of the particulate removal efficiency as a function of particle size are presented for the specific economizers used in this work. The following assumptions are made in this analysis:

a. The gas flowing around the heat exchanger tube can be treated as a continuum.

b. The heat-exchanger tube is treated as an isolated cylinder in cross-flow, with forced-convection dominating natural convection.

c. The sticking efficiency is unity because of the condensate on the wall of the heat exchanger tubes.

d. The slip correction factor is introduced into the Brownian diffusivity (Stokes-Einstein relation).

e. To obtain efficiencies of tube bank heat exchangers we can determine the particle collection efficiency of an isolated single cylinder and then combine these for the entire exchanger.

* Inertial impaction

Inertial impaction occurs when a particle is unable to follow the rapidly curving streamlines around an obstacle and, because of its inertia, continues to move toward the obstacle along a path of less curvature than the flow streamlines. Thus, collision occurs because of the particle's momentum. Note that the mechanism of inertial impaction is based on the premise that the particle has mass but no size, whereas interception is based on the premise that the particle has size but no mass.

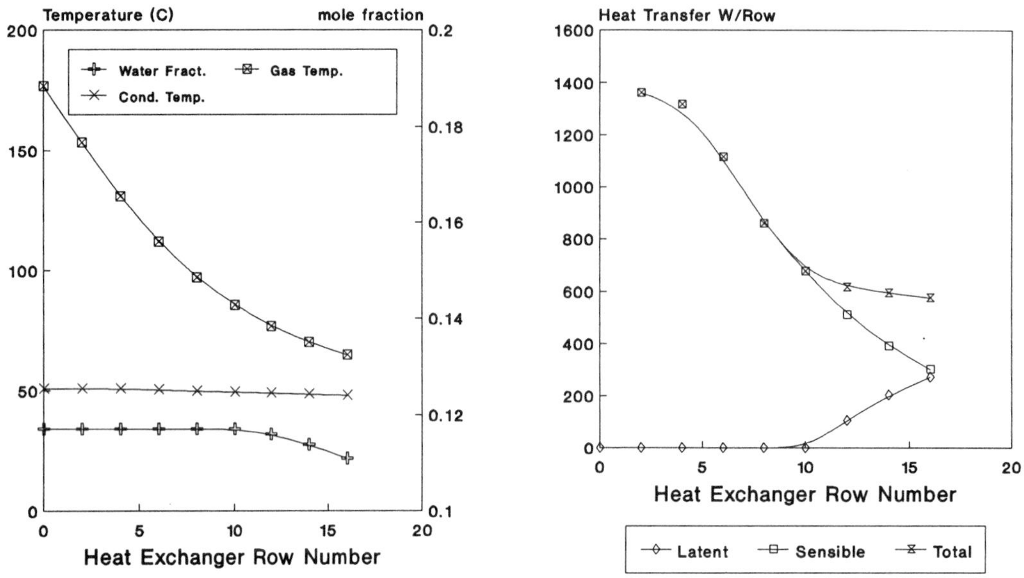

Figure 6. Predicted Performance of Air-Cooled Condensing Heat Exchanger - Case I

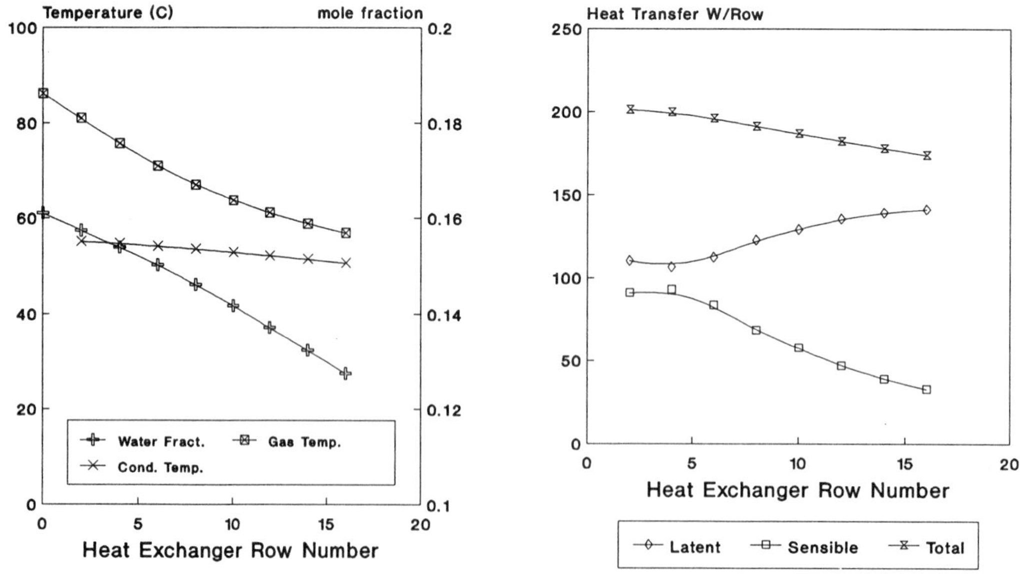

Figure 7. Predicted Performance of Air-Cooled Condensing Heat Exchanger - Case II

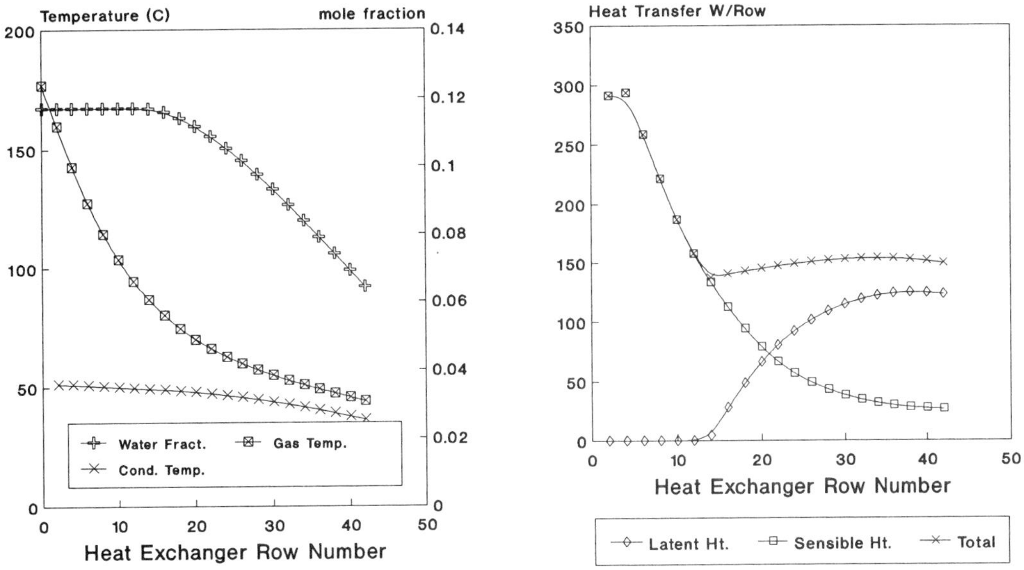

Figure 8. Predicted Performance of Water-Cooled Condensing Heat Exchanger - Case I

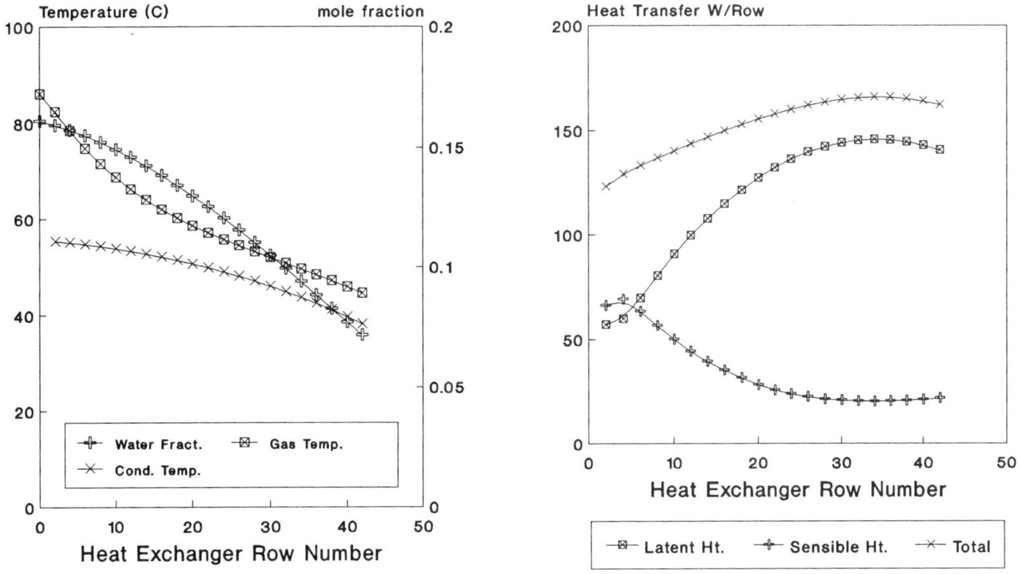

Figure 9. Predicted Performance of Water-Cooled Condensing Heat Exchanger - Case II

* Interception

Interception takes place when a particle, following the streamlines of flow around a cylinder, is of size sufficiently large that its surface and that of the cylinder come into contact. Thus, if the streamline on which the particle center lies is within a distance $d_p/2$ of the cylinder, interception occurs.

* Brownian diffusion

Particles in a gas undergo Brownian diffusion that will bring some particles in contact with the cylinder due to their random motion as they are carried past the cylinder by the flow. A concentration gradient is established after the collection of a few particles and acts as a driving force to increase the rate of deposition over that which would occur in the absence of Brownian motion. Because the Brownian diffusivity of particles increases as particle size decreases, the removal mechanism will be most important for very small particles. When analyzing collection by Brownian diffusion, the particles are treated as diffusion massless points.

* Thermophoresis

Heavy molecules and particles in a temperature gradient experience a force (usually in the direction from hot to cold, which produces a net transport called thermophoresis. Thermophoretically driven mass transport is important in some gaseous systems (e.g., flame "structure" and surface combustion of hydrogen (air) and most aerosol systems (soot and ash deposition on cooled heat-exchanger surfaces, sampling targets, and/or turbine blades) as shown in Rosner and Fernandez dela Mora [13].

For each of the mechanisms discussed above a collection efficiency can be defined. For a single cylinder this efficiency is the rate of particle collection divided by the mass flow rate of particulates through an area defined by the tube diameter and length. Table 1 presents the summary of relations used for isolated single cylinder collector efficiencies for four different collection mechanisms. In this analysis the effects of Stefan flow and particlegrowth due to condensation have not been included. These would be expected to provide some increase in particle collection.

For a single cylinder, the combined collection efficiency due to all of the effects considered is:

$$\eta_c = 1-(1-\eta_B)\cdot(1-\eta_{TH})\cdot(1-\eta_{IN})\cdot(1-\eta_{IM}) \quad (2)$$

Using this the collection efficiency for the first row of tubes is:

$$\eta_{1st} = \eta_c \cdot \frac{total\ tube\ proj.\ area}{total\ cross\ section\ area} \quad (3)$$

For a heat exchanger with n rows the collection efficiency can be calculated as:

$$\eta_{nrows} = 1-((1-\eta_{1st})\cdot(1-\eta_{2nd})\cdot\ldots\ldots(1-\eta_n)) \quad (4)$$

The predicted collection efficiency for both the air and the water cooled economizers used in this work have been estimated using the approach above as a function of particle size. The heat transfer conditions used are those shown in Figures 6 and 8. Based upon these conditions, the particle collection efficiency for four different mechanisns as shown in Table 1 were evaluated. Figure 10 shows the collection efficiency as a function of the particle size for the air cooled case. The particle collection efficiency from interception is not shown because the value is negligebly small. For larger particles inertial impaction is clearly dominant. For the smaller particles thermophoresis is shown to greatly increase the particle collection efficiency over isothermal diffusion although the removal efficiency for the small particles is very small. The water-cooled condenser as shown in Figure 11 clearly presents the greater potential for particle removal and this is primarily due to the increased gas velocities around the tubes.

EXPERIMENTAL RESULTS

Results are presented for the air and water-cooled units in Tables 2 and 3, respectively. Along with the measured outlet temperatures and flue gas moisture contents values for these parameters predicted using the methods discussed earlier are presented.

With regard to heat transfer performance the calculation method discussed in the previous section did fairly well in terms of predicting the outlet gas temperature and the condensation rate. The particulate removal efficiency was found to range between 42 and 66% for the air cooled case and was found to be 89% in the water cooled case. The particle removal efficiency shows no significant trend with the inlet flue gas moisture content or the total amount of water vapor condensed. Increasing the flue gas water vapor content does not appear to yield any significant benefit in particulate collection.

DISCUSSION

An estimate of the particle removal efficiency which might be expected in these heat exchangers can be made using a simple "cut diameter" approach. In this approach the particle diameter for which a 50% particle removal efficiency is expected is identified and then it is assumed that all particles larger than this size are removed and all small particles pass through. In the case of the air-cooled condenser the cut diameter is about 23 microns (Figure 10). Based on typical size distributions for pulverized coal flyash [18] 50 to 70% of the ash is smaller than this size. Removal efficiency then would only be expected to be about 30 to 50%. The present results show somewhat higher removal efficiencies.

For the water-cooled condenser the cut diameter is about 7 microns. Based on this and typical flyash size distribution the removal efficiency would be expected to range from 60 to 80%. Tests results presented above indicated 89%.

The predictions of particle removal efficiency presented in this paper have neglected the consideration of some mechanisms, including particle growth due to condensation and Stefan flow. These may be leading to somwhat greater particle removal than predicted. Generally, however, the removal of particles appears to be still dominated by simple inertial impaction and consideration of this alone can be used for at least preliminary estimates of performance.

Table 1. Summary of Isolated Single Collection Efficiencies

	Mechanism	Efficiency	References	Remarks
1.	Brownian diffusion	$\eta_B = 3.68 A_f^{1/3} P_e^{2/3}$	Friedlander [15]	Diffusing mass less points
2.	Thermophoresis	$\eta_{TH} = \eta_B \cdot F(\text{Suction})$	Rosner [16]	Hot to Cold
3.	Interception	$\eta_{IN} = 2A_f \left(\frac{d_p^2}{d_w}\right)$	Flagan et al. [17]	Neglect of particle inertia $(d_p/d_w \ll 1)$
4.	Inertial Impaction	$\eta_{IM} = 0.01978749 \ln(8\,Stke) + 0.5136545 (Stke-0.125) - 0.0482858 (Stke-0.125)^2$	Wessel et al. [14]	0.125<Stke<0.5
		$\eta_{IM} = \{1+1.54424 (Stke-0.125)^{-1} - 0.538013 (Stke-0.125)^{-2} + 0.2020116 (Stke-0.125)^{-3}\}^{-1}$		0.5<Stke

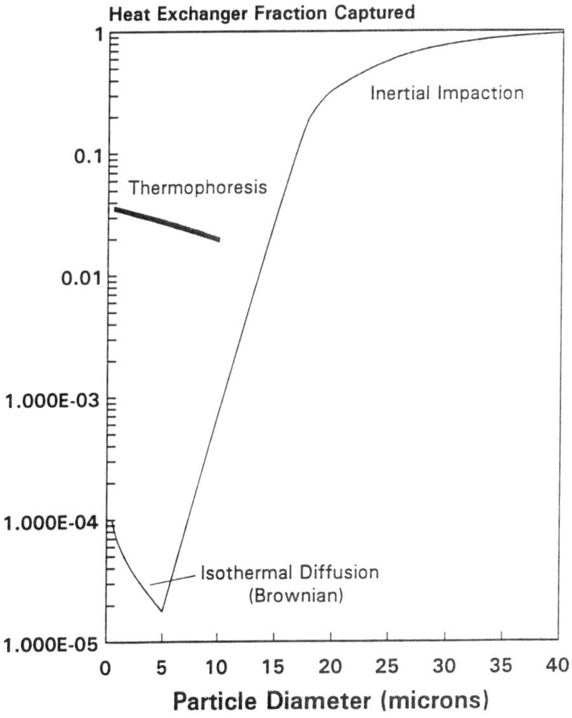

Figure 10. Results of Prediction of Particulate Capture with Air-Cooled Condenser

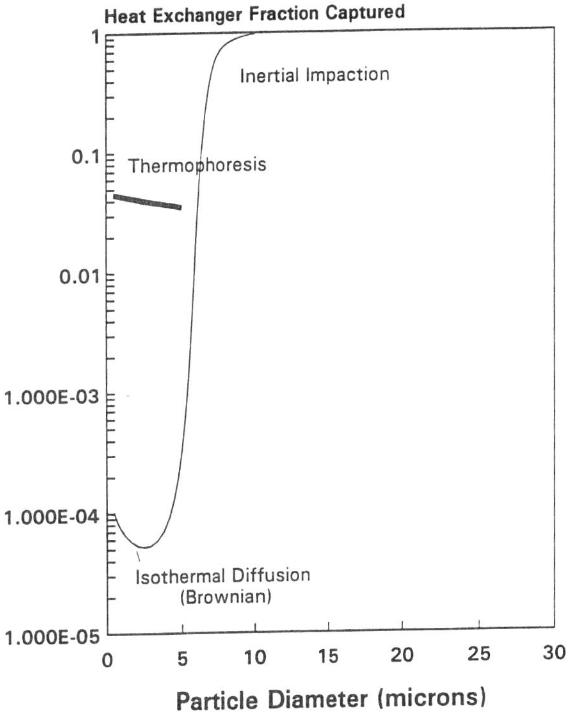

Figure 11. Results of Prediction of Particulate Capture with Water-Cooled Condenser

TABLE 2 - AIR COOLED CONDENSING ECONOMIZER

EXPERIMENTAL DATA

Run	1	2	3	4	5	6	7	8	9
Firing rate, W	82,000	82,000	82,000	62,000	62,000	62,000	62,000	62,000	62,000
% Ash in slurry	10	10	10	10	10	10	10	10	10
Slurry Spray Rate cm^3/s	1.0	1.0	1.0	1.08	1.05	1.05	1.05	1.37	1.26
Particulate Conc. Inlet, gm/dscm	0.706	0.847	0.777	0.742	0.600	0.494	0.512	1.45	1.59
Particulate Conc. Outlet, gm/dscm	0.388	0.494	0.332	0.247	0.212	0.212	0.282	0.586	0.561
Moisture, Inlet % (measured)	24	15	14	ND	22	22	20	14	13
Moisture, Outlet % (measured)	22	10	11	ND	20	17	15	12	12
Moisture, Outlet % (calculated)	20.4	14.3	13	___	18.9	16.5	14.7	12.1	
Particle Removal Efficiency	45	42	57	66	63	55	45	60	65
Flue gas temperature Inlet, C (measured)	177	191	193	127	143	81	81	127	133
Flue gas temperature Outlet, C (measured)	74	71	74	54	66	56	52	53	71
Flue gas temperature Outlet, C (calculated)	77	71	69	___	71	61	58	60	

ND - No data

TABLE 3 - WATER COOLED CONDENSING ECONOMIZER

EXPERIMENTAL DATA

Run	1	2
Firing rate, W	62,000	62,000
% Ash in slurry	10	10
Slurry Spray Rate, cm^3/s	0.74	0.84
Particulate Concentration Inlet, gm/dscm	0.625	0.516
Particulate Concentration Outlet, gm/dscm	0.0660	0.0576
Moisture, Inlet % (measured)	11	19
Moisture, Outlet % (measured)	6	7
Moisture, Outlet % (calculated)	5.1	7.6
Particle Removal Efficiency, %	89	89
Flue gas temperature Inlet, C (measured)	146	97
Flue gas temperature Outlet, C (measured)	45	44
Flue gas temperature Outlet, C (calculated)	39	46

CONCLUSIONS

The thermal performance and particulate removal efficiency of condensing economizers applied to flue gas from coal water slurry-fired boilers or furnaces has been evaluated. With regard to heat transfer performance, the calculation method developed by Colburn and Hougan was found to reasonably predict outlet temperature and water vapor content. The mechanisms which could be expected to influence particulate removal across a condensing economizer are discussed and estimates of the particle removal efficiency as a function of particle size are presented for the specific economizers studied. Isolated single cylinder collector efficiencies for diffusion, thermophoresis, interception, and inertial impaction are evaluated. Inertial impaction is found to be the dominant mechanism for flyash with particle diameters in the range of 1 to 50 microns. For heat exchangers with multiple tube rows, the collection efficiency can be approximated by extending the single tube results. Collection efficiencies were found to be somewhat higher than predicted for typical flyash, indicating that some mechanisms not included in the prediction may be contributing. Inertial impaction, however, can still be used to at least roughly predict performance.

REFERENCES

1. Applications Manual-Condensing Boilers, The Chartered Institution of Building Services Engineers, London, 1989.

2. Thompson, R., Goldstick, R., Taback, H. and Vogt, R., Flue Gas Condensation Heat Recovery for Industrial Boilers, International Symposium on Industrial Fuel Technology, American Flame Research Committee, Chicago, Oct. 1981.

3. Proceedings-Symposium on Condensing Heat Exchangers, Atlanta, Georgia, 1982, Gas Research Institute Report GRI-82/0009.3.

4. Ball, D.L., White, E., Lux, J. Jr., and Locklin, D. Condensing Heat Exchanger Systems for Oil-Fired Residential/Commercial Furnaces and Boilers, Phase I and II, Prepared by Battelle Columbus Laboratories, Brookhaven National Laboratory Report BNL 51617, Oct. 1982.

5. Dietz, R., Butcher, T. and Wieser, R., Emissions and Thermal Efficiency at a Flue Gas Condensing Economizer, Brookhaven National Laboratory Report BNL 51456. June 1981.

6. U.S. Code of Federal Regulations, Title 40, Part 60, Appendix A. U.S. Government Printing Office, Washington, D.C., 1990.

7. Colburn, A.P. and Hougen, O.A. Design of Cooler Condensers for Mixtures of Vapors with Non-Condensing Gases, Inc. Eng. Chem., Vol. 26, pp. 1178-82, 1934.

8. Kern, D.Q. Process Heat Transfer, pp. 340-51, McGraw-Hill, New York, 1950.

9. Heat Transfer and Fluid Flow Data Books, Gemini Publishing Co., 1990.

10. Razgatis, R., Ball, D., Lux, J., Jr., White, E., Markle, R., Bradbury, E., Sekreioglu, I. and Locklin, D. Condensing Heat Exchanger Systems for Residential/Commercial Furnaces and Boilers, Phase III. Prepared by Battelle Columbus Laboratory. BNL 51770, Feb. 1984.

11. Henley, E.J. and Seader, J.D. Equilibrium-Stage Separation Operations in Chemical Engineering, John Wiley & Sons, New York, 1981.

12. Kays, W. and London, A. Compact Heat Exchangers, Third Ed., McGraw-Hill, New York, 1984.

13. Rosner, D.E. and Fernandez de la Mora, J. Small particle transport across turbulent nonisothermal boundary layers," ASME Trans.-J. Engineering for Power, Vol. 104, pp. 885-894, 1982.

14. Wessel, R.A. and Righi, T. "Generalized correlations for inertial impaction of particles on a circular cylinder," Aerosol Science and Technology, Vol. 9, pp. 29-60, 1988.

15. Friedlander, S.K. "Smoke, dust and haze: Fundamentals of aerosol behavior," Wiley, New York, 1977.

16. Rosner, D.E. "Transport processes in chemically reacting flow systems," Butterworth- Meinemann, Boston, 1986.

17. Flagan, R.C. and Seinfeld, J.H. "Fundamentals of air pollution engineering," Prentice Hall, New Jersey, 1988.

18. Lowe, A. "The radiative properties and effects of fly ash particles," in Combustion of Pulverized Coal the Effect of Mineral Matter, The University of Newcastle, 1979.

HTD-Vol. 197, Two-Phase Flow and Heat Transfer
ASME 1992

CRITICAL HEAT FLUX OF TWO-PHASE THERMOSYPHONS WITH AN EXTERNAL DOWN-FLOW CHANNEL

I. Pioro
Institute of Engineering Thermophysics
Ukrainian Academy of Sciences
Kiev, Ukraine

A. Panek and Y. Lee
Department of Mechanical Engineering
University of Ottawa
Ottawa, Ontario, Canada

ABSTRACT

To improve the maximum heat transfer capacity of a two-phase closed thermosyphon, additional structural components such as internal separate channels, external down flow channels, etc., are often added to thermosyphons to separate the opposing flows of vapour and condensate.

There are several studies made on the critical heat flux on such thermosyphons with additional structural elements but they are mostly limited to the cases with those having internally inserted separate flow channels.

In the present paper, we present the experimental study made on the critical heat fluxes of a two-phase closed thermosyphons with an external down flow channel.

NOMENCLATURE

c_{pf}	specific heat, J/(kg K)
D	internal diameter of thermosyphon, m
G_f	mass velocity of liquid phase alone, kg/(m^{2}s)
g	acceleration due to gravity, m/s^2
i_{fg}	latent heat of vaporization, J/kg
L	length, m
q	heat flux, W/m^2
T	temperature, C (K)
u_f	velocity in tube assuming total flow is liquid, m/s
θ	title angle
ρ	density, kg/m^3
σ	surface tension, N/m
Fr	Froude number, $u_f^2/(g\{\sigma/[g(\rho_f - \rho_g)]\}^{0.5})$
K	Kutateladze number, $q_{cr}/\{i_{fg}\rho_g^{0.5}[\sigma g(\rho_f - \rho_g)]^{0.25}\}$

SUBSCRIPTS

c	condenser section
cr	critical
e	evaporator section or heating zone
f	liquid
g	vapour
sub	subcooling

INTRODUCTION

Two-phase closed thermosyphons are widely used in many engineering practices as heat transfer devices of high efficiency (Negishi et al., 1985; Vasilliev, 1990). The most of such systems used in the industry are of counter-flow configuration, and its thermal capacity is mainly limited by the relatively low values of its critical heat fluxes (CHF).

To maximize the heat transfer capacity of such counter-flow two-phase closed thermosyphons, additional structural components such as internal separate flow channels, external down flow channels, etc., are often added to thermosyphons to separate the opposing flows of vapour and condensate (both of these thermosyphons with additional flow channels are called in this paper as **separate-flow two-phase thermosyphons**).

The thermosyphons with external down flow channels have a number of advantages over those with the internally inserted flow channels. The formers are more adaptable to cost-effective production and require less materials because they need only one external condensate flow line.

There are many engineering thermal devices which utilize the principle of fluid circulation by natural buoyancy forces in a closed loop. These include some boilers of special configuration and various equipments used in the chemical industry. Because of this historical background, CHFs for thermosyphons with additional structural components have been calculated, assuming that the case of boiling of liquids circulating in tubes is directly applicable to their systems (Bezrodny et al., 1984; Kai, 1990). In practice, however, this assumption is seldom correct, especially for the cases where fluid velocity and pressure are high. CHF is also strongly affected by many parameters which can vary widely for different devices (Collier, 1980; Kutateladze, 1990; Whalley, 1987).

There are several studies made on the critical heat flux on such thermosyphons with additional structural elements to separate flows of vapour and condensate within the thermosyphons but they are mostly limited to the cases with those having internally inserted separate flow channels (Pioro et al., 1988).

In this paper, we present the results of an experimental study on the critical heat flux of two-phase closed thermosyphons with external down flow channels.

EXPERIMENTAL APPARATUS AND PROCEDURES

A schematic diagram of the experimental system used in the study is shown in Fig. 1 and the physical dimensions of the four thermosyphons used in the experiment as well as the ranges of the experimental parameters are given in Table 1. These values were chosen to reflect the practical aspect of the application of such thermosyphon systems.

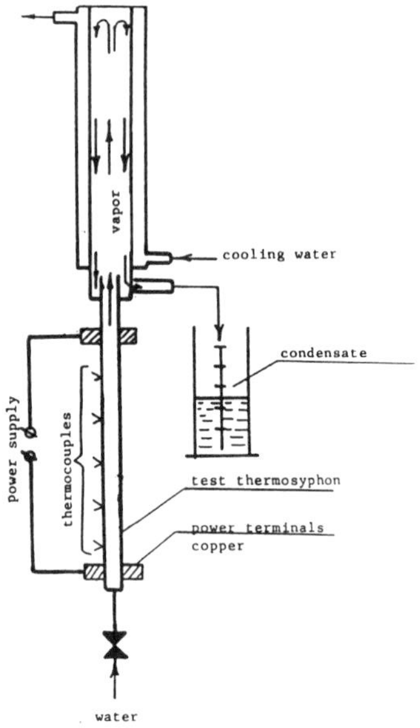

Figure 1 Experimental apparatus

The evaporator section was directly heated by the electrical resistance of the evaporator section. Provision was made so as to vary the length of heating section, L_e. The whole apparatus was completely covered with thick insulation material. The working fluid was delivered to the evaporator section by a pump. The condensate from the condenser section was removed and its flow rate was measured.

TABLE 1 Physical Characteristics of Test Section and the Ranges of Experimental Parameters

D_c = 24 mm; L_c = 0.975m; Tube Material = s.s. 304

T_f = 283 K; ΔT_{sub} < 90 K;

D, mm	3	7	14	19
L_e, m	0.153 - 0.31	0.19 - 0.73	0.262	0.6 - 1.13
u_f, m/s	0.09 - 0.46	0.007 - 0.25	0.0026 - 0.048	0.0028 - 0.073
ρu_f, kg/(m²s)	90 - 460	100 - 250	2.6 - 48	2.8 - 73

Working fluid - deminelized water

Condenser pressure - atmospheric

θ (with respect to the horizontal) - 90°

A number of chromel-alumel (K-type) thermocouples were used for the measurement of the outer surface temperature of the test thermosyphons and for the outer surface temperature of the thermal insulation of the apparatus.

DC electric power input was controlled by a variable transformer and determined from separately measured voltage and current.

The usual experimental procedure for the measurement of the critical heat flux was used; i.e. the power input to the evaporator section was increased stepwise until a sudden sharp increase in the wall temperature was noticed and the power input at the moment was recorded. The circulating flow rate of water was calculated from the measurement of the condensate flow rate which was removed from the condenser section at the time as close to the moment the dry-out was noticed.

ANALYSIS OF THE EXPERIMENTAL DATA

The analysis of the experimental data requires appropriate selection of variables which, for the present case, would affect the critical heat flux. From the many well known correlations, obtained from theoretical and experimental studies of the critical heat flux for the flow boiling in tubes and especially for the cases of counterflow two-phase closed thermosyphons, it was relatively easy to deduce the main parameters which would affect the critical heat flux for the closed two-phase thermosyphons with an external down-flow channel. These include the physical characteristics of the evaporator section of the test thermosyphons, the circulation flow velocity and some thermodynamic properties of the working fluid such as latent heat of evaporation, densities of liquid and vapour, surface tension etc. The system pressure was taken into account by changes in the respective thermodynamic properties of the working fluid.

The functional relationship among these variables of the present experimental data were processed through the following procedure.

If we assume that a class of function is given by f(x, p) where x is a vector of an independent variable and p a vector of a unknown parameter, then we can write the following:

$$J = \sum_{i=1}^{n} [y_i - f(x_i, p)]^2 \qquad (1)$$

where y_i are the measured values of the function $f(x_i, p)$ and n the number of measurements. Now the minimalization procedure can be applied to Eq. (1) in order to find the parameters. An iteration process that unconditionally converges to the global minimum can de used to solve this problem. The iteration procedure may be described by an equation:

$$p^{(s)} = p^{(s-1)} - q^{(s-1)} \upsilon \, [\, \nabla_p J \,]^{(s-1)} \qquad s = 1, 2, \ldots \qquad (2)$$

and begins with some number q^0 which is decreased until the following condition is met:

$$| J^{(s)} - J^{(s-1)} | < q^{(s-1)} \upsilon \, [\, \nabla_p J^{(s-1)} \,]^2 \qquad (3)$$

where υ is a constant between 0 and 1, and ∇_p denotes the gradient of Eq. (1) with respect to the parameters p. We terminate the iterations when one of the following conditions is satisfied:

$$| p^{(s)} - p^{(s-1)} | < \Delta p$$

or

$$| J^{(s)} - J^{(s-1)} | < \Delta J$$

where Δp and ΔJ are the pre-selected constants for the accuracy of calculation. This general algorithm was used to identify all the constants in the empirical equations presented here. The first stage of the calculation was Monte Carlo minimalization method and the algorithm just described above was used as the second stage.

DISCUSSION AND CONCLUSIONS

The experimental results are shown in Figures 2 to 6. It was noticed from Fig. 2 that at low circulation rate, i.e. ($u_f <$ 0.004 m/s), the critical heat fluxes of the present separate-flow thermosyphons are almost identical to those of the counter-flow thermosyphons (Negishi et al., 1985; Pioro et al., 1988), given as:

$$K_{cr} = 0.16\{1 - \exp[-(D_e/L_e)(\rho_f/\rho_g)^{0.13}\mathrm{Cos}^{1.8}(\theta - 55)]\}^{0.86} \qquad (4)$$

This may be due to the fact that the two-phase flow involved in both thermosyphons at very low heat flux are similar because the vapour velocity is extremely low which induces a separate-flow thermosyphon to act as if it were a counter-flow thermosyphon. Therefore, for low circulation rates, e.g., ($u_f < 0.004$ m/s), q_{cr} seems to be independent of u_f and dependent on (D_e/L_e). And the critical heat flux for such cases may be calculated from the well known correlations available for the counter-flow two-phase thermosyphons.

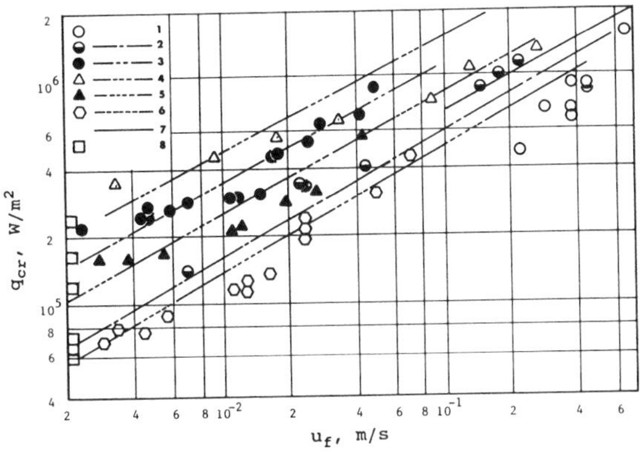

Experimental Results, Present Study:

1 D_e = 0.003 m; L_e = 0.153 m
2 D_e = 0.007 m; L_e = 0.35 m
3 D_e = 0.014 m; L_e = 0.262 m
4 D_e = 0.019 m; L_e = 0.25 m
5 D_e = 0.019 m; L_e = 0.60 m
6 D_e = 0.019 m; L_e = 1.136 m

7 Equation (5)
8 Equation (4)

Figure 2 Critical Heat Fluxes

As illustrated in Fig. 2, for the circulation rate greater than $u_f > 0.5$ m/s, the critical heat flux is seen to depend primarily on the circulation rate, the influence of which becomes dominant over that of the physical characteristics of the thermosyphons. This is also deduced from the empirical relation obtained for counter-flow two-phase thermosyphons (Negishi et al., 1985; Pioro et al., 1988) which does not contain any physical dimensions of the thermosyphons, given as:

$$q_{cr} = 0.23 i_{fg} \rho_g^{0.5} u_f^{0.5} \rho_f^{0.25} [\sigma g(\rho_f - \rho_g]^{0.125}$$

$$\times [1 + 0.057 c_{pf} \Delta T_{sub} (\rho_f/\rho_g)^{0.8}/i_{fg}] \qquad (5)$$

It is safe to state, therefore, that the boiling process in separate-flow thermosyphons can be regarded as a transitional phase whose beginning and end may be described by the respective empirical equations for the critical heat fluxes of the counter-flow thermosyphons (Negishi et al., 1985; Pioro et al., 1988), and of the forced flow boiling in tubes (Collier, 1980; Kutateladze, 1990; Whalley, 1987).

To express the experimental data in a dimensional equation including only those parameters that directly affect the CHF, we obtained the following equation from the present experiment:

$$q_{cr} = 2.5 \times 10^6\, u_f^{0.5} L_e^{-0.76} \tanh(54 D_e) \qquad (6)$$

In a dimensionless number, we also obtained:

$$K_{cr} = 0.9\{1 - \exp[-(D_e/L_e)(\rho_f/\rho_g)^{0.13} Fr^{0.25}\} \qquad (7)$$

where D_e = 3 - 19 mm; L_e = 0.15 -1.14 m; u_f = 0.003 - 0.46 m/s and ΔT_{sub} < 90 °C.

In Figures 3 to 6, the appropriate correlation lines given by Eqs. (6) and (7) are shown with the experimental results.

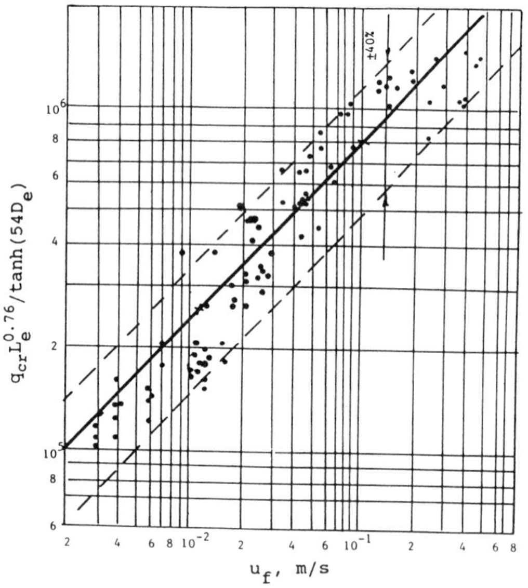

Figure 3 Effect of Fluid Circulation Velocity

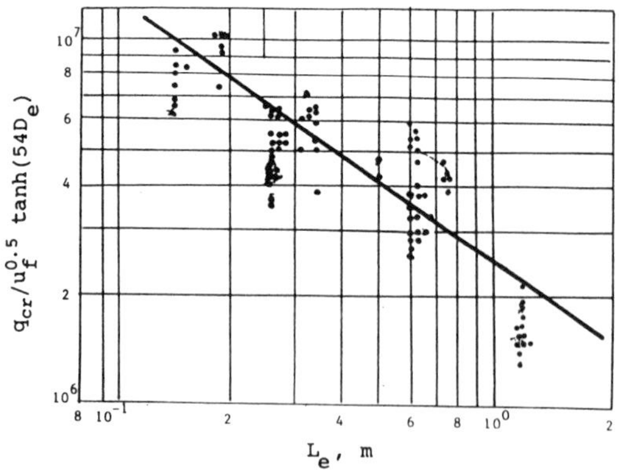

Figure 4 Effect of Evaporator Length

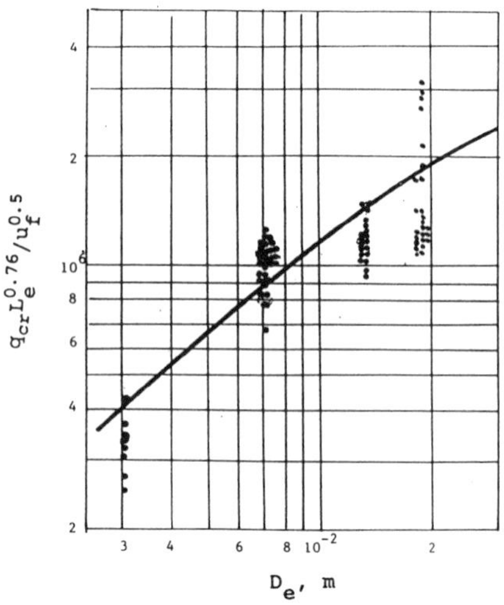

Figure 5 Effect of Evaporator Diameter

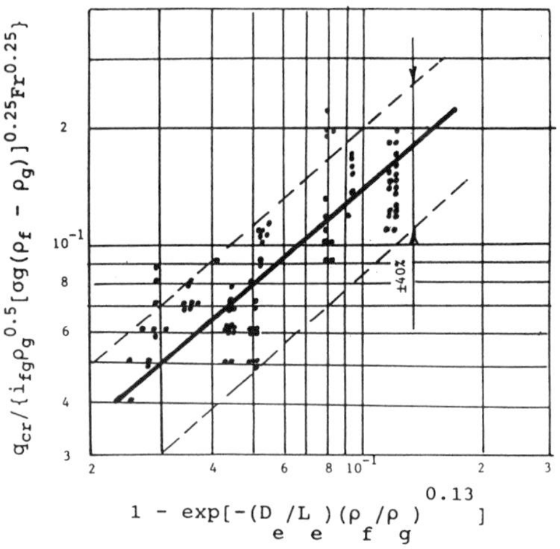

Figure 6 Correlation in Dimensionless Parameters

ACKNOWLEDGEMENT

The experimental part of this collaborative study was carried out at the Institute of Engineering Thermophysics, Ukrainian Academy of Science in Kiev and the study was partially funded by the Natural Science and Engineering Research Council of Canada, NSERC No. A-5175.

REFERENCES

Bezrodny, M.K., Vokov, S.S., Ivanov, V.B., 1984, "Development and Perfection of Cooling Systems for Energy Technological Machines in Nonferrous Metallurgy based on the Closed Two-phase Thermosyphons", Promyshlennaya Energetika (Industrial Energetics), No. 2, pp. 34-37.

Collier, J.G., 1980, *Convective Boiling and Condensation*, McGraw-Hill, New York, 2nd ed., p. 435.

Kai, M., 1990, "Investigation of the Dryout Limit in Evaporator of the Separate Type Heat Pipe", *Pre-print B-5, 7th Int. Heat Pipe Conf.*, Minsk.

Kutateladze, S.S., 1990, *Heat Transfer and Hydrodynamic Resistance: Reference Handbook*, Moscow, Energoatomizdat, p. 367.

Negishi, K. et al. (editors), 1985, *Practical Heat Pipes*, Nikan Ind. Pub. Co., Tokyo.

Pioro, L.S., Pioro, I.L., 1988, *Two-Phase Thermosyphons and Their Application in Industry*, Kiev, Naukova dumka, p. 136.

Vasilliev, L.L. (editor), *Proceedings of 7th Inter. Heat Pipe Conf.*, Minsk, May 1990.

Whalley, P.B. *Boiling, Condensation and Gas-Liquid Flow*, Clarendon Press, Oxford, pp. 174-190, 1987.

HTD-Vol. 197, Two-Phase Flow and Heat Transfer
ASME 1992

SUDDEN RELEASE OF SUPERHEATED LIQUIDS

J. Schmidli and G. Yadigaroglu
Nuclear Engineering Laboratory
Swiss Federal Institute of Technology
Zurich, Switzerland

S. Banerjee
Department of Chemical and Nuclear Engineering
University of California
Santa Barbara, California

ABSTRACT

A series of experiments were conducted with Refrigerants-114, R-12 and propane (b.pts of 3.4, -30.1 and -42.4 ^{0}C at 1 bar) to investigate phenomena arising from catastrophic failure of vessels containing liquefied gases. The experiments were initiated by shattering spherical glass flasks of different sizes (100, 1000 and 2000 ml). The parameters varied were height of the flasks above ground, the filling level and the superheat (which in turn governs the thermodynamic flash fraction), and the relative humidity of the surrounding air.

From high-speed video (1000 fps) and movie (up to 4000 fps) recordings it was possible to determine the shape and the expansion velocity of the release during the initial stage during which gravity plays no important role. For sufficient released internal energy, the early evolution of the release was always spherical with constant expansion velocity during the first few milliseconds until Rayleigh-Taylor type instabilities appeared at the surface of the droplet/vapor cloud.

For the propane experiments, the pressure history of the release was also recorded; the local pressure decayed continously and very rapidly to atmospheric without an intermediate plateau. From very fast responding thermocouples and traces on droplet sensitive paper located at different distances from the center of the release, information about droplets, temperatures and the condensed water vapor in the cloud were gathered.

For all the different test conditions pool formation was significant and therefore has to be taken into account for the later stages of dispersion.

INTRODUCTION

Liquefied gases are often stored or processed under pressure. If a failure of the containment envelope occurs, some portion of the released material will promptly flash and entrain with it fine droplets - the rest will impinge on the ground to form a pool. The evaporation of the pool is relatively slow compared to flashing.

The release scenario can be roughly divided into three partly overlapping stages: inertial expansion driven by internal energy, gravitational slumping of the vapor/droplet mixture if it is denser than air, and subsequent turbulent dispersion due to the wind and related effects.

A large number of dispersion models for the last (and perhaps the second stage) are presently available. All calculations suffer, however, from the uncertainty in the "source term" i.e. the situation at the end of the first stage which occurs very rapidly. In the main, this is due to lack of information about aerosol formation from a flashing liquid i.e. what fraction of the initial liquid mass forms an aerosol and the thermophysical conditions in the vapor/droplet mixture. There is also little known about pool formation and the resulting delayed source for the dispersion of reactive, toxic or flammable substances. Therefore, the main aim of our work is to understand the mechanisms governing the first stage of dispersion of superheated liquid releases, the inertially controlled expansion stage. The emphasis is on catastrophic failures of the containing vessels, though directed jets will also be considered at a later stage.

One of the main reasons there was not a lot research in this field in the past, is the short time period of the first stage. In our small scale experiments it was less then a tenth of a second.

To develop the model for the initial stage of such releases, as mentioned before required for definition of the source term, a large number of experiments need to be performed. Ideally the parameters studied should include thermophysical properties of the contained fluid, the superheat, the height of the release above ground and the failure mode of the containment.

Because of the importance of the problem for the many commonly handled, hazardous liquefied gases (e.g. ammonia, hydrogen fluoride, propane, chlorine), a few studies have already been conducted - though the coverage of parameters and detailed measurements are sparse. Notable amongst these are the early investigations of Hess et al. (1974) and Maurer et al. (1977) using propane and propylene in cylindrical tanks. Their combined data were used to provide models for the inertially driven expansion based on a diffusion type equation. The mechanical energy that would be released if the pressurized material were to be expanded isentropically to ambient conditions is the main parameter in this type of model. The models are based on an inner, uniformly mixed zone and an outer one with a Gaussian concentration profile. The zones are considered to be hemispherical in shape, since inertial effects dominate those due to gravity and atmospheric turbulence in the first stage. These ideas form the basis for the TNO (1978) and the essentially similar World Bank (1985) models. In both cases, pool formation is ignored and therefore the initially released mass is overpredicted. According to Schmidli et al. (1990) and Nolan et al. (1991) there is also evidence that the initial expansion velocity is overpredicted by the such models.

To be useful, a model for the initial expansion should provide the concentration of the vapor phase in the cloud, the droplet concentration and composition, the phase temperature and the dimensions of the pool (if one is formed). While the predictions need not to be accurate during the rapid inertial expansion, they must be reasonably reliable for the end of this stage, since gravitational slumping and atmospheric dispersion calculations start with this prediction as initial conditions. The Schmidli et al. (1990) and Nolan et al. (1991) results indicate that the TNO/World Bank models may predict, for the fluids used, too large and a too dilute a cloud at the end of the inertial expansion, which could lead in some cases to significantly non-conservative results. For example, too dilute a cloud of flammable material may suggest that much of the release spreads rapidly to a concentration below the lower flammability limit. Similar nonconservative predictions may be obtained for the concentration of toxic releases. Because of these uncertainties an extensive set of experiments were undertaken, as discussed below, to define the situation and provide the physical understanding needed for modeling it.

EXPERIMENTAL APPARATUS

The equipment used for the current experiments resulted from the experience gained from first series of experiments by Schmidli et al. (1990) and is sketched in Figure 1. Refrigerants 12 and 114 were used for the first tests, propane for those that followed. Butane will be used in the future, too. Table 2 summarizes the experimental parameters investigated.

The liquefied gases were held in spherical glass flasks that could be suspended at various heights above the ground to study ground- as well as elevated releases. The flasks were shattered with a remotely-operated pointed hammer. The initial condition of the liquid, which was always in its saturated state, was measured and controlled by a thermocouple inside the vessel. A fast-responding piezoelectric pressure transducer installed at the center of the vessel provided the pressure decay history of the release.

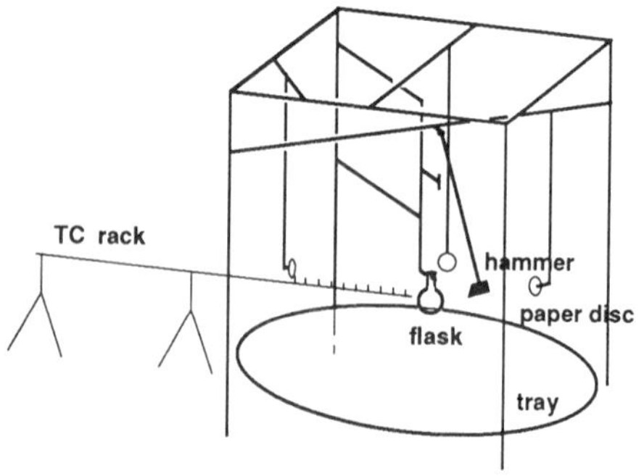

Figure 1: Experimental setup

The evolution of the aerosol cloud that was formed was followed using a standard video system taking 50 half-frames per second (hfps), a Kodak Ektapro system at 1000 frames per second (fps) or a Hycam 16 mm movie camera with speed up to 4000 fps. Very high speed visualisation for the first stage was required because the initial cloud expansion was very rapid. For the propane experiments, 10 specially-developed very fast responding K-type thermocouples were used. Their noninsulated wires (and junction) had a diameter of 50 microns and the thermocouples had a time constant of less then 1 millisecond under our experimental conditions. They were mounted on a radial rack, starting 25 cm from the release center with spacing of 10 cm. The thermocouple signals provided information about the location of the cloud front. A rough local concentration distribution could also be calculated.

For the small refrigerant releases a tray with a diameter of 1.5 m covered with an insulated layer was mounted on a high- precision electronic balance to gather data about pool formation. The balance was a Mettler PE 4000 having a resolution of 0.1 g. For the propane experiments, this system was replaced by a tray of 1 m diameter mounted on 3 load cells from Hottinger Baldwin. The accuracy of this system was of the order of 1 g.

To investigate the droplet characteristics, the propane was colored and its droplets were caught on paper mounted on discs protected with a second, slitted, rotating disc. The speed of rotation could be adjusted to have the surface of the paper exposed only once during passage of the cloud. The paper used was a high quality ordinary copy paper. The intensity of the colored spot on the

Substance	Boiling point at 1.0 bar [^{0}C]	Sat. pressure at 25 ^{0}C [bar]	Latent heat at 1.0 bar [kJ kg^{-1}]	Liquid specific heat at 20 ^{0}C [kJ kg^{-1} K^{-1}]	Isentropic quality for 25 ^{0}C release
R-114	3.4	2.1	136	0.984	0.15
R-12	-30.1	6.5	166	0.967	0.28
Propane	-42.4	9.5	426	2.57	0.35

Table 1: Thermophysical properties of test fluids

paper is a measure of the initial droplet size, while the size of the mark is related to the droplet size at the moment of the impact on the disc. To detect any water vapor condensation the ordinary paper was replaced later by a water sensitive paper. Water droplets on the paper lead to a chemical reaction on the surface of the paper and resulted in a blue spot whose dimensions are related to the water quantity collected.

A calibration experiment, presently under progress, will allow to relate the size of the marks made on the paper by impinging droplets to their actual size. Thus the largest drop diameters will be obtained. To more accurately determine the droplet size distribution, more advanced techniques are required, e.g. laser holography.

The signals of the thermocouples and of the pressure transducer were sampled with a Keithley K500 data acquisition system connected to a 286/12 MHz personal computer. The combination of hardware and self-written software produced using the data acquisition language Asyst, allowed collection of 30000 data points per experiment. The maximal total sampling frequency was 4 kHz for 12 channels or 333 Hz for each channel. This sampling speed provided an approximate time resolution and mixture temperature information, but should be increased to follow more precisely the cloud front propagation. The signal of the pressure transducer was further sampled at a much higher frequency of up to 50 kHz with a digital oscilloscope Nicolet NC-310.

TEST MATRIX

The very first experiments were performed using Refrigerant-114 in 50 ml and 100 ml vessels at relativly low superheats. Results from these experiments were published by Schmidli et al. (1990) and are not presented here again. Later, additional R-114, R-12 and propane releases were produced, and the test matrix will be further extended in the near future with runs at different superheats of propane and butane. Table 1 summarizes the relevant thermophysical properties of the test fluids and Table 2 the test conditions.

The parameter of primary interest is the initial superheat of the liquefied, pressurized fluid. The superheat can also be expressed in terms of a flash fraction, i.e. the quality calculated assuming isentropic flashing down to ambient pressure. The isentropic quality x is defined as

$$x = (s_{f1}-s_{f2})/s_{fg2}$$

where: s is the entropy, (J kg^{-1} K^{-1}), f stands for liquid; g for vapor;, fg their difference; 1 and 2 are the initial and final (athmospheric pressure) states, respectively.

The R-114 experiments were performed first, followed by R-12 runs, which were carried out to compare the behaviour of different refrigerants under conditions leading to the same isentropic quality. The refrigerant experiments were stopped due to new governmental regulations prohibiting the release of such refrigerants, and propane was used next.

Parameters that were further varied were the vessel size and its filling level, as well as the release height, since it was felt that all these could influence the expansion velocity, the aerosol formation and the fraction of the release that would form a pool. However, measurements of the pool mass fraction could so far only be obtained for refrigerants in the 100 ml vessels and with an improved system for propane for the 1000 ml vessels.

Fluid	Flask size [ml]	Height of release	Filling level	Superheat [^{0}C]	Isentropic flash fraction
	50	Ground	full	6.6	0.05
R-114	or	or	or	14.6	0.10
	100	5 dia.	0.5	21.6	0.15
	100	Ground	full	26.6	0.19
	or	or	36.6	0.25	
	1000	5 dia.	0.5	46.6	0.32
	100	Ground	full		
R-12	or	or	or	52.1	0.26
	1000	5 dia.	0.5		
	1000	Ground	full	47.4	0.23
Propane	or	or	or		
	2000	1 Meter	0.5	60.4	0.30

Table 2: Test matrix

For the large releases the measurement system used started oscillating violently due to high initial momentum of the release impacting on the tray. Because of these strong oscillations, we define the pool fraction as the mass detected at the end of the oscillation period of the system (about 1.5 s after the release for the load cell system). This practical definition makes sense, since the part of the pool that evaporated rapidly before this time is mostly taken along in the initial cloud.

EXPERIMENTS

First we present the results gathered by visualisation. To achieve a better understanding of the phenomena we mainly focus in this paper on elevated releases for which there was no interference with the ground during the very early phase of the release.

Delay In Nucleation And Flashing

Figures 2 and 3 show the initial phases of the release of R-12 from a height of 5 initial diameters above ground level with different fill levels. In both cases the cloud expanded significantly between the 1000 fps video frame in which the hammer impacts the flask and the very next one. Thus only an upper limit for any delay in nucleation that exists could be established and the exact time needed for significant flashing at this pressure could not be determined with the recording speed of 1000 fps used. Since the cloud expands volumetrically, a relatively large change in volume (say a factor of 2) implying significant vapor generation - a void fraction of 0.5 - produces only a 26 percent change in diameter. It can be, however, said that bubble nucleation and significant flashing occured in less then 1 ms.

Propane experiments were recorded with high speed movies at up to 4000 fps and a similar situation was experienced. It can only be said again that the bubble nucleation and onset of significant flashing is less than 250 microseconds. Thus, in the cases of practical interest, we can assume almost instantaneous bubble nucleation and very rapid flashing after a catastrophic failure of the pressure vessel.

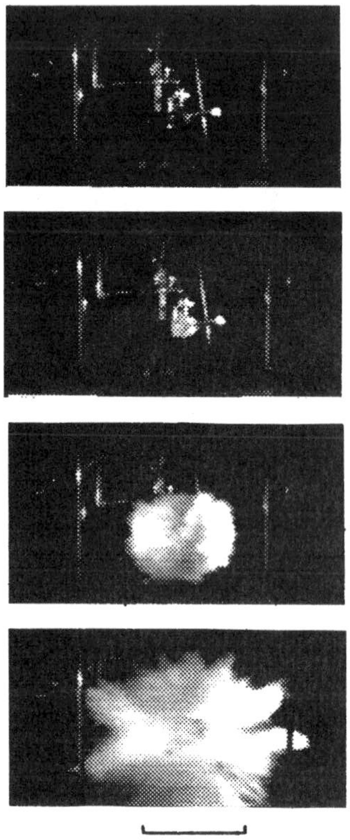

50 cm

Figure 2: R-12, 1 liter full, 52.1 ^{0}C superheat 1000 fps; frames: 0, 1, 12, 25

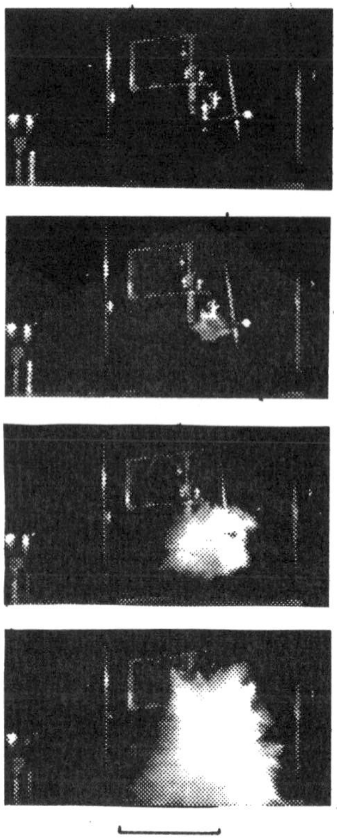

50 cm

Figure 3: R-12, 1 liter half-full, 52.1 ^{0}C superheat 1000 fps; frames: 0, 1, 12, 25

Cloud Shape And Pool Formation

Refrigerant-114 experiments were conducted at 3 different superheats. For the lower superheats of 26.6 ^{0}C and 36.6 ^{0}C, the released fluid behaved differently than for all the other fluid and superheat conditions tested. The release did not evolve as an explosion-like expansion which leads to a roughly spherical cloud. The cloud expansion velocity was much smaller and the liquid did not break up as much. The immediately-released internal energy was not large enough to form a spherical vapor/droplet cloud and thus the release was immediately dominated by gravity (Fig. 4). Even if a large part of the liquid did not break up into droplets, some of it flashed and an aerosol cloud did form, however.

R-12 and propane showed a different behavior at conditions leading to a similar or even smaller isentropic flash fraction then R-114 at 36.6 ^{0}C superheat and expanded spherically like R- 114 at 46.6 ^{0}C superheat. A theoretical expansion velocity v_{theo} can be calculated assuming that the internal energy available after isentropic flashing (quality x) converts to kinetic energy:

$$\frac{v_{theo}^2}{2} = h_{f1} - (h_{f2} + x * h_{fg2})$$

where: h is the enthalpy (J kg^{-1}). Comparing the measured cloud expansion velocities to v_{theo} leads to the suggestion that it is not the flash fraction that determines the shape of the release but the order of magnitude of the fraction of internal energy convertible to kinetic energy (Table 3).

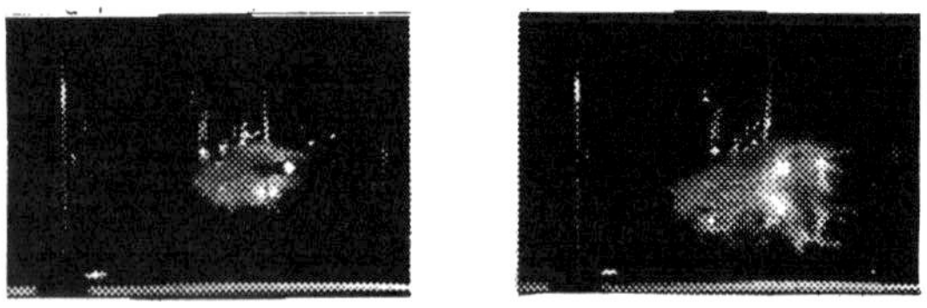

Figure 4: R-114 release, 26.6 ^{0}C superheat, 1 liter flask, full, after 25 and 80 milliseconds

The R-12, propane, and R-114 at 46.6 ^{0}C superheat releases behaved similarly. For the first 8 - 20 milliseconds the shape of the expanding cloud, consisting of liquid, any initially existing vapor and vapor generated by flashing, were for all the different conditions mentioned above, roughly spherical; the clouds grew with radial constant speed. The length of this period was mainly determined by the vessel size (Table 4). The length of this period and the initial constant velocity discussed below determine the size of the initial cloud at the end of the rapid expansion period.

After this period which led to an increase in radius of the order of 5 - 10 of the initial flask radius, which is equal to an increase in volume by a factor larger then 100, "fingers" were formed at the surface of the cloud and the cloud expansion velocity slowed down dramatically.

Fluid	Pressure [10^5 Pa]	Superheat [^{0}C]	Isentropic quality	v_{theo} [m/s]
R-114	2.5	26.5	0.19	69.7
	3.4	36.5	0.25	88.6
	4.4	46.5	0.32	107.9
R-12	6.0	52.1	0.26	120.4
Propane	5.4	47.4	0.23	142.2
	7.8	60.4	0.30	181.8

Table 3: Theoretical expansion velocity

Flask Filling	1000 ml		2000ml	
	Full	Half	Full	Half
Fluid				
R-12	11.8	10.5	19.8	19.0
R-114	10.0	6.5	21.0	18.5

Table 4: Average time [ms] after release for the beginning of finger formation,
(R-12 at 22 ^{0}C, R-114 at 50 ^{0}C, both elevated)

Due to flashing and vapor generation the fluid, initially a liquid continuum, changed to a vapor continuum and the liquid chunks formed underwent aerodynamical breakup into droplets. It seemed that these droplets thrown outwards determined the expansion rate up to the stage of finger formation. Thus the initial dilution of the cloud happened by inclusion in the cloud of the air through which the droplets were travelling. Up to this point the releases for half-filled and full vessels look similar in shape, though the releases of half-filled vessels had, obviously, a denser bottom part (Figs. 2 and 3).

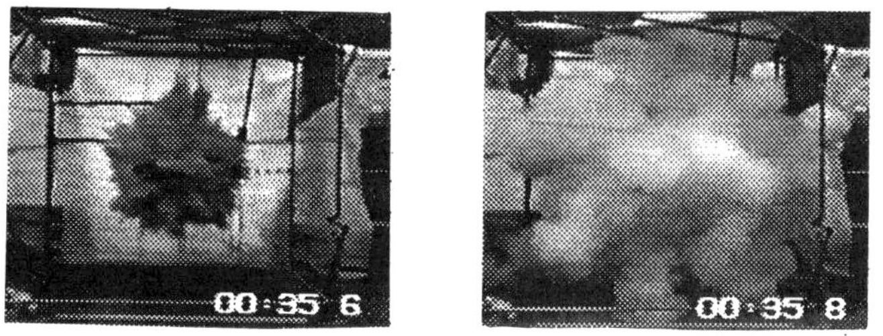

Figure 5: Propane release, 60.4 ^{0}C superheat, 2 liter flask, full, 50 hfps; frames: 1 and 10

The fingers that form in the later stages of the expansion are probably due to Rayleigh-Taylor type instabilities at the accelerating cloud/air interface. After their appearance, the aerosol clouds behave differently according to the different filling level. The 100 percent filled vessels led to clouds that retained their spherical shape (Fig. 5) and expanded further with decreasing horizontal and vertical speeds until gravitational forces dominated the inertial ones and caused the cloud to start slumping. The clouds from 50 percent filled vessels, after the initial spherical growth period at constant expansion velocity, changed shape from spherical to one characterized by an upwards directed jet coupled to what appeared to be a denser part of the release, thrust downwards towards the ground (Figs. 3,6).

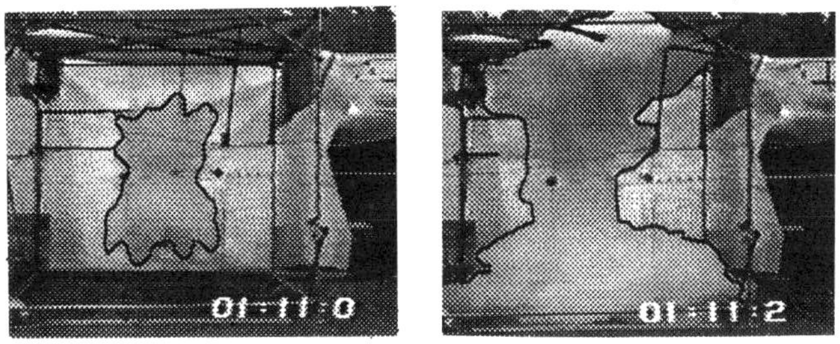

Figure 6: Propane release, 60.4 ^{0}C superheat, 2 liter flask, half-full, 50 hfps; frames: 1 and 10

At first sight, this was rather surprising. However, even though the delay time for flashing could not be determined with our instrumentation, the vapor mass above the liquid starts certainly to expand before bubbles can nucleate and should impart a downward initial thrust to the liquid mass before it expands and breaks up into fine droplets. A second factor that may occur is a prefential flashing of the liquid at the existing vapor/liquid interface for a partially filled flask. This would also account for the downward thrust of the liquid (much like a rocket effect) and the upward directed, less dense, jet. This is of course speculative at present and needs further experimental confirmation.

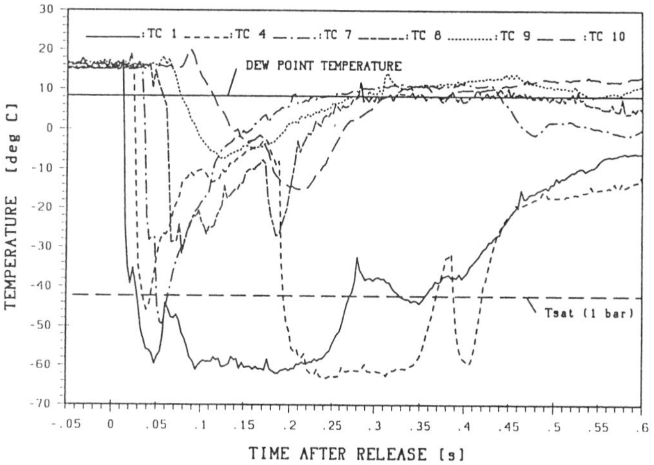

Figure 7: Cloud tempereature of propane
(Run 097, 5 ^{0}C, 2 liter flask, full, release height 1 m)

Water vapor droplets become visible, at least in the later stages of the expansion. This is because the cloud temperature falls well below the dewpoint of the entrained air (Fig 7). Thus, for the same temperature, a vessel failed in a higher humidity environment should lead to a larger visible cloud than at lower humidity, which was indeed observed (Table 5).

Time after release [ms]	Cloud Radius [cm] Exp R077 38 % rel. Hum.	Cloud Radius [cm] Exp R079 83 % rel. Hum.
0	6	6
5	14	15
10	21	21
15	31	30
20	36	40
25	42	49

Table 5: Effect of humidity on the visible cloud size
(R-12, 1000 ml, 5 dia., elev., 22 ^{0}CAmbient Temp.)

The visual evidence for a denser bottom part of the cloud for a half-filled vessel is confirmed by the data from the pool measurements. Figure 8 shows the pool fractions for R-114 releases at different superheats. It is obvious that for half- filled vessels a larger part of the initial mass hits the ground and forms a liquid pool. The pool fractions for the two lower superheats of R-114 are much higher than for the high superheat release due to the formation of the pool by liquid chunks. The pool fractions of half-filled vessels for R-12 were also larger than for full ones (Table 6). For large enough released internal energy, the pool was formed not by liquid chunks but by impingment of the droplets on the ground. The visible part of the pool had a smaller diameter than the trays used.

Release Height	Filling Full	Half
Ground	21	36
Elevated	23	39

Table 6: Average pool fraction [%] of R-12 released at 22 ^{0}C, 100 ml flask

Droplets hit the ground also further away from the release center, but because their volume was small at this distance, they evaporated immediately.

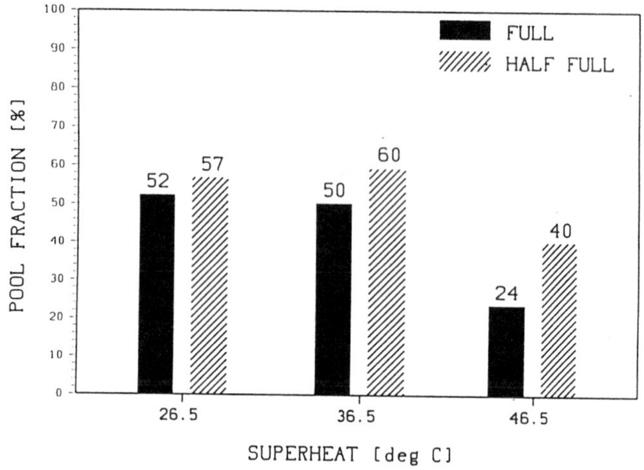

Figure 8: Average pool fractions
(R-114, 100 ml vessels, release height 5 diameters)

Pool history

A typical temperature history of the pool is shown in Figure 9 for a propane release. The measured temperature dropped well below the boiling point of -42.4 ^{0}C at ambient pressure of the propane and stayed at this temperature longer than the recording time of 200 seconds. It is clear that the decrease of the pool mass as function of time depends very strongly on the thermophysical properties of the ground, its initial temperature and the ambient conditions. The pool boils due to heat transfer from the ground and evaporates due to mass transfer into the air/wind.

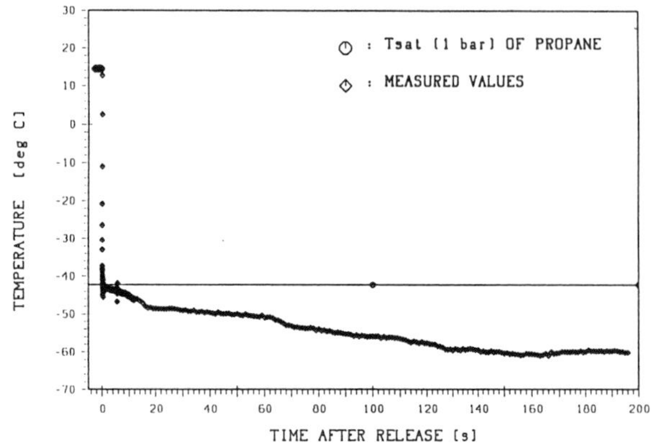

Figure 9: Pool temperature
(Propane, Run 016, 18 ^{0}C, 2 l flask, full, ground level release)

After a catastrophic release of tons of dangerous materials the boiling/evaporating pool could be a source of vapor/aerosol for long periods of time. In the experiments done, the tray was covered with an insulation, and therefore the pool regression was very slow and mainly due to evaporation into the air and not due to boiling with heat conducted from the tray.

Cloud Expansion

Quantitative results of the initial, roughly spherical evolution of the release were gathered by analyzing the high speed videos and movies frame by frame. As long as the cloud surface can be approximated by a sphere, the radius as a function of time grows linearly (Fig. 10). The growth slows down when the "fingers" start to evolve.

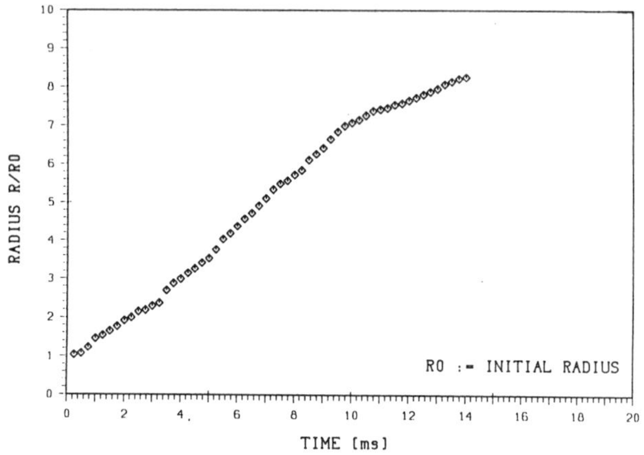

Figure 10: Cloud radius r of propane
(Run 142, 18 ^{0}C, 2 liter flask, full, release height 1m)

The measured initial velocities are presented in Tables 7 and 8. The superheat or the release energy that is convertible to kinetic energy appears to determine the release velocity. For propane, an increase in superheat from 47.4 to 60.4 ^{0}C, or in terms of isentropic flash fraction from .23 to .30, doubled the initial expansion velocity. For refrigerants no comparisons could be made with our data, but Nolan et. al (1991) obtained similar results with R-11. The filling level and the flask sizes used had only a minor effect on the initial speed.

Flask		1000 ml		2000ml	
Superheat [^{0}C]	Filling	Full	Half	Full	Half
47.4		17.4 (0.12)	18.0 (0.13)	19.4 (0.14)	17.6 (0.12)
60.4		38.0 (0.21)	35.5 (0.20)	41.3 (0.23)	26.9 (0.15)

Table 7: Average expansion velocity in m/s and as fraction of the theoretical velocity v_{theo} for elevated propane releases

The two-liter releases at high superheat do show some diminution of the expansion velocity for the half-filled flask. Whether this is a significant effect or experimental scatter remains to be determined.

Flask	Filling Superheat [°C]	1000 ml Full	Half	2000ml Full	Half
R-12	52.1	14.7 (0.12)	13.1 (0.11)	13.9 (0.12)	13.1 (0.11)
R-114	26.6	*	*	*	*
	36.6	*	*	*	*
	46.6	12.7 (0.12)	13.4 (0.12)	11.7 (0.11)	12.4 (0.11)

* no significant spherical expansion velocity

Table 8: Average expansion velocity in m/s and as fraction of the theoretical velocity v_{theo} for elevated refrigerants releases

Once fingers form, they behave initially like jets and therefore the temperature recorded depends very strongly on position in the aerosol cloud. If the tip of a finger moves along the thermocouple line, the temperature recorded falls below the saturation temperature at ambient pressure of the released fluid; indeed droplets cool the cloud to obtain heat for their vaporization (Fig. 7).

Overall, thermocouple recordings presented no surprise. The further away the thermocouples, the later the temperature drop observed took place. The magnitude of the temperature drop decreased with increasing distance from the release center, very likely because of the lower concentration of evaporating droplets in the cloud. During the finger growth stage, the concentration of released material dropped not due to the aerosol cloud front capturing air within it as it moved uniformly outward in the initial stage, but by entrainment of air at the surface of the fingers due to shear layers between the jet-like fingers and the surrounding air.

Droplet characteristics

Information regarding the droplets in the cloud was obtained from the paper discs used. Droplet impact marks down to 50 microns could be measured. The discs were mounted at horizontal distances of 25, 50, 75 and 100 cm from the release center. Since the paper was protected by rotating slitted discs, information about droplet history could be gathered. The first paper was usually impacted by a liquid mass during about 1 ms. During this period no single droplets could be distinguished; the liquid was essentially a continuum. Following this period, few marks having diameters up to 6 millimeters appeared on the paper.

On the second paper, less continuous liquid was caught and more spots in the range of millimeters in diameter appeared, and the first sharp, distinct droplet marks could be observed. On the third and fourth pieces of paper mainly spots of the order of 3 millimeters and less were detected, if any at all.

The largest 5 spots of propane and condensed water vapor from a particular run are presented in Figure 11. The size of the marks is related to the actual droplet sizes.

The following sequence of events appears to summarize our qualitative findings, but must still regarded as speculative: Initially the release expands with the surface layer of liquid rapidly vaporizing and fragmenting. However the main part of the liquid nucleates bubbles more slowly which grow till the liquid disintegrates and is thrown outwards. After a certain distance the liquid breaks into large droplets of several millimeters in diameter. Since the Weber number of these large droplets exceeds the critical one, the droplets break up further into much smaller droplets. All this appears to take place within the first 50 to 75 cm. The droplets subsequently evaporate quite rapidly due to the large increase in surface area by break up.

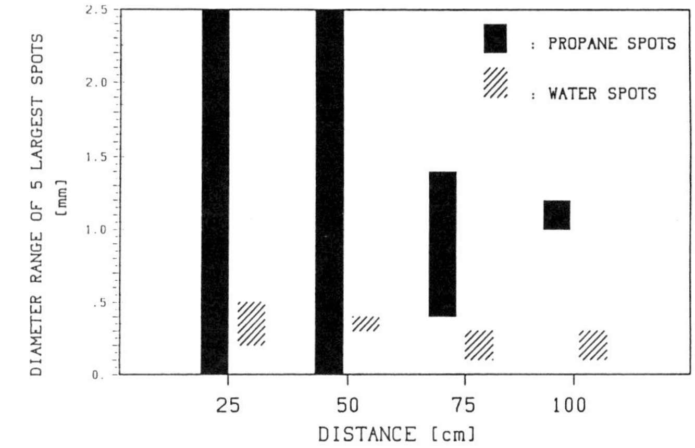

Figure 11: Spot size observed in an experimental run (Run 129, 5 ^{0}C, 1 liter flask, half-full, release height 1m)

Using water sensitive paper in the last series of experiments, water droplets could be detected (Fig. 11). The marks made by the water droplets were mainly less then 1 millimeter in diameter, the largest were usually of the order of 200-500 microns. Propane at a temperature of lower than -12 ^{0}C at 1 bar can form hydrates with water; water vapor could also simply condense on the propane droplets and therefore could lead to a blue spot, connected to a red one from the colored propane on the water sensitive paper. In all the experiments conducted we could detect sporadically few such double spots. Their formation occurs, however, seldom and should therefore not be important. Zumsteg (1988) performed experiments releasing a cold dense gas cloud (not a pressurised liquid) and the measured condensed water vapor particles had a maximum diameter of 5 microns.

The ratio of water marks to propane marks increases with increasing distance from the release center. This is consistent with the fact that the cloud remains visible after the initial expansion due to the formation of a mist containing mainly condensed atmospheric water vapor.

Pressure History

Figure 12 shows a typical pressure history measured with the piezoelectrical transducer placed inside the vessel. First the pressure increased a little for about .1 to .15 milliseconds and then dropped down continously and very rapidly. The strong oscillations after the release are likely an effect of the very rapid depressurisation on the transducer and its support system.

With more sophisticated electronics and instrumentation currently being installed we should be able to obtain information about the pressure undershoot, if any. No evidence of the intermediate plateau in pressure (lasting for about 30 ms) reported by Nolan et al. (1991) appeared in our experiments. This difference is probably due to the fact that Nolan et al. used a different, mechanically opened vessel with a much longer opening time than ours. This suggests that it may be sufficiently accurate in any model to assume that the pressure decays quasi-instantaneously to ambient pressure with a propagation speed equal to the speed of sound through the liquid. The pressure undershoot, if any, is likely to be of little consequence for the subsequent behaviour of the cloud.

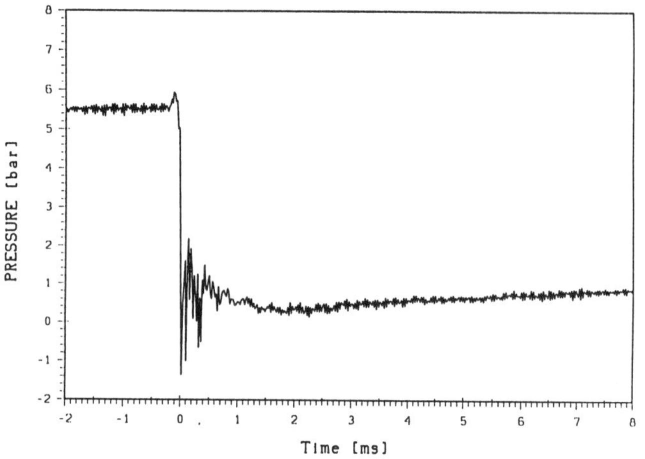

Figure 12: Pressure history
(Propane, 5 ^{0}C, 1 liter flask, half-full, release height 1m)

CONCLUSIONS

From these experiments, information needed for a model describing the inertial-driven stage of a catastrophic release of superheated liquid was gathered. Initially, following failure of a vessel containing pressurized liquid, the expansion is spherical and expands radially with constant velocity determined by the initial superheat and the thermophysical properties of the fluid. This period ends with the growth of fingers due to what appears to be Rayleigh-Taylor type instabilities. When these instabilities start growing like fingers, the expansion rate of the cloud decelerates enormously. The processes that lead to the initial breakup of the liquid mass are still not clear. Perhaps the liquid is thrown outwards due to bubble nucleation and growth in the early stages - initially in a shell like structure that breaks up later into droplets. These droplets undergo further breakup into a much larger number of fine droplets until they have a Weber number less than the critical one.

The filling level has a strong effect on the shape of the cloud after the linear radial growth period, whose duration seems to depend from the initial radius. The filling is an important parameter governing the mass of the pool formed, too. This pool formed by the impingement of droplets on the ground is quite significant and has to be taken into account. Water vapor condensation should also be considered in calculating properly the concentrations in the cloud for the later stage of the releases.

Acknowledgements:

We wish to thank the Swiss National Science Foundation for partly supporting this work under Contract No. 21.25277.88. The experiments were performed at the Paul Scherrer Institute and at the Casaccia Research Center of ENEA in Italy, the latter with the collaboration and guidance of Dr. G.P. Celata. Both centers provided valuable technical assistance, which is gratefully acknowledged. Furthermore this work is part of the CEC STEP Program CLOUD project.

REFERENCES

Hess, K., Hoffmann, W., and Stoeckl, A., "Propagation Process after Bursting of Tanks Filled with Liquid Propane," International Loss Prevention Symposium, The Hague, 1974.

Maurer, B., Hess, K., Giessbrecht, H., and Leuckel, W., "Modelling of Vapor Cloud Dispersion and Deflagration after Bursting of Tanks Filled with Liquified Gases," International Loss Prevention Symposium, Heidelberg, 1977.

Nolan, P.F., Pettitt, G.N., Hardy, N.R., and Bettis, R.J., "Release conditions following loss of containment, J. Loss Prevention Process Ind., 3, 1990.

Nolan, P.F., Pettitt, G.N., Hardy, N.R., "The Physical Modelling of Two Phase Releases following the Sudden Failure of Pressurised Vessels," Proceedings of the International Conference and Workshop on Modeling and Mitigation the Consequences of Accidential Releases of Hazardous Materials, New Orleans, 1991.

Opschoor, U., "Evaporation," TNO Yellow Book, 1978.

Schmidli, J., Banerjee, S., and Yadigaroglu, G.,"Effects of vapor/aerosol and pool formation on rupture of vessels containing superheated liquid," J. Loss Prevention Process Ind., 3, 1990.

World Bank,"Manuel of Industrial Hazard Techniques", 1985.

Zumsteg, F., "Laborversuche zur instationären Ausbreitung kalter Gaswolken", PhD Thesis Nr. 8644, ETH Zurich, 1988

HTD-Vol. 197, Two-Phase Flow and Heat Transfer
ASME 1992

THREE-DIMENSIONAL TEMPERATURE DISTRIBUTION AND LOCAL TRANSFER COEFFICIENT OF COOLING IN A PACKED COLUMN

K. T. Yu, X. J. Yuan, and G. Q. Dong
Chemical Engineering Research Center
Tianjin University
Tianjin, People's Republic of China

ABSTRACT

Hot water was cooled by countercurrent air in an experimental packed column of one meter in diameter filled with corrugated metal sheet structured packing. Forty eight temperature transducer were inserted into the packing for simultaneous measurement of water temperature distribution under various operating conditions. Three-dimensional temperature profiles at different cross sections of the column are reported.

Mathematical derivation shows the similarity between the dimensionaless temperature and the residence time of flowing liquid. Thus the flow pattern of the liquid phase along the column can be obtained from the experimental temperature profiles.

A set of partial differential equations is established to modeling the cooling process. The local coefficients of mass transfer along the whole column can be evaluated by the model computation together with the experimental temperature profiles. The significance of the variation of the local coefficients is discussed.

I. INTRODUCTION

As a result of developing many high efficient packings in recent years, the use of packed column nowadays for cooling and other processes such as distillation and absorption becomes very popular in industries. Since the gas-liquid contacting process in a packed column can be regarded essentially as a countercurrent interacting two-phase flow accompanied simultaneously heat and mass transfer, the basic problems concerned are primary the flow distribution and the transfer efficiency. Much effords have been devoted to study these problems experimentally from the overall point of view, such as finding the overall heat and mass transfer coefficients under various operating conditions for a specific packing and representing them in an empirical dimensionless equation. The problem of flow distribution is often treated implicitly into the transfer coefficient. However, when one deals with the scale up of a packed column to a larger diameter, such simple approach by using an overall coefficient is often in failure because the effect of maldistribution in a large column always becomes so serious that the transfer efficiency may be lowered considerably than that measured in laboratory. On the other hand, a packed column can be in its highest transfer efficiency only under the perfect condition of the plug flow. Thus, the problem of flow distribution, or specifically the deviation from the plug flow, is a topic of great interest in recent research work on packed column.

The usual method of studying liquid flow distribution in a packed column is to collect the amount of liquid flowing down by a number of receivers evenly distributed at the bottom of the column. Such data, however, do not reflect the true distribution along a column. Recently, by the use of tracer technique [1,2], the three dimensional mean residence time distribution of liquid in a packed column was reported for random packings. Such distribution simulates the qualitative flow distribution, or in other words, the flow pattern. For the sole purpose of investigating the flow pattern, the tracer method is preferable since a relatively simple partial differential equation need be solved for estimating the model parameters. However, the type of flow pattern in a packed column has not yet been well established. Stichlmair et al.[3] investigated the effect of maldistribution on flow pattern by measuring the temperature profiles along the whole packed column. Inthis paper, a method is developed by using the experimental measurement of temperature distribution for evaluating at once both flow pattern and the local transfer coefficient. Furthermore, another advantage of using this method is that the size of temperature probe can be greatly reduced to avoid any appreciable interruption of flow. The influence of gas rate to the liquid flow pattern is not covered in this paper, although such effect has been demonstrated experimentally to be relatively unimportant.

II .EXPERIMENTAL WORK

The test column is one meter in diameter filled with 1.7 meters corrugated stainless steel sheet structured packing which is similar to Sulzer Mellapak 250Y as shown in Figs. 1A and 1B. The main reason for adopting such structured packing is due to its low pressure drop and high transfer efficiency, as had been found in many industrial absorption and distillation proesses. Warm water at constant temperature was feeding dowmward from the top of the column by a specially designed distributor to insure practically uniform inlet distribution. Atmospheric air was blowing up countercurrently from the bottom through a distrbutor of sieve tray type. Forty eight probes (temperature transducer) of type AD590JH produced by American AD Company with accuracy of 0.1℃ were installed in three sections of the packed bed (see Fig. 2). The electrical

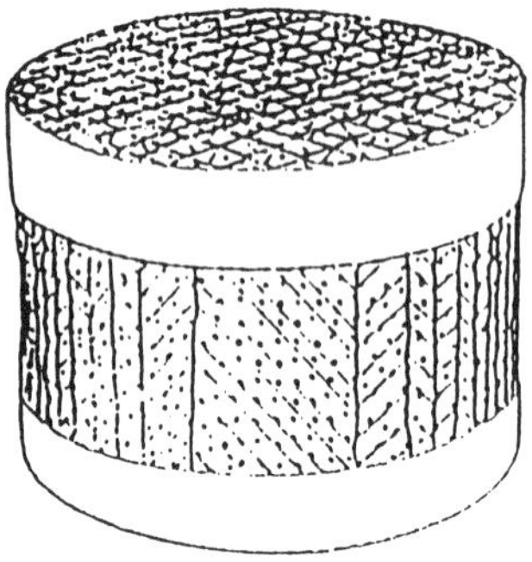

Fig.1A Metallic corrugated sheet packing of 250Y type

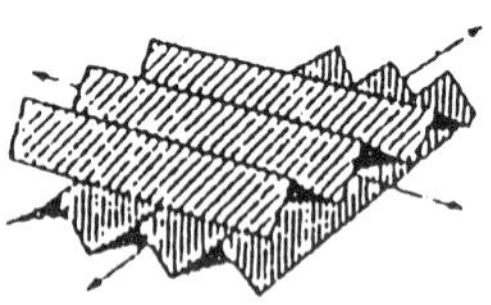

Fig.1B Flow Channels in packing element

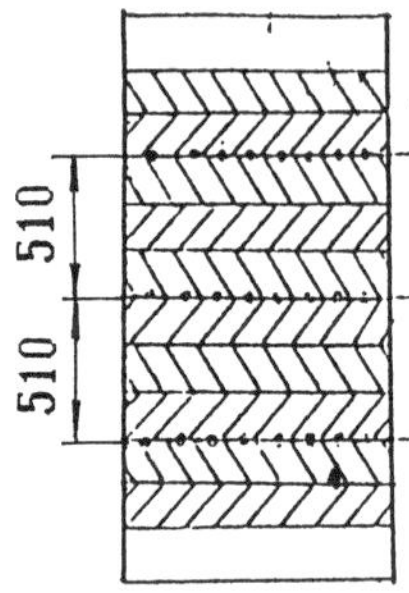

Fig.2 Location of AD590JH in packing bed

signals from all probes were collected and analyzed by a on-line microcomputer.

In order to compare the temperature field under different operating conditions, the following dimensionless temperature T_R is used instead of the true temperature T:

$$T_R = (T - T_{gin}) / (T_0 - T_{gin}) \qquad (1)$$

where T_{gin} and T_0 are the inlet temperatures of air and water respectively. Figs. 3 and 4 are the typical experimental dimensionless temperature profiles in a section. Fig. 5 shows the instantaneous profiles in three sections of the packed column.

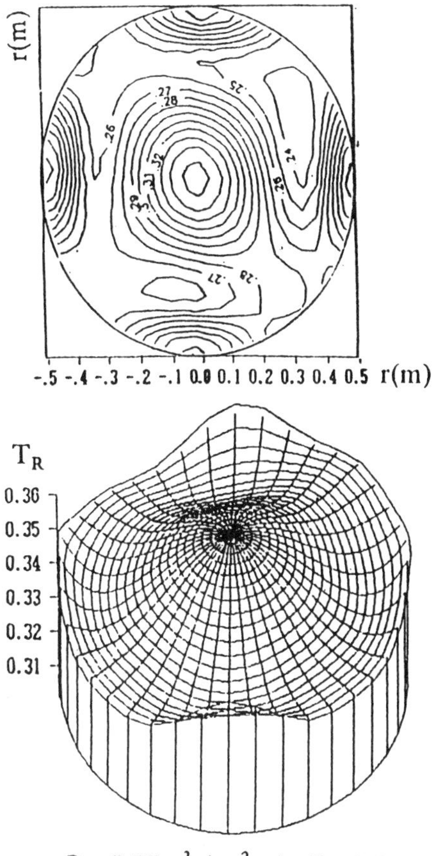

$Q = 8.97m^3/m^2 \cdot h$; $F = 2.1$

Fig.3 Temperature(T_R) profiles

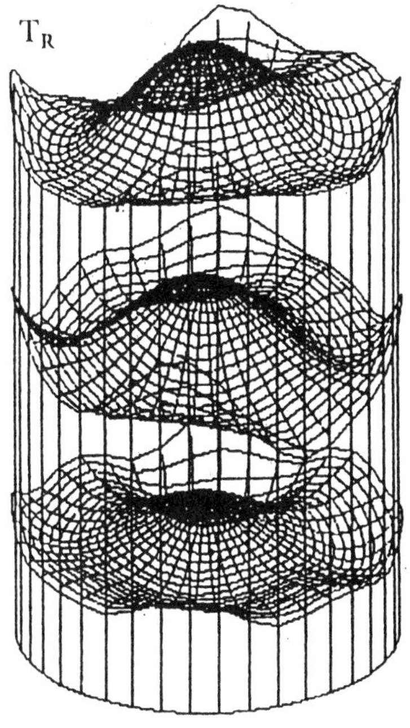

$Q = 8.97m^3/m^2 \cdot h$; $F = 2.1$

Fig. 5 Temperature(T_R) profiles on three cross sections

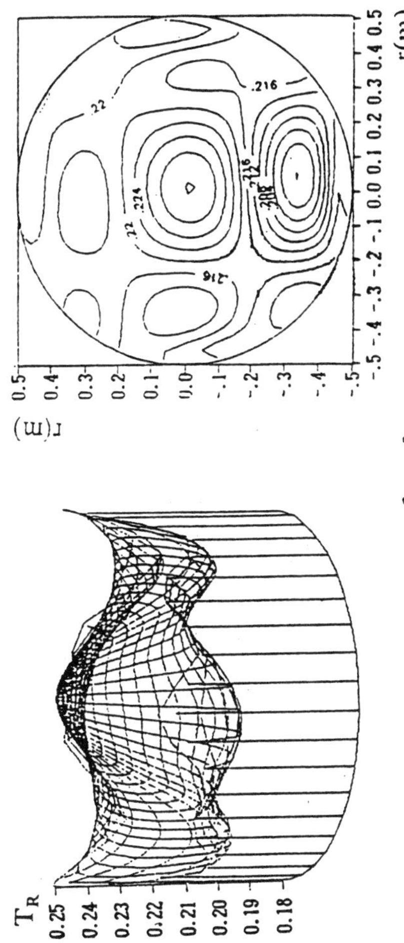

$Q = 3.85m^3/m^2 \cdot h$; $F = 1.2$

Fig.4 Temperature(T_R) profiles

III. SIMULARITY BETWEEN TEMPERATURE FIELD AND FLOW PATTERN

Assuming the gas and liquid phases flowing through the packed bed are well mixed and referring to Fig. 6, it is possible to take a differential

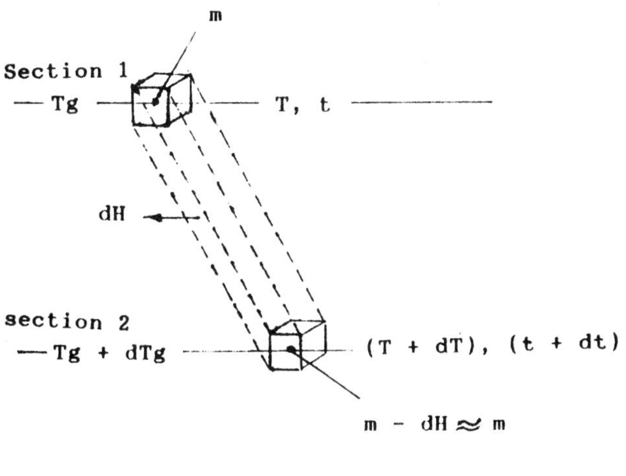

Fig. 6 The moving of liquid element

distance in a random path where a liquid element with mass m (approaching infinitesimal) is moving downward from section 1 to section 2 and contacting with the surrounding gas. The temperature and residence time of the liquid mass entering the section 1 are designated as T and t and that of the liquid mass leaving the section 2 are (T+dT) and (t+dt) respectively. During the cooling process, a very small amount of liquid is vaporized which equals to the increase of humidity in the surrounding gas. The amount of heat necessary for such vaporization, however, is very small in comparison with the sensible heat loss of the liquid. Therefore, the amount of heat transferred from liquid phase is approximately equal to:

$$\bar{q} = mC_p dt \tag{2}$$

Such amount of heat should be absorbed by the gas phase and can be represented by the following conventional equation:

$$\bar{q} = h\bar{A}(T - T_g)dt \tag{3}$$

where h and A are the heat transfer coefficient and effective area respectively. Equating eqs. (2) and (3) yields

$$dT = (h\bar{A} / mC_p)T_g dt$$

or

$$dT / (T - T_{gin}) = (h\bar{A} / mC_p)[(T - T_g) / (T - T_{gin})]dt$$

$$= (h\bar{A} / mC_p)kdt \tag{4}$$

where k is a function of t, ranging usually from (0.5–1.0) in common, cooling process. If an average value of k, i.e. k_m, can be taken, then integrate eq. (4) from t=o to t=t to give

$$ln[(T - T_{gin}) / (T_0 - T_{gin})] = (hA / mC_p)k_m t$$

or

$$lnT_R = Kt \tag{5}$$

Although K is changing along the height of the packed bed, eq. (5) indicates that, if the logarithm of the dimensionless temperature T_R representing a contour line (isotherm) is multiplied by the value of K which is estimated in the column section concerned, a line of constant residence time can be obtained. Such conversion means the existence of similarity, or a certain relationship, between the dimensionless temperature and the residence time of the flowing liquid. In other words, the flow pattern of the liquid which is commonly represented by the residence time distribution profiles can be mapped out, at least qualitatively, from the experimental temperature contours.

IV. EVALUATION OF LOCAL TRANSFER COEFFICIENT

Even though the experimental temperature profile indicates qualitatively the liquid flow pattern, yet the numerical values of three dimensional liquid velocity can only be computed by solving a set of partial differential hydrodynamic equations. Once the velocity distribution is known, the local, or regional, coefficients of heat and mass transfer can be calculated if the temperature distribution is known from the experimental measurement.

Since the flowing liquid is in the state of turbulence, we may start with the well known Reynolds equation, viz.

$$u_j\frac{\partial u_i}{\partial x_j} = \frac{1}{\rho}\frac{\partial}{\partial x_j}\left[- p\delta_{ij} + D\left(\frac{\partial u_i}{\partial x_j} + \frac{\partial u_j}{\partial x_i}\right) - \rho\overline{u'_i u'_j}\right] \tag{6}$$

In the present countercurrent two phase flow, the resistance created by the rising gas against the downward flowing liquid results in the increase of turbulence, or in other words, the increase of Reynolds stress. Thus a multiplication factor f may be added to the term of Reynold stress in eq. (6).

In finding Reynolds stress, one of the common practice is by introducing Boussinesq's concept which gives

$$- f\ (\rho\overline{u'_i u'_j}) = D_e\left(\frac{\partial u_i}{\partial x_j} + \frac{\partial u_j}{\partial x_i}\right) \tag{7}$$

where the eddy diffusivity D_e is function of liquid load, gas rate and the type of packed bed concerned.

Substituting eq. (7) into eq. (6) and noting that in most turbulent flow as well as in present study $D_e >> D$, the following equation of motion in three dimensions is obtained:

$$u_j\frac{\partial u_i}{\partial x_j} = \frac{1}{\rho}\frac{\partial}{\partial x_j}\left[- p\delta_{ij} + D_e\left(\frac{\partial u_i}{\partial x_j} + \frac{\partial u_j}{\partial x_i}\right)\right] \tag{8}$$

Similarly, the corresponding equation of motion for gas phase may be written. For the sake of simplicity, the gas flow is assumed to be one dimensional and uniform.

Besides the hydrodynamic equations, an equation representing the heat and mass transfer can also be established. Consider a differential element with the dimension of dx, dy and dz. The liquid is flowing in x, y, z directions and the gas flows only in z direction. Furthermore, heat is transferred from the liquid phase to the gas phase, i. e. equivalent to a heat generation source, at a rate of q per unit volume of the packed bed. Assuming the coefficient of thermal diffusion α is constant in three directions, an energy balance yields the following equation:

$$u_r \frac{\partial T}{\partial r} + u_z \frac{\partial T}{\partial z} = \frac{\alpha}{r}\frac{\partial}{\partial r}\left(r\frac{\partial T}{\partial r} \right) + \frac{\partial^2 T}{\partial z^2} + \frac{q}{\rho C_p} \tag{9}$$

The rate of heat transfer q can be expressed in terms of temperature driving force $(T-T_g)$as follows:

$$q = h_a(T - T_g) \tag{10}$$

where h_a is the volumetric coefficient of heat transfer in the unit of Joules per unit volume per unit time and per degree temperature difference. Since the evaporation of liquid is accompanying in a liquid–gas direct contacting process, eq. (10) can also be written in the form of mass transter equation, viz.

$$q = k_a(H^* - H)\varphi \tag{11}$$

where the driving force is represented by the difference of saturated humidity H* and the humidity of the gas, all measured at the same temperature; k_a is the volumetric mass transfer coefficient. In using eq. (11), the evaluation of H should be started from the bottom of the column by adding the amount of liquid vaporized in each element to the humidity of the incoming gas, and proceed similar calculation to the next neighboring element. The humidity difference is then evaluated numerically. In this paper, eq. (11) is used for evaluating the mass transfer characteristics of the packing.

In our experimental work, if the initial distribution of both liquid and gas are closely uniform, the temperature profiles or the flow pattern of the liquid in all cross sections may be considered to be symmetical with respect to the centerline of the column for practical purpose as it can be seen from Figs. 3, 4 and 5.

In the computation concerning a packed column, it is more convenient to use the cylindrical coordinate rather than the Cartesian. Thus, the foregoing partial differential equations together with the equation of continunity is converted into the following form:

$$u_r \frac{\partial u_r}{\partial r} + u_z \frac{\partial u_r}{\partial z} = -\frac{1}{\rho}\frac{\partial p}{\partial r} + D_e \left\{ \frac{\partial}{\partial r}\left[\frac{1}{r}\frac{\partial}{\partial r}(r u_r) \right] + \frac{\partial^2 u_r}{\partial z^2} \right\} \tag{12}$$

$$u_r \frac{\partial u_z}{\partial r} + u_z \frac{\partial u_z}{\partial z} = -\frac{1}{\rho}\frac{\partial p}{\partial z} + D_e \left[\frac{1}{r}\frac{\partial}{\partial r}\left(r\frac{\partial u_z}{\partial r} \right) + \frac{\partial^2 u_z}{\partial z^2} \right] \tag{13}$$

$$u_j \frac{\partial T}{\partial x_j} = \alpha\left(\frac{\partial^2 T}{\partial x_j^2} \right) + \frac{K_a (H^* - H)_T \varphi}{C_p \rho} \tag{14}$$

$$\frac{1}{r}\frac{\partial}{\partial r}(\rho r u_r) + \frac{\partial}{\partial z}(\rho u_z) = 0 \tag{15}$$

The boundary conditions of the foregoing equations are: At the top of the column:

The inlet liquid is uniform and unidirectional, i.e.

$$u_z = Const = (u_z)_0$$

At the centerline of the column:

The liquid flow is symmetrical flow, i.e.

$$\partial u_r / \partial r = 0$$

At the wall of the column:

Due to the partial flow of liquid along the column wall and the exchange flow between the wall region and the bulk of packing, the authors suggested the following boundary conditions [4]:

$$\partial u_r / \partial z = -(A / D)(u_r - BV) \tag{16}$$

$$\partial V / \partial z = -(D / C)(du_r / dr) \tag{17}$$

where V is the liquid flow rate along the wall; A, B, C and D are constants depending on the type of packing. For 250Y structured packing, they are 0.0015971, 0.3079, 0.0127 and 0.034 respectively.

At the bottom of the column

It is expected that $\partial u_z / \partial z = 0$

In addition, the summation of liquid leaving the column should be equal to the amount of entering.

The method of finite grid netword was used to solve numerically the foregoing set of partial differential equations. The value of eddy diffusivity D_e was taken from our experimental measurement [5]. As the first step of our computation, eqs. (12), (13) and (15) were solved to give the local directional liquid velocities at each grid point, and then superimposed the corresponding local experimental temperature. By means of eq. (14), the local coefficient of mass transfer at each point can be obtained.

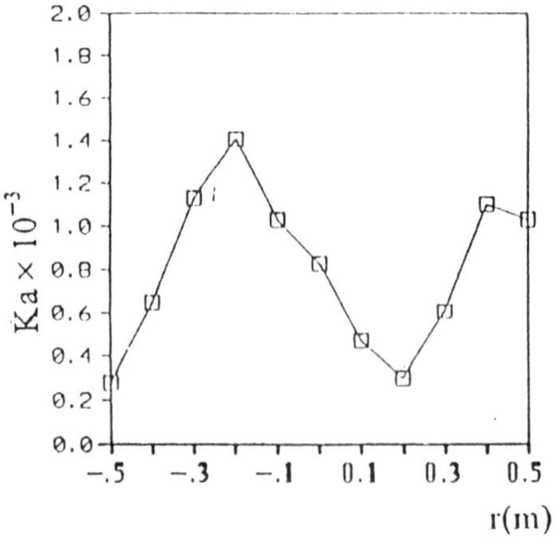

$Q = 8.97m^3/m^2 \cdot h, F = 2.1, z = 1.36m$

Fig. 8 Mass transfer coefficient (ka) profiles

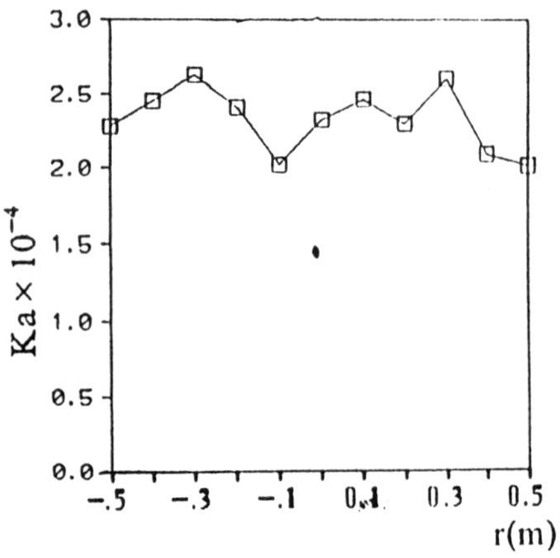

$Q = 8.97m^3/m^2 \cdot h, F = 2.1, z = 0.68m$

Fig. 7 Mass transfer coefficient (ka) profile

Figs. 7 and 8 show the variation of the calculated local mass transfer coefficients with the column radius and column height. It is worthy to note the following result from these figures:

(1) The variation of K_a along radial direction is not so serious as expected for random packing in a larger diameter column. This is due to the fact that 250Y structured packing possesses very good radial flow distribution. Thus this kind of packing can be scaled up to industrial column with good transfer efficiency.

(2) The values of K_a drop down seriously at the top section of the column because the air in this region is nearly saturated so as to reduce considerably the amount of mass and heat transfer. It may inspire the idea that the local rather than overall coefficient can give more reliable information especially in the scaling up of a packed column.

IV. CONCLUSION

For studying the flow pattern and transport characteristics in gas–liquid countercurrent two–phase flow through a packed bed, the method of measuring the three dimensional temperature profiles is demonstrated to be successful. This method is based on the simularity of liquid temperature profile and its residence time distribution, and also on the computation of transport coefficient by combining the basic hydrodynamic and heat transfer differential equations with the expermental temperature measurement.

The effectiveness of a gas–liquid contacting process can be assessed by the extent of deviation of flow pattern from the plug flow as well as the radial and axial variation of local transport coefficients. With this idea, the method presented in this paper may be useful in future research and development work.

NOMENCLATURE

A	constant
$\bar{A}$	heat transfer area
B	constant
C	constant
D	constant in eqs. (16) and (17)
D	diffusivity
D_e	eddy diffusivity
f	multiplication factor
h	coefficient of heat transfer
h_a	volumetric coefficient of heat transfer
H	humidity of the gas (air)
H_s^*	saturated humidity of the gas (air)
k	equals to $(T-T_g)/(T-T_{gin})$
k_m	average value of k
K	equals to $(hA/mC_p)k_m$
k_a	volumnetric coefficient of mass transfer
m	mass of liquid (water) element
p	pressure
$\bar{q}$	total amount of heat transfer
q	rate of heat transfer per unit volume
r	radius of the packe column
t	residence time of the liquid (water)
T	liquid (water) temperature
Tg	gas (air) temperature
T_{gin}	inlet gas (air) temperature
T_R	dimensionless temperature
To	inlet liquid (water) temperature
u_r	liquid (water) velocity in r (radial) direction
u_z	liquid (water) velocity in z (axial) direction
V	liquid (water) velocity along the wall
w	equals to (G+H)
x y	zcoordinates
α	thermal diffusivity
φ	latent heat of vaporization
ρ	density of liquid (water)

REFERRENCE

(1) Z. T. Zhang, Ph. D. Dissertation, Chemical Engineering Research Center, Tianjin Universtiy (1986)

(2) Z. T. Zhang, and K. T. Yu, J. Chem. Ind. & Eng. (China), 39, No.2, 162 (1988)

(3) Stichlmair, J. and S. Ulbrich, I. Chem. E. Symposium Series No. 104, B 213 (1987)

(4) X. J. Yuan, F. S. Li and K. T. Yu, J. Chem. Ind. & Eng. (China), 40, No.6, 686 (1989)

(5) H. Y. Li, M. S. Dissertation, Chemical Engineering Research Center, Tianjin University (1987)

AUTHOR INDEX

Two-Phase Flow and Heat Transfer – 1992